Brazilian Marine Biodiversity

Series Editor

The book series Brazilian Marine Biodiversity was designed to communicate to a broad and international readership the diversified marine and coastal habitats along the large Brazilian coast.

* * *

The diversity of marine habitats found in Brazil is astonishing and includes estuaries, coral reefs, rocky shores, sandy beaches, rhodolith beds, mangroves, salt marshes, deep-sea habitats, vegetated bottoms, and continental shelf. These habitats are addressed from an ecosystem perspective across the series, and characterized in terms of distribution and peculiarities along the Brazilian coast, records of relevant species, and information on the prevailing structuring ecological and oceanographic processes governing biodiversity.

The series also presents an analysis of the role of biodiversity and the importance of ecosystem services, and discusses the threats to each habitat, such as pollution, habitat loss, invasive species, overfishing, and global environmental changes. Conservation efforts are also considered as well as gaps in scientific knowledge and science-policy interface.

This series is an initiative of the UNESCO Chair for Ocean Sustainability, hosted by the Oceanographic Institute and the Advanced Studies Institute of the University of São Paulo, within the Brazilian Network for Monitoring Coastal Benthic Habitats (ReBentos), which is supported by the Brazilian National Council for Scientific and Technological Development (CNPq), the Research Program on Biodiversity Characterization, Conservation, Restoration and Sustainable Use of the São Paulo Research Foundation (BIOTA-FAPESP), the Coordination for the Improvement of Higher Education Personnel (CAPES) and the Brazilian Innovation Agency (FINEP). ReBentos is part of the Brazilian Network on Global Climate Change Research (Rede Clima) and the Science and Technology National Institute on Climate Changes (INCT Mudanças Climáticas) at the Ministry of Science, Technology, and Innovation (MCTI).

Yara Schaeffer-Novelli
Guilherme Moraes de Oliveira Abuchahla
Gilberto Cintrón-Molero
Editors

Brazilian Mangroves and Salt Marshes

Springer

Editors
Yara Schaeffer-Novelli
Instituto Oceanográfico
Universidade de São Paulo USP
São Paulo, SP, Brazil

Guilherme Moraes de Oliveira Abuchahla
Leibniz Centre for Tropical Marine
Research (ZMT)
Bremen, Germany

Gilberto Cintrón-Molero
Instituto BiomaBrasil
São Paulo, SP, Brazil

ISSN 2520-1077 ISSN 2520-1085 (electronic)
Brazilian Marine Biodiversity
ISBN 978-3-031-13485-2 ISBN 978-3-031-13486-9 (eBook)
https://doi.org/10.1007/978-3-031-13486-9

This Springer imprint is published by the registered company Springer Nature Switzerland AG
The registered company address is: Gewerbestrasse 11, 6330 Cham, Switzerland

Coordination

Support

Foreword

These authors selected an unusual example to expose their philosophy. Tropical and too unstable for most land uses and growing on mud too salty for even other vegetation, we come to the mangroves. Commonly located near busy seaports where they may be abandoned, leaving the authors whom I know and expect will request readers to join with compassion and conviction to save and conserve the mangroves for their many benefits for us all.

Frank Howard Wadsworth, in memoriam
Director of the Institute of Tropical Forestry, US Forest Service (1942 to 1978), International Institute of Tropical Forestry, USDA Forest Service, San Juan, Puerto Rico
Puerto Rico, 2019

Preface

The coast of Brazil was born from the tectonic divorce of Gondwanaland, but geomorphic forces and mangroves have acted over time to smooth out rifts and create settings that in turn transformed the coast and hid the marks of the separation. In fact, the smooth shoreline of Brazil is the manifestation of nature's own healing power and self-organization that in the process creates unexpected novelty. Jagged rifting sediments and mangroves combined to create a coastline dominated by beaches, sheltered coastal lagoons, and estuaries that in turn have become habitats for life that through animal migrations have transhemispheric importance.

The study of mangroves covers a wide domain. For convenience, we have subdivided the book into parts that correspond roughly to levels of organization. Just like a Russian nesting doll or a fractal, knowledge comes in layers, so we have attempted to present those layers as well from the largest broadest view in four parts.

The first part, Tropical and Subtropical Brazilian Coastal Zone, introduces the reader along the coastline through different scales of observation. For instance, the term "mangrove" refers to individual trees or whole ecosystems. This apparent ambiguity reflects their scaleless fractal nature. They may occur as individuals, ecosystems, landscapes, and whole coasts as in northern Brazil. Along the country's coastline, we meet various transformations and local versions of mangrove consortia with local factors to enhance the diversity of the coast itself while coupling with the continental shelf and the offshore ecosystem. Here, we provide a closer look at the responses of mangroves and salt marshes, where the reader can catch a glimpse of their secret, their ability to deal with extreme dynamism. The entire part ends up revisiting a 1990 perspective on broad generalizations based on rudimentary technology available at the time, now through modern tools and perspectives. This highlights the importance of how trained eyes and technology make for a great partnership. Not much changed through the decades in terms of interpretation, but the depth of knowledge provided for a much better understanding and appreciation of the complexity of natural processes and their openness.

Part II, The Mangrove Ecosystem, tackles core principles in the understanding of mangroves as individuals, populations, and communities. The power of keen observation is highlighted but the need for scope is identified as a requisite for detecting long-term trends and ascertaining possible responses to environmental changes as well as establishing phenological patterns of mangrove forests. Genetic and epigenetic studies provide clues on the orthogenesis of mangrove ecosystems, taking into consideration their importance as information processors and historical entities. Furthermore, other aspects of mangrove responses to the environment reflect the close link between species, ecosystem, and settings. We give special attention to litterfall data from all along the Brazilian coast, reflecting once more the diversity and potential of our coast to hosting and providing new life.

The third part, Mangrove and Salt Marsh Associates, focuses on diverse organisms and communities that are essential aspects of ecosystem function. Generally overlooked by mangrove specialists due to its structural complexity, microorganism communities drive geochemical cycles and ecosystem services. Macroalgae can also be found in mangrove and salt marsh ecosystems along the entire coast; their temporal and spatial variability is driven by changes and fluctuations in environmental factors describing survival strategies that involve the production of a diverse array of bioactive compounds. Zooming out, meio- and macrofauna inhabit the usually low oxygen, highly saline sediment of mangroves and salt marshes. They are a set of silent but crucial actors that process organic matter and route it among multiple pathways. On the other side of the spectrum is the iconic community of Brachyuran crabs. Their roles span whole domains, ecological, social, and even hemispheric by supporting human societies and trans-hemispheric migratory birds. With effect, this is one of the most well-known study cases of ecological connectivity. The same is true for the ichthyofauna, which act as a bridge between the massive productivity in mangroves and salt marshes and the adjacent oligotrophic high seas. Last, but not least, we highlight the importance of flagship species, such as the Antillean manatee, as a powerful ecosystem conservation tool.

Part IV, Conservation Strategies, deals with an emerging level of organization that is related to the integration of social and ecological processes to shape a new domain, that of an integrated social-ecological system (i.e., biogeocoenosis). Throughout Brazilian history, mangroves played an important role in the shaping of the coastline; initially perceived as barriers in the eyes of the arriving Europeans, they were nevertheless exploited in diverse ways, facilitating the establishment of coastal cities. This ambiguity is a manifestation of the human ecological shortsightedness that can only be changed by blending or learning from traditional and ethnic practices that were formed through centuries of cohabitation. In the national context, specific extractive reserve protected areas take traditional communities into consideration, and the system that regulates them forms the backbone of the country's conservation strategy. Among important conservation strategies, we highlight the importance of education for a citizen-driven, transformative practice. In the same way, we suggest that a setting-landscape perspective is the most effective way of perceiving mangrove and salt marsh ecosystems. These settings are both transformative and self-organizing through different scales in space and time, creating a densely woven fabric that can only be managed in an integrated fashion. The complexity emerging from this fabric breeds irreducible complexity related to the outcomes of interventions. This suggests that environmental management should be multiscale adaptive through dynamic framing in order to achieve compatibility with the hierarchic complexity that managing mangroves and salt marshes require.

We hope you enjoy the book and do not forget to get your feet muddy as well!

São Paulo, SP, Brazil — Yara Schaeffer-Novelli
São Paulo, SP, Brazil — Gilberto Cintrón-Molero
Bremen, Germany — Guilherme Moraes de Oliveira Abuchahla

Acknowledgments

This book is dedicated to all those pioneers that raised awareness about the importance of scientifically based education in wetland conservation. The editors want to specially acknowledge some key players as Dr. Manuel Vagas-Velez (UNESCO ROSTLAC, Uruguay), Dr. Carlos Chiriboga (OEA, Washington, D.C.), Marilza Cordeiro-Marino (IBt/SMA, São Paulo, Brazil), Norma Crud Maciel (FEEMA/IMA, Rio de Janeiro, Brazil), Tânia Mara Simões do Carmo (UFES, Espírito Santo, Brazil), Waldemar Londres Vergara Filho (ICMBio, Pará, Brazil), and finally Sr. Gilinho (Cananéia fisherman). They represent many others who have devoted their lives to understand and properly cherish the mangrove ecosystem. Also, we must mention the decisive support of entities such as the World Wildlife Fund – USA, Organization of the American States, Conselho Nacional de Desenvolvimento Científico e Tecnológico (CNPq), Coordernação de Aperfeiçoamento de Pessoal de Nível Superior (CAPES), Brazilian state research agencies, and the Rede de Monitoramento de Habitats Bentônicos Costeiros (ReBentos). We are very thankful for all authors that offered their expertise to contribute to the comprehensive chapters in this volume.

Contents

About the Editors

Yara Schaeffer-Novelli is a naturalist with a PhD in zoology and specialization in biological oceanography from the University of São Paulo, Brazil. She is a senior professor in the Oceanographic Institute at the University of São Paulo where she taught and mentored graduate students since 1970. Her main research interests are the tropical marine ecosystems with emphasis on mangrove forests, and her experiences span from ecosystem studies to environmental impact assessment, teaching, and outreach.

Gilberto Cintrón-Molero is a marine biologist with a large experience with ecological studies on coral reefs and mangroves all along the Caribbean and Latin America. He worked for the Oceanography Program in the Department of Natural Resources, Government of Puerto Rico. During the past more than 20 years, he worked as an international affairs specialist in the U.S. Department of the Interior's Fish and Wildlife Service International Division. His main research interests are coastal ecology and management.

Guilherme Moraes de Oliveira Abuchahla is a biologist with a Master of Science degree from the Christian-Albrechts University in Kiel, Germany. He is currently a doctorate candidate in Mangrove Ecology at the Leibniz Centre for Marine Tropical Research in Bremen, Germany. His main research interests are coastal tropical ecology, integrated coastal zone management, teaching, and outreach.

About the Editors

Part I
Tropical and Subtropical Brazilian Coastal Zones

Chapter 1
The Brazilian Coastal Landscapes: A Narrative

Cláudia Câmara do Vale, Gilberto Cintrón-Molero, and Yara Schaeffer-Novelli

1.1 Introduction

The length of Brazil's coastline is reported to be from 7408 km (CEMBRA 2012) to 9200 km (Dominguez 2009). However, adding the shoreline length of the coastal states using IBGE data (2016), the total length adds to 10,959.52 km. This discrepancy is not due to lack of information or measurement error but instead reflects a fundamental property of all coastlines; the length of any coastline can vary drastically depending on the unit of length and the scale of the map. This implies that there is no explicit answer to the question of the length of a given shoreline. Rather length can only be estimated; the fractal-like properties of coastlines make the notion of length inapplicable.

Length is a function of resolution and detail captured by measuring. However, interesting aspects of the Brazilian coastline emerge from the consideration of its fractal dimension (D) (Mandelbrot 1967). This dimension allows the assessment of the complexity or roughness of a coastline and, we suggest, its potential diversity. A fractal dimension approaching 1 corresponds to a nonfractal monotonous straight line. This suggests a high diversity of shoreline habitats is associated with a higher

C. C. do Vale (✉)
Departamento de Geografia, Universidade Federal do Espírito Santo, Vitória, ES, Brazil

G. Cintrón-Molero
Instituto BiomaBrasil, São Paulo, SP, Brazil
e-mail: gilcintron@earthlink.net

Y. Schaeffer-Novelli
Instituto BiomaBrasil, Recife, PE, Brazil

Instituto Oceanográfico, Universidade de São Paulo, São Paulo, SP, Brazil

IUCN, SSC, Mangrove Specialist Group, Gland, Switzerland
e-mail: novelliy@usp.br

Y. Schaeffer-Novelli et al. (eds.), *Brazilian Mangroves and Salt Marshes*, Brazilian Marine Biodiversity, https://doi.org/10.1007/978-3-031-13486-9_1

(>1) fractal dimension (> rugosity). The fractal dimension of Brazil's coast has been estimated considering coastal lengths from CEMBRA (2012), Dominguez (2009), and IBGE (2016). Estimates varied from 1.2 to 1.5, depending on the rugosity of each coastal sector.

A clue to the fractal dimension of the Brazilian coast is provided by the observation that although globally this coast ranks 12th in length among the planet's shorelines (IBGE 2016), it significantly ranks second in terms of mangrove area (FAO 2007; Giri et al. 2010). Another interesting fact related to their fractal nature is that coastlines self-organize to dissipate the greatest amounts of energy. This reveals another important feature of the Brazilian coast, its dynamic adaptability, and constant transformability. The coast of Brazil is characterized by barrier structures typical of a trailing coast where sediments are abundant and strongly influenced by sea-level changes and wave action. Energy dissipation is related to work done in moving sediments and building structures such as sandy barriers and geomorphic structures that provide shelter for mangrove forests and salt marshes among other ecosystems. This results in low dimensional landforms such as straight beaches and locally dominated by wave regimes with low incidence angles.

Thus, along the length of the Brazilian coast, we observe a highly diverse suite of features spanning various spatiotemporal process scales, reflecting the endless shaping of coastal landforms by ecogeomorphologic factors acting since the splitting of the supercontinent of Pangea (Dominguez 2009), during the Mesozoic. This variability is manifested in terms of geographical position as well as landform types, sizes, endurance, and orientation.

What we see today along the coast of Brazil reflects the combined interaction of geological, oceanographic, and ecogeomorphic processes interacting with sediment supply, climate, and oceanographic factors, and antecedent topography to produce the modern dynamic coastal landscapes. Mangroves play important roles in such dynamic landscapes, because they can rapidly colonize and stabilize new intertidal areas. Sediment supply and antecedent geology are fundamental controls of the first-order hierarchy of coastal landscapes, but in another sense, each landscape is unique and distinctive, "Perfect Landscapes" (Phillips 2007), that results from a combination of processes, interactions, and events unlikely to be duplicated. Landscapes are evolving entities and coastal evolutionary processes are open ended. This entails the potential to produce an endless diversity of forms, unbounded complexity, novelty, and dynamism (Ab'Sáber 2001).

The Brazilian coast shows good records of sea-level oscillations in its sedimentary rocks, reliably documenting temporal and spatial variations in terms of features that developed when sea-level stood at different elevations in the past (Souza 2015). The Brazilian sea-level history is complex; at least three high sea level episodes are recognized; a high still stand some 5600 years BP (+5 m), a high still stand some 12,000 years BP (+8 m), and perhaps others dated around 200,000 years BP (+8 m) (Souza 2015).

On trailing edge continental shorelines wave-dominated coasts, sea-level behavior is the most important factor controlling coastal sedimentary features such as sandy beaches, deltas, and barriers, producing changes in aerial features such as beach ridges, cheniers, dunes, and complex combinations of such landforms. The Postglacial Marine Transgression and sediment supply have been important forcing functions shaping coastal geomorphology. About 18,000 years BP ice sheets were close to the maximum extent and the sea stood some 120–130 m below present level (Dominguez 2009). Later sea-level rose, submerging the continental shelf, reaching up to 5 m above present level before falling to the present level about 6000 years ago (Dominguez 2009). The text refers to the actual coastline in relation to the sea level. The processes of progradation and recession refer to the coastline itself during the different sea-level rising and lowering across the eras. However, as a result of these changes in sea level, the present coast is characterized by its relative absence of barrier island lagoonal systems (Dominguez 2009). That is, today's barrier islands and estuaries are, in a geological sense, rare features. These oscillations in sea-level highlight the dynamism and malleability of the Brazilian coastal features and help comprehend its rich diversity. Coasts evolve, adapt, and transform at multiple scales; they are not static entities and are best understood within the framework of complex adaptive systems (CASs) and Earth Surface Systems (ESSs) (for more information, see Chaps. 3 and 18). The Brazilian coast is full of signs that if properly interpreted have valuable lessons for contemporary planners, policymakers, and coastal communities.

In order to better understand the configuration of a coast dynamically shaped by short- and long-term spatiotemporal processes, we adopted a typology originally developed by Silveira (1964) and summarized by Ab'Sáber (2001), which divides the coast into six distinct sectors: Amazonian Equatorial Coast; Septentrional Northeastern Coast, Oriental Northeastern Coast, Eastern Coast, Southeastern Coast, and Southern Cost (Fig. 1.1). There have been many attempts to classify coastal environments since Hartt's "Geology and Physical Geography of Brazil" (1870), but Silveira's (1964) typology was here adopted as a practical geographic tool to illustrate the variability and diversity of landforms within coastal sectors. Coastal classifications have received little attention until recently when geographical information systems (GIS) have enabled the association of numerous attributes for susceptibility and vulnerability assessments. The Brazilian coast is eminently Tropical (77%) but has a significant Subtropical length (23%) along its southernmost states below 23° 26′ S, within the Southeastern and Southern sectors.

Along the coast of Brazil, mangrove forests find numerous favorable environments to colonize, develop, and persist; except conspicuously along the coast of Rio Grande do Sul (RS; Fig. 1.1) state where unfavorable climatic factors interact with physiography to create oligohaline environments that inhibit and preclude their establishment in what at first seems to be favorable conditions in terms of available geomorphic features (Duke et al. 1998; Cintrón-Molero and Schaeffer-Novelli 2019). After Indonesia, Brazil has the largest mangrove area of any other country with 13,989.66 km^2, about 8.5% of the global total (Spalding et al. 2010; Simões and Oliveira 2018). According to Schaeffer-Novelli et al. (2000), mangrove forests are widely distributed over some 90% of the Brazilian coast, generally confined to sheltered intertidal mudflats, particularly at the larger coastal indentations, such as bays, estuaries and behind coastal barriers and barrier islands, and in coastal lagoons

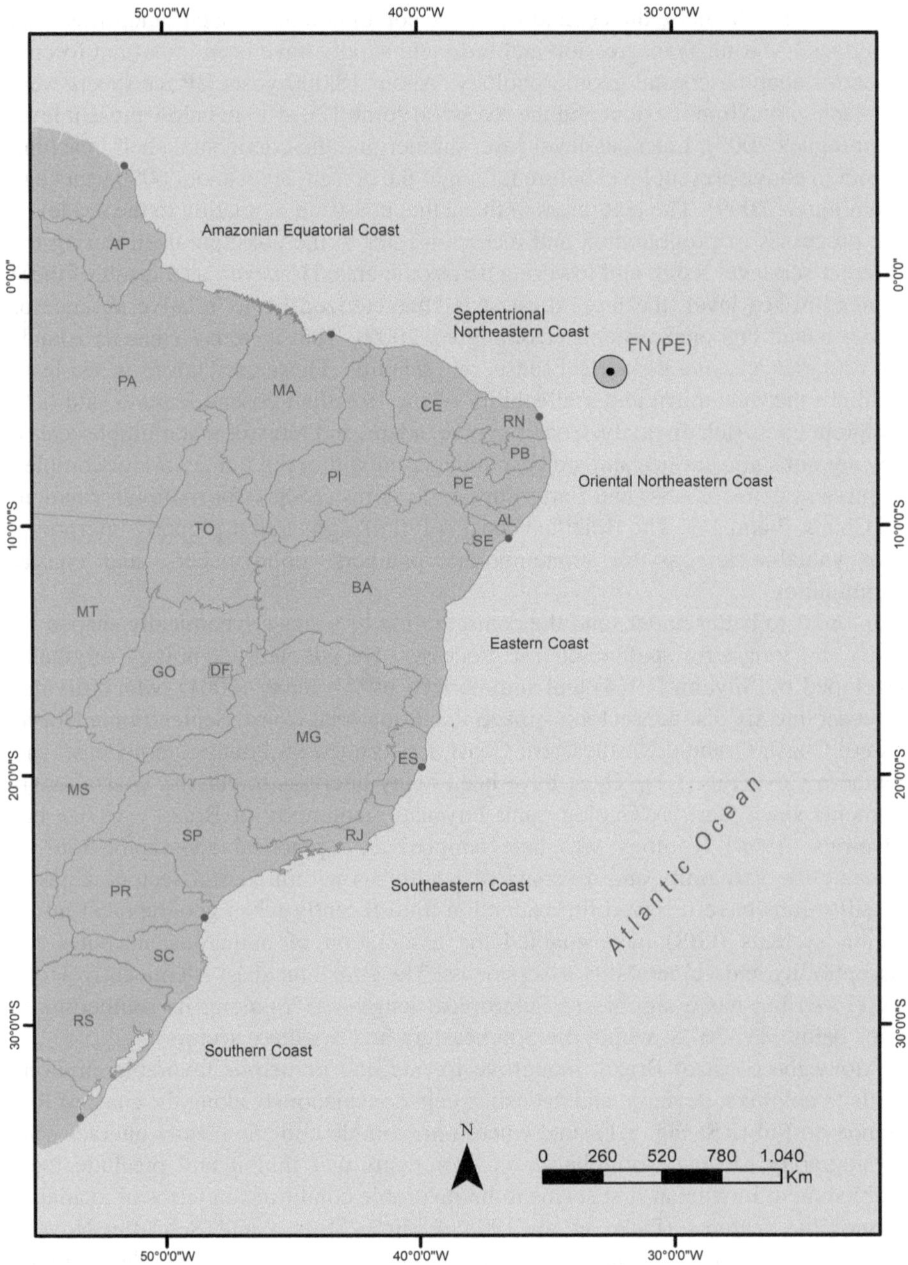

Fig. 1.1 Distribution of coastal sectors of Brazilian coast according to Ab'Sáber (2001). The 17 Brazilian coastal states are: *AP* Amapá, *PA* Pará, *MA* Maranhão, *PI* Piauí, *CE* Ceará, *RN* Rio Grande do Norte, *PB* Paraíba, *PB* Pernambuco, *AL* Alagoas, *SE* Sergipe, *BA* Bahia, *ES* Espírito Santo, *RJ* Rio de Janeiro, *SP* São Paulo, *PR* Paraná, *SC* Santa Catarina, *RS* Rio Grande do Sul, *FN* Fernando de Noronha Archipelago

subject to frequent flooding by tides and subject to saline intrusions. During the Quaternary changes in sea level and climate the most recent morphological elements of the Brazilian coast were added; strand plains (prograded barriers), tidal flats, wetlands, and coastal dune fields (Dominguez 2009). From 04° 26′12″ N, in the state of Amapá, on the Amazonian Equatorial Coast, mangrove forests are present intermittently along the coast to their latitudinal limit 28° 30′ S at Laguna (Santa Catarina State).

Conversely, salt marshes are predominant south from the mangrove latitudinal limit (Costa et al. 2009). Nevertheless, they may occur throughout the tropical domain in the lower edges of mangrove ecosystems and in hypersaline areas and salt flats within or at the terrestrial-most margin of mangroves. They will colonize the lands within the spring tide reach under direct sunlight. Despite their prominence in some areas of the coast, there is little consensus on what is the actual area of occurrence and occupation by salt marshes (Junk et al. 2013). This fact can be explained by the seasonal variation in aerial structure (leaves or blades) that these plants naturally show and that is responsible for the high export of particulated organic matter to surrounding environments (Long and Mason 1983; Adam 1990; Silva et al. 1993a; Abreu et al. 2006; Marangoni and Costa 2009).

1.2 Amazonian Equatorial Coast (04° 26′12″ N to 02° 20′ 43″ S)

The Amazonian Equatorial Coast is 4250 km long (IBGE 2016) and encompasses a tide-mud dominated (Amapá-Guianas sector) to the west, and the tide-dominated Pará-Maranhão unit in the east. Temperatures do not vary greatly during the year and only two distinctive seasons are recognized (summer and winter); the latter is characterized by intense rainfall rather than low temperatures. Over 80% of Brazil's mangroves are found within the deltaic complex of Brazil's three northernmost coastal states (Amapá, Pará, and Maranhão) (see Chap. 3, Maps 1, 2, and 3, respectively).

The mangrove coast to the east encompasses two world geoecomorphic records; the largest continuous mangrove system, with 480 km in length and 7500 km^2 in coverage (Souza Filho 2005) as well as the gorge of the largest river in length and water and sediment discharge; the Amazon River (or Amazonas). This river, along with the Tocantins and the Parnaíba rivers, brings to the coast the largest sediment load of the entire South American continent.

Two contrasting sedimentary regimes characterize this coast defined by the down-drift (East) and up-drift (West). The different style of the western sector sedimentation produces a muddy shoreline lacking reentrants and the hinterland is dominated by a Precambrian basement with very short rivers. The presence of muddy sediments even increases the tidal amplitude by decreasing bottom friction.

Mangroves can become tenuously established over transient mudflats while persisting in more stable features as in well-developed mangrove forests lining the larger rivers: The Oiapoque and the Caciporé rivers, whose lower reaches are subjected to saltwater wedges that enter deep inland into a wide and low coastal plain. This riverine-marine lowland is interrupted in places by vegetated beach ridges within a dissected paleo-coastline.

From the extreme northern limits, the coast is less incised, with fewer prominent estuaries such as the Cunani, Calçoene, and Flechal rivers but many creeks ("igarapés") that reach the coastal shallows, often occupied by mangroves, or interrupted by beach ridges marking old coastlines and beaches now occupied by "restinga"[1] formation.

From the offshore island of Maracá to the Araguari River Delta, flooded fields and small, shallow, muddy lagoons characterize the coast, with mangroves growing over fine sediments deposited at the delta.

Along the Amazon right margin, the amount of fine suspended sediments and the volume of freshwater are so huge that mangrove forests are found only as dispersed elements in a mosaic composed of mixed floodplain forests and patches of herbaceous vegetation that, like mangroves, are created and destroyed seasonally due to instability of the sediments and cyclic erosive events generated by periods of greater precipitation and river discharge. At the mouth of the Amazon River, a huge fluvial island has formed, the Marajó Island, considered the world's largest island of this type, covering 48,000 km^2 (Ab'Sáber 2001).

The eastern side of the Marajó Island is lined by mangrove forests where saline influence is greater, including the edges of igarapés and tidal creeks that drain them. Although tidal influence reaches inland as far as Óbidos (some 1000 km upstream the Amazonas River) saline influence is present only near the mouth of the river facing the east.

Between Belém (PA) and São Luís (MA), the coast is extremely rugged and unstable; known as "coasta de ria" (inlet coast), it is characterized by narrow drowned river valleys cut into the Barreiras Formation[2] and flooded by the Holocene sea-level transgression and its subsequent regression, which exposed fine sediments suitable for colonization by mangroves. In this setting, mangroves find perfect conditions for colonization and persistence, including tropical humid climate, ample fresh water, and sediment supply as well as tidal mixing driven by a macrotidal regime. This combination of factors contributes to its lush development with trees reaching 30–45 m in height.

The low and flat coastal plain allows saltwater and mangroves to penetrate more than 25 km inland. From Marajó Bay to Pirabas River, south of Belém, the coastal

[1] The *restinga* is an elongated shoreline feature of sandy nature and low amplitude that tends to close recesses (Souza et al. 2008), at times covered in vegetation of diverse nature (e.g., halophilous, psammophilous, xerophytes, and hygrophytes).

[2] The Barreiras Formation is a series of Cenozoic segments of diverse origins and complex genesis, forming barriers, tablelands, and abrupt cliffs that separate the interior of the country from the Atlantic Ocean (Moura-Fé 2014).

plateau reaches the coastline, where active bluffs are being carved by erosion into the Tertiary sediments of the Barreiras and Pirabas formations, which are exposed to the action of waves and tidal currents, as in the Gurupi River, which separates the states of Pará and Maranhão (Souza Filho 2005). Here, estuarine channels extend for approximately 60 km.

The margins of São Marcos Bay (MA), an estuary 100 km long and 16 km wide, shelter exuberant mangrove forests, especially in its western portion. The São Marcos Bay is one of the largest indentations on the Amazonian Equatorial Coast of Brazil.

1.3 Septentrional Northeastern Coast (02° 20′43″ S to 05° 04′12″ S)

This segment can be analyzed in two sectors: one to the west is a dune field that constitutes the Lençóis Maranhenses National Park (LMNP), and the other to further east, between the Piauí State and the city of Natal (RN) (see Chap. 3, Maps 3, 4, 5, and 6).

The most characteristic feature of this coastal setting is the LMNP, the largest dune field in Brazil and one of the largest in the world. At its widest portion, the dune field is 31 km wide, while its total width, including vegetated portions, exceeds 120 km (Ab'Sáber 2001). This sand barrier extends along the coast for some 76 km, migrating landwards along a NE-SW direction (Ab'Sáber 2001). To the east of the Lençóis Maranhenses, there are active prograded barriers dominated by active transgressive dune fields, which become smaller from west to east and include deflation plains.

Parts of the coast are dominated by dune fields and mangrove forests are restricted to few existing rivers, such as the Preguiças River that flows into the Atlantic and the Da Fome River, which flows into Tutóia Bay, in the Parnaíba River Delta and eastern boundary of Lençóis Maranhenses National Park. This area is notable because of the prominent dune fields and the common occurrence of three red mangrove species (*Rhizophora mangle* L., *Rhizophora racemosa* Meyer, and *Rhizophora harrisonii* Leechman) (Prance et al. 1975).

Further east is the only true active delta found on the Brazilian coast: the Parnaíba River delta, which borders the state of Piauí. This delta is characterized by its environmental heterogeneity, creating habitats for several types of vegetation in a setting rich in faunal and floral biodiversity. In addition to the mangrove forests that colonize the sedimentary landforms in coastal depressions, there are other types of vegetation, such as pioneer psamophilic vegetation and subperennial dune vegetation, among others (Costa and Cavalcanti 2010). The Parnaíba River delta is the southernmost limit of the occurrence of *R. harrisonii* and *R. racemosa* (Schaeffer-Novelli 1991).

At the border of Piauí with Ceará states, the estuary of the Timonha River is flanked by moving coastal dunes. This coast faces directly into the trade winds, rainfall is nil in the dry season, and extensive dune fields form and eventually drift, filling estuaries, lagoons, and mangrove ecosystems. Aridity, high sediment supply, and strong winds are reasons for the dominance of dune fields in Maranhão, Piauí, Ceará, and Rio Grande do Norte states. Beyond the Timonha River, mangroves are found in Camocim, Acaraú, Aracati-Mirim, and in other smaller estuaries. The coast of Ceará State is almost rectilinear (W-E), with extensive sandy beaches, and the occurrence of mangrove forests is limited. Within the metropolitan area of Fortaleza, mangrove forests have resisted massive urban expansion in the two important estuaries of Ceará and Cocó rivers. The Ceará River estuary (~03° 44′ S) presents 1158 ha of mangroves (Reis-Neto et al. 2011).

Along the semiarid section is the coast of Rio Grande do Norte State, which presents similar characteristics to the coast of Ceará in terms of prevailing aridity with 750 mm year^{-1}. The exception along the coast of Rio Grande do Norte is the Apodi-Mossoró estuary, carved into the Barreiras Formation. In its margins, several relict salt marshes are being filled by sediments and recolonized by mangrove forests, where salinity is not too high (Costa et al. 2014). Saltworks are characteristic in the northern coast of Rio Grande do Norte, where high temperatures, low precipitation, and high evaporation favor the activity.

1.4 Oriental Northeastern Coast (05° 40′12″ S to 10° 30′ 13″ S)

Cape Calcanhar (RN) is the northernmost extremity of the Oriental Northeastern Coast, which covers part of this state and the Paraíba, Pernambuco, and Alagoas states (see Chap. 3, Maps 6, 7, 8, and 9).

Here a new physiographic element appears along the coastline; the coral reefs forming linear stretches of beach rock pavements marking paleo-shorelines (Santos et al. 2007). The beach rocks in front of Natal run parallel to the shore and are composed of two lines that dip slightly seaward. Carbon dating of the more continuous outer "reef" dates from 4700 years BP, whereas the inner one is older dating from 6250 years BP (Oliveira et al. 1990). These cemented iron oxide sand grain barriers were formed during a period of rapid sea-level rise and high sea-level still stands. Their exposure and erosional features are due to the present low sea level.

The Barreiras Formation, however, remains the most striking element of this coast, whose promontories reach the coast forming occasional subdeveloped beach arches. The Rio Grande do Norte coast is on a sediment-starved coast; the rivers are small and do not contribute with significant sediment amounts to the coast. Only small- and medium-sized rivers reach the coast. The rivers with the largest discharge (Piranhas-Açu and Apodi-Mossoró) are dammed and sediments are trapped in their reservoirs and do not reach the sea. As a result, sediment loss to the dune fields, longshore transport, and spit barrier formations contribute to a negative sediment budget. Sediment is deposited only in the head of small estuaries behind sand strips, forming narrow *restingas*.

Riverine forests are only found on the head of extensive estuaries. Coastal dune fields are present along the Natal coast just landward from the beach overlaying the Barreiras Formation and extending inland several kilometers. The county of Tibau do Sul is the limit of occurrence of the active dunes of the Oriental Northeastern Coast. The Curimataú River, next to the Paraíba State, is the last large estuary of Rio Grande do Norte State. At its margins, the mangrove forests are well developed although shrimp farming is also extensive.

The coast of Paraíba State, which is almost rectilinear in the first few kilometers from (N-S), is lined by active cliffs and only a few rivers are incised into the Barreiras Formation plateau. Among them is the estuary of the Mamanguape River, which cuts inland flanked by the plateau and extending inland approximately 14 km, with extensive mangrove forests surrounded by sugarcane crops. The mouth of the Paraíba do Norte River to the north of the capital city of João Pessoa is an estuary of considerable dimensions, containing the mangrove forest of Cabedelo, which is considerably altered at the margins by human impact.

From the city of João Pessoa toward the south to the border with Pernambuco State, the coast remains almost rectilinear, with few small rivers reaching the Tertiary cliff-dominated coast. Shrimp farms and mangroves are found inside the estuary formed by the Tracunhaém and Siriji rivers. Shrimp farming activity along this coast has taken a great toll on the salt flats and marshes, which are essential habitats for migratory birds along an important Neotropical migratory route.

From the metropolitan region of Recife-Olinda to the south, mangroves occupy Holocene terrain as in the case of the Capibaribe-Jordão-Pina-Tejipió estuarine complex. The Abreu e Lima industrial complex of the Suape Port was built over reclaimed former mangroves taking advantage of the shelter created by the Cape Santo Agostinho and the estuary of the Ipojuca River. This has intensely modified the landscape as well as the coastal dynamics, suppressing coral reefs and resulting in the elimination of a large part of the mangrove forests there (499.83 ha), besides causing multiple indirect impacts (Tavares 2015).

The Sirinhaém River basin covers an area of 2000 km^2, which drains into a large estuary with well-preserved mangrove forests, particularly upstream from the city of Barra do Sirinhaém on the right margin of the river. This river is also known as Formoso River, an important touristic spot in the coast of the state.

Within Pernambuco State, the only oceanic insular mangroves of the South Atlantic stand in Fernando de Noronha Archipelago and deserve attention despite their small size (see Chap. 3, Map 18). The island's main water body, the Maceió Creek, supports the Sueste's 0.89 ha *Laguncularia racemosa* (L.) Gaertn. f. mangrove stand that surrounds a lagoon behind dune barriers (Ridley 1890; Pessenda et al. 2008; Barcellos et al. 2017). This small stand is fed by freshwater from a small watershed drained by the Maceió Creek and saltwater percolation across the porous dune and the intermittent seawater intrusions across the inlet (Barcellos et al. 2017).

The coastline of Alagoas State is 220 km long and runs along a NE-SW direction. The border with Pernambuco State follows the Persinunga River between the counties of São José da Coroa Grande and Maragogi. There, the coast is characterized by active cliffs undercut by waves, in a clear process of recession, in proximity to the shore of the coral reefs. Many rivers are similar to the Manguaba River, with a narrow upper course that widens at the mouth. Among these are the Tatuamunha, Camaragibe, Jirituba, Paripueira, and Prataji rivers, all mangrove-lined in their lower course. These are just a few of the most significant estuaries along this wave-dominated coast.

Despite the predominance of an extensive sandy plain on the coast of Alagoas State, the Mundaú-Manguaba estuarine lagoon complex (09° 35′ S) stands out as the largest mangrove forest. It is formed by the Mundaú River, which flows into the lagoon of the same name (27 km^2), and by the Paraíba do Meio River, which flows into the Manguaba Lagoon (42 km^2); the two largest lagoons in the state (Lins et al. 2013). The Mundaú-Manguaba riverine complex experiences a rainy and dry regime and dampened tidal influence. River discharge controls salinity during the rainy season, whereas tides move seawater into the system in the dry season (summer).

Another important part of the coast of Alagoas State is the estuarine lagoon system on the Roteiro municipality, associated with the mouth of São Miguel River involving the counties of Barra de São Miguel to the north, and Roteiro, to the south. The Coruípe River, 40 km south of Barra de São Miguel, has its mouth between the seaside towns of Pontal de Coruípe and Batel. Here mangroves are well developed on the left margin of the Coruípe River and entirely altered on the right margin, where the town of Batel is located, right between the mangrove fringe and beach ridges. From there, the Quaternary plain widens and is truncated by beach ridges formed by regressive Pleistocene and Holocene shoreline movements. An extensive field of dunes is superimposed on the ridges, reaching the mouth of the São Francisco River.

The southern border of Alagoas State is geographically limited by the São Francisco River, which defines the border with the state of Sergipe. The São Francisco River delta (800 km^2) has been the topic of an interesting debate by geomorphologists. Galloway (1975) considered this delta an icon for a wave-dominated delta as his 1975s tripartite model. However, Dominguez et al. (1982) questioned the validity of Galloway's model in the case of the São Francisco, since this delta, as well as others along the Brazilian coast, has been influenced by a 5 m transgressive event that has exerted a major control on the shore.

Galloway's classification, although thought-provoking, is simplistic in that it neglects sea-level fluctuations in shaping deltas. Thus, the São Francisco River mouth may not be a "classical" delta but is unique in terms of the complexity of the processes that have shaped it. It can be considered as in Phillips' (2007) contention about complex systems, a "perfect landscape." Landscapes are a circumstantial contingent result of several factors acting within a specific environmental setting. Models are inevitable simplifications of reality; a tripartite model neglects many formative processes that, for more detailed understanding, must be considered. For example, greater detail can shed light on the sediment dynamics that drove mangrove colonization on the different *facies* formed and reformed during this deltaic development.

The southernmost point in the Alagoas State marks the boundary between the Oriental Northeastern Coast and the Eastern Coast sectors. Along this coastal segment, the Barreiras Formation approaches the coast and forms cliffs up to 15 m high that rise abruptly from the foreshore; narrow and rugged beaches predominate.

1.5 Eastern Coast (10° 30′13″ S to 19° 40′29″ S)

The Eastern Coast sector extends from Sergipe State, at the São Francisco River southward to the left margin of the mouth of the Doce River in Espírito Santo State (see Chap. 3, Maps 10, 11, and 12).

This coast is smooth and backed by the Barreiras cliffs generating an abrasive and aligned character. The Sergipe State has a straight coastline that extends for about 163 km (Vilar and Santos 2011), with four large estuaries: the São Francisco, Sergipe, Vaza-Barris, and Piauí-Real-Fundo, which interrupt the straightness of the coastline and open large estuaries where mangrove forests can develop. This part of the Eastern coast presents typical features of marine transgression and regressions, such as marine terraces, coastal ridges, coastal dunes, fluvial-lagoonal plains, and estuaries.

From the right margin of the São Francisco River delta, a tributary flows south parallel to the coastline crossing 15 km between Pleistocene and Holocene ridge depressions and reaching the dune formations. This tributary is locally known by various names such as Parapuca Canal, soon after its branching from the São Francisco River, also Carapitanga, Poço, and Boca do Poço canals. The latter has its outlet in the Atlantic Ocean (Santos et al. 2014) where 21.68 km^2 of mangrove forests are found according to Carvalho and Fontes (2007), with *R. mangle* as the dominant species.

Toward the south is the Japaratuba River, which, in its lower course, becomes sluggish and supports a mixed mangrove forest containing *R. mangle*, *L. racemosa*, and *Avicennia germinans* (L.) L., which occupies soils rich in organic matter. The Sergipe River estuary is one of the four largest of the state. According to Carvalho and Fontes (2006), the mangrove area of the Sergipe River occupies 54.96 km^2. However, it is in the broader and substantially larger Vaza-Barris River estuary where mangroves are better preserved and occupy a considerable area of the coastal plain. At the mouth itself, marine influence on coastal hydrodynamics is remarkable, with waves and coastal currents being active processes. The more open portion of the estuary is not conducive to the establishment of mangrove seedlings, due to the instability of the sandy margins and erosive processes. However, it is on several islands within this estuary that mangrove forests are better developed and preserved.

The 132 km long Piauí River flows along the geographic microregions of Agreste do Lagarto and Sergipe's southern coast, flowing between the counties of Estância, Sergipe, and Jandira, in the Mangue Seco village, already in the Bahia State, downstream from its confluence with the Fundo River on the left margin and Real River on its right margin. The mangroves of the Piauí-Real-Fundo rivers estuarine complex can be evaluated jointly according to various authors because of their geographic proximity and the sharing of similar environmental characteristics (ADEMA 1984; Carvalho and Fontes 2006, 2007). In this estuarine complex, several tidal channels reach inland, having their margins occupied by narrow mangrove forests.

The coast of Bahia State, with 800 km in length, is mostly N-S oriented but markedly segmented, presenting different compartments that can be individualized by climate, geological, geomorphological, oceanographic, regional, and local conditions. The Geodiversity Survey Map of Bahia State (Carvalho and Barreto 2010) identifies and classifies 71 geological environmental units, distributed in different domains. Along the coast, this includes plateaus, cliffs, low hills and domes, structural steps, erosive ridges, fluvial and coastal features, and reefs.

The Barreiras Formation is present along the entire coast of Bahia except where sedimentary rocks of the Mesozoic rifts outcrop along the coast. The oldest geological elements on this coast, from north to south, are the São Francisco Craton and the Araçuaí Fold Belt. These two provinces have exerted a major control on coastal development since the South America-Africa breakup (Dominguez 2009). The outcome is that the continental shelf in the fold belt is much wider than in the cratonic section characterized by a narrower shelf. Mangue Seco, the northernmost spot on the coast of Bahia, is located on a fluvial bar currently occupied by an extensive coconut plantation, on the right margin of the Real River, on the border with Sergipe State. From this point south, there are nearly 34 km of extensive and uninhabited sandy beaches fringing a narrow coastal plain sometimes with regressive coastal lagoons and different physiognomic patterns of *restingas*. In this section of the Brazilian coast, mangroves are absent.

The Itapicuru River, which starts in the piedmont of the Chapada Diamantina, drains an approximate area of 36,440 km^2 and flows into the Atlantic, in the area known as Costa dos Coqueiros, on the northern coast of Bahia. There the flow is influenced by tides and most of the sediment is deposited, creating a substrate for the establishment, growth, and maintenance of extensive mangroves. An interesting feature of this area is the paleo-bay head delta associated with the Itapicuru River and its extensive lower river mangroves. Research shows a 10-meter-deep estuary existed during the last sea transgression (Dominguez 2009). Bayhead deposits reach 6 m in thickness and *Crassostrea rhizophorae* middens dated around 5100 years BP indicates the presence of mangroves during that time (Dominguez 2009).

Between the estuary of the Itapicuru River and the mouth of Itariri River, there are sandy beaches and coastal lagoons without direct contact with the Atlantic Ocean, besides the different physiognomic patterns of the *restinga* vegetation. The Itariri River meanders in its lower estuarine reach and spreads over a coastal depression, where the occurrence of mangrove forests is restricted to its margins, but better developed near its mouth.

Beyond this section is the Todos os Santos Bay within the Recôncavo Baiano basin, forming an embayment covering some 1100 km^2 and 200 km in the perimeter. Its geologic origins as a rifted coast date back to the rifting of the present African and South American continents, with the emergence of an aborted rift and successive processes of subsidence and uplift, influenced by the processes of marine transgression and regression, and to the exposure to a past semiarid climate (Dominguez et al. 2009). This is a sediment-starved coast with few rivers of small size within small drainage basins and low precipitation values reaching the bay or the coast. In this sector, the largest river is the Paraguaçu; which also drains the plateau of the Chapada Diamantina and has its waters dammed at the Pedra do Cavalo Dam. The dam furnishes 60% of the water supply to the metropolitan region of the state's capital Salvador (Hadlich 2009). When it reaches the calm and sheltered waters of the Iguape and the Todos os Santos bays, its sediments are deposited at the foot of the hills and cliffs, creating favorable environments for mangrove formation (Hadlich 2009).

Further south, from the Garcez Inlet on the left margin of the Jequiriçá River to the Ponta do Curral Bar, is a coastal plain with wide regressive beach ridges and wide marine-built terraces separating the Todos os Santos Bay from the islands of Tinharé and Boipeba. Beyond Pratagi Beach wide regressive beach ridges, sometimes covered by *restinga* vegetation or exposed, sometimes containing mangroves and coastal lagoons separated by internal marine-built terraces, face the Camamu Bay, another inlet of considerable size incorporating several rivers, bays, and tidal channels covered by mangrove forests. This bay is located on the southern coast of Bahia, where extensive areas of mangrove forests are present.

The Santarém, Igrapiúna, Orojó, and Serra rivers are the largest rivers that flow into Camamu Bay forming an intricate and diverse estuarine complex: some 6 km further south in the county of Itacaré, on the right margin of the Contas River (Ilhéus municipality). The coast of Itacaré is characterized by sandy beaches anchored by promontories of the Barreiras Formation. From its southern limit, the coast is straight, interrupted by small mangrove-lined estuaries. From Ilhéus to the mouth of the Una River, and for about 50 km, there are straight sandy beaches interrupted by marine-cut terraces, remnants of old razed Pleistocene cliffs. At least four river bars and tidal channels make up an intricate estuarine complex where mangrove forests are found. From the mouth of the Salsa River to the mouth of Jequitinhonha River, several environmental, geological, and geomorphological processes interact to produce a variety of features and strand plain evolutionary sequences. Among the morphological features in the wide strand plain are Pleistocene and Holocene marine terraces, beach ridges, alluvial fans, paleo-cliffs, and residues of marshes and mangroves.

From the south of the Jequitinhonha River plain, the occurrence of offshore reefs such as the Coroa Alta Reef is reflected in the coastal morphodynamics, which becomes influenced by wave refraction. In this part of the coast, the Santo Antônio and Braço do Norte rivers merge to flow into the Atlantic Ocean, in narrow watercourses lined by mangroves, near the village of Santo Antônio. Another mangrove-lined estuary is the João da Tiba River that empties into the Atlantic along the right margin of the city of Santa Cruz de Cabrália. Along this sector of the coast, offshore reefs play an important morphodynamic role in inducing the formation of tombolos.

Along the coast of Porto Seguro, mangrove forests are restricted to the estuary of the Buranhém River, which is considerably altered and polluted. Several small mangrove-lined estuaries occur south of Porto Seguro, such as those of the Frades River, in Itaquena, the Caraíva River, by the namesake city, and the Corumbau River, that flows into Ponta de Corumbau. A coastal feature that is not related to any large river is the presence of beach ridges of the coastal plain of Caravelas, which are old Pleistocene and Holocene formations that become truncated at Ponta da Baleia. This coastline reflects a complex interaction among the offshore reef growth (the Abrolhos Reef-Bank), wave refraction, sea-level changes, and long-shore drift.

Offshore reef development led to the formation of a low-energy zone in the vicinity of Caçumba Island that favored the deposition of fine sediments and extensive mangrove and tidal flat development (Dominguez 2009). The left margin of the Doce Stream is the border between the states of Bahia and Espírito Santo, which is the southerly end of the Eastern Coast sector. A striking feature of this stretch of coastline is the presence of the Barreiras Formation near the coast. At Conceição da Barra (ES), the Doce River hosts a small mangrove stand. The total mangrove area along Espírito Santo is 70.35 km^2, and the species found include *R. mangle*, *L. racemosa*, *Avicennia schaueriana* Stapt & Leechm., and *A. germinans* (Vale and Ferreira 1998). The wide strand plain of the Doce River is related to the transgression and regression of the sea level during the Quaternary and comprises beach ridge terraces, lagoons, freshwater, and mangrove swamp deposits. The strand plain is the result of the interaction of sea-level with fluvial sediment inputs, wave reworking, and drowning of the paleo-delta.

The area receives several rivers that flow to the Atlantic, but some could not break through the beach ridges forcing diversions, to the north or the south. In these rivers, the development of mangroves is extensive, with the São Mateus estuary being the largest with 10.1 km^2 of mangrove forest. The mouth of the São Mateus River reflects both erosive and depositional processes, possibly cyclical, that destroyed in 1991 part of the city of Conceição da Barra (Vale 2010; Vale and Ross 2011).

1.6 Southeastern Coast (19° 40′19″ S to 25° 39′53″ S)

The Southeastern Coast (see Chap. 3, Maps 12, 13, 14, and 15) begins along the paleo-cliffs at the right margin of Doce River originating from wave erosion during the Quaternary marine transgression. Because of the large volume of freshwater and sediments carried by the Doce River into the Atlantic Ocean, mangroves are not found here. The abrupt narrowing of the Piraquê coastal plain of the Doce River is due to a smaller river, the Doce Stream that separates the plain from the Barreiras Formation at Aracruz.

From the Barra do Sahy (ES), a sequence of beaches occurs where mangroves are absent. About 10 km separates the Barra do Saí from the mouth of the Piraquê River, which is formed by the Piraquê-Mirim and the Piraquê-Açu rivers (Vale and Ferreira 1998). Completely inserted within the Barreiras Formation, both have well-developed mangrove forests on their margins, occupying an area of 15.80 km^2 (Vale and Ferreira 1998). In the section that stretches from the mouth of the Piraquê River to the entrance of Vitória Bay, mangrove forests are found in small rivers, such as Reis Magos and Jacaraípe.

The Barreiras Formation retreats and reappears as paleo-cliffs and marine erosional terraces, which constitute laterites, on which some red mangroves (*R. mangle*) are seen, but they do not form groves of "rock mangroves," as in the coast of Bahia (Vale and Schaeffer-Novelli 2018).

The coastal landscape of the city of Vitória has both continental and island portions, a high heterogeneity, and geological-geomorphological diversity. The granitic-gneiss outcrops that characterize the relief of the Espírito Santo State coast, especially those found in the proximity of the Vitória Bay, have acted as traps for sediments brought by the several rivers that drain into the bay (Ferreira 1989). This process of sedimentation provides for the accretion of mangroves, a fact visible in the Santa Maria da Vitória River Delta.

The Vitória Bay is one of the most prominent features of the Espírito Santo coast, where the Santa Maria da Vitória, Bubu, and Aribiri rivers, as well as the Passagem and Lameirão channels, add up to 18 km^2 of mangrove landscape around the bay (Vale and Ferreira 1998). A striking feature of these mangrove forests is the occurrence of salt flats, particularly in transition areas as well as in the inner part of the larger mangrove forests.

From the entrance of the Vitória Bay toward the south, the coast presents its morphology dominated by a small inlet anchored in Precambrian outcrops that reach the coast and continue to the south at Anchieta; dominated by broad truncated sand cords due to transgression and regression processes or dominated by the Barreiras Formation, which, to the south, forms active cliffs (Vale and Schaeffer-Novelli 2018). Although the coast has more inlets toward its southern end, the occurrence of mangroves is lower than on the estuaries and rivers of the northern portion. Mangroves are found in Guarapari, on the margin of the estuaries of the Una and Perocão rivers, which break through to the sea at Santa Mônica, and along the margin of the Jabuti, Aldeia Velha, and Lameirão rivers, which together with other smaller ones open into Guarapari Bay where mangroves occupy 5.7 km^2 (Vale and Ferreira 1998). Furthermore, mangrove forests are found in the Benevente River estuary, in Anchieta, where the most structurally developed and best-conserved forests in the state are. This mangrove forest occupies 4.57 km^2 along the river with the predominance of *R. mangle* and *Avicennia* spp.

The rivers Iconha, Itapemirim, and Itabapoana are characterized by the occurrence of riverine mangrove forests. These features extend from the right margin of the Itabapoana River until reaching the Paraíba do Sul River, at the northern coast of Rio de Janeiro State.

The northern coast of Rio de Janeiro State lacks mangroves. Extensive areas of *restinga* vegetation are the distinctive feature of this coastline. Further south mangroves occur in the Paraíba do Sul River delta, near the cities of Atafona and São João da Barra. Here, several of the paleo-channels can be observed reflecting the dynamics under which the delta was formed. There are truncated beach ridges on both sides of the river, marine terraces, *restinga* vegetation, and mangrove deposits. A coastal feature that stands out at the mouth of the Paraíba do Sul River is the Convivência Island, which supports the interpretation that this feature was part of a cuspate destructive delta, dominated by waves Dominguez et al. (1981). Mangroves are found on both margins of the delta associated with interridge depressions that form a mangrove-lined channel that runs parallel to the coastline as a branch of the Paraíba do Sul River with a considerable mangrove coverage.

Southward, coastal aeolian dunes occur in the vicinity of Cabo Frio because of the local arid microclimate generated by a periodic but intense oceanic upwelling (Dias and Kjerfve 2009). The inlet between the peninsulas of Búzios, Cabo Frio, and Arraial do Cabo constitutes an interesting stretch of the coast of Rio de Janeiro, but the occurrence of mangrove is restricted to the small Una River. From Arraial do Cabo to the entrance of Guanabara Bay, the coastline presents a succession of coastal lagoons that reflect their origin as features shaped by marine transgressions and regressions. The largest hypersaline coastal lagoon in the country is the Araruama Lagoon, surrounded by *restinga* and halophytes such as *Sesuvium portulacastrum* (L.) L., *Salicornia gaudichaudiana* Mog., and *Blutaparon portulacoides* (St. Hil.), as well as the Poaceae *Sporobolus virginicus* (L.) Kunth (FEEMA 1988; Debenay et al. 2001). This coastal stretch takes an east-west orientation, marking the end of the Barreiras Formation.

The Guanabara Bay is an iconic Brazilian feature. Satellite images dating back to 2018 show continuous mangrove areas, the largest one located in the north part of the bay, drained by the Macacu, Guapi, Guaraí, Cacerebu, and Guaxindiba rivers (Zee et al. 2017). According to Pires (2010), this complex encompassed a mangrove area of 10.36 km^2 in 2002. A smaller area of remnant mangrove forest is found close to Jardim Gramacho, where a fringe of variable width from the water line to the urbanized littoral ranges from 700 to 1300 m in width (Vale and Schaeffer-Novelli 2018).

Southward to the Guanabara Bay, a set of E-W barriers enclosing lagoons characterizes the southern coast of Rio de Janeiro where the occurrence of mangroves is limited. The larger occurrence is in Sepetiba Bay, where they are widely distributed, forming fringe forests or islands, such as in the Guaratiba Inlet. Further south the Bocaina Mountain Range defines the border between the states of Rio de Janeiro and São Paulo. In Ubatuba municipality, mangroves occur in the estuary of Escuro River at the Dura Beach (Cunha-Lignon et al. 2009), and in front of Picinguaba Bay on the margins of the namesake river that flows along a narrow fluvial-marine plain in direct contact with Precambrian Bocaina Mountain Range outcrops (Vale and Schaeffer-Novelli 2018).

Within the wider fluvial-marine plain of the city of Caraguatatuba, on the margin of the Juqueriquerê River, mangroves are found within an urban setting. In the county of São Sebastião, there are urban mangroves by the ferry boat terminal and in Araçá Bay (Schaeffer-Novelli et al. 2018). At the southernmost part of this county, mangroves also are found on the margins of the Sahy River, which ends at Barra do Sahy Inlet. Here, the development of holiday homes is a significant conservation and land-use issue. Further south the mangroves at the Juquehy River persisted despite the river rectification. In Bertioga municipality, mangroves occur along both margins of the Bertioga Channel and at Itaguaré and Guaratuba estuaries.

A wide fluvial plain has developed in the central portion of the São Paulo State, where the Serra do Mar Mountain Range retreats inland. One of the largest archipelagos on the Brazilian coast lies right behind the coastline comprising the municipalities of Santos e São Vicente; named the Santista-Vicentino Archipelago (Vale and Schaeffer-Novelli 2018). The Santos Port, the largest South American harbor complex, is in this complex of rivers that flow toward the Atlantic. Colocated is the greatest concentration of industries in the country shaping an industrial and urban complex that includes the city of Cubatão. Here mangroves are found in a sheltered environment, with an ample contribution of freshwater and terrigenous sediments, that together with a suitable climate guarantees their establishment, development, and persistence. The original extent of the Baixada Santista mangroves was estimated by Silva et al. (1993b) as 131 km^2, with only 53 km^2 in relatively good condition by 1991. Within this estuarine complex, the largest extents of mangroves persist on Barnabé, Bagres, and Piaçaguera islands. Among the rivers that flow into the area are the Piaçabuçu, Santana, Cubatão, Casqueiro, Jurubatuba, Moji, Quilombo, Itapanhaú, and the Bertioga Channel. Mangroves still thrive in some of these rivers (Menghini et al. 2011).

Beyond this estuarine complex, toward the south, the coastal plain of São Paulo becomes mostly rectilinear, but mangroves are found in the county of Itanhaém, on the margins of the river of the same name. As a historical note, the mangroves of Itanhaém were described by Lamberti (1969) in one of the first modern mangrove studies in Brazil. The waters of Jequiá, Aguapé, and Preto rivers form the Itanhaém estuary that shows well-developed mangroves. Here the Precambrian outcrops reach the coastline, forming high outcrops, still covered by Atlantic Forest vegetation. The Guaraú Outcrop (about 500 m high) separates two rivers in which mangroves occur along their margins. The Peruíbe River rectified before the lower estuary runs along the north slope of Guaraú Outcrop and has altered mangroves because of the proximity to an urban area. On the southern slope of the Guaraú Outcrop mangroves are much better preserved. About 40 km south is the coastal plain of Cananéia-Iguape, another notable geomorphic feature of the Eastern Coast sector in the state of São Paulo (Ab'Sáber 2001).

Before the *restinga* barriers had been formed, the sea made shallow penetrations into the entire regional coast, reaching, during the main Holocene transgression, the foothills of the Serra do Mar Mountain Range. The Cananéia-Iguape coastal system is composed of three natural main islands: Cardoso, Comprida, and Cananéia, separated by rivers and channels. Also, the artificial Iguape Island delimitates the system to the southwest by the Valo Grande Canal opened in 1852 (see Cunha-Lignon et al. 2011). The ca. 64-km-long Comprida Island, between the Cananéia and Icapara inlets, integrates the Iguape-Cananéia Lagoonal System, the result of the latest and faster transgression and marine regression of the Holocene. The island is an elongated sequence of regressive sandy ridges, whose width varies between 3 and 4 km, limited to the west by a lagoon that because of its length has different names: Mar de Cananéia, or Mar de Fora in the southwest, adjacent to Cananéia, Mar

Pequeno or Mar de Iguape in the northeast adjacent to Iguape Island. To the east, the system is bordered by the Atlantic. The Cardoso Island is separated from the mainland by the Ararapira Channel and the Trapandé Bay, with dense mangrove colonization.

The Varadouro Channel connects the Cananéia-Iguape Lagoonal System to Pinheiros Bay, in the state of Paraná (Schaeffer-Novelli et al. 1990a). Here, mangroves cover islands of various dimensions. Another feature that stands out in this part of the Southeastern Coast is the Paranaguá Bay (PR), formed by a combination of small bays, inlets, and islands. The Mel and Peças islands are found at the entrance of the bay.

The last portion of the Southeastern Coast to be described comprises an area from the Antonina Bay, where mangroves are already found in a subtropical setting. To the south, the Paranaguá Bay opens within a wide coastal plain, in which several rivers flow parallel to the coastline draining into the sheltered waters of the bay. The Guaratuba Bay, close to the limit with the state of Santa Catarina, is another large indentation where mangroves have developed along the margin of tidal channels on the edges of small hills.

1.7 Southern Coast (25° 39′53″ S to 33° 45′07″ S)

The last section of the coast starts at Paranaguá Bay and extents to Chuí (RS), the southernmost tip of the country (see Chap. 3, Maps 15, 16 and 17). This coast rests against the crystalline massifs that form the Serra do Mar Mountain Range that stretches from the state of Rio de Janeiro (21° S) to Santa Catarina (28° S) (Almeida and Carneiro 1998). Its most prominent feature is the scraped coastal range that when it intersects the coastline forms embayments, strand plains, and estuaries. Mangroves occur along the coast of the state of Santa Catarina, including on the margins of the Saí-Guaçu and Saí-Mirim rivers on the northern coast of the state (Ab'Sáber 2001).

Among the most significant estuaries is the São Francisco Bay (also known as Babitonga Bay) that extends over 130 km^2, where the adjacent lands include stretches of older outcrop formations, sometimes in direct contact with the Atlantic or inland reaching the scarped coastal range of the Serra do Mar (Vale and Schaeffer-Novelli 2018; França et al. 2019). Three sections can be distinguished in this bay: One being the bay itself, which receives water from two other sectors; the Palmital Channel to the north and another to the south composing the sector that separates the São Francisco do Sul Island from the mainland, the Linguado Canal. This sector of the Santa Catarina State's coast shelters about 75% of all the mangroves of the state, covering 59.94 km^2 (Vale and Schaeffer-Novelli 2018).

Southward a combination of beach arches anchored in the protuberances of the Precambrian basement and colder waters do not propitiate the colonization by mangroves. At the northern end of Santa Catarina Island, there is a small stretch of fringed mangroves at the entrance of the Norte Bay. At the Itacorubi River, mangroves and associated salt marshes are well preserved (Sierra de Ledo and Soriano-Sierra 1998). In the Sul Bay or Sul Channel, the mangroves of Tavares River correspond to the second largest mangrove area of the island covering 7.44 km^2 (Espinoza 2008).

The Southern latitudinal limit for *R. mangle* is at Sonho Beach (27° 50′ S), Palhoça municipality, in the mainland of Santa Catarina State. The white mangrove *L. racemosa* and the black mangrove *A. schaueriana* reach their southern limit at the Ponta Grossa River (28° 30′S) within the Santo Antônio Lagoon, Laguna municipality (Cintrón-Molero and Schaeffer-Novelli 1981; Schaeffer-Novelli et al. 1990b; Soares et al. 2012) (see Chaps. 3 and 18). Beyond this point is Brazil's largest Quaternary sandy progradation. The coastal plain is made up of a series of barriers separated by low line areas occupied by fresh and salt marshes and large water bodies, some connected to the sea (through the Rio Grande and the Tramandaí inlets), some not.

The Patos Lagoon is a coastal lagoon of relevance to migratory birds coming from both North and South pole, depending on the season (see Chap. 12). The most prominent lagoon is the Patos Lagoon. There, salt marshes colonize mud flats, mainly *Spartina densiflora* Brong and *Juncus* spp. (Marangoni and Costa 2009). Nevertheless, other marsh species can be found in the area, namely, *Spartina alterniflora* Loisel., *Scirpus maritimus* L., *Scirpus olneyi* A. Gray, *Juncus acutus* L., *Fimbristylis spadicea* L., *Myrsine parviflora* A. DC, *Acrostichum danaefolium* Langsd. Fisch (also a mangrove-associate), *Vigna luteola* (Jacq.) Benth, and *Salicornia gaudichaudiana* Mog. (Marangoni 2003). The Southern Coast segment ends at Chuí (33° 45′07″ S).

1.8 Summary

The Brazilian coast is characterized by a broad suite of features spanning several spatiotemporal processes and scales, reflecting the sculpturing of coastal landforms by ecogeomorphic factors. Climatic, geological, oceanographic, and ecological processes interact with sediment supply and antecedent topography to produce the modern coastal landscape that incorporates a record of sea-level change in terms of erosional and depositional features as well as biological indicators. Sea-level oscillations highlight the dynamism of coastal features, helping to comprehend its malleability and diversity. Coasts evolve and adapt at multiple scales; they are not static entities and are best understood within the framework of complex adaptive systems and Earth Surface Systems. Over most of Brazil's coast, mangrove forests and salt marshes are widely distributed although generally confined to sheltered upper intertidal mudflats in the larger coastal indentations, such as bays, estuaries, and behind coastal barriers and barrier islands, in lagoons subject to frequent flooding by tides and saline intrusions.

References

Ab'Sáber AN (2001) Litoral do Brasil. Metalivros, São Paulo

Abreu PC, Costa CSB, Bemvenuti C, Odebrecht C, Granéli W, Anésio AM (2006) Eutrophication processes and trophic interactions in a Shallow Estuary: preliminary results based on stable isotope analysis (d13C and d15N). Estuaries Coasts 29(2):277–285

Adam P (1990) Saltmarsh ecology. Cambridge University Press, New York

Administração Estadual do Meio Ambiente – ADEMA (1984) Levantamento da flora e caracterização dos bosques de mangues do estado de Sergipe – ADEMA Report, Aracaju

Almeida FFM, Carneiro CDR (1998) Origem e evolução da serra do mar. Rev Bras Geocienc 28(2):135–150

Barcellos RL, Lins SRRM, Coelho-Jr C, Travassos PEPF (2017) Processos sedimentares sazonais e análise da feição arenosa no sistema ambiental do Sueste, Fernando de Noronha, Estado de Pernambuco. Arq Cienc Mar 50(1):42–71

Carvalho LM, Barreto MA (2010) Levantamento da Geodiversidade do Estado da Bahia. CPRM, Salvador, 184pp

Carvalho MES, Fontes AL (2006) Caracterização geomorfológica da zona costeira do Estado de Sergipe. In: Geomorfologia Tropical e Subtropical: processos, métodos e técnicas: Proceedings of the VI Simpósio Nacional de Geomorfologia/Regional Conference on Geomorphology, 6–10 September. Universidad Autónoma del Estado de México, Goiânia, 15pp

Carvalho MES, Fontes AL (2007) A carcinocultura no espaço litorâneo sergipano. Rev Fapese 3(1):87–112

Centro de Excelência para o Mar Brasileiro – CEMBRA (2012) Direito do Mar. In: Fernandes LPC (ed) O Brasil e o Mar Brasileiro no Século XXI: Relatório aos tomadores de decisão do País. CEMBRA, Rio de Janeiro, pp 19–35

Cintrón-Molero G, Schaeffer-Novelli Y (1981) Los manglares de la costa brasileña: revisión preliminar de la literatura. Abuchahla GMO (elect rev ed). [S.I] Informe Técnico preparado para la Oficina Regional de Ciéncia y Tecnologia para America Latina y el Caribe de UNESCO y la Universidad Federal de Santa Catarina, 47 pp

Cintrón-Molero G, Schaeffer-Novelli Y (2019) The role of atmospheric-tropospheric rivers in partitioning coastal habitats and limiting the poleward expansion of mangroves along the southeast coast of Brazil. Int J Hydro 3(2):92–94

Costa JLPO, Cavalcanti APB (2010) Fitogeografia da planície deltaica do rio Parnaíba, Piauí/ Maranhão-Brasil: Análise da distribuição das espécies e interferência antrópica. Rev Eletr Geogr 2(4):88–92

Costa CSB, Iribarne OO, Farinac JM (2009) Human impacts and threats to the conservation of South American salt marshes. In: Silliman BR, Grosholtz T, Bertness MD (eds) Salt marshes under global Siege, 1st edn. University of California Press, Berkeley

Costa DFS, Rocha RM, Cestaro LA (2014) Análise fitoecológica e zonação de manguezal em estuário hipersalino. Mercator 13:119–112

Cunha-Lignon M, Menghini RP, Santos LC, Niemeyer-Dinola C, Schaeffer-Novelli Y (2009) Estudo de caso dos manguezais do Estado de São Paulo (Brasil): aplicação de ferramentas com diferentes escalas espaço-temporais. Rev Gerenc Cost Integr 9:79–91

Cunha-Lignon M, Kampel M, Menghini RP, Schaeffer-Novelli Y, Cintrón G, Dahdouh-Guebas F (2011) Mangrove forests submitted to depositional processes and salinity variation investigated using satellite images and vegetation structure surveys. J Coast Res SI64:344–348

Debenay JP, Geslin E, Eichler BB, Duleba W, Sylvestre F, Eichler P (2001) Foraminiferal assemblages in a hypersaline lagoon, Araruama (RJ) Brazil. J Foramin Res 31(2):133–151

Dias GTM, Kjerfve B (2009) Barrier and beach ridge systems of the Rio de Janeiro coast. In: Dillenburg S, Hesp PA (eds) Geology and geomorphology of Holocene coastal barriers of Brazil. Springer, Berlin, pp 226–252

Dominguez JML (2009) The coastal zone of Brazil. In: Dillenburg S, Hesp PA (eds) Geology and geomorphology of Holocene coastal barriers of Brazil. Springer, Berlin, pp 17–51

Dominguez JML, Bitencourt ACSP, Martin L (1981) Esquema evolutivo da sedimentação quaternária nas feições deltaicas dos rios São Francisco (SE/AL), Jequitinhonha (BA), Doce (ES) e Paraíba do Sul (RJ). Rev Bras Geocienc 11(4):227–237

Dominguez JML, Martin L, Bittencourt ACSP (1982) Evolução paleogeográfica do delta do rio Jequitinhonha durante o Quaternário, influência das variações do nível do mar. In: Atas do IV Simpósio do Quaternário no Brasil. ABEQUA, Rio de Janeiro, pp 69–92

Dominguez JML, Andrade ACS, Almeida AB, Bitencourt ACSP (2009) The Holocene barrier strand plains of the state of Bahia. In: Dillenburg S, Hesp PA (eds) Geology and geomorphology of Holocene coastal barriers of Brazil. Springer, Berlin, pp 254–288

Duke NC, Ball MC, Ellison JC (1998) Factors influencing biodiversity and distributional gradients in mangroves. Global Ecol Biogeogr 7:27–47

Espinoza HCF (2008) Evolução temporal da cobertura vegetal do manguezal do rio Tavares (Florianópolis-SC) empregando sensoriamento remoto e SIG. Master thesis, Universidade do Vale do Itajaí, 70 pp

Ferreira RD (1989) Os Manguezais da Baia de Vitória (ES): Um Estudo de Geografia Física Integrada. PhD dissertation, Universidade de São Paulo, 302 pp

Food and Agriculture Organization – FAO (2007) The world's mangroves 1980–2005. FAO, Rome, 77 pp

França MC, Pessenda LCR, Cohen MCL, Azevedo AQ, Fontes NA, Silva FB, Melo JCF Jr, Piccolo MC, Bendassolli JA, Macario K (2019) Late-Holocene subtropical mangrove dynamics in response to climate change during the last millennium. Holocene 29(3):445–456

Fundação Estadual de Engenharia do Meio Ambiente – FEEMA (1988) Perfil Ambiental do Município de Araruama, Rio de Janeiro, Brasil. FEEMA, Araruama

Galloway WE (1975) Process framework for describing the morphologic and stratigraphic evolution of deltaic depositional systems. In: Broussard ML (ed) Deltas, models for exploration. Houston Geological Society, Houston, pp 87–98

Giri CE, Ochieng E, Tieszen LL, Zhu Z, Singh A, Loveland T, Masek J, Duke NC (2010) Status and distribution of mangrove forests of the world using earth observation satellite data. Global Ecol Biogeogr 20:154–159

Hadlich GM (2009) Distribuição de apicuns e de manguezais na Baía de Todos os Santos, Bahia, Brasil. In: Anais do XIV Simpósio Brasileiro de Sensoriamento Remoto, 25–30 April. INPE, Natal, pp 4607–4614

Hartt CF (1870) Geology and physical geography of Brazil. Osgood, Boston, 620 pp

Instituto Brasileiro de Geografia e Estatistica – IBGE (2016) Caracterização do território: posição e extensão. In: Anuário Estatístico do Brasil. IBGE 1, Rio de Janeiro, 457 pp

Junk WJ, Piedade MTF, Lourival R, Wittmann F, Kandus P, Lacerda LD, Bozelli RL, Esteves F, Nunes da Cunha AC, Maltchik L, Schöngart J, Schaeffer-Novelli Y, Agostinho AA (2013) Brazilian wetlands: their definition, delineation, and classification for research, sustainable management, and protection. Aquat Conserv 24(1):5–22

Lamberti A (1969) Contribuição ao conhecimento da ecologia das plantas do manguezal de Itanhaém. Bol Bot 317:23

Lins RC, Fragoso CR, Cirilo JA (2013) Simulações da hidrodinâmica do Complexo Estuarino-Lagunar Mundaú-Manguaba (CELMM/AL). In: XX Simpósio Brasileiro de Recursos Hídricos, 17–22 November. ABRH, Bento Gonçalves, RS, pp 1–8

Long SP, Mason CF (1983) Saltmarsh ecology, Tertiary level biology series. Blackie, Glasgow

Mandelbrot B (1967) How long is the coast of Britain? Statistical self-similarity and fractional dimension. Science 156(3775):636–638

Marangoni JC (2003) Caracterização da paisagem de uma área de preservação: estudo de caso da Lagoinha (Rio Grande, RS). Atlântica 25(2):163–169

Marangoni JC, Costa CSB (2009) Natural and anthropogenic effects on salt marsh over five decades in the Patos Lagoon (Southern Brazil). Braz J Ocean 57(4):345–350

Menghini RP, Coelho C, Rovai AS, Cunha-Lignon M, Schaeffer-Novelli Y, Cintrón G (2011) Massive mortality of mangrove forest in Southeast Brazil (Baixada Santista, State of São Paulo) as a result of harboring activities. J Coastal Res SI 64:1793–1797

Moura-Fé MM (2014) Barreiras: série, grupo ou formação? Rev Bras Geogr Fís 7(6):1055–1061

Oliveira MIM, Bagnoli E, Farias CC, Nogueira AMB, Santiago M (1990) Considerações sobre a geometria, petrografia, sedimentologia, diagênese e idade dos beachrocks do Rio Grande do Norte. In: Congresso Brasileiro de Geologia, Natal-RN 33-33 September, pp 621–634

Pessenda LCR, Gouveia SEM, Ledru M-P, Aravena R, Ricardi-Branco FS, Bendassolli JA, Ribeiro AS, Saia SEMG, Sifeddine A, Menor EA, Oliveira SMB, Cordeiro RC, Freitas AMM, Boulet R, Filizola HF (2008) Interdisciplinary paleovegetation study in the Fernando de Noronha Island (Pernambuco state), northeastern Brazil. An Acad Bras Cienc 80(4):677–691

Phillips JD (2007) The perfect landscape. Geomorphology 84:159–169

Pires IO (2010) Manguezais da região do Recôncavo da Baía de Guanabara: revista através dos mapas. Rev Gerenc Cost Integr 8:5–22

Prance GT, Silva MF, Albuquerque BW, Araújo IJS, Carreira LMM, Braga MMN, Macedo M, Conceição PN, Lisbôa PLB, Braga PI, Lisbôa RCL, Vilhena RCQ (1975) Revisão taxônomica das espécies amazônicas da Rhizophoraceae. Acta Amaz 5(1):5–22

Reis-Neto AS, Cunha-Lignon M, Arruda-Reis MCC, Meireles AJA (2011) The Ceará river mangrove's landscape (northeast Brazil) – comparative analyses of 1968 and 2009. J Coastal Res 2:1802–1805

Ridley HN (1890) Notes on the Botany of Fernando Noronha. Bot J Linn Soc 27:1–95

Santos CLA, Vital H, Amaro VE, Kikhuchi RKP (2007) Mapeamento de Recifes Submersos na costa do Rio Grande do Norte, RN Brasil: Macau e Maracajaú. R Bras Geocienc 25:27–36

Santos M, Cavalcanti L, Matos HR, Schaeffer-Novelli Y, Cunha-Lignon M, Bitencourt MS, Koedam N, Dahdouh-Guebas F (2014) Anthropogenic activities on mangrove areas (São Francisco River Estuary, Brazil Northeast): a GIS-based analysis of CBERS and SPOT images to aid in local management. Ocean Coast Manage 89:39–50

Schaeffer-Novelli Y (1991) Manguezais brasileiros. Full Professorship dissertation, Universidade de São Paulo, 42 pp

Schaeffer-Novelli Y, Mesquita HSL, Cintrón-Molero G (1990a) The Cananéia Lagoon estuarine system, São Paulo, Brazil. Estuaries 13(2):193–203

Schaeffer-Novelli Y, Cintrón-Molero G, Adaime RR, Camargo TM (1990b) Variability of mangrove ecosystems along the Brazilian coast. Estuaries 13(2):204–218

Schaeffer-Novelli Y, Cintrón-Molero G, Soares MLG, Tognella-de-Rosa MMP (2000) Brazilian mangroves. Aquat Ecosyst Health 3:561–570

Schaeffer-Novelli Y, Cintrón-Molero G, Reis-Neto AS, Abuchahla GMO, Neta LCP, Lira-Medeiros CF (2018) The mangroves of Araçá Bay through time: An interdisciplinary approach for conservation of spatial diversity at large scale. Ocean Coast Manage 164:60–67

Sierra de Ledo B, Soriano-Sierra EJ (1998) Contribución para el manejo integrado de sistemas estuarinos y sus recursos en la Isla de Santa Catarina, Brasil. In: Soriano-Sierra EJ, Sierra de Ledo B (eds) Ecologia e Gerenciamento Manguezal de Itacorubi. FEPEMA/UFSC-CCB/NEMAR, Florianópolis, pp 379–388

Silva CP, Pereira CMP, Dorneles LPP (1993a) Espécies de gramíneas e crescimento de *Spartina densiflora* Brong. em uma marisma na Laguna dos Patos, RS, Brasil. Cad Pesq Ser Bot 5(1):95–108

Silva IX, Moraes RP, Santos RP, Pompéia SL, Martins SE (1993b) A degradação dos ecossistemas da Baixada Santista, São Paulo. In: III Simpósio de Ecossistemas da Costa Brasileira: subsídios a um gerenciamento Ambiental. ACIESP, Serra Negra, pp 30–38

Silveira JD (1964) Morfologia do litoral. In: Azevedo A (ed) Brasil, a terra e o homem, I As Bases Físicas. Companhia Editora Nacional, São Paulo, pp 253–305

Simões E, Oliveira AL (2018) Brazilian Blue Initiative – Structure, Definition and Guidelines. Brasília: MMA, ICMBio, WWF-Brazil

Soares ML, Estrada GCD, Fernandez V, Tognella MMP (2012) Southern limit of the Western South Atlantic mangroves: assessment of the potential effects of global warming from a biogeographical perspective. Estuar Coast Mar Sci 101:44–53

Souza CR (2015) The Bertioga coastal plain: an example of morphotectonic evolution. In: Vieira BC, Salgado AAR, Cordeiro LJ (eds) Landscapes and landforms of Brazil. Springer, Berlin, pp 115–134

Souza Filho PWM (2005) Costa de manguezais de macromaré da Amzônia: cenários morfológicos, mapeamento e quantificação de áreas usando dados de sensores remotos. Rev Bras Geof 23(4):427–435

Souza CRG, Hiruma ST, Sallun AEM, Ribeiro RR, Sobrinho JMA (2008) "Restinga": Conceitos e Empregos do Termo no Brasil e Implicações na Legislação Ambiental. Instituto Geológico, São Paulo, 104 pp

Spalding M, Kainuma M, Collins L (2010) World Atlas of Mangroves. Eathscan, New York, 319 pp

Tavares PF (2015) Leis autorizativas para a supressão de manguezais e as consequências socioambientais para o estado de Pernambuco. Master thesis, Universidade Federal de Pernambuco, 86 pp

Vale CC (2010) Correlação entre os processos erosivos e sedimentares e o comportamento das espécies vegetais dos manguezais da foz do rio São Mateus, litoral norte no Estado do Espírito Santo. Rev GEOUSP 27:113–134

Vale CC, Ferreira RD (1998) Os manguezais do litoral do Espírito Santo. In: Anais do IV Simpósio de Ecossistemas Brasileiros. ACIESP, São Paulo, pp 88–94

Vale CC, Ross JLS (2011) As transformações morfológicas e fitogeográficas do estuário do rio São Mateus, litoral norte do Estado do Espírito Santo entre 1970–2008. Rev GEOUSP 21:3–23

Vale CCV, Schaeffer-Novelli Y (2018) A zona costeira do Brasil e os manguezais. In: Instituto Chico Mendes de Conservação da Biodiversidade Atlas dos Manguezais do Brasil. Instituto Chico Mendes de Conservação da Biodiversidade, Brasília, pp 34–53

Vilar JWC, Santos MAN (2011) As áreas litorâneas de Sergipe (Brasil): da análise geográfica à gestão integrada do território. Rec geogr Am Cent EGAL:1–19

Zee D, Medeiros R, Scarano FR, Klabin I (org) (2017) Baía de Guanabara: Passado, Presente, Futuros. Andrea Jackobsson Estúdio Editorial Ltda, Rio de Janeiro, 157 pp

Chapter 2
Environmental and Biotic Factors Driving Distributional Patterns in Mangrove and Salt Marsh Ecosystems

Sarah Charlier-Sarubo, Marília Cunha-Lignon, and Eduardo J. Soriano-Sierra

2.1 Introduction

Mangrove distribution is regulated by a complex combination and interaction of biotic and environmental factors, including salinity, soil type, degree of anoxia, nutrient availability, physiological tolerances, predation, and competition, at different scales (Ellison 2002). Seemingly, according to Alongi (2008), both composition and structure (also called phytosociology) of mangrove forests are the results of a complex interplay of physiological tolerances and competitive interactions leading to a mosaic of interrupted or arrested succession sequences in response to physical/chemical gradients and changes in geomorphology.

Zonation theories in mangroves have a rich experimental and observational history (Krauss et al. 2008). Mangrove forests have been described as having bands of vegetation, typically arranged along tidal gradients which are dominated by one or two species (Feller et al. 2010). It is related to physical processes such as tidal frequency and edaphic factors that change along a gradient from seaward to landward (Friess et al. 2012).

Although several hypotheses have been put forth to explain zonation, most publications report only dominant species and describe zones without quantifying them (Ellison et al. 2000). It is essential to consider that mangrove and salt marsh

S. Charlier-Sarubo (✉)
Centro de Estudos do Mar, Universidade Federal do Paraná, Pontal do Paraná, PR, Brazil

M. Cunha-Lignon
Campus de Registro, Universidade Estadual Paulista, Registro, SP, Brazil

E. J. Soriano-Sierra
Núcleo de Estudos do Mar, Universidade Federal de Santa Catarina, Campus Reitor João David Ferreira Lima, Florianópolis, SC, Brazil

Y. Schaeffer-Novelli et al. (eds.), *Brazilian Mangroves and Salt Marshes*, Brazilian Marine Biodiversity, https://doi.org/10.1007/978-3-031-13486-9_2

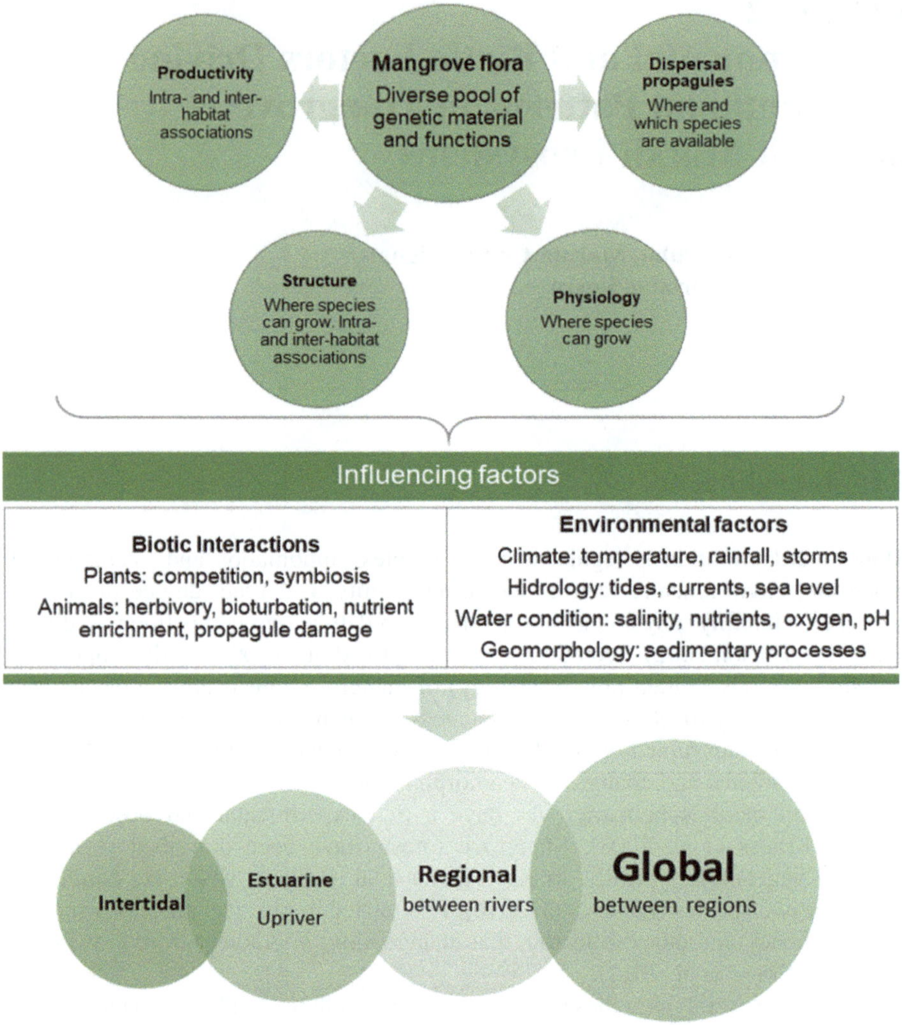

Fig. 2.1 Relationship between mangrove floristic and its main strategies to succeed under a range of biotic and environmental factors, considered at four geographic scales. (Based on Duke et al. 1998)

ecosystems respond to biotic and abiotic factors, considering different time and spatial scales (Fig. 2.1).

Physical form and structure, physiological capabilities, productive capacity and growth, and reproductive development with the dispersal of propagules are the main strategies mangrove species developed to live in such a harsh environment. Each attribute is then influenced by a range of biotic and environmental factors, which merge to determine the distributional patterns of mangrove at different scales (Ball

1996; Duke et al. 1998; Krauss et al. 2008; Feller et al. 2010), even though biotic interactions are mostly reported at intertidal scales.

2.2 Biotic Factors Influencing Species Richness and Distribution of Mangroves

According to Duke et al. (1998), the distribution of mangrove species across the intertidal profile is influenced by associated fauna, which may indirectly assist in the establishment of propagules. Predation on propagules is also essential for the distributional patterns of some taxonomic groups of mangroves and in specific geographic regions (Smith 1996).

Cannicci et al. (2008) have reviewed many studies documenting the impacts of fauna on forest development, productivity, and structural complexity, concluding that the ecological role played by insect and crab herbivory is not restricted to tissue lost or damage but also at an ecosystem level on functioning and performance.

Other authors (Berger et al. 2006; Peters et al. 2020) have proposed modeling approaches to explain plant competition and succession on mangrove forests. They initially state dominating species will be gradually replaced in the canopy due to tree competition for light and use crown height, crown shape, and tree size as main competition parameters. Berger et al. (2006) argued that it is essential to assume that the fast growth rate of the pioneer species slows down relative to those of the following species within succession. A decrease in nutrient availability may be responsible for these relative changes in the growth rates of different species.

2.2.1 Predation on Mangrove Propagules and Its Effects on Forest Structure

Cannicci et al. (2008) summarized three models that have been proposed in the literature in order to evaluate and explain the impact of crab propagule predation on vegetation structure: (i) The dominance-predation model, which suggests an inverse ratio between the rate of predation of particular species and its dominance in the grove canopy; (ii) the canopy-gap-mediated model, which considers that predation could be more intense under closed canopies than in adjacent gaps, and (iii) the flooding regime model, which considers the time available for crabs to forage at low intertidal and upper intertidal belts, suggesting that propagule predation may be related to inundation period.

Clarke and Kerrigan (2002) conclude that a dominance-predation effect may occur in some mangrove stands, but it is not a general phenomenon among the tested species. Their assessment has also corroborated the canopy-gap-mediated model, suggesting that crabs can clear high-density stands and enhance establishment and growth of new seedlings. Krauss et al. (2008) observed that the impact of arthropod consumers on the survival of mangrove propagules or seedlings varies with light and temperature conditions in several different systems. While some studies in Australia presented that canopy gaps could afford a refuge from crab predation for mangrove propagules, a related study on the Caribbean coast of Panama, where the crab species are different, found no difference in crab predation rates between understory and gap environments (Krauss et al. 2008).

Farnsworth and Ellison (1997) demonstrated the significant presence and the ecological importance of predispersal propagules predation on mangroves world-wide. They have found the most frequent predators are scolytid beetles, lepidopteran larvae, and crabs. While crabs and beetles were detected in propagules throughout the world, burrowing moth larvae were more common in the southern hemisphere.

2.2.2 *Bioturbation Processes Influencing Ecosystem Engineering*

Crabs are considered the engineer species of mangrove ecosystems by their habitat and their bioturbation activity, which significantly decreases ammonium and sulfide concentrations in mangrove soil, playing a crucial ecological role in the structural and functional stability of mangrove ecosystems (Cannicci et al. 2008).

They physically modify the structure of sediments, favor their oxygenation and their enrichment in organic matter, breaking and burying leaves and propagules into their burrows. They also improve the circulation of water and organic matter, thus positively benefiting mangrove productivity (Capdeville et al. 2019).

Kristensen and Alongi (2006) proved that the activity of fiddler crabs (*Uca vocans*) affected redox-sensitive elements, such as iron (Fe) and sulfur (S), down to a depth of 2 cm, not only during feeding but also during other activities involving movement, such as walking when the legs sink into the sediment. Further, the deposition of feeding pellets and burrow maintenance activities probably enhances the effective mixing depth, resulting in increased productivity of *Avicennia marina* saplings associated with those crabs.

2.2.3 *The Importance of Dispersal Ability and Factors Regulating Propagule Establishment*

Most mangrove species are typically dispersed by water-buoyant propagules, which allow them to take advantage of estuarine, coastal, and ocean currents both to restore existing stands and to establish new ones (Duke et al. 1998). The direction they travel depends on sea currents and land barriers, but the distance they reach stands on the time propagules remain buoyant and viable, which differ for each species (Duke et al. 1998).

According to the previous authors, it seems that a range of factors may influence propagule dispersal on a local scale (intertidal and estuarine scale), possibly including propagule size, but also water quality, substrate condition, tidal position, immerse propagule durability, germinating and rooting capacity, resistance of propagules to damage caused by predators, and the general viability of propagules.

Although dispersal potential alone does not correlate with zonation, the combination of buoyancy and early growth may influence the distribution of some high shore species, which float for long periods and initiate roots and shoot more slowly than other floating species (Clarke et al. 2001). Delgado et al. (2001) suggested that factors as propagule abundance, dispersal, buoyancy, and sensitivity to flooding and mechanical stress were essential in determining mangrove species' distribution in the environmental setting of an estuary on the Pacific coast of Costa Rica.

According to Sousa et al. (2007), dispersing propagules are consistently transported downward toward lower tidal heights; this result strongly suggests that dispersal limitation plays an essential role in structuring higher elevation, inland forest stands. On the other hand, agitation by waves may prevent the smaller propagules from establishing at the lower, seaward boundary of the forest. In the zone between these upper and lower boundaries, there is no limitation by propagules supply nor water depth, and a variety of interacting biotic and abiotic processes determine forest composition and species distribution (Sousa et al. 2007). Such local dispersal may also influence the considerable genetic variation that has been observed among local populations of *Avicennia germinans*, *A. marina*, and *Rhizophora mangle* (Sousa et al. 2007).

2.3 Environmental Factors and Physiological Adaptation

The distribution of mangroves and salt marshes is related to physical, environmental, and climatic factors at different scales (Fig. 2.1), as proposed by Ball (1996), Smith (1996), Duke et al. (1998), Pennings and Moore (2000), Krauss et al. (2008), and Feller et al. (2010). If a species is present, then the environment must be suitable for it, not the opposite. Hence, the presence of certain species in mangrove environments depends first on the proximity of source populations (Duke et al. 1998), while physiological attributes of mangrove species also play critical roles in the development of forest structure, and certain environmental factors can lead to adaptive responses (Ball 1996, 2002; Feller et al. 2010; Arrivabene et al. 2014).

Factors like physiological tolerance (e.g., salinity, flood, and temperature), growth responses, competitive abilities toward other species, availability of suitable places for the establishment, and establishment ability regulate each mangrove plant in some way; from the individual to its position along the intertidal profile, from their upstream location within an estuary to their coastal range and throughout the world in one or more biogeographic regions (Duke et al. 1998; Krauss et al. 2008).

2.3.1 Climate Parameters and Its Relation to the Early Development

Krauss et al. (2008) identified the importance of not frequently studied environmental factors, such as temperature, atmospheric CO_2 concentration, and sea-level rise (SLR) as important drivers not only to mangrove establishment on a global scale but also to seedling growth and persistence on an intertidal scale.

Winter air temperature and precipitation regimes greatly influence mangrove physiology and literature indicates that freeze and drought sensitivity are range limit dependents (Lovelock et al. 2016; Osland et al. 2017). Although mangroves may face either high or low temperatures extremes, most studies have emphasized their sensitivity to freezing temperatures, since this factor limits its distribution latitudinally on a regional scale (Krauss et al. 2008). Those authors also cite some effects, which high temperature may cause, such as physiological processes limitation and death of the tissues or the whole plant.

By contrast, salt marshes emerge to be less constrained by climate compared to mangroves, which is demonstrated by their occurrence in the subtropics and tropics in restricted areas where mangrove development is excluded. Thus, traditional mapping of coastal wetland distribution probably underestimates the overall area of salt marshes found in the tropics, leading to the presumption that latitude is the overriding constraint controlling marsh (temperate) versus mangrove (subtropical to tropical) (Friess et al. 2012).

While increased atmospheric CO_2 is likely to improve the potential growth of mangrove species, some factors may delay, limit, or prevent mangrove seedling and growth, such as competition from other species, presence of maternal reserves, or other phenological aspects (Krauss et al. 2008). They also mention that CO_2 response will usually depend on other growth-limiting factors such as salinity and nutrient availability.

2.3.2 Effects of Salinity Gradients on Mangrove and Salt Marsh Species Distribution

To succeed on a substrate that contains high concentrations of soluble salt, mangrove trees present salt-eliminating mechanisms that have been classified into three groups: (1) Salt excluders, (2) salt secretors, and (3) salt accumulators (Parida and Jha 2010). The salt-excluding *R. mangle* keeps the xylem sap almost free of salt by ultrafiltration at the membranes of root cells. Salt secretors regulate internal salt levels by secreting excess salt through foliar glands, like *Laguncularia racemosa* and *A. schaueriana* do. Salt accumulators, on the other hand, accumulate high concentrations of salt by efficient isolation of ions to the vacuoles in the leaf, translocation outside the leaf, possible cuticular transpiration, and efficient leaf turnover to salt shedding (Parida and Jha 2010), such as the genera *Rhizophora* and *Avicennia*. Both genera present more than one mechanism of salt adaptation in different plant tissues: *Rhizophora*, salt exclusion and accumulation, while *Avicennia* presents all three mechanisms.

As reported by Costa and Davy (1992), in the tropical and subtropical coastal regions of Brazil, salt marshes occur as fringes in front of mangroves and as spots in the interior of these. In the first case, the smooth cordgrass *Spartina alterniflora* generally dominates salt marshes. When salt marshes occur in the interior of mangroves, in oligohaline places, the species generally dominant is *Spartina densiflora*, accompanied by several other plants, like those of the genera *Limonium* and *Salicornia*.

In well-structured salt marshes, as is often the case in temperate zones, the zoning of vegetation on the intertidal scale is characteristic, forming distinct bands according to the salinity, frequency, and duration of the floods (Soriano-Sierra 1999). The porewater salinity accompanies the flood gradient, presenting the same salinity of the waters of the estuary at the more frontal parts of salt marshes, as it decreases to the highest part. In the low and medium salt marsh, only a few halophytic herbs occur; however, in the upper salt marsh, more than 40 species can occur that are only tolerant of low salinity.

According to Ball (1996), mangroves species differ in their ability to grow in extremely low or high salinities, although maximum growth of most species occurs under moderate salinity regimes. Ball (1998) concluded that three factors seem to promote the highest species richness in moderate saline sites: (i) salinities are within the range of tolerance of most species; (ii) species richness is greatest where the diversity of functional types is high; and (iii) seasonal variation in salinity, within tolerance limits, may cause temporal variations in resource use, which may support species coexistence.

Cunha-Lignon et al. (2011a) demonstrated that although high sedimentation rates promote mangrove colonization, the low salinity overcomes it by creating suitable conditions to macrophytes reproduction, which has a superior adaptation to salt-limited environments (Schaeffer-Novelli et al. 2016).

2.3.3 Sediment Fertility and Its Influence on Mangrove Structure

Sediment nutrient availability can vary spatially along tidal gradients and temporally with seasonal and interannual variation in nutrient delivery and cycling (Feller et al. 2002, 2010).

Feller et al. (2002) proved that essential nutrients such as nitrogen (N) and phosphorus (P) are not uniformly distributed within mangrove forests. Sediment fertility can switch from conditions of N to P limitations across fringe, dwarf (or stunted), and transition zone, and the same nutrient limitation does not affect all ecological processes within an ecosystem (e.g., plant growth, internal nutrient cycling, decomposition) consistently. On the other hand, Reis et al. (2019) registered in fringe mangroves higher N inputs and higher N losses to the atmosphere, compared to basin mangroves in polluted sites (N-enriched areas) and conserved sites (non-N-enriched areas). According to the authors, the current and future N pollution scenarios via water pollution and N deposition from the atmosphere, mangroves could become a more significant source of nitrous oxide to the atmosphere as fringe mangroves continue to cycle higher inputs of N.

To understand how nutrient availability affects within-stand nutrient-cycling processes for *R. mangle*-dominated forests with high P-deficient soils, fertilization experiments indicated that P deficiency is a significant factor limiting the primary productivity of dwarf red mangroves (Feller et al. 1999; Lovelock et al. 2006).

McKee (1993) concluded that spatial and temporal variation in sediment redox potential and sulfide levels influences both mangrove seedling dynamics and root systems of adult trees. Seedling recruitment patterns across the intertidal zone may also differ locally and regionally depending on nutrient input and canopy structure (McKee 1995).

2.4 Response of Mangrove and Salt Marsh Species to Sedimentary Processes

The distribution of mangrove species corresponds with the intertidal accommodation space, which progressively supplies through sediment accumulation (Woodroffe et al. 2016). The vegetation structure of mangroves forests is a good indicator of sedimentary processes and environmental changes (Cunha-Lignon et al. 2009a).

Confirming the importance of these processes to mangrove dynamics, Thomas et al. (2017) produced a global map using seven categories of mangrove changes. According to these authors, natural processes of deposition and erosion contribute to 32.4% and 20.3% of mangrove area changes around the globe, respectively; moreover, both processes were commonly observed simultaneously. Although these observations cannot be readily managed as with human-induced alterations, it is known that mangrove forest extent is dynamic and strongly influenced by external pressures (Schaeffer-Novelli et al. 1990; Schaeffer-Novelli et al. 2000; Thomas et al. 2017).

2.4.1 Depositional Process Regarding Sea Level Rise

The origin and evolution of salt marshes result from processes of sedimentation of fine particles from the sea and dry land in sheltered coastal sectors (Boorman 1999). The substrates are initially colonized by algae and bacteria, which form a biofilm that contributes to the retention of sediments. It is then the turn of halophilic macrophytes, which retain more sediment due to their high stem and radicular density.

The elevation of the substrate level progressively decreases flooding by tides and salinity, favoring succession for plants less tolerant to these factors. This substrate becomes increasingly higher and more compact, hence vegetated by species progressively less tolerant to salinity (Soriano-Sierra 1993; Cunha-Lignon et al. 2009b; Schaeffer-Novelli et al. 2016).

In depositional areas, succession takes place with *S. alterniflora* colonizing sediment banks. This species of smooth cordgrass helps the establishment of propagules and the development of *L. racemosa* seedlings and saplings (Cunha-Lignon et al. 2009a). This vertical evolution should accompany the rise in average sea level if the sediment supply continues. The growth of marsh and mangrove species always occurs until the highest level is reached by the tide. This geomorphic structure is dynamic and capable of adapting to sea level rise (SLR) by local deposition and erosion processes; thus, expansion toward the mainland is possible when no obstacles are present landward (Arasaki et al. 2008; Schaeffer-Novelli et al. 2016).

Marsh and mangrove establishment and growth provide positive feedback, since increased deposition stabilizes roots and contributes to elevating the substrate diminishing flooding frequency and increasing deposition (Schaeffer-Novelli et al. 2016). Chen et al. (2018) considered sediment trapping mechanisms based on the hydrodynamic related to sediment settling and direct trapping by vegetation and confirmed that salt marsh grasses are more efficient than mangrove trees at inducing sediment trapping over a tidal-cycle scale. Structural vegetation characteristics will be established based on the site conditions (stable or in progradation) (Cunha-Lignon et al. 2011b).

The SLR rate associated with the sedimentation balance, oceanographic conditions, coastal geomorphology, and topography are factors that can determine mangrove dynamics (Soares 2009).

2.4.2 Erosional Process Concerning Extreme Events

Shoreline stabilization and coastal protection are some of the security services provided by mangrove and salt marsh ecosystems (Alongi 2008; UNEP 2014) by reducing wave energy, increasing sedimentation, reducing erosion, and mitigating the effects of storm surges and debris movement (Alongi 2008; Spalding et al. 2014; Di Nitto et al. 2014).

Catastrophic erosional storms are rare along the Brazilian coast (Ward et al. 2016). On the other hand, climate change will lead to a higher frequency of extreme storm events (IPCC 2013), which may increase the volume of water carried by channels, maintaining reduced salinity, and intensifying erosional processes along the fringes. While in sites where the sediment supply is limited from marine sources, storms can provide valuable input to maintain surface elevation (Ward et al. 2016).

Erosive processes on salt marsh and mangrove ecosystems are expected because of SLR and consequent changes in water circulation patterns. McLoughlin et al. (2015) found a stronger correlation between wave energy flux and volumetric erosion rates along the marsh edges than with lateral erosion rates. On the other hand, lateral erosion over affluent estuaries is likely due to increased precipitation. Erosion will enforce the retreat of the frontal banks of salt marshes, by removing the tufts of macrophytes, and may even cause loss of the banks locally (Soriano-Sierra 1993). In erosional areas, *R. mangle* and/or *A. schaueriana,* with high structural development, dominate the mangrove fringe.

2.4.3 Sedimentary Processes Across Scales

The characteristics of mangrove forests are strongly influenced by coastal types and process scales (Schaeffer-Novelli et al. 2000). These authors cite that the first-order control over the extensive coastal zones in which mangroves develop is provided by plate tectonics, which has little influence on the Brazilian coast resulting in a contribution of large amounts of sediments from the rivers to the continental shelves.

Second-order controls are associated with sedimentary processes driven by waves, tides, and fluvial energies that shape the systems morphodynamically and make diversification of mangrove forests possible (Schaeffer-Novelli et al. 2000; Cunha-Lignon et al. 2009b). Those forcing functions led to regional trends in structural development and function. Schaeffer-Novelli et al. (1990, 2000) proposed the division of the Brazilian coast into eight mapping units (segments) based on similar environmental conditions, geomorphic processes, disturbance regimes, and landscape mosaics predominate (see Chap. 3).

At the broadest geomorphological macroscale, mangroves occupy settings created or modified by geomorphic forces, including climate and SLR (Woodroffe et al. 2016). At a lower-level component, regional, extremely active geomorphic processes occur in the coasts of Amapá and Pará, submitted to the Amazon River energy, while barrier-lagoon systems are considered as relatively stable areas (Schaeffer-Novelli et al. 2000). At the microscale, a variety of geomorphological and biological surface and subsurface processes run, such as the hydroperiod and surface elevation changes, which regulate forest structure, although response times may be slow (see Chap. 3, Maps 1 and 2).

Although zonation has been described as evidence of plant succession in the past, there have been few investigations of the replacement of mangrove species over time (Woodroffe et al. 2016). Krauss et al. (2014) assumed that accounts of peat deposition from mangrove root growth over the Holocene infer a capacity for root zone impact on elevation change associated with water level changes over geological time frames.

2.5 Climate Change and Its Effects on Mangrove Forests and Salt Marshes

Historically, mangroves have responded not only to sea-level oscillations but also to changes in long-term shifts in temperature and precipitation (Alongi 2015). Local variability and local geomorphology are important factors envisioning whether mangroves of a specific region will adapt or survive (Alongi 2015). The capacity of mangroves to adapt to SLR may reside not only in the properties of individual species or forests but also in processes that operate at the landscape and regional levels (Schaeffer-Novelli et al. 2002).

Climate change is likely to have an extensive impact on mangrove and salt marsh ecosystems (Saintilan et al. 2014; Ellison 2015), through processes including SLR, changing ocean currents, increased storminess, increased temperature, changes in precipitation, and increased atmospheric CO_2 (Fig. 2.2) (McKee et al. 2012; Ward et al. 2016).

Mangroves are likely to be less affected by SLR in areas with high sediment availability, uplifting or stable coasts, high productivity, and large tidal ranges (Ward et al. 2016), along wet tropical coasts, and in areas adjacent to significant river input (Alongi 2008), such as the estuary of the Amazon River and Parnaíba River delta (see Chap. 3, Map 4).

The abovementioned factors combined with increased temperatures at the latitudinal extremes of mangrove distribution distribution and the predicted increase in the strength and frequency of El Niño events could promote the advance of mangroves in higher latitudes in South America. However, in the semiarid regions, where mangroves typically occur in estuaries, and irrigation and damming are prevailing, mangroves are expected to suffer from the increase in salt stress and resultant decreases in productivity, combined with decreases in sediment input (Alongi 2015; Ward et al. 2016).

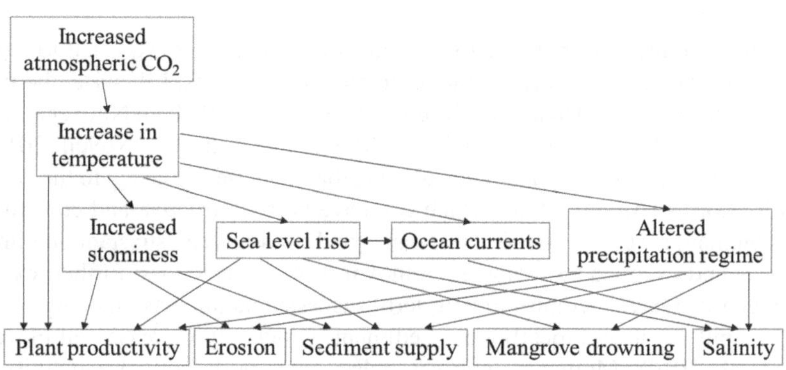

Fig. 2.2 Conceptual framework of the main impacting factors of climate change, the shifts in ocean-atmosphere system, and their influence in mangrove aspects, which influence their distribution. (Based on Ward et al. 2016)

Saintilan et al. (2014) indicate that mangrove species have proliferated at or near their global poleward limits over the past half-century to the detriment of salt marsh species. Soares et al. (2012) confirmed that mangroves' southern limit on the Atlantic coast is Santo Antônio Lagoon (28° 28′ S; 48° 50′ W) in the municipality of Laguna, Santa Catarina State, as stated earlier by Schaeffer-Novelli et al. (1990) (see Chap. 3, Map 16). At this site *L. racemosa* (dominant species) is stunted, a common trait to peer species globally at their southern limit, although *A. schaueriana* may reach up to 10 m, suggesting a vigor well out of its range. It is expected that the mangrove forests will expand toward regions south of their present latitudinal limit considering the changes observed in the climate and oceanographic characteristics in the western South Atlantic (Soares et al. 2012).

2.5.1 Mangrove and Salt Marsh Vulnerability to Climate Change

The tidal range is likely to affect the mangrove vulnerability to SLR directly; mangrove forests located in microtidal areas are generally more vulnerable than those settled in macrotidal environments (Ward et al. 2016). On the other hand, these ecosystems can modify their environment actively through surface elevation change processes by promoting sediment trapping and retention (Krauss et al. 2014).

Mangroves migration landward via seedling recruitment and vegetative reproduction depend on the ability of individual mangrove species to colonize newly available habitat at a rate that keeps pace with the rate of relative SLR (Duke et al. 1998; Gilman et al. 2008; Di Nitto et al. 2014; Alongi 2015), the slope of adjacent land, and the presence of obstacles to landward migration of the landward mangrove boundary (Gilman et al. 2008; Krauss et al. 2014; Di Nitto et al. 2014; Feller et al. 2017). The intertropical marshes that usually develop as fringes in front of mangroves (Lana 2003) are expected to be locally extinct by permanent submersion caused by SLR.

Mangrove latitudinal limits are to expand as mangroves are expected to replace marshes with the increase of temperature in temperate regions. Likewise, the marsh is expected to also expand to higher latitudes (Krauss et al. 2014; Di Nitto et al. 2014; Saintilan et al. 2014; Alongi 2015; Cintrón-Molero and Schaeffer-Novelli 2019).

Mangrove species show physiological traits that increase resiliency to the damage caused by storms (Alongi 2008), which can have both destructive and constructive impacts on mangrove ecosystems. Hurricanes and cyclones can substantially impact mangroves and even lead to complete removal or large-scale loss; further, extreme storm events can have medium- to long-term successional impacts on mangroves by giving a rapid input of allochthonous sediment, which can increase soil elevation (Ward et al. 2016). Moreover, nutrient pulses may push both productivity and mangrove growth (Lovelock et al. 2011).

2.6 Final Remarks

It would be unreliable to assume simple assessments of one or two factors to have an overall descriptive representation of how the structure and functioning, as well as zonation and succession work in the mangrove and salt marsh ecosystems. Since mangrove plants are related overall by their ability to grow in the intertidal zone, they have naturally developed different attributes and strategies to persist in this environment (Duke et al. 1998; Krauss et al. 2014; Woodroffe et al. 2016).

A variety of biotic and environmental factors influence each attribute which couples to define the distributional patterns of each species associated with others at global, regional, estuarine, and intertidal scales (Duke et al. 1998). Given the mangrove species resilience, their condition reflects environmental changes, highlighting the adaptability of resident species that can be used as bioindicators of forcing functions (Cunha-Lignon et al. 2009a; Arrivabene et al. 2014). Mangroves are expanding their latitudinal range as global temperatures keep rising (Alongi 2015) and the shift from salt marsh to mangrove dominance on subtropical and temperate shorelines has important implications for ecological structure, function, and global change adaptation (Saintilan et al. 2014).

Around 70% of Brazilian mangroves are in preserved areas, but the effectiveness of this progress continues weakened by bureaucracy, the lack of enforcement on conservation policy violations, and economic affairs (Ferreira and Lacerda 2016). Sadly, human impacts have often critically reduced the area occupied by mangrove forests (Giri et al. 2011) or monopolized areas into which they might have expanded. Land-use conversions not only directly affect mangrove forests, but also reduce the opportunities for mangroves to extend farther landward as the sea rises (Woodroffe et al. 2016).

References

Alongi D (2008) Mangrove forests: resilience, protection from tsunamis, and responses to global climate change. Estuar Coast Shelf S 76:1–13

Alongi D (2015) The impact of climate change on mangrove forests. Curr Clim Change Rep 1:30–39

Arasaki E, Alfredini P, Amaral RF, Lamparelli CC (2008) Os Efeitos no Ambiente Marinho da Elevação do Nível do Mar em Regiões da Baixada Santista. Brasil. Rev Bras Recurs Hidric 13(2):165–175

Arrivabene HP, Souza I, Có WLO, Rodella RA, Wunderlin DA, Milanez CR (2014) Functional traits of selected mangrove species in Brazil as biological indicators of different environmental conditions. Sci Total Environ 476:496–504

Ball MC (1996) Comparative ecophysiology of mangrove forest and tropical lowland moist rainforest. In: Mulkey SS, Chazdon RL, Smith AP (eds) Tropical forest plant ecophysiology. Chapman and Hall, New York, pp 461–496

Ball MC (1998) Mangrove species richness in relation to salinity and waterlogging: a case study along the Adelaide River floodplain, northern Australia. Global Ecol Biogeogr 7:73–82

Ball MC (2002) Interactive effects of salinity and irradiance on growth: implications for mangrove forest structure along salinity gradients. Trees 16:126–139

Berger U, Adams M, Grimm V, Hildenbrandt H (2006) Modelling secondary succession of neotropical mangroves: causes and consequences of growth reduction in pioneer species. Perspect Plant Ecol 7:243–252

Boorman LA (1999) Salt marshes – present functioning and future change. Mangrove Salt Marsh 3: 227–241

Cannicci S, Burrows D, Fratini S, Smith TJ, Offenberg J, Dahdouh-Guebas F (2008) Faunal impact on vegetation structure and ecosystem function in mangrove forests: a review. Aquat Bot 89(2): 186–200

Capdeville C, Pommier T, Gervaix J, Fromard F, Rols J-L, Leflaive J (2019) Mangrove facies drives resistance and resilience of sediment microbes exposed to anthropic disturbance. Front Microbiol 9:3337

Chen Y, Li Y, Thompson C, Wang X, Cai T, Chang Y (2018) Differential sediment trapping abilities of mangrove and saltmarsh vegetation in a subtropical estuary. Geomorphology 318: 270–282

Cintrón-Molero G, Schaeffer-Novelli Y (2019) The role of atmospheric-tropospheric rivers in partitioning coastal habitats and limiting the poleward expansion of mangroves along the southeast coast of Brazil. Int J Hydro 3(2):92–94

Clarke PJ, Kerrigan RA, Westphal CJ (2001) Dispersal potential and early growth in 14 tropical mangroves: do early life history traits correlate with patterns of adult distribution? J Ecol 89(4): 648–659

Clarke PJ, Kerrigan RA (2002) The effect of seed predators on the recruitment of mangroves. J Ecol 90:728–736

Costa CSB, Davy AJ (1992) Coastal saltmarsh communities of Latin America. In: Seeliger U (ed) Coastal plant communities of Latin America. Academic, San Diego, pp 179–199

Cunha-Lignon M, Coelho-Jr C, Almeida R, Menghini RP, Correa F, Schaeffer-Novelli Y, Cintrón-Molero G, Dahdouh-Guebas F (2009a) Mangrove forests and sedimentary processes on the South Coast of São Paulo State (Brazil). J Coastal Res SI 56:405–409

Cunha-Lignon M, Mahiques M, Schaeffer-Novelli Y, Rodrigues M, Klein D, Goya S, Menghini RP, Tolentino C (2009b) Analysis of mangrove forest succession, using cores: a case study in the Cananéia-Iguape Coastal System, São Paulo – Brazil. Braz J Oceanogr 57(3):1–13

Cunha-Lignon M, Kampel M, Menghini RP, Schaeffer-Novelli Y, Cintrón G, Dahdouh-Guebas F (2011a) Mangrove forests submitted to depositional processes and salinity variation investigated using satellite images and vegetation structure surveys. J Coastal Res SI 64(1):344–348

Cunha-Lignon M, Coelho-Jr C, Almeida R, Menghini RP, Schaeffer-Novelli Y, Cintrón G, Dahdouh-Guebas F (2011b) Characterisation of mangrove forest types in view of conservation and management: a review of mangals at the Cananéia region, São Paulo State. Brazil. J Coastal Res SI 64(1):349–353

Delgado P, Hensel PF, Jiménez JA, Day JW (2001) The importance of propagule establishment and physical factors in mangrove distributional patterns in a Costa Rican estuary. Aquat Bot 71:157–178

Di Nitto D, Neukermans G, Koedam N, Defever H, Pattyn F, Kairo JG, Dahdouh-Guebas F (2014) Mangroves facing climate change: landward migration potential in response to projected scenarios of sea level rise. Biogeosciences 11(3):857–871

Duke NC, Ball MC, Elisson JC (1998) Factors influencing biodiversity and distributional gradients in mangroves. Global Ecol Biogeogr 7:27–47

Ellison AM (2002) Macroecology of mangroves: large-scale patterns and processes in tropical coastal forests. Trees 16:181–194

Ellison J (2015) Vulnerability assessment of mangroves to climate change and sea-level rise impacts. Wetl Ecol Manag 23:115–137

Ellison AM, Mukherjee BB, Karim A (2000) Testing patterns of zonation in mangroves: scale dependence and environmental correlates in the Sundarbans of Bangladesh. J Ecol 88:813–824

Farnsworth EJ, Ellison AM (1997) Global patterns of pre-dispersal propagule predation in mangrove forests. Biotropica 29(3):318–330

Feller IC, Whigham DF, O'Neill JP, McKee KL (1999) Effects of nutrient enrichment on within-stand cycling in a mangrove forest. Ecology 80(7):2193–2205

Feller IC, Whigham DF, McKee KM, O'Neill JP (2002) Nitrogen vs. phosphorus limitation across an ecotonal gradient in a mangrove forest. Biogeochemistry 62:145–175

Feller IC, Lovelock CE, Berger U, McKee KL, Joye SB, Ball MC (2010) The biocomplexity of mangrove ecosystems. Annu Rev Mar Sci 2:395–416

Feller IC, Friess DA, Krauss KW, Lewis RR (2017) The state of the world's mangroves in the 21st century under climate change. Hydrobiologia 803(1):1–12

Ferreira AC, Lacerda LD (2016) Degradation and conservation of Brazilian mangroves, status and perspectives. Ocean Coast Manage 125:38–46

Friess DA, Krauss KW, Horstman EM, Balke T, Bouma TJ, Galli D, Webb EL (2012) Are all intertidal wetlands naturally created equal? Bottlenecks, thresholds and knowledge gaps to mangrove and saltmarsh ecosystems. Biol Rev 87(2):346–366

Gilman EL, Ellison J, Duke NC, Field C (2008) Threats to mangroves from climate change and adaptation options: a review. Aquat Bot 89:237–250

Giri C, Ochieng E, Tieszen LL, Zhiu Z, Singh A, Loveland T, Masek J, Duke N (2011) Status and distribution of mangrove forests of the world using earth observation satellite data. Global Ecol Biogeogr 20:154–159

Intergovernmental Panel for Climate Change – IPCC (2013) Climate Change 2013: the physical science basis. Contribution of Working Group I. In: Stocker TF, Qin D, Plattner G-K, Tignor M, Allen SK, Boschung J, Nauels A, Xia Y, Bex V, Midgley PM (eds) Fifth assessment report of the Intergovernmental Panel on Climate Change. Cambridge University Press, New York, 1523 pp

Krauss KW, Lovelock CE, McKee KL, López-Hoffman L, Ewe SM, Sousa WP (2008) Environmental drivers in mangrove establishment and early development: a review. Aquat Bot 89(2): 105–127

Krauss KW, McKee KL, Lovelock CE, Cahoon DR, Saintilan N, Reef R, Chen L (2014) How mangrove forests adjust to rising sea level. New Phytol 202:19–34

Kristensen E, Alongi DM (2006) Control by fiddler crabs (*Uca vocans*) and plant roots (*Avicennia marina*) on carbon, iron, and sulfur biogeochemistry in mangrove sediment. Limnol Oceanogr 51:1557–1571

Lana PC (2003) As marismas da Baía de Paranaguá, características gerais, modo de apropriação e implicações para a Legislação Ambiental. Desenvolvimento e Meio Ambiente 8:11–23

Lovelock CE, Ball MC, Feller IC, Engelbrecht BM, Ling EM (2006) Variation in hydraulic conductivity of mangroves: influence of species, salinity, and nitrogen and phosphorus availability. Physiol Plantarum 127(3):457–464

Lovelock CE, Feller IC, Adame M, Reef R, Penrose H, Wei L, Ball M (2011) Intense storms and the delivery of materials that relieve nutrient limitations in mangroves of an arid zone estuary. Funct Plant Biol 38:514–522

Lovelock CE, Krauss KW, Osland MJ, Reef R, Ball MC (2016) The physiology of mangrove trees with changing climate. In: Goldstein G, Santiago LS (eds) Tropical tree physiology: adaptations and responses in a changing environment. Springer, New York, pp 149–179

McKee KL (1993) Soil physicochemical patterns and mangrove species distribution: reciprocal effects? J Ecol 81:477–487

McKee KL (1995) Seedling recruitment patterns in a Belizean mangrove forest: effects of establishment ability and physico-chemical factors. Oecologia 101(4):448–460

McKee KL, Rogers K, Saintilan N (2012) Response of salt marsh and mangrove wetlands to changes in atmospheric CO_2, climate and sea level. In: Middleton BA (ed) Global change and the function and distribution of wetlands. Springer, Dordrecht, pp 63–96

McLoughlin SM, Wiberg PL, Safak I, McGlathery KJ (2015) Rates and forcing of marsh edge erosion in a Shallow Coastal Bay. Estuar Coast 38:620–638

Osland M, Feher LC, Griffith KT, Cavanaugh KC, Enwright NM, Day RH, Stagg CL, Krauss KW, Howard RJ, Grace JB, Rogers K (2017) Climatic controls on the global distribution, abundance, and species richness of mangrove forests. Ecol Monogr 87(2):341–359

Parida AK, Jha B (2010) Salt tolerance mechanisms in mangroves: a review. Trees 24:199–217

Pennings SC, Moore DJ (2000) Zonation of shrubs in western Atlantic saltmarshes. Oecologia 126: 587–594

Peters R, Walther M, Lovelock C, Jiang J, Berger U (2020) The interplay between vegetation and water in mangroves: new perspectives for mangrove stand modelling and ecological research. Wetlands Ecol Manage 28:697–712

Reis CRG, Reed SC, Oliveira RS, Nardoto GB (2019) Isotopic evidence that nitrogen enrichment intensifies nitrogen losses to the atmosphere from subtropical mangroves. Ecosystems 22(5): 1126–1144

Saintilan N, Wilson NC, Rogers K, Rajkaran A, Krauss KW (2014) Mangrove expansion and salt marsh decline at mangrove poleward limits. Global Change Biol 20:147–157

Schaeffer-Novelli Y, Cintrón-Molero G, Adaime RR, Camargo TM (1990) Variability of the mangrove ecosystem along the Brazilian Coast. Estuaries 13(2):204–218

Schaeffer-Novelli Y, Cintrón-Molero G, Soares MLG, De Rosa T (2000) Brazilian mangroves. Aquat Ecosyst Health 3:561–570

Schaeffer-Novelli Y, Cintrón-Molero G, Soares MLG (2002) Mangroves as indicators of sea level change in the muddy coasts of the world. In: Healy T, Wang Y, Healy J-A (eds) Muddy coasts of the world: processes, deposits and function. Elsevier, Amsterdam, pp 245–262

Schaeffer-Novelli Y, Soriano-Serra EJ, Vale CC, Bernini E, Rovai AS, Pinheiro MAA, Schmidt AJ, Almeida R, Coelho-Jr C, Menghini RP, Martinez DI, Abuchahla GMO, Cunha-Lignon M, Charlier-Sarubo S, Shirazawa-Freitas J, Cintrón-Molero G (2016) Climate changes in mangrove forests and salt marshes. Braz J Oceanogr 64(SP2):37–52

Smith-III TJ (1996) Mangrove forest structure. In: Feller IC, Sitnik M (eds) Mangrove ecology: a manual for a field course. Smithsonian Institution, Washington, DC, pp 65–74

Soares MLG (2009) A conceptual model of the responses of mangrove forests to sea level rise. J Coast Res SI 56:267–271

Soares MLG, Estrada GCD, Fernandez V, Tognella MMP (2012) Southern limit of the Western South Atlantic mangroves: assessment of the potential effects of global warming from a biogeographical perspective. Estuar Coast Mar Sci 101:44–53

Soriano-Sierra EJ (1993) Etude écologique des marais salés du bassin d'Arcachon: structure et évolution des schorres, production et dégradation de leur végétation et échanges de matières particulaires entre les schorres et le bassin. PhD dissertation, Université Bordeaux I, 254 pp

Soriano-Sierra EJ (1999) Ecossistemas de marismas da Lagoa da Conceição. II. Fitocenosis. In: Sierra B, Soriano-Sierra EJ (eds) O Ecossistema da Lagoa da Conceição v4. FEPEMA, Florianópolis, pp 171–184

Sousa WP, Kennedy PG, Mitchell BJ, Ordóñez L, Benjamín M (2007) Supply-side ecology in mangroves: do propagule dispersal and seedling establishment explain forest structure? Ecol Monogr 77(1):53–76

Spalding MD, Ruffo S, Lacambra C, Meliane I, Hale LZ, Shepard CC, Beck MW (2014) The role of ecosystems in coastal protection: adapting to climate change and coastal hazards. Ocean Coast Manage 90:50–57

Thomas N, Lucas R, Bunting P, Hardy A, Rosenqvist A, Simard M (2017) Distribution and drivers of global mangrove forest change, 1996–2010. PLoS One 12(6):1–14

United Nations Environment Programme – UNEP (2014) The importance of mangroves to people: a call to action. In: van Bochove J, Sullivan E, Nakamura T (eds) United Nations environment programme. World Conservation Monitoring Centre, Cambridge

Ward RD, Friess DA, Day RH, MacKenzie RA (2016) Impacts of climate change on mangrove ecosystems: a region by region overview. Ecosyst Health Sustain 2(4):1–25

Woodroffe CD, Rogers K, McKee KL, Lovelock CE, Mendelssohn IA, Saintilan N (2016) Mangrove sedimentation and response to relative sea-level rise. Annu Rev Mar Sci 8:243–266

Chapter 3
Variability of Mangroves Along the Brazilian Coast: Revisiting

Gilberto Cintrón-Molero, Yara Schaeffer-Novelli, André S. Rovai, Clemente Coelho-Jr., Ricardo P. Menghini, Renato de Almeida, Cláudia Câmara do Vale, Elaine Bernini, Guilherme Moraes de Oliveira Abuchahla, and Marília Cunha-Lignon

---

Supplementary Information The online version contains supplementary material available at https://doi.org/10.1007/978-3-031-13486-9_3.

G. Cintrón-Molero (✉)
Instituto BiomaBrasil, São Paulo, SP, Brazil

Y. Schaeffer-Novelli
Instituto BiomaBrasil, Recife, PE, Brazil

Instituto Oceanográfico, Universidade de São Paulo, São Paulo, SP, Brazil

IUCN Mangrove Specialist Group, São Paulo, SP, Brazil

A. S. Rovai
Department of Oceanography and Coastal Sciences, College of the Coast and Environment, Louisiana State University, Baton Rouge, LA, USA

C. Coelho-Jr.
Instituto BiomaBrasil, Recife, PE, Brazil

Instituto de Ciências Biológicas, Universidade de Pernambuco, Recife, PE, Brazil

R. P. Menghini
Instituto BiomaBrasil, Recife, PE, Brazil

Complexo Educacional das Faculdades Metropolitanas Unidas, Mestrado Profissional em Saúde Ambiental, São Paulo, SP, Brazil

R. de Almeida
Instituto BiomaBrasil, Recife, PE, Brazil

Centro de Ciências Agrárias, Ambientais e Biológicas, Universidade Federal do Recôncavo da Bahia, Cruz das Almas, BA, Brazil

C. C. do Vale
Departamento de Geografia, Universidade Federal do Espírito Santo, Vitória, ES, Brazil

Y. Schaeffer-Novelli et al. (eds.), *Brazilian Mangroves and Salt Marshes*, Brazilian Marine Biodiversity, https://doi.org/10.1007/978-3-031-13486-9_3

3.1 Introduction

Considering the high dynamism of coastal systems, the advances in the concepts of complex systems theory to illuminate contemporaneous coastal management issues related to multiple spatial and temporal scales, and the recent knowledge produced on Brazilian mangroves and saltmarshes in this chapter, we revisit and expand the contribution presented by Schaeffer-Novelli et al. (1990) (hereafter referred to as the *1990 paper*) to understand mangrove and salt marsh patterns and processes along the Brazilian coast and to discuss the originally proposed macroecological concepts.

The 1990 paper was written at a time when interest in mangroves on a global scale was beginning to blossom triggered mainly by a series of outstanding publications, that is, Chapman's (1975) book on mangrove biogeography, Walsh's (1974) seminal work on mangrove zonation, and the concept of "outwelling" by Odum (1971) and Odum and Heald (1975). Seminal papers such as The Ecology of Mangroves by Lugo and Snedaker (1974) were shifting attention from vegetation to ecological processes. Other major contributions were the UNESCO's release of the world's most thorough mangrove bibliographic survey since 1614 (Rollet 1981), and the Handbook for Mangrove Area Management (Hamilton and Snedaker 1984). The Association of Marine Science Researchers (ALICMAR) was created in 1974 and provided the first social-technical platform for sharing ecological knowledge in the Americas. At about the same time, the Organization of American States (OEA) and UNESCO's Regional Office for Latin America subsidized the participation of researchers in various meetings in Hawaii, United States (1974), and Cali, Colombia (1978). Simultaneously, several academic institutions initiated local activities supported by national research organizations, which also sponsored academic exchanges, planting the seed of mangrove ecology on increasingly fertile ground.

Thus, the 1990 paper was an outgrowth of the convergence of attention from multiple Brazilian organizations being directed at coastal systems and the emerging recognition that ecological knowledge would contribute significantly to administer systems that were being recognized for their ecologic importance for coastal fisheries, whereas just a decade earlier, they had been misjudged as useless wastelands suitable for reclamation – a mindset that had prevailed since colonial times and was firmly entrenched in society. What was emerging since the 1960s and early 1970s (Odum 1969; Odum 1970) was the realization that perhaps these lands were not wastelands after all, and that they merited to be managed more rationally, in terms of what today is termed an ecosystem services-based approach (Gregory and Goudie

E. Bernini
Departamento de Engenharia e Meio Ambiente, Centro de Ciências Aplicadas e Educação, Universidade Federal da Paraíba, Rio Tinto, PB, Brazil

G. Moraes de Oliveira Abuchahla
Leibniz Centre for Tropical Marine Research (ZMT), Bremen, Germany

M. Cunha-Lignon
Instituto BiomaBrasil, Recife, PE, Brazil

Campus Experimental de Registro, Universidade Estadual Paulista, Registro, SP, Brazil

2011). Changing deeply seated mindsets requires considerable efforts, such as increasing research levels for understanding the behavior of these systems and building a cadre of human resources that could increase societal awareness to promote their conservation rather than their reclamation and transformation.

The 1990 paper adopted a mesoscale (landscape) perspective for convenience. Landscapes are the most tangible ecological criteria and remain an appropriate level of observation for broad-scale management as well as for focusing on smaller scales. In this respect, we propose a refinement of the mesoscale classification scheme used in the 1990 paper and argue that the application of the coastal environmental setting (CES) framework (Thom 1982; Woodroffe 1992; Twilley et al. 2018;) improves our capacity to appropriately understand and scale mangroves' macroecological attributes and responses to natural and anthropogenic stressors at larger spatial and temporal scales.

In addition, later in this chapter, we discuss an approach we call *dynamic framing* as complementary for adopting a landscape perspective coherent with nature's investment and endurance strategy. However, of critical importance is that the mesoscale landscape approach merges human social systems and geomorphic systems into a unitary system, and recognizes multiple interactive and intercausal scales, geomorphological, social, and ecological processes (Huggett 1995) that are interdependent and vital for sustainability.

3.2 Scales and Variability in Mangrove Macroecology

The Brazilian coast is characterized by mangrove forests along most of its 10,959.52 km length (IBGE 2016), between the latitudes 04°20′12″N and 33°45′07″S. The measured length of any coast is a function of the scale and resolution of the measurement (Mandelbrot 1983), so it is not surprising that various length estimates exist in the literature (see Chap. 1 for more details). On a global scale, Brazil's coastline length ranks 12th (IBGE 2016), but shelters the second-largest mangrove area cover in the world (FAO 2007; Giri et al. 2011; Hamilton and Casey 2016). This suggests that from a broad perspective, this coast contains landscape features that are particularly favorable for mangrove development. Because a broad view subdues detail, processes, and structures, it is not surprising that a closer look often reveals unexpected variability at different levels.

Reality is complex and stratified into characteristic scales, dynamics, and patterns. These tend to be bundled into discrete scales of interaction (Rowe 1961; Simon 1962). Variability is the result of complexity; the diversity of components compounded by the spatiotemporal diversity of factors influences landscape responses and development at various scales, leaving distinct signatures that reveal dominant influences. The interaction between factor regimes and scales results in relatively distinct landforms. Thus, the original paper intended to examine zonality, that is, Brazilian coastal patterns, in terms of features and regional environments. The outcome was a proposed division of the coast into segments within which similar broad climatic, geomorphologic, and oceanographic features and comparable management needs are found. Mangroves were accounted as mostly azonal perhaps because of the dominance of local (site) factors in influencing development. Complexity defies any attempt of classifying any coast where the diversity of landscape elements is high and where these elements and forcing functions act in combination and interact in complex ways.

In the context of revisiting the 1990 paper, we find it desirable to review some aspects of the notion of variability, considering that coasts are the most dynamic places on the planet. It is misleading to consider coastal features as static or perceive variability as problematic. Variability is a manifestation of complexity and although it presents obstacles to generalizing and identifying clear-cut patterns in nature, it is part of it and is present at all scales driven by external and internal factors such as self-organization. Variability paradoxically entails the iterative power of order, of system-level responses that eventually can lead to adaptive change and the capture of environmental energies to gradually perform increasingly more complex geoecologic work that makes more complete use of all available energies.

Furthermore, categorizations are based on generalizations and variability tends to obscure categories. In fact, all categorization schemes are simplifications and ignore variability at some scale. In nature, absolute categories do not exist, because categories exist only as (human) ideas, whereas reality is a *continuum*; change, variability, and transformation are pervasive but until very recently we have perceived reality in terms of static components and neglected processes and change. A shift taking place in ecology is the increasing adoption of a process-based perspective. The relationships between ecological processes and spatiotemporal patterns on a variety of scales are one of the most relevant research topics for most unresolved questions in ecology. Even climate was accounted as constant until very recently. What is pertinent in this appreciation is that temporal variability is an important area of concern, because the temporal scope of human observation is often very limited when considering the extended endurance of many geomorphic features. Focusing on the narrow spatiotemporal window of human experience inevitably provides a "keyhole" or partial view of coastal systems that ignoring its limitations can undermine management efforts no matter how well-intended they might be. In general, ecological events have a characteristic frequency and a corresponding spatial scale, and an ecological study of the landscape must conform to these scales (Turner 1998; Blackburn and Gaston 2002).

The Brazilian coast has been divided into different segments by several authors for different purposes, highlighting certain features and processes. The complexity of responses from a complex system approach is obviously overwhelming. In our 1990 paper, the purpose was to highlight geographic variability in the context of settings for mangrove establishment and development. That paper did not intend to describe causal factors or system dynamics but was limited to describing patterns of mangrove structure along the coast in very broad terms. To do that we conveniently divided the coast into eight broad segments oriented to mangrove abiotic drivers (Fig. 3.1) such as climatological (temperature, precipitation, and potential evapotranspiration), hydrographic (river order rank), and oceanographic (tidal amplitude),

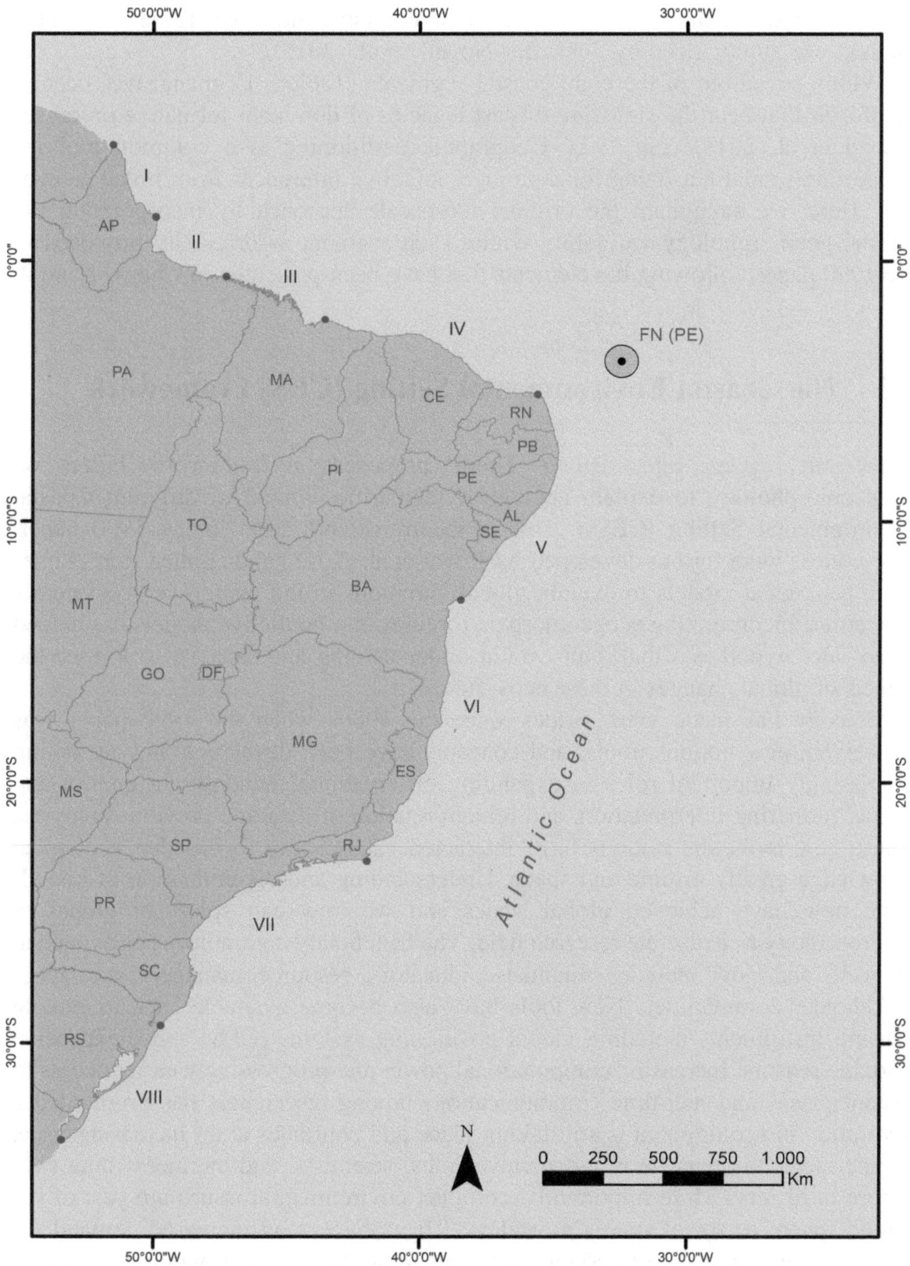

Fig. 3.1 Coastal segments proposed in Schaeffer-Novelli et al. (1990), divided by dots: I – Cape Orange (04°30′N) to Cape Norte (01°40′N), II – Cape Norte to Ponta Curuçá (00°36′S), III – Ponta Curuçá to Ponta Mangues Secos (02°15′S), IV – Ponta Mangues Secos to Cape Calcanhar (05°08′ S), V – Cape Calcanhar to Recôncavo Baiano (13°00′S), VI – Recôncavo Baiano to Cabo Frio (23° 00′S), VII – Cabo Frio to Torres (29°20′S), and VIII – Torres to Chuí (22°35′S)

including the resulting mangrove development (structure; see Table 3.1). That method was summarized by Schaeffer-Novelli et al. (2016).

Within seven out of the eight coastal segments (Table 3.1), mangroves occupy landforms that bear the signature of past legacies of dominant formative processes (Rovai et al. 2018) (Fig. 3.2). Geographic partitioning is a common tool for supporting spatial reasoning for deriving qualitative inferences from broad categories. Here, we strengthen the original mesoscale approach by incorporating the coastal geomorphology variability within each segment as originally proposed in the 1990 paper, following the elements that have been presented in Chaps. 1 and 2.

3.3 The Coastal Environmental Setting (CES) Framework

Some fifty years ago, Bruce Thom proposed a framework based on ecogeomorphology to explain ecological regularities linked to different Coastal Environmental Setting (CESs). This is the framework used in the 1990 paper. These ideas were further developed by Rovai et al. (2018) and applied to multiple-scale ecological models to explain global variations in the mangrove ecosystem's properties. Incorporating ecogeomorphic forcings into predictive models has helped to advance hypotheses that improve our understanding and capacity to foresee the effects of global changes in these ecosystems.

Ecology has made great strides since the 1980s when the 1990 paper was conceived; new notions, tools, and concepts have been developed and are taking increasingly important roles in expanding observational windows in quality and scope, furthering interpretation, and reinterpretation of data and previous analyses. These new tools and notions have interacted catalytically to broaden ecological knowledge greatly in time and space. Understanding and dissemination of knowledge now have achieved global scales and we now can speak of global or macroecology as a discrete research field. The beneficiary community has expanded as well, and now includes scientists, educators, resource managers, and large stakeholder communities. New tools have also become available such as remote sensing instruments including global positioning systems (GPS) and inexpensive portable sensors. Increasing computational power and progressively easier access to distant places and real-time communications among researchers has propitiated a revolution in ecology that is still taking place and continues at an increasing pace. Furthermore, there are now more universities, scientists, and engineers than ever before in history. More importantly, complex environmental issues are part of the public sphere or social spaces nowadays. Thus, there is an increased demand for scientific communication to nonprofessionals to promote greater public understanding and engagement by educated constituencies.

The 1990 paper's perspective remains relevant as an appropriate level of observation for revealing mesoscale order as a starting point for a triadic approach that pays attention to events that take place at other levels: the focal level, the next higher level, and the level immediately below (Salthe 1985). Here we will demonstrate that

Table 3.1 Eight coastal segments proposed in Schaeffer-Novelli et al. (1990)

Segment	Description
I	From Cape Orange (04°30′N) to Cape Norte (01°40′N), at the northern limit of the mouth of the Amazon River. It is characterized by homogeneous forests dominated by the *Avicennia*. Mangroves colonize coastal rivers, extending to considerable distances inland. The genus *Rhizophora* occupies the estuarine portion of the rivers, where the marine influence is direct. In these areas, *Montricardia* and *Laguncularia* occupy the inner parts of the forests.
II	From North Cape (01°40′N) to Ponta Curuçá (00°36′S). Mangrove development and cover are sparse in this segment due to the influence of the fluvial discharge of the Amazon River. The forests are mixed, with freshwater mudbanks dominating the northern part of the Amazon River mouth. The mangroves are mostly formed by the genus *Avicennia* at sites of low elevation and low salinity, while the *Rhizophora* occurs at sites with more significant marine influence or periodically inundated by tides.
III	From Ponta Curuçá (00°36′S) to Ponta Mangues Secos (02°15′S). The genus *Rhizophora* dominates the forest fringes. The higher ground behind the fringes is colonized by *Avicennia* and *Laguncularia*. Low-energy, depositional environments are colonized by *Spartina*. The genus *Conocarpus* is found in transition zones to upland.
IV	Ponta Mangues Secos (02°15′S) to Cape Calcanhar (05°08′S). Mangroves are poorly developed along this stretch of coast due to the lack of freshwater input associated with prolonged dry seasons. High salt concentrations limit mangroves to river mouths.
V	Cape Calcanhar (05°08′S) to Recôncavo Baiano (13°00′S). Due to the high energy of this section of the coast, mangroves develop in sheltered areas, associated to estuaries and coastal lagoons. *Rhizophora* and *Laguncularia* appear as pioneers. In the inner parts of the forests, *Avicennia* and *Laguncularia* form mixed forests.
VI	From Recôncavo Baiano (13°00′S) to Cabo Frio (23°00′S). Relatively extensive mangroves are commonly found behind *restingas*. All three genera of mangroves are found, either mixed or in monospecific stands. In the Todos os Santos Bay, *Laguncularia* is dominant, colonizing sandy-clay soils. *Rhizophora* is found predominantly on the margins, forming a narrow strip on the fringes. When dominant, they form monospecific stands frequently flooded by the tides. *Avicennia* and *Laguncularia* may also form mixed marginal forests.
VII	From Cabo Frio (23°00′S) to Torres (29°20′S). The tallest mangrove trees border estuaries, channels, and some river downstream. Forests can be monospecific or mixed of the three genera. Recent sediments of barrier islands may be colonized by *Spartina*, forming saltmarshes completely flooded by high tides. *Rhizophora* colonizes muddy sediments with large amounts of organic matter, while *Avicennia* is found in higher deposits forming extensive forests. Landward transitional zones are often colonized by *Hibiscus*, *Crinum*, and *Acrosthicum*. The latitudinal limit for real mangrove species is on the coast of Santa Catarina, at 27°30′S (*R. mangle*) and 28°30′S (*A. schaueriana* and *L. racemosa*).
VIII	From Torres (29°20′S) to Chuí (33°45′S). This stretch of coastline is formed by extensive beach deposits, associated with dune ridges and sandy ridges. Abundant lagoons, isolated from the ocean by multiple barriers, result from successive transgressive and regressive geologic events. Low winter temperatures and a wide temperature range inhibit the growth of typical mangrove species, promoting the development of salt marshes.

Adapted from Schaeffer-Novelli (1999)

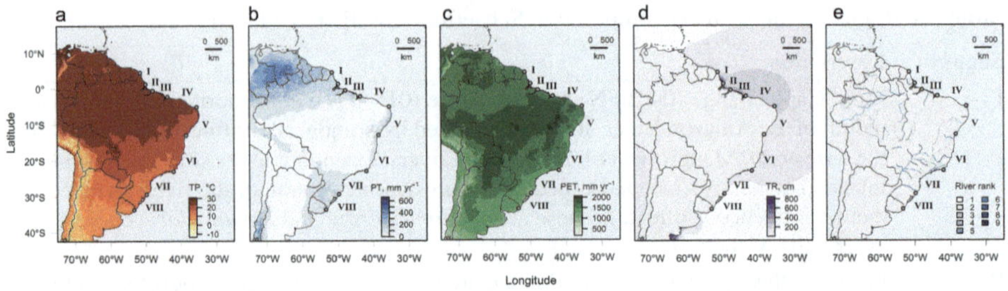

Fig. 3.2 Coastal segments proposed in Schaeffer-Novelli et al. (1990) highlighting dominant environmental forcing's and climate-driven threats. Mangroves are present in segments I–VII, salt marshes are predominant at segment VIII. (**a**) TP-mean annual temperature, (**b**) PT-mean annual precipitation, (**c**) PET-mean annual potential evapotranspiration, (**d**) TR-mean tidal range, and (**e**) River order rank. Sources: TP and PT data from Hijmans et al. (2005), PET from Title and Bemmels (2018), TR from Vestbo et al. (2018), and river order rank from Patterson and Kelso (2018)

the CES framework[1] provides the most obvious and tangible ground for improvement in spatial resolution, accessing constraints, and moving toward higher fidelity scales. All ecological processes and structures are multiscales (Allen and Hoekstra 1992). We provide a reanalysis of mangrove structural (biomass) and functional (primary productivity, carbon sequestration) attributes discussed in the 1990 paper. We used global compilations on climatic and oceanographic variables to predict mangrove ecological traits at a continental scale, expanding models proposed in our original 1990 paper from a conceptual to an empiric perspective. Particularly, we explored how the relative contribution of rivers, tidal range, along with regional climate, shapes distinct CES, reflected in substrate conditions to which plants respond (Thom 1982; Woodroffe 1992; Twilley et al. 2018) (Fig. 3.3). Distinct CES, for example, deltas, estuaries, and lagoons, are formed by the relative contribution of geophysical variables (e.g., river discharge, tidal amplitude, wave energy). Along with regional climatic drivers, these geophysical forcings constrain carbon partitioning among ecosystem compartments (soil, above- and belowground biomass). CES types include large rivers, small deltas (grouped as deltaic by Thom 1982), tidal systems (estuaries, bedrock as defined by Thom 1982), lagoons (including composite settings as defined by Thom 1982), carbonate coastal settings, and arheic or dry coastlines.

The following brief, yet comprehensive overview on dominant global types of CES was originally summarized by Rovai et al. (2018). However, we suggest consultation of the original sources (Thom 1982; Woodroffe 1992) for additional information. One of the major factors defining the different CES is sediment source (i.e., river-borne), which represents a combination of geophysical processes and

[1]The term Coastal Environmental Settings (CESs) refer to a typology of mangrove-occurring localities that share certain composed by geophysical, geomorphic, and biologic characteristics.

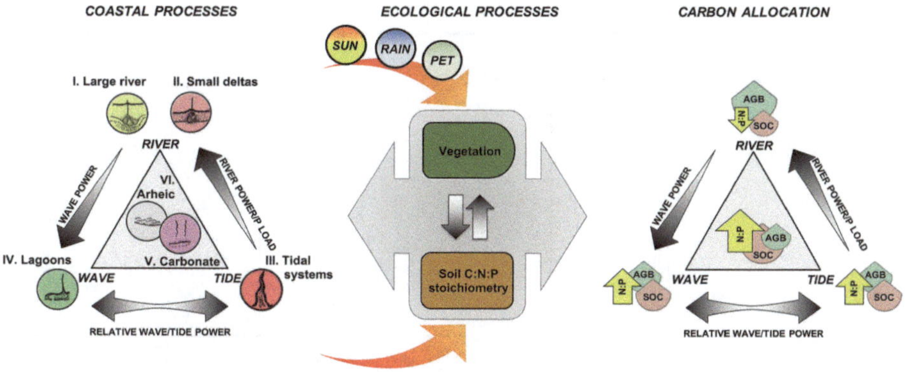

Fig. 3.3 Coastal ecogeomorphology conceptual framework, showing how bidirectional fluxes between abiotic and biotic components control nutrient stoichiometry and carbon storage in mangroves. CES types: I – large rivers; II – small deltas; III – tidal systems; IV – lagoons; V – carbonate coastal settings; and VI – arheic, or dry coastlines. PET: potential evapotranspiration; C:N:P: carbon-to-nitrogen-to-phosphorus ratio. Adapted from Twilley et al. (2018)

local geology influencing mangrove dynamics (Thom 1982; Woodroffe 1992; Woodroffe 2002).

The CES framework provides an alternative to the latitude gradient paradigm, and its use has advanced our capacity to predict mangrove ecological attributes such as aboveground biomass (Rovai et al. 2016), litterfall production (Ribeiro et al. 2019), and soil organic carbon (Rovai et al. 2018) at larger scales with a high confidence level. This is particularly useful for coastlines that lack such information. Here, we focus on the variability of mangrove aboveground biomass (AGB), litterfall (NPP_L, or Net Primary Productivity Litterfall), and soil organic carbon (SOC) along the Brazilian coastline, as these ecosystem attributes constitute the largest long-term (>100 years), perennial carbon pools in mangrove forests. However, a pressing need remains for generating estimates of belowground biomass (roots) and productivity as these are significant components of ecosystem-level C stock and budget, respectively. CESs provide an ecological/terrain conceptual unit for management that is easily geographically defined.

3.4 Aboveground Biomass

Previous attempts to predict continental-scale mangrove aboveground biomass (AGB) include latitude (Saenger and Snedaker 1993; Twilley et al. 1992) and climate-based models (Hutchison et al. 2014). Although latitude-based models can indirectly encompass critical climatic and geophysical variables, their individual contribution to explain AGB value spatial patterns is unknown, since their explanatory power is not explicitly weighted in the statistical analysis. Although a climatic modeling approach explicitly includes climate variables such as temperature and

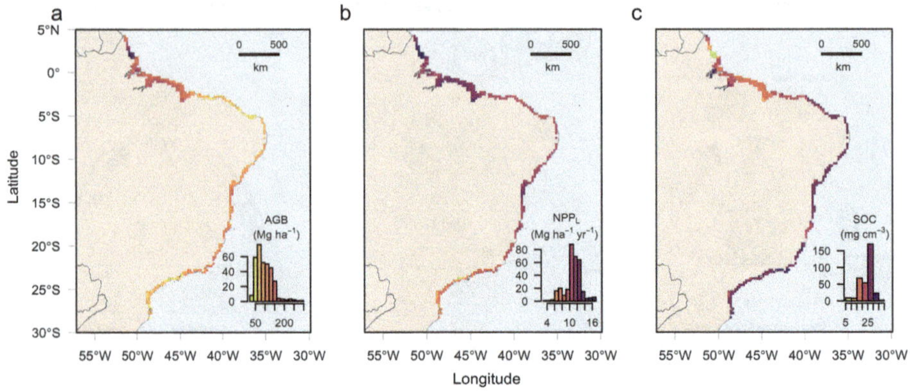

Fig. 3.4 Predicted mangrove aboveground biomass (**a**) (AGB in Mg ha^{-1}), litterfall productivity (**b**) (NPP_L in Mg ha^{-1} year^{-1}), and soil organic carbon density (**c**) (SOC in mg cm^{-3}) in Brazilian mangroves. Histograms depict the frequency of modeled values for each mangrove attribute. AGB data extracted from Rovai et al. (2016), NPP_L, from Ribeiro et al. (2019), and SOC from Rovai et al. (2018)

precipitation to explain mangrove AGB at the global scale (Hutchison et al. 2014), this analysis is limited not only by the number of climatic variables included in the model but also by the lack of other environmental variables that directly influence mangrove structural and functional properties at regional and local scales (Twilley 1995; Twilley and Rivera-Monroy 2009).

A literature review to assemble a global dataset containing information on published mangrove AGB and forest structure data is summarized in a review by Rovai et al. (2016). The inclusion of other geophysical variables in the climatic-geophysical model significantly improves AGB estimates at the latitudinal scale as demonstrated for the neotropics. As in the conceptual model proposed in the 1990 paper, the review by Rovai et al. (2016) shows that at continental scales, higher tidal amplitudes contributed to high forest biomass associated with warm temperatures, abundant rainfall, and low potential evapotranspiration (Figs. 3.1a–c and 3.3a). For the Brazilian coast, this model corroborates the mangrove forest structural development described for each segment proposed in the 1990 paper (see Chaps. 4 and 6), with higher AGB values predicted for low latitude, deltaic and macrotidal coastlines (Segments I–III, Fig. 3.1; Table 3.1), and lower values along increasingly austral latitudes, tide- and wave-dominated, or dry coastlines (Segments IV and VII, Table 3.1; Fig. 3.4a).

Mangrove AGB values in Brazil range from 25.3 to 284.8 Mg ha^{-1} (mean = 95.8 Mg ha^{-1}), within the range estimated for the neotropics (16.6–627.0 Mg ha^{-1}, mean = 88.7 Mg ha^{-1}) (Rovai et al. 2016). Using a biomass-to-carbon conversion factor of 0.475 (Hamilton and Friess 2018) and a mangrove forest cover of 7675 km^2 (Hamilton and Casey 2016), the total C stored in mangroves' AGB in Brazil is estimated at 0.04 PgC, which corresponds to 7.3% of global C stocks in mangrove AGB (Rovai et al. 2016).

Rovai et al. (2016) show that CES represents a major determinant on mangrove wetland development, configuration, and realized maximum biomass, particularly considering the diversity of mangrove geoecological settings and associated dynamics (Thom 1982; Woodroffe 1992; Twilley 1995). This energy signature is strongly influenced by the local tidal range and river discharge, critical geophysical variables explaining a significant percentage of the AGB total variance (Rovai et al. 2016). Indeed, tidal amplitude, a component of the hydroperiod regime in coastal regions, significantly influenced mangrove structural development. Higher tidal amplitude promotes nutrient exchange and aeration of soil layers, which reduces sulfide production and accumulation, allowing higher growth rates and forest development (Lugo and Snedaker 1974; Castañeda-Moya et al. 2013).

3.5 Net Primary Productivity – Litterfall (NPP_L)

Ribeiro et al. (2019) provided the first model that accounts for continental-scale variability in mangrove Net Primary Productivity [Litterfall] (NPP_L) in response to climatic and geophysical variables combined. Their results advance the current understating of mangrove NPP_L variability across latitudinal and longitudinal gradients, considering that previous studies did not account for the role of geophysical forces in driving large-scale NPP_L variability. Instead, correlations were usually performed using absolute variation in latitude degrees as a predictor of mangrove primary productivity (e.g., Twilley et al. 1992; Saenger and Snedaker 1993; Bouillon et al. 2008).

The model by Ribeiro et al. (2019) addresses a core question in mangrove macroecology, clarifying the role of factors that control mangrove NPP_L at larger spatial scales. The authors show that mangrove NPP_L is controlled by a combination of climatic (temperature and precipitation) and geophysical forces, such as tidal range. Here we used the model results by Ribeiro et al. (2019) for the neotropics to estimate NPP_L for Brazilian mangroves (Fig. 3.4b). The predicted NPP_L values for Brazilian mangroves ranged from 3.79 to 16.97 Mg ha^{-1} year^{-1} (mean = 10.92 Mg ha^{-1} year^{-1}), and the range reported for the neotropics is 1.66–28.81 Mg ha^{-1} year^{-1} (mean = 10.25 Mg ha^{-1} year^{-1}) (see Ribeiro et al. 2019 for details). Using a biomass-to-carbon conversion factor of 0.475 (see Hamilton and Friess 2018), the predicted mean NPP_L for Brazilian mangroves corresponds to 5.5 Mg C ha^{-1} year^{-1}. Using Hamilton and Casey's (2016) estimative for Brazilian mangrove forest cover of 7675 km^2, the annual rate of C removed from the atmosphere by mangrove NPP_L in the country is estimated at 4 Tg C, which corresponds to 30% of total NPP_L in the neotropics (Ribeiro et al. 2019).

Higher NPP_L rates were predicted for mangrove forests influenced by large river systems, such as along the Amazon River coastline. These patterns of high NPP_L rates predicted for river-dominated coastlines are consistent with observed values reported for other deltaic coastal settings in the neotropics such as in the San Juan River delta (Colombia), Orinoco River delta (Venezuela), and Essequibo River

(Guyana) (see Ribeiro et al. 2019 for details). These regions with high NPP_L are located in tropical regions subjected to low annual variability in temperature, high rates of rainfall (>2000 mm year^{-1}) (Hijmans et al. 2005), and macrotidal regimes (Carrère et al. 2012). Conversely, the low rates of NPP_L in Brazil were predicted for mangroves subjected to lower winter temperatures, reduced tidal amplitude (i.e., Segment VII, Table 3.1; Fig. 3.2), as well as reduced annual precipitation and reduced river discharge (Segment IV, Table 3.1; Fig. 3.2), which altogether constrain high primary productivity and forest development.

Ribeiro et al. (2019) showed that the interaction between precipitation and temperature accounted for most of the variability in mangrove NPP_L across the neotropics. Temperature and precipitation regimes have long been described as important drivers of mangrove NPP_L (Pool et al. 1975; Twilley 1995; Day et al. 1996; Feher et al. 2017). Temperature affects plants' vital processes from photosynthesis and respiration to reproductive success and carbon storage (Duke 1990; Lovelock 2008). Similarly, rainfall also influences mangrove growth and primary production (Day et al. 1996; Twilley et al. 1997; Agraz-Hernández et al. 2015). Lower primary production has been reported for mangrove forests along dry coastlines, whereas the highest NPP_L rates were related to areas with rainfall regimes over 2000 mm year^{-1} (Hernández and Mullen 1975; Félix-Pico et al. 2006; Lema and Polanía 2007). The synergism between temperature and precipitation regimes plays a major role in determining mangrove development and distribution (Spalding et al. 2010; Osland et al. 2016; Feher et al. 2017).

The results in Ribeiro et al. (2019) also highlighted the role of tidal regimes in mangrove NPP_L variability at larger scales. These findings support previous studies that show a strong positive influence of tidal amplitudes in primary production (Cintrón and Schaeffer-Novelli 1981; Alongi 2002). Tides are an energy subsidy to mangroves' primary production (Odum et al. 1982) and as this energy increases, so is the amount of organic matter exchanged between mangroves and adjacent environments (Twilley et al. 1986, 1992). Periodic tidal inundation promotes nutrient exchange and soil aeration, which reduces the accumulation of toxic substances (e.g., sulfides) and enhances forest development (Lugo and Snedaker 1974; Castañeda-Moya et al. 2013). In addition, earlier studies have shown tides to be a major driver of carbon allocation between above- and belowground compartments in mangrove forests. For instance, higher tides are frequently associated with well-developed mangrove forest stands (Cintrón and Schaeffer-Novelli 1981; Twilley 1995; Rovai et al. 2016). Conversely, mangrove root biomass was found to be higher in sites subjected to infrequent inundation (Castañeda-Moya et al. 2011; Adame et al. 2017). Similarly, higher soil organic carbon stocks have been negatively correlated with tides (Rovai et al. 2018). Also, the tidal amplitude is an important component of hydroperiod influencing mangrove species zonation (Crase et al. 2013) as well as the vertical range of suitable environment for mangrove establishment (Hutchings and Saenger 1987).

Although not selected as a significant term in the model by Ribeiro et al. (2019), potential evapotranspiration (PET) has been acknowledged as one of the major climatic factors determining the distribution of life zones on Earth (Holdridge

1967). PET represents the amount of water that could potentially be used by plants, but it is transferred back to the atmosphere through evaporation, thus, being an important regulator of forest water balance (Holdridge 1967). The interaction between PET and precipitation is especially important for mangroves, due to soil water content and salinity balance (Clough 1992; Wolanski et al. 1992). Indeed, PET has been shown to play a major role in the continental-scale variability of aboveground biomass and soil organic carbon stocks in mangroves (Rovai et al. 2016, 2018).

In equatorial climates, where temperatures are constantly high throughout the year, precipitation rates are moderate to high and the ratio between precipitation and PET is low, so mangrove forests can allocate more energy to their aboveground biomass and thus are better developed (Schaeffer-Novelli et al. 1990; Clough 1992). Where PET exceeds rainfall, the water deficit leads to decreased soil moisture, and consequently higher soil salinities, water stress on mangrove trees, and restricted forest development (Schaeffer-Novelli et al. 1990; Day et al. 1996; Castañeda-Moya et al. 2006). Moreover, the upper limit of distribution and survival of particular mangrove species is very often determined by soil salinity and soil water content, which are regulated by the conjunction of PET, rainfall, and tidal amplitude (Wolanski et al. 1992; Castañeda-Moya et al. 2006).

Furthermore, the influence of river discharge on mangrove ecosystems functioning is also indubitable. Nevertheless, excessive freshwater discharges act as a constraint by promoting competition by glicophytes that limits mangrove colonization. This is true in the Amazon estuary as well as south of Laguna (28°30′S) where freshwater habitats prevail displacing mangroves and favor freshwater marsh development. Overall, rivers are responsible for most of the freshwater input in mangroves, acting as a source of nutrients (phosphorus) and decreasing interstitial salinity (Pool et al. 1975; Castañeda-Moya et al. 2013). Riverine mangroves are characterized by optimal structural growth, with high values of aboveground biomass and NPP_L resulting from high nutrient availability, abundant freshwater drainage, and reduced soil salinity levels, which are controlled by river discharge (Cintrón et al. 1978; Castañeda-Moya et al. 2006). River discharge is particularly important in dry (or arheic) coastlines such as in Northeast Brazil (see Chap. 1). In these dry climates, evapotranspiration exceeds the moisture supplied by precipitation, and river discharge becomes an important source of freshwater that controls salinity within limits that are not stressful for mangrove survival, forming extensive hypersaline flats (or “apicuns”).

The *apicum* (in singular) is a spatial-temporal ecogeomorphic feature of the mangrove ecosystem; it is a morphoclimatic hydrosere, a dynamic feature of the high intertidal zone, and technically a high salt marsh feature. The high salt marsh is influenced by precipitation, runoff, or seepage (Costa and Davy 1992; Hadlich et al. 2010). In dry coastlines with minor river discharge, massive mangrove diebacks can occur triggered by inland droughts, multidecadal fluctuations in sea level such as the 18.6-year Metonic Cycle (Munk et al. 2002), reductions in rainfall, and abnormally high air temperatures (Duke et al. 2017; Lovelock et al. 2017). During these events, only mangroves fringing estuary channels and upstream riverine stands remained healthy and mostly intact (Duke et al. 2017).

3.6 Soil Organic Carbon (SOC) Stocks

In the present work, we include a new mangrove ecological feature not covered in the 1990 paper, the continental-scale variability of Soil Organic Carbon (SOC) stocks in response to climatic and geophysical drivers. Mangroves have long been recognized for their potential role as a significant global carbon sink that may mitigate atmospheric CO_2 enrichment (Twilley et al. 1992). They were recently recognized as the most carbon-dense forests in the tropics (Donato et al. 2011), culminating in an increase in research papers reporting mostly on local and regional carbon stocks. Few studies have attempted to deliver global mangrove carbon budgets (Chmura et al. 2003; Bouillon et al. 2008). Only recently have specific models been developed to account for global variation in mangrove SOC stocks (Jardine and Siikamäki 2014; Atwood et al. 2017; Rovai et al. 2018). Attention has been driven to SOC stocks, because most of the carbon in mangroves ecosystems is stored in this compartment (Twilley et al. 1992; Hamilton and Friess 2018), where it remains stable for much longer compared to AGB.

Rovai et al. (2018) demonstrated how local and regional estimates of SOC linked to CES can render a more realistic spatial representation of global mangrove SOC stocks. They combined 107 published and unpublished studies conducted worldwide to yield a dataset consisting of depth-integrated (top meter) mangrove SOC density values, reporting on 551 sites from 43 countries. In contrast to previous studies (e.g., Jardine and Siikamäki 2014; Atwood et al. 2017), this dataset included exclusively soil profiles that were at least 0.3 m in depth (which were then normalized to a depth of 1 m), and mangrove SOC density values obtained from elemental analyses or chemical determination (i.e., wet oxidation). Rovai et al. (2018) showed that the diversity of CESs can contribute to the global integration of complex geomorphological, geophysical, and climatic responses that explain the contribution of mangroves to global carbon sequestration. Their approach improved our capacity to predict the global contribution of coastal systems such as mangroves to carbon dynamics in the Earth system. Although their global mangrove SOC budget estimate was similar to early ones, for example, 2.3 PgC (Rovai et al. 2018) and 2.6 PgC (Atwood et al. 2017), they showed that mangrove SOC stocks vary markedly across different types of CESs, increasing from river- to tide/wave-dominated to carbonate coastlines. For example, a global estimate, recently provided by Atwood et al. (2017), used a country-level mean mangrove SOC stock of 283 Mg C ha^{-1} based on values from 48 countries to extrapolate global patterns for the remaining 57 countries that lack data on mangrove SOC. Results in the study by Rovai et al. (2018) indicate that for those countries, many of which comprise mostly carbonate CESs, the global mean reference value of mangrove SOC stocks suggested by Atwood et al. (2017) is about 50% lower than values based on distinct CESs. Moreover, their analysis showed that the CES framework has the potential to resolve unexpected patterns observed between carbonate and river-dominated coastal landforms identified in former global mangrove SOC budgets (Jardine and Siikamäki 2014). They showed that mangrove SOC stocks have been underestimated by up to 44%

(a difference equivalent of roughly 200 MgC ha^{-1}) and overestimated by up to 86% (around 400 MgC ha^{-1}) in carbonate and deltaic settings, respectively, likely due to the omission of geomorphological and geophysical drivers in accounting for the large-scale variability of mangrove SOC stocks.

Here we used Rovai et al. (2018) results to compute estimates of SOC for Brazilian mangroves (Fig. 3.4c). Lower SOC density values were predicted for deltaic and macrotidal (Segments I–III, Table 3.1; Fig. 3.2) and arid (Segment IV, Table 3.1; Fig. 3.2) CESs. Higher SOC values were consistent along tide- and wave-dominated coastlines (Segments V–VII, Table 3.1; Fig. 3.2). Mangrove SOC stocks in the soil top meter in Brazil ranged from 72.1 to 388.3 MgC ha^{-1} (mean = 240.4 MgC ha^{-1}), within the global range of 33.8 to 464.1 MgC ha^{-1} (mean = 296.6 MgC ha^{-1}) (Rovai et al. 2018). Using the mangrove forest area of 7675 km^2 (Hamilton and Casey 2016), the total carbon stored in mangroves soils in Brazil is estimated at 0.15 PgC, which corresponds to 6.5% of global SOC stocks in contrast to the 9.3% suggested earlier (Hamilton and Friess 2018).

3.7 Advancing the CES Framework: Challenges for Mangrove Macroecologists

Tremendous advances have been made recently in terms of mapping the global mangrove forest cover. The two most recent mangrove forest cover estimates range from nearly 82,000 (Hamilton and Casey 2016) to 132,000 km^2 (Giri et al. 2011). This difference of approximately 40% in mangrove forest cover is due to different methodologies used to classify mangrove occurrence within each degree-cell. While the database in Hamilton and Casey (2016) (CGMFC-21)[2] estimates the percent cover for each degree-cell within a mangrove forest, the earlier database in Giri et al. (2011) (MFW)[3] uses a presence approach. Despite these methodological aspects, both CGMFC-21 and MFW databases have a very high resolution of approximately 900 m^2 (30 × 30 m at the equator).

The parameters on which we based most of the discussion in this chapter (that is, AGB, NPP_L, and SOC) were conveyed using the mangrove forest cover provided by the CGMFC-21 database but adjusted to a much lower fidelity (approximately 625 km^2 or 25 × 25 km at the equator) than the original spatial resolution. As pointed out in the original sources, we based our analyses on Rovai et al. (2016 and 2018) and Ribeiro et al. (2019).

There are essentially two main reasons that may be preventing the development of robust higher-resolution large-scale mapping of mangrove ecological attributes. First, the attempt to balance the loss of information during the trade-off process of down- and upscaling data with different resolutions (Blackburn and Gaston 2002).

[2]CGMFC-21 (project): Continuous Global Mangrove Forest Cover for the Twenty-first Century.

[3]MFW (dataset): Mangrove Forest Cover Loss dataset.

Indeed, recent efforts in macroecology strived to consolidate a database of environmental variables that are thought to be relevant to species' ecology and geographic distribution at a reasonable spatial resolution (0.08333° or approximately 8.3 km at the equator) (Title and Bemmels 2018). Even the WorldClim database (Hijmans et al. 2005), which has over 3000 citations, has resolutions that range from 1 to 340 km^2. In this respect, some of the predictors used in the analyses we present here have a coarse native resolution, such as river discharge (0.5°). Thus, it is reasonable to work with an intermediary cell size (e.g., 0.25°) that is spatially representative of most CES domains, which the modeling framework is based on. Second, although the integration of information on mangrove typology based on local hydrology and topography (e.g., fringe vs. interior sites) would potentially allow for more robust local and global estimates, most papers in which the analyses presented here are based on do not include accurate information on hydroperiod. Accordingly, the spatial resolution of most global compilations on marine and terrestrial environmental variables (Title and Bemmels 2018) does not reflect the variability compatible with neither the CMFGC-21 nor the MFW database native resolution.

In order to perform a multiscale spatial analysis, both dependent and independent variables would have to be available at differing resolutions. Moreover, the set of environmental variables would have to hold ecological meaning across different spatial scales, which is unlikely as variables that control SOC formation in coastal wetlands differ at different scales (check Holmquist et al. 2018; Osland et al. 2018; and Rovai et al. 2018). While the scale-dependent issues discussed here are perhaps one of the major challenges mangrove ecologists will face when upscaling ecological traits from site-level observations, the CES framework resolved much of the dramatic difference in mangrove SOC estimates, particularly in terms of spatial variability with mangrove soil properties following close the energetic signature of distinct coastline types (Rovai et al. 2016, 2018; Twilley et al. 2018; Ribeiro et al. 2019).

3.8 CES Restrict the Atlantic South American Mangrove Limit

Laguna is an interesting threshold and is currently considered the southernmost limit of mangroves in Brazil (Cintrón and Schaeffer-Novelli 1981; Soares et al. 2012). However, it is attention-grabbing, because suitable habitats further south in the country seem to be present (Ximenes et al. 2018), yet mangroves as an ecosystem stop abruptly at Laguna (28°48′S). The mean sea surface temperatures here vary from 18.2 °C in summer to 16 °C in winter (Ximenes et al. 2018).

In a biogeographical terminus, this is a particularly interesting transitional zone, because it appears not only to be a limit to a species but to an ecosystem; at this geographic location, a regime or phase shift takes place. The discontinuity apparent in Laguna is a topic of great relevance to climate change research and the understanding of the future of mangroves in the region. Rather than mangrove expansion, the region may have been experiencing a contraction due to increasing freshwater dominance that might have resulted in a freshwater barrier blockage that now limits further mangrove expansion to the south beyond Laguna (Cintrón-Molero and Schaeffer-Novelli 2019) (Map 16).

Furthermore, because of potential conflict with agricultural land use in the Patos Lagoon region, it is likely that manmade attempts to restrict saline intrusions could further limit mangrove expansion to the south in the near future. Below Laguna is the 620 km coastal tract of Rio Grande do Sul State (Map 17), which encompasses South America America's longest barrier structure, running almost uninterruptedly except near Cassino and Tramandaí inlets; the former is the inlet to the Patos Lagoon. Both are permanent openings due to the high freshwater discharges of the coastal lagoons behind the barrier.

Patos Lagoon's extensive marshes are dominated by the genera *Spartina*, *Juncus*, *Cyperus*, *Typha*, *Scirpus*, *Paspalum*, and *Sesuvium* (Delaney 1962), which prevail in an eminently freshwater environment promoted by abundant rainfall water ($P \approx 1500$ mm year^{-1}), reduced potential evapotranspiration, high percolation rates, seepage, river flows, and microtidal regime (Hijmans et al. 2005; Carrère et al. 2013; Cohen et al. 2013). These occupy the biogeographic changeover zone that extends to northeastern Argentina (Costa and Davy 1992). The coast of the Rio Grande do Sul State, south of 34°S, is well known to receive rain throughout the year, including the passage of Mesoscale Convective Systems (MCS) (Houze Jr 2004), severe frontal systems as well as sporadic severe hail and frost events. The larger continental landmass at 10–25°S is conducive to the development of deep convective activity fed by Amazon moisture transport by a low-level jet into the area. This makes this area the most active MCS region in the world (Nesbitt and Zipser 2003).

Mangroves are documented to persist at the latitude of 38°45′S at Westport, Australia, where the mean annual atmospheric temperature is 18 °C and the coldest is 17 °C and where humid subtropical (Cfa) and maritime (Cfb) climate prevails (Peel et al. 2007). At Corner Inlet, Australia, they are found at 38°54′S. So, the abrupt phase shift at Laguna (Brazil) is a prominent feature that merits further and more detailed attention in the context of environmental change prediction. In any case, climate change is perhaps one of the most active research areas in present times, and southeastern Brazil and mangrove ecosystem dynamics offer fruitful research possibilities that would lead to understanding how climate influences coastal vegetation.

The southern domain is an area where planetary, regional, and local processes interact but where it is realistic to locate instrumentation to provide local-level data recordings and frequent site-level vegetation and interstitial salinity monitoring. This is a region where active climatological research is taking place and where climatology is of great interest because of its impact on agriculture and the local economy. This research is bound to help support new hypotheses about the distribution and abrupt limit of mangrove occurrence in this coastal segment.

3.9 Dynamic Framing and the Three Coastal Domains

The segments described, in the 1990 paper, are embedded within three broad domains that span the whole coast; they remain relevant to serve as guiding posts for versatile back-and-forth shifting of observation scales, an approach we have designated as *dynamic framing*. The three domains we identified are (Fig. 3.5)

- The *Northernmost Domain* is highly moisture- and tide-subsidized and extends from the Guyanas and Amapá (Brazil, Cape Orange, Oiapoque River) to Cape São Roque).
- The *Central Deltaic Coast Domain* extends from below Cape São Roque to Cabo Frio, as a domain characterized by warm temperatures but strong lateral constraints due to high levels of wave/energy.
- The *Cabo Frio to Laguna Domain*, largely below the Tropic of Capricorn and increasingly influenced by cold frontal systems and the convective activity of the South Atlantic Convergence Zone. This portion of the Brazilian coast is periodically and strongly influenced by local, regional (South American Monsoon System, Robertson et al. 2005), and global forcings (e.g., ENSO).

An apparent paradox by which muddy coasts act simultaneously as outwelling sources of biological organic matter while being geological sinks is resolved by recognizing a dialectical perspective between scales. In the short term, outwelling is notable and characteristic but over long temporal scales, deposition and accumulation prevail. This suggests that the CES scale integrates equilibrial and nonequilibrial dynamics at the scale of the whole system.

3.10 Final Remarks

It is misleading to consider coastal features as static or perceive variability as a disturbing feature. Variability is representative of complexity and although it presents obstacles to generalizing and identifying clear-cut patterns in nature, it is part of it and is present at all scales driven by external and internal factors, especially climate and self-organization. The emerging awareness about mangrove systems in sequestering carbon emissions and their contribution to climate regulation increases the relevance of continued research for education, developing robust conservation policy and for suggesting future research grounded in the emerging field of complexity science.

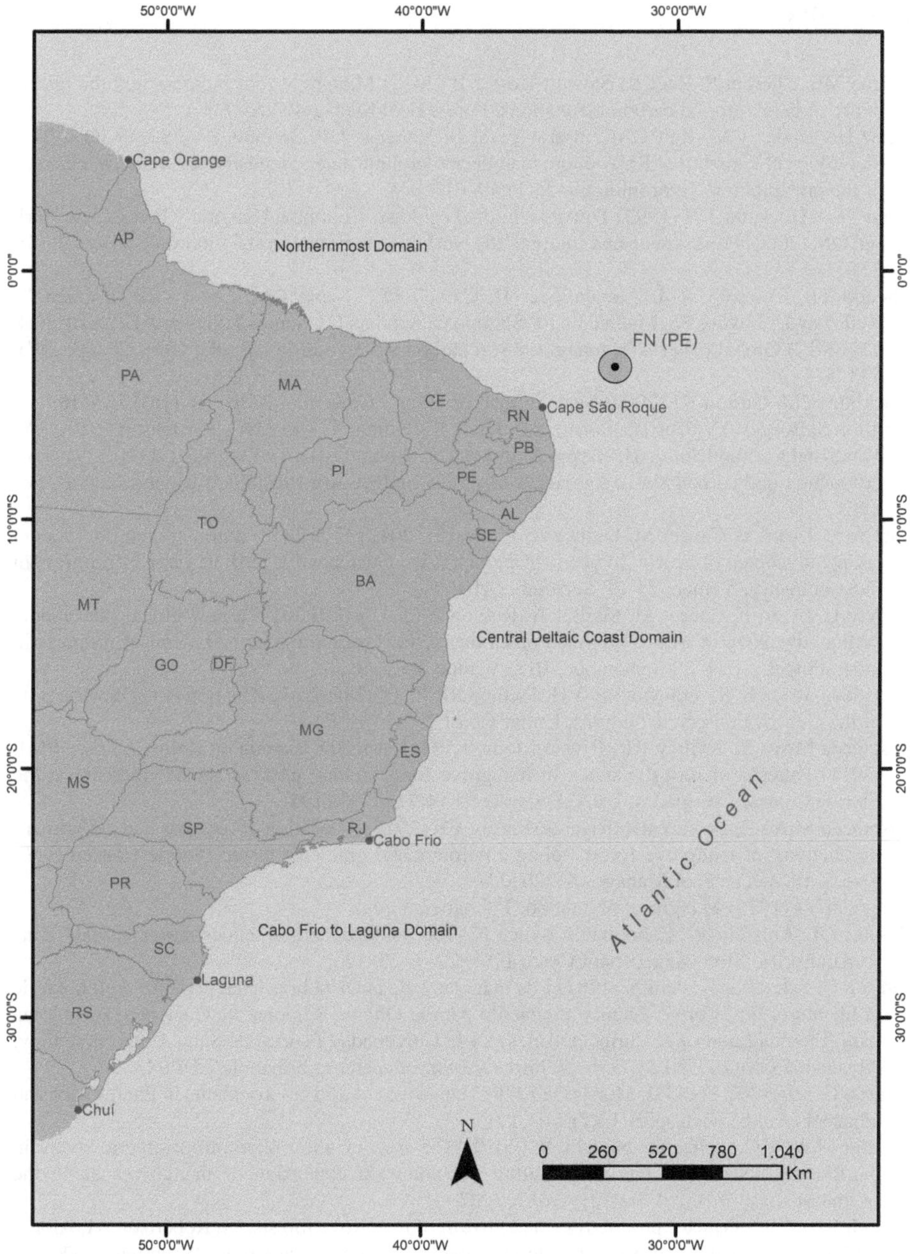

Fig. 3.5 Northernmost, Central Deltaic, and Cabo Frio to Laguna Coastal Domains

References

Adame MF, Cherian S, Reef R, Stewart-Koster B (2017) Mangrove root biomass and the uncertainty of belowground carbon estimations. Forest Ecol Manag 403:52–60

Agraz-Hernández CM, Keb CAC, Iriarte-Vivar S, Venegas GP, Serratos BV, Sáenz JO (2015) Phenological variation of Rhizophora mangle and ground water chemistry associated to changes of the precipitation. Hydrobiologia 25(1):49–61

Allen TFH, Hoekstra TW (1992) Toward a unified ecology. Columbia University Press, New York

Alongi DM (2002) Present state and future of the world's mangrove forests. Environ Conserv 29(3): 331–349

Atwood TB, Connolly RM, Almahasheer H, Carnell PE, Duarte CM, Lewis CJE, Irigoien X, Kelleway JJ, Lavery PS, Macreadie PI, Serrano O, Sanders CJ, Santos I, Steven ADL, Lovelock CE (2017) Global patterns in mangrove soil carbon stocks and losses. Nat Clim Change 7(7): 523–528

Blackburn TM, Gaston KJ (2002) Scale in macroecology. Glob Ecol Biogeogr 11(3):185–189

Bouillon S, Borges AV, Castañeda-Moya E, Diele K, Dittmar T, Duke NC, Kristensen E, Lee SY, Marchand C, Middelburg JJ, Rivera-Monroy VH, Smith TJ III, Twilley RR (2008) Mangrove production and carbon sinks: a revision of global budget estimates. Glob Biogeochem Cycles 22:2

Carrère L, Lyard F, Cancet M, Guillot A, Roblou L (2012) FES 2012: a new global tidal model taking advantage of nearly 20 years of altimetry. In: Ouwehand L (ed) 20 years of progress in radar altimetry, Venice, 24–29 September 2013

Carrère L, Lyard F, Cancet M, Guillot A, Roblou L (2013) FES 2012: a new global tidal model taking advantage of nearly 20 years of altimetry. In: Proceedings of '20 years of progress in radar altimetry', 24–29 September 2012, Venice, Italy

Castañeda-Moya E, Rivera-Monroy VH, Twilley RR (2006) Mangrove zonation in the dry life zone of the Gulf of Fonseca, Honduras. Estuar Coast 29(5):751–764

Castañeda-Moya E, Twilley RR, Rivera-Monroy VH, Marx BD, Coronado-Molina C, Ewe SML (2011) Patterns of root dynamics in mangrove forests along environmental gradients in the Florida Coastal Everglades, USA. Ecosystems 14(7):1178–1195

Castañeda-Moya E, Twilley RR, Rivera-Monroy VH (2013) Allocation of biomass and net primary productivity of mangrove forests along environmental gradients in the Florida Coastal Everglades, USA. For Ecol Manag 307:226–241

Chapman VJ (1975) Mangrove vegetation. J. Cramer, Vaduz

Chmura GL, Anisfeld SC, Cahoon DR, Lynch JC (2003) Global carbon sequestration in tidal, saline wetland soils. Glob Biogeochem Cycle 17(4):22–1–22–12

Cintrón G, Schaeffer-Novelli Y (1981) Los manglares de la costa brasileña: revisión preliminar de la literatura. In: Informe Técnico preparado para la Oficina Regional de Ciencia y Tecnología para América Latina y el Caribe de Unesco y la Universidad Federal de Santa Catarina (ed/rev: Abuchahla GMO). UNESCO, p 47. http://www.producao.usp.br/handle/BDPI/43826

Cintrón G, Lugo AE, Pool DJ, Morris G (1978) Mangroves of arid environments in Puerto Rico and adjacent islands. Biotropica 10(2):110–121

Cintrón-Molero G, Schaeffer-Novelli Y (2019) The role of atmospheric-tropospheric rivers in partitioning coastal habitats and limiting the poleward expansion of mangroves along the southeast coast of Brazil. Int J Hydrol 3(2):92–94

Clough BF (1992) Primary productivity and growth of mangrove forests. In: Robertson AI, Alongi DM (eds) Tropical mangrove ecosystems. American Geophysical Union, Washington, DC, pp 225–249

Cohen S, Kettner AJ, Syvitski JPM, Fekete BM (2013) WBMsed, a distributed global-scale riverine sediment flux model: model description and validation. Comput Geosci 53:80–93

Costa CSB, Davy AJ (1992) Coastal Saltmarsh communities of Latin America. In: Seeliger U (ed) Coastal plant communities of Latin America. Academic, San Diego, pp 179–199

Crase B, Liedloff A, Vesk PA, Burgman MA, Wintle BA (2013) Hydroperiod is the main driver of the spatial pattern of dominance in mangrove communities. Glob Ecol Biogeogr 22:806–817

Day JW Jr, Coronado-Molina C, Vera-Herrera FR, Twilley R, Rivera-Monroy VH, Alvarez-Guillen H, Day R, Conner W (1996) A 7 year record of above-ground net primary production in a southeastern Mexican mangrove forest. Aquat Bot 55:39–60

Delaney PJV (1962) Quaternary geologic history of the coastal plain of Rio Grande do Sul, Brazil. S Am Coast Stu Techn Rep 10(A):1–63

Donato DC, Kauffman JB, Murdiyarso D, Kurnianto S, Stidham M, Kanninen M (2011) Mangroves among the most carbon-rich forests in the tropics. Nat Geosci 4(5):293–297

Duke NC (1990) Phenological trends with latitude in the mangrove tree *Avicennia marina*. J Ecol 78(1):113–133

Duke NC, Kovacs JM, Griffiths AD, Preece L, Hill DJE, van Oosterzee P, Mackenzie J, Morning HS, Burrows D (2017) Large-scale dieback of mangroves in Australia's Gulf of Carpentaria: a severe ecosystem response, coincidental with an unusually extreme weather event. Mar Freshwater Res 68(10):1816–1829

Feher LC, Osland MJ, Griffith KT, Grace JB, Howard RJ, Stagg CL, Enwright NM, Kraus KW, Gabler CA, Day RH, Rogers K (2017) Linear and nonlinear effects of temperature and precipitation on ecosystem properties in tidal saline wetlands. Ecosphere 8(10):e01956

Félix-Pico E, Holguín-Quiñones O, Hernández-Herrera A, Flores-Verdugo F (2006) Mangrove primary production at El Conchalito Estuary in La Paz Bay (Baja California Sur, Mexico). Cienc Mar 32(1A):53–63

Fisheries and Aquaculture Department – FAO (2007) The world's mangroves 1980–2005. FAO Romefos

Giri CE, Ochieng E, Tieszen LL, Zhu Z, Singh A, Loveland T, Masek J, Duke NC (2011) Status and distribution of mangrove forests of the world using earth observation satellite data. Glob Ecol Biogeogr 20:154–159

Gregory KJ, Goudie AS (2011) The Sage handbook of geomorphology. Sage, Los Angeles

Hadlich GM, Celino JJ, Ucha JM (2010) Diferenciação físico-química entre apicuns, manguezais e encostas na Baía de Todos os Santos, Nordeste do Brasil. Geociências 29(4):633–641

Hamilton SE, Casey D (2016) Creation of a high spatio-temporal resolution global database of continuous mangrove forest cover for the 21st century (CGMFC-21). Glob Ecol Biogeogr 25(6):729–738

Hamilton SE, Friess DA (2018) Global carbon stocks and potential emissions due to mangrove deforestation from 2000 to 2012. Nat Clim Change 8(3):240–244

Hamilton SE, Snedaker SC (eds) (1984) Handbook for mangrove area management. Environment and Policy Institute East-West Center, International Union for the Conservation of Nature and Natural Resources & United Nations Educational, Scientific and Cultural Organization

Hernández A, Mullen K (1975) Produtividad primaria neta en un manglar del Pacífico Colombiano. In: Memorias del Seminario Sobre El Pacífico Colombiano. Universidad del Valle, Cali

Hijmans RJ, Cameron SE, Parra JL, Jones PG, Jarvis A (2005) Very high-resolution interpolated climate surfaces for global land areas. Int J Climatol 25:1965–1978

Holdridge LR (1967) Life zone ecology. Tropical Science Center, San José

Holmquist JR, Windham-Myers L, Bliss N, Crooks S, Morris JT, Megonigal JP, Troxler T, Weller D, Callaway J, Drexler J, Ferner MC, Gonneea ME, Kroeger KD, Schile-Beers L, Woo I, Buffington K, Breithaupt J, Boyd BM, Brown LN, Dix N, Hice L, Horton BP, MacDonald GM, Moyer RP, Reay W, Shaw T, Smith E, Smoak JM, Sommerfield C, Thorne K, Velinksy D, Watson E, Grimes KW, Woodrey M (2018) Accuracy and precision of tidal wetland soil carbon mapping in the conterminous United States. Sci Rep-UK 8(1):9478

Houze RA Jr (2004) Mesoscale convective systems. Rev Geophys 42(RG4003)

Huggett RJ (1995) Geoecology: an evolutionary approach. Routledge, New York City

Hutchings P, Saenger P (1987) Ecology of mangroves. University of Queensland Press, St Lucia

Hutchison J, Manica A, Swetnam R, Balmford A, Spalding M (2014) Predicting global patterns in mangrove forest biomass. Conserv Lett 7(3):233–240

Instituto Brasileiro de Geografia e Estatistica – IBGE (2016) Caracterização do território: posição e extensão. In: Anuário Estatístico do Brasil, vol 1. IBGE, Rio de Janeiro

Jardine SL, Siikamäki JV (2014) A global predictive model of carbon in mangrove soils. Environ Res Lett 9(10):104013

Lema LF, Polanía J (2007) Estructura y dinámica del manglar del delta del río Ranchería. Caribe colombiano. Rev Biol Trop 55(1):11–21

Lovelock CE (2008) Soil respiration and belowground carbon allocation in mangrove forests. Ecosystems 11(2):342–354

Lovelock CE, Feller IC, Reef R, Hickey S, Ball MC (2017) Mangrove dieback during fluctuating sea levels. Sci Rep-UK 7(1):1–8

Lugo AE, Snedaker SC (1974) The ecology of mangroves. Annu Rev Ecol Syst 5(1):39–64

Mandelbrot BB (1983) The fractal geometry of nature, 1st edn. WH Freeman, New York City

Munk W, Dzieciuch M, Jayne S (2002) Millennial climate variability: is there a tidal connection? J Clim 15:370–385

Nesbitt SW, Zipser EJ (2003) The diurnal cycle of rainfall and convective intensity according to three years of TRMM measurements. AMS J Clim 16:1456–1475

Odum EP (1969) The strategy of ecosystem development. Science 164:262–270

Odum WE (1970) Pathways of energy flow in a South Florida Estuary. PhD dissertation, University of Miami

Odum WE, Heald EJ (1975) Mangrove forests and aquatic productivity. In: An introduction to land-water interactions. Springer, Berlin/Heidelberg, pp 129–136

Odum WE, McIvor CC, Smith TJ III (1982) The ecology of the mangroves of South Florida: a community profile. US Fish and Wildlife Service – Office of Biological Services, Washington, DC

Osland MJ, Enwright NM, Day RH, Gabler CA, Stagg CL, Grace JB (2016) Beyond just sea-level rise: considering macroclimatic drivers within coastal wetland vulnerability assessments to climate change. Glob Change Biol 22:1–11

Osland MJ, Gabler CA, Grace JB, Day RH, McCoy ML, McLeod JL, From AS, Enwright NM, Feher LC, Stagg CL, Hartley SB (2018) Climate and plant control on soil organic matter in coastal wetlands. Glob Change Biol 24(11):5361–5379

Patterson T, Kelso NV (2018) Natural earth. https://www.naturalearthdata.com/downloads/10m-physical-vectors/10m-rivers-lake-centerlines/

Peel MC, Finlayson BL, McMahon TA (2007) Updated world map of the Köppen-Geiger climate classification. Hydrol Earth Syst Sci 5:439–473

Pool DJ, Lugo AE, Snedaker SC (1975) Litter production in mangrove forests of southern Florida and Puerto Rico. In: Proceedings of the international symposium on biology and management of mangroves, Honolulu, 8–11 October 1974

Ribeiro RA, Rovai AS, Twilley RR, Castañeda-Moya E (2019) Spatial variability of mangrove primary productivity in the neotropics. Ecosphere 10(8):e02841

Robertson AW, Mechoso CR, Ropelewski CF, Grimm AM (2005) The American monsoon systems. In: Chang C-P, Wang B, Lau N-CG (eds) The global monsoon system: research and forecast. Secretariat of the World Meteorological Organization, Geneva, pp 197–206

Rovai AS, Riul P, Twilley RR, Castañeda-Moya E, Rivera-Monroy VH, Williams AA, Simard M, Cifuentes-Jara M, Lewis RR, Crooks S, Horta PA, Schaeffer-Novelli Y, Cintrón G, Pozo-Cajas PPR (2016) Scaling mangrove aboveground biomass from site-level to continental-scale. Glob Ecol Biogeogr 25(3):286–298

Rovai AS, Twilley RR, Castañeda-Moya E, Riul P, Cifuentes-Jara M, Manrow-Villalobos M, Horta PA, Simonassi JC, Fonseca AL, Pagliosa PR (2018) Global controls on carbon storage in mangrove soils. Nat Clim Change 8(6):534–538

Rowe JS (1961) The level-of-integration concept and ecology. Ecology 42(2):420–427

Saenger P, Snedaker SC (1993) Pantropical trends in mangrove above-ground biomass and annual litterfall. Oecologia 96(3):293–299

Schaeffer-Novelli Y (1999) Situação atual do grupo de ecossistemas: manguezal, marisma e apicum, incluindo os principais vetores de pressão e as perspectivas para sua conservação e usos sustentável. Avaliação e ações prioritárias para a conservação da biodiversidade da zona costeira e marinha. Projeto de Conservação e Utilização Sustentável da Diversidade Biológica Brasileira – PROBIO. Report BDT_mangue-1999

Schaeffer-Novelli Y, Cintrón-Molero G, Adaime RR, Camargo TM (1990) Variability of mangrove ecosystems along the Brazilian coast. Estuaries 13(2):204–218

Schaeffer-Novelli Y, Soriano-Sierra EJ, Vale CC, Bernini E, Rovai AS, Pinheiro MAA, Schmidt AJ, Almeida R, Coelho-Jr C, Menghini RP, Martinez DI, Abuchahla GMO, Cunha-Lignon M, Charlier-Sarubo S, Shirazawa-Freitas J, Cintrón-Molero G (2016) Climate changes in mangrove forests and salt marshes. Braz J Oceanogr 64(sp2):37–52

Simon HA (1962) The architecture of complexity. Proc Am Phils Soc 106(6):467–482

Soares MLG, Estrada GCD, Fernandez V, Tognella MMP (2012) Southern limit of the Western South Atlantic mangroves: assessment of the potential effects of global warming from a biogeographical perspective. Estuar Coast Shelf S 101:44–53

Spalding M, Kainuma M, Collins L (2010) The world Atlas of mangroves. Earthscan, London

Thom BG (1982) Mangrove ecology – a geomorphological perspective. In: Clough BF (ed) Mangrove ecosystems in Australia: structure, function and management. Australian National University Press, Canberra, pp 3–17

Title PO, Bemmels JB (2018) ENVIREM: an expanded set of bioclimatic and topographic variables increases flexibility and improves performance of ecological niche modeling. Ecography 41(2): 291–307

Turner MG (1998) Landscape ecology. In: Dodson SI, Allen TFH, Carpenter SR, Ives AR, Jeanne RL, Kichell JF, Langston NE, Turner MG (eds) Ecology. Oxford University Press, New York/ Oxford, pp 77–122

Twilley RR (1995) Properties of mangrove ecosystems related to the energy signature of coastal environments. In: Hall CAS (ed) Maximum power: the ideas and applications of H. T. Odum. University Press of Colorado, Denver, pp 43–62

Twilley RR, Rivera-Monroy VH (2009) Ecogeomorphic models of nutrient biogeochemistry for mangrove wetlands. In: Perillo GME, Wolanski E, Cahoon DR, Brinson MM (eds) Coastal wetlands: an integrated ecosystem approach. Elsevier BV, Dordrecht, pp 641–683

Twilley RR, Lugo AE, Patterson-Zucca C (1986) Litter production and turnover in basin mangrove forests in southwest Florida. Ecology 67(3):670–683

Twilley RR, Chen R, Hargis T (1992) Carbon sinks in mangroves and their implications to carbon budget of tropical coastal ecosystems. Water Air Soil Poll 64:265–288

Twilley RR, Pozo M, Garcia VH, Rivera-Monroy VH, Zambrano R, Bodero A (1997) Litter dynamics in riverine mangrove forests in the Guayas river estuary, Ecuador. Oecologia 111(1):109–122

Twilley RR, Rovai AS, Riul P (2018) Coastal morphology explains global blue carbon distributions. Front Ecol Environ 16(9):1–6

Vestbo S, Obst M, Fernandez FJQ, Intanai I, Funch P (2018) Present and potential future distributions of Asian Horseshoe crabs determine areas for conservation. Front Mar Sci 5:164

Walsh GE (1974) Mangroves: a review. In: Reimhold RJ, Queen WH (eds) Ecology of halophytes. Academic, New York, pp 51–174

Wolanski E, Mazda Y, Ridd P (1992) Mangrove hydrodynamics. In: Tropical mangrove ecosystems. American Geophysical Union, Washington, DC, pp 43–62

Woodroffe CD (1992) Mangrove sediments and geomorphology. In: Robertson AI, Alongi DM (eds) Tropical mangrove ecosystems. American Geophysical Union, Washington, DC, pp 7–41

Woodroffe CD (2002) Coasts: form, process and evolution. Cambridge University Press, Cambridge

Ximenes AC, Ponsoni L, Lira CF, Koedam N, Dahdouh-Guebas F (2018) Does sea surface temperature contribute to determining range limits and expansion of mangroves in Eastern South America (Brazil)? Remote Sens-Basel 10:1787

Part II
The Mangrove Ecosystem

Chapter 4
Reproductive Phenology of the Brazilian Mangrove Species

Elaine Bernini, Frederico Lage-Pinto, and Yara Schaeffer-Novelli

4.1 Introduction

Phenology is the study of the occurrence of repetitive biological phenomena and their causes in terms of environmental factors as well as the interrelationship between phases of the same or different species (Lieth 1974). The phenological events studied in plants include vegetative phases (buds and leaf fall) and reproductive phases (flowering and fruiting).

Plant reproductive phenology is controlled by complex interactions between biotic and abiotic factors (Wolkovich et al. 2014). The biotic factors include interactions with pollinators, dispersers, and morphological and physiological adaptations (Van Schaik et al. 1993; Liebsch and Mikich 2009; Wolkovich et al. 2014), while abiotic factors include precipitation, air temperature, photoperiod, solar radiation, and soil water availability (Morellato et al. 2000; Engel and Martins 2005; Couralet et al. 2013; Wolkovich et al. 2014; Borchert et al. 2015). The reproductive patterns of tropical plants and their relationship with abiotic factors have been described for various ecosystems (Ballestrini et al. 2011; Nadia et al. 2012; Morellato et al. 2013; Rodríguez-Gallego and Navarro 2015; Ulsig et al. 2017).

The study of the reproductive phenology of plant species is fundamental to understanding the dynamics of ecosystems (Fournier 1974) since the periodicity and synchrony of phenological events influence the structure and functioning of

E. Bernini (✉) · F. Lage-Pinto
Laboratório de Ecologia de Costeira e Oceânica, Centro de Ciências Aplicadas e Educação, Universidade Federal da Paraíba, Rio Tinto, PB, Brazil

Y. Schaeffer-Novelli
Instituto Oceanográfico, Universidade de São Paulo, São Paulo, SP, Brazil

Instituto BiomaBrasil, Recife, PE, Brazil

IUCN Mangrove Specialist Group, São Paulo, SP, Brazil

Y. Schaeffer-Novelli et al. (eds.), *Brazilian Mangroves and Salt Marshes*, Brazilian Marine Biodiversity, https://doi.org/10.1007/978-3-031-13486-9_4

communities (Williams et al. 1999; Encinas-Viso et al. 2012; Revilla et al. 2015). In addition, monitoring reproductive phenomena provides data on the quantity and quality of available wildlife resources (Bullock and Sollis-Magallanes 1990; Araújo et al. 2011), while also informing projects aiming at the recovery of degraded areas (Zamith and Scarano 2004; Garcia et al. 2009). Furthermore, phenological studies are currently being used to monitor climate change (Morellato 2008; Cleland et al. 2012) and can contribute to assessing and mitigating the consequences of land-use changes and other anthropogenic disturbances, such as fragmentation, fire, and invasive species (Morellato et al. 2010; Morellato et al. 2016).

Mangroves are highly ecologically important ecosystems (Adame et al. 2010; Mumby 2006; Donato et al. 2011) whose area has declined alarmingly in recent decades (Valiela et al. 2001; Giri et al. 2011; Hamilton and Casey 2016). The loss of mangrove forests has been caused by tourism, aquaculture, urban development, overexploitation of resources, agriculture, and industrialization (Alongi 2002), and the reminiscent mangroves remain under intense anthropogenic pressure, with an estimated global deforestation rate between 0.16% and 0.39% per year (Hamilton and Casey 2016) (see Chaps. 2 and 16).

Currently, mangroves are being affected by global climate changes that further aggravate anthropogenic pressures (Wong et al. 2014). Global changes in precipitation rates and air temperature as well as rising sea levels may modify reproductive phenology and reduce the production of flowers and fruits in addition to altering seed dispersal and seedling establishment (Alongi 2008; Ellison 2012; Van der Stocken et al. 2017). Thus, climate change is expected to have consequences for population dynamics and the biogeographical distribution of mangrove species (Perry and Mendelssohn 2009; Van der Stocken et al. 2017).

This chapter aims to assess the state-of-the-art research on reproductive phenology in mangroves in Brazil. We address methodological issues, describe patterns of flowering and fruiting, and highlight knowledge gaps on the phenology of mangrove species. The review was carried out by conducting a bibliographical survey of scientific journals and books as well as theses, dissertations, and monographs, since such references cover a large portion of the studies of the phenology of mangrove species in Brazil, namely, *Avicennia germinans* (L.) Stearn, *Avicennia schaueriana* Stapf & Leechm. ex Moldenke, *Laguncularia racemosa* (L.) Gaertn. f., *Rhizophora harrisonii* Leechman, *Rhizophora mangle* L., and *Rhizophora racemosa* G. F. Mayer.

4.2 Methodological Approaches

In Brazil, most studies of reproductive phenology monitor flowering (Fig. 4.1) and fruiting (Fig. 4.2), but the definition of phenophases varies according to the species. For *Rhizophora mangle*, for example, the recorded phenophases may be flower buds, flowers at anthesis, fruits, and propagules, while for *Avicennia schaueriana* and *Laguncularia racemosa*, they may be buds, flowers, and propagules (Nadia et al. 2012). Some authors also record immature and mature fruits (Matni 2007; Lima

Fig. 4.1 Flowers of (**a**) *Avicennia germinans*, (**b**) *A. schaueriana*, (**c**) *Laguncularia racemosa*, and (**d**) *Rhizophora mangle*. (Photos: Clemente Coelho-Jr)

2012). The detailed monitoring of flowering (buds and flowers) and fruiting (immature fruits, mature fruits, and propagules) is useful to better plan recovery actions for degraded areas.

The research methodology can influence the analysis and interpretation of the phenological patterns, making it difficult to compare results among studies, so the choice of the evaluation method is of relevance (Bencke and Morellato 2002; D'Eça-Neves and Morellato 2004). A total count of flowers and fruits is virtually impossible for trees, so direct and indirect methods have been developed to monitor reproductive phenophases.

In Brazil, direct observation procedures for mangrove species include the semi-quantitative methods of canopy counting (Fernandes 1999) and Fournier intensity (Fournier 1974), in which the number of objects (e.g., flower buds or fruits) in different categories is recorded on an ordinal scale. Direct observation can be carried out by the quantitative method of branch counting (Christensen 1978), by which all the flowers or fruits on a branch are counted. Direct methods are faster than the indirect method (Stevenson et al. 1998; D'Eça-Neves and Morellato 2004).

The indirect method is conducted by quantifying the dry weight of phenophases, with the aid of litterfall collectors, which are baskets that come in different shapes and sizes. For mangrove species, the studies that have been carried out with collectors mainly emphasize the importance of phenological patterns in the

Fig. 4.2 Fruits of (**a**) *Avicennia germinans*, (**b**) *A. schaueriana*, (**c**) *Laguncularia racemosa*, and (**d**) *Rhizophora mangle*. (Photos: Elaine Bernini)

production of forest biomass. Such studies do not always separate reproductive structures into categories (flowers and fruits) and/or species. We found only one study (Mehlig 2006) that specifically used collectors to evaluate phenology in Brazil. However, studies that employed collectors to quantify primary production and that specified the occurring phenophases/mangrove species were also included in our data compilation.

Studies comparing direct and indirect observation methods in tropical forests have found different results. Morellato et al. (2010) reported differences in seasonal patterns, while Stevenson et al. (1998) reported similar seasonal patterns for both methods. Morellato et al. (2010) drew attention to the following deviations when estimating phenology using collectors: (1) species composition may vary between methods; (2) there is a time interval between the direct observation of a phenophase in a tree and the record of it in the collectors; and (3) there is sensitivity to local effects, such as the presence of large quantities of flowers and fruits of a certain species in the collectors or flowers or heavy fruits that overestimate production. As mangrove species exhibit significant differences in propagule weight, the evaluation of phenology using collectors should be avoided.

Fernandes et al. (2005) studied the reproductive phenology of *L. racemosa* through the direct methods of canopy and branch counting. The authors noted the difficulty of recording more than 100 items in higher trees using the canopy counting method. It is possible to quantify the absolute number of each item more easily using the branch counting method, but this is logistically more complicated in forests with very high trees. However, the results of the two methods showed a significant correlation with both the flower phenophase ($r = 0.94$, $p < 0.001$) and the fruit phenophase ($r = 0.93$, $p < 0.001$), indicating that the indirect method is efficient for monitoring and describing reproductive phenological patterns.

The Fournier intensity (1974) has been the most-used method for mangrove species in Brazil followed by canopy counting, branch counting, and collectors (see Tables 4.1, 4.2, 4.3 and 4.4). In the northern region of the country, the most used method is canopy counting, while the Fournier method (1974) is more common in the other regions.

Sample size may also influence the characterization of phenological patterns. Fournier et al. (1975) suggested a sample size of 10 individuals to analyze the phenology of tropical tree species, but Morellato et al. (2010) recommend at least 15 trees to better estimate the pattern of a sampled population. Considering the most-used evaluation methods in Brazilian mangrove forests (Fournier 1974; Fernandes 1999), the sample number was 15 or more trees in 70% and 10 trees in 30% of the surveys analyzed, indicating an adequate sample number.

The frequency of observations is another important aspect of phenological studies. Morellato et al. (2010) showed that larger sample numbers describe phenological patterns with increasing accuracy as the frequency of observations increases, while small samples lose information and accuracy. Thus, biweekly observations (i.e., every 15 days) provide reasonable accuracy regardless of sample size. For the studies analyzed here, only Mehlig (2006) and Bernini and Rezende (2010) made biweekly observations, while the other studies performed monthly observations (see Tables 4.1, 4.2, 4.3 and 4.4). This result can be partly attributed to a lack of financial resources and logistical issues but considering that the number of sampled individuals has been adequate in most studies, the monthly observation frequency would not represent a severe loss of information (Morellato et al. 2010) that might compromise the phenological characterization by the studies here presented.

Phenological patterns may vary over time due to climate variations and differential flower and fruit production, so long-term studies (three years or more) are recommended. For Brazilian mangroves, most of the data refer to one year of observation with sampling periods exceeding two years in rare cases (see Tables 4.1, 4.2, 4.3 and 4.4). The reasons for the short study duration can be the limitation of resources, as above, as well as a short period of time available to perform the monitoring and/or achieve the study objectives since most studies are linked to dissertations and theses. Monitoring for three or more annual cycles is mainly important for studies aimed at investigating climate change, as it will have

Table 4.1 Data on reproductive phenology of *Avicennia germinans* in Brazilian mangroves in different locations (coastal state and coordinates)

Location	Method and sample number	Periodicity and duration (months)	Flowering duration (months)	Fruiting duration (months)	Flowering pattern	Fruiting pattern	Period of highest flowering intensity				Period of highest fruiting intensity			
							D	R	DRT	RDT	D	R	DRT	RDT
Amapá[a] 02°10′ N	Semiquantitative[i] ($n = 50$)	Monthly (12)	8	9–11	Annual	Annual	•					•		
Pará[b] 00°50′ S	Semiquantitative[i] ($n = 15$)	Monthly (12)	5–7	1–4	Annual	Annual	•				•			
Pará[c] 00°50′ S	Semiquantitative[i] ($n = 20$)	Monthly (36)	12	9–10	Continuous	Annual	•					•		
Pará[d] 00°53′ S	Semiquantitative[i] ($n = 20$)	Monthly (12)	11	11	Continuous	Continuous	•					•		
Pará[e] 01°00′ S	Semiquantitative[i] ($n = 20$)	Monthly (12)	9	8	Annual	Annual	•				•			
Paraíba[f] 06°49′ S	Semiquantitative[j] ($n = 16$)	Monthly (12)	8	7	Annual	Annual	•					•		
Rio de Janeiro[g] 21°17′ S	Quantitative[k] ($n = 4$)	Monthly (12)	10	8–9	Annual	Annual		•				•		
Rio de Janeiro[h] 21°36′ S	Quantitative[k] ($n = 7$)	Biweekly (24)	5–6	4–5	Annual	Annual		•				•		

D dry, *R* rainy, *DRT* dry and rainy transition, *RDT* rainy and dry transition

[a]Fernandes (1999); [b]Matni (2007); [c]Rodrigues (2015); [d]Silva and Fernandes (2011); [e]Gardunho (2009); [f]Lage-Pinto et al. (unplublished data); [g]Bernini et al. (2014); [h]Bernini and Rezende (2010); [i]Fernandes (1999); [j]Fournier (1974); [k]Baskets. See Chap. 3, Maps: 1 – Amapá, 2 – Pará, 7 –Paraíba, 8 – Pernambuco, 13 – Rio de Janeiro, 14 – São Paulo, and 15 – Paraná

Table 4.2 Data on reproductive phenology of *Avicennia schaueriana* in Brazilian mangroves in different locations (coastal state and coordinates)

Location	Method and sample number	Periodicity and duration (months)	Flowering duration (months)	Fruiting duration (months)	Flowering pattern	Fruiting pattern	Period of highest flowering intensity				Period of highest fruiting intensity			
							D	R	DRT	RDT	D	R	DRT	RDT
Pernambuco[a] 07°47′S	Quantitative[h] ($n = 7$)	Monthly (12)	5	–	Annual	–	•							
Pernambuco[b] 07°50′S	Semiquantitative[i] ($n = 10$)	Monthly (48)	4–7	5–6	Annual	Annual	•				•			
Rio de Janeiro[c] 22°42′S	Semiquantitative[i] ($n = 20$)	Monthly (24)	10	9	Annual	Annual	•					•		
Rio de Janeiro[d] 23°00′S	Quantitative[j] ($n = 6$)	Monthly (36)	–	6–9	–	Annual						•		
São Paulo[e] 25°00′S	Quantitative[j] ($n = 5$)	Monthly (38)	8	4	Annual	Annual		•				•		
Paraná[f] 25°29′S	Semiquantitative[i] ($n = 10$)	Monthly (13)	13	12	Sub-continuous	Sub-continuous	•	•			•			
Paraná[f] 25°49′S	Semiquantitative[i] ($n = 10$)	Monthly (13)	13	12	Sub-continuous	Continuous	•	•			•			
Paraná[g] 25°49′S	Semiquantitative[i] ($n = 20$)	Monthly (12)	8	3	Sub-annual	Sub-annual	•	•				•		

D dry, *R* rainy, *DRT* dry and rainy transition, *RDT* rainy and dry transition

[a]Medeiros and Sampaio (2013); [b]Nadia et al. (2012); [c]Rodrigues (2015); [d]Cardoso et al. (2015); [e]Adaime (1985); [f]Lima (2012); [g]Alvarenga (2015); [h]Christensen (1978); [i]Fournier (1974); [j]Baskets

Table 4.3 Data on reproductive phenology of *Laguncularia racemosa* in Brazilian mangroves in different locations (coastal state and coordinates)

Location	Method and sample number	Periodicity and duration (months)	Flowering duration (months)	Fruiting duration (months)	Flowering pattern	Fruiting pattern	Period of highest flowering intensity				Period of highest fruiting intensity			
							D	R	DRT	RDT	D	R	DRT	RDT
Amapá[a] 02°10′N	Semiquantitative[p] ($n = 50$)	Monthly (12)	9–12	9–12	Sub-annual	Sub-annual	●					●		
Pará[b] 00°50′S	Semiquantitative[p] ($n = 15$)	Monthly (12)	1–5	2–9	Annual	Annual	●					●		
Pará[c] 00°50′S	Semiquantitative[p, q] ($n = 20$; $n = 5$)	Monthly (12)	11	12	Continuous	Continuous		●				●		
Pará[d] 01°00′S	Semiquantitative[p] ($n = 20$)	Monthly (12)	12	12	Continuous	Continuous		●				●		
Pernambuco[e] 07°47′S	Quantitative[r] ($n = 7$)	Monthly (12)	5	–	Annual	Annual		●						
Pernambuco[f] 07°50′S	Semiquantitative[p] ($n = 10$)	Monthly (48)	6–9	5–10	Sub-annual	Sub-annual		●				●		
Pernambuco[g] 08°43′S	Quantitative[s] ($n = 8$)	Monthly (12)	7	7	Annual	Annual		●				●		
Paraíba[h] 06°49′S	Semiquantitative[r] ($n = 16$)	Monthly (12)	8	3	Annual	Sub-annual		●				●		
Rio de Janeiro[i] 21°17′S	Quantitative[s] ($n = 4$)	Monthly (12)	3	2–4	Annual	Annual		●				●		
Rio de Janeiro[j] 21°36′S	Quantitative[s] ($n = 7$)	Biweekly (24)	4	4	Annual	Annual		●				●		
Rio de Janeiro[k]	Semiquantitative[r] ($n = 20$)	Monthly (24)	8	8–10	Annual	Annual		●				●		

22°42′S														
Rio de Janeiro[l] 23°00′S	Quantitative[s] ($n = 6$)	Monthly (36)	–	2–6	–	Sub-annual						●		
São Paulo[m] 25°00′S	Quantitative[s] ($n = 5$)	Monthly (38)	3	7	Annual	Annual		●				●		
Paraná[n] 25°29′S	Semiquantitative[r] ($n = 10$)	Monthly (13)	12	7	Sub-continuous	Annual	●	●				●		
Paraná[n] 25°49′S	Semiquantitative[r] ($n = 10$)	Monthly (13)	11	7	Sub-continuous	Annual	●	●				●		
Paraná[o] 25°49′S	Semiquantitative[r] ($n = 20$)	Monthly (12)	8	3	Annual	Annual		●				●		

D dry, *R* rainy, *DRT* dry and rainy transition, *RDT* rainy and dry transition

[a]Fernandes (1999); [b]Matni (2007); [c]Fernandes et al. (2005); [d]Gardunho (2009); [e]Medeiros and Sampaio (2013); [f]Nadia et al. (2012); [g]Longo (2009); [h]Lage-Pinto et al. (unplublished data); [i]Bernini et al. (2014); [j]Bernini and Rezende (2010); [k]Rodrigues (2015); [l]Cardoso et al. (2015); [m]Adaime (1985); [n]Lima (2012); [o]Alvarenga (2015); [p]Fernandes (1999); [q]Christensen (1978); [r]Fournier (1974); [s]Baskets. See Chap. 3, Maps: 1 – Amapá, 2 – Pará, 7 – Paraíba, 8 – Pernambuco, 13 – Rio de Janeiro, 14 – São Paulo, and 15 – Paraná

Table 4.4 Data on reproductive phenology of *Rhizophora mangle* in Brazilian mangroves in different locations (coastal state and coordinates)

Location	Method and sample number	Periodicity and duration (months)	Flowering duration (months)	Fruiting duration (months)	Flowering pattern	Fruiting pattern	Period of highest flowering intensity				Period of highest fruiting intensity			
							D	R	DRT	RDT	D	R	DRT	RDT
Amapá[a] 02°10′N	Semiquantitative[o] ($n = 50$)	Monthly (12)	12	12	Continuous	Continuous	•					•		
Pará[b] 00°50′S	Semiquantitative[o] ($n = 15$)	Monthly (12)	6–7	6–12	Annual	Annual	•				•			
Pará[c] 00°50′S	Semiquantitative[o] ($n = 20$)	Monthly (36)	12	12	Continuous	Continuous				•		•		
Pará[d] 00°52′S	Quantitative[p, q] ($n = 15$; $n = 14$)	Biweekly (24)	12	12	?	?				•		•		
Pernambuco[e] 07°47′S	Quantitative[p] ($n = 7$)	Monthly (12)	5	–	Annual	Annual			•					
Pernambuco[f] 07°50′S	Semiquantitative[r] ($n = 10$)	Monthly (48)	9–12	11–12	Annual	Continuous		•			•			
Pernambuco[g] 08°43′S	Quantitative[p] ($n = 8$)	Monthly (12)	8–12	8–12	Annual	Annual		•			•	•		
Rio de Janeiro[h] 21°17′S	Quantitative[p] ($n = 4$)	Monthly (12)	12	10–12	Continuous	Annual	•	•				•		
Rio de Janeiro[i] 21°36′S	Quantitative[p] ($n = 7$)	Biweekly (24)	7–8	6–11	Annual	Annual		•				•		
Rio de Janeiro[j] 22°42′S	Semiquantitative[r] ($n = 20$)	Monthly (24)	5–7	11	Annual	Continuous	•					•		
		Monthly (36)	–	10	–	Annual						•		

Rio de Janeiro[k] 23°00′S	Quantitative[p] ($n = 6$)													
São Paulo[l] 25°00′S	Quantitative[p] ($n = 5$)	Monthly (38)	5	11	Annual	Continuous				•		•		
Paraná[m] 25°29′S	Semiquantitative[r] ($n = 10$)	Monthly (13)	11	13	Sub-continuous	Sub-continuous	•	•			•	•		
Paraná[m] 25°49′S	Semiquantitative[r] ($n = 10$)	Monthly (13)	13	12	Sub-continuous	Sub-continuous	•	•			•	•		
Paraná[n] 25°49′S	Semiquantitative[r] ($n = 20$)	Monthly (12)	9	10	Annual	Annual		•				•		

D dry, *R* rainy, *DRT* dry and rainy transition, *RDT* rainy and dry transition

[a]Fernandes (1999); [b]Matni (2007); [c]Rodrigues (2005); [d]Mehlig (2006); [e]Medeiros and Sampaio (2013); [f]Nadia et al. (2012); [g]Longo (2009); [h]Bernini et al. (2014); [i]Bernini and Rezende (2010); [j]Rodrigues (2015); [k]Cardoso et al. (2015); [l]Adaime (1985); [m]Lima (2012); [n]Alvarenga (2015); [o]Fernandes (1999); [p] Baskets; [q] Christensen (1978); [r] Fournier (1974). See Chap. 3, Maps: 1 – Amapá, 2 – Pará, 7 – Paraíba, 8 – Pernambuco, 13 – Rio de Janeiro, 14 – São Paulo, and 15 – Paraná

long-term effects that may alter the phenological patterns of mangrove plants (Alongi 2008; Ellison 2012).

One variable that may change over time is the synchrony of reproductive phenophases. Synchrony refers to a simultaneous occurrence of a given phenophase between individuals and populations, but it has received little attention in phenological studies of mangrove species. In Brazil, synchrony has mainly been evaluated without the use of quantitative methods. Nadia et al. (2012) and Lage-Pinto et al. (2021) evaluated synchrony using the index proposed by Augspurger (1983). This synchrony index is a quantitative method that estimates the overlap in the flowering or fruiting period between individuals of the same species, but it does not consider the differences in the intensities of the phenophases. Freitas and Bolmgrem (2008) suggested an index that includes the total duration of the phenophase of an individual and the variation in the number of flowers and fruits within this interval. This index corrects the overestimation of synchrony when the measurement does not account for the differences in the intensity of the phenophase. The use of quantitative methods to evaluate the flowering and fruiting synchrony of mangrove species would facilitate data comparison, especially when long-term monitoring is required to evaluate phenological responses to climate change.

4.3 Phenological Patterns

Adaime (1985) was the first researcher to describe the flowering and fruiting phenological patterns of mangrove species in Brazil (São Paulo state), even though the focus was on the primary productivity of the ecosystem. Since then, studies of reproductive phenology have become more frequent but have been unevenly distributed along the Brazilian coast (Tables 4.1, 4.2, 4.3 and 4.4). This scenario reflects the distribution of both species and researchers.

In Brazil, the northern limit of the distribution of mangrove species is in Amapá State, and the southern boundary varies according to the species (Cintrón and Schaeffer-Novelli 1992) (see Chap. 1). From the northernmost portion of Brazil, *Rhizophora harrisonii* and *R. racemosa* occur until Maranhão State (Santos 1986), and *Avicennia germinans* occurs until Rio de Janeiro State (Maciel and Soffiati-Netto 1998). The state of Santa Catarina is the southern limit of occurrence of *A. schaueriana*, *L. racemosa*, and *R. mangle* (Cintrón-Molero and Schaeffer-Novelli 1981, 1992) (see Chap. 3, Maps).

Mangroves in the westmost part of the Amazonian Equatorial Coast (Amapá and Pará states) (see Chap. 1) are the subject of most of the studies since they have been the focus of a research group that aims to determine the reproductive phenological patterns of mangroves in that region (Fernandes 2016). However, there continues to be a lack of information on the phenology of the mangroves in Maranhão State, at the eastmost part of the coastal segment. Concerning the other studied mangrove areas in the country, the states of Pernambuco, Rio de Janeiro, and Paraná are the ones presenting more data on flowering and fruiting (Tables 4.1, 4.2, 4.3 and 4.4).

In terms of species, most of the data correspond to *L. racemosa* and *R. mangle*, reflecting their wider distributions, but there are no records for several Brazilian states. Despite the wide geographic distribution of *A. schaueriana*, the number of studies on this species is relatively low. There are fewer studies of *A. germinans*, *R. harrisonii*, and *R. racemosa*, reflecting the smaller biogeographic range of these species in Brazil.

The reproductive phenology data of the typical mangrove species in Brazil are summarized in Tables 4.1, 4.2, 4.3 and 4.4. The frequency of the phenological patterns was classified according to Newstrom et al. (1994). The phenological patterns of *R. harrisonii* and *R. racemosa* are described throughout the text, but *Conocarpus erectus* was not included in this chapter, because, according to Lugo (1998), this species is erroneously listed as a typical mangrove species as it can tolerate salt but not flooding.

Avicennia germinans presents marked seasonality in its formation of flowers and fruits, whose pattern is mainly annual (Table 4.1). According to 75% of the studies analyzed, the presence of flowers and fruits extends for eight or more months of the year. Flowering shows a tendency toward greater intensity during the dry season, and the fruiting peak mainly occurs during the rainy season. *Avicenna schaueriana* exhibits a seasonal pattern and is mainly characterized by the annual pattern (Table 4.2). Flower production is typically highest during the dry season and was recorded as occurring over eight or more months of the year in 71% of the studies. Fruiting may occur in the dry or rainy season, but the duration of this phenomenon varies widely throughout the year along the Brazilian coast (3–12 months).

Flowering and fruiting in *L. racemosa* show mainly annual patterns, being characterized as seasonal or not (Table 4.3). Reproductive phenophases exhibit peaks in the rainy season. In most studies, flowering occurs in more than seven months, while fruiting occurs in less than eight months throughout the year.

Rhizophora mangle exhibits continuous or annual patterns for flowering and fruiting (Table 4.4). Peak flower production mainly occurs during the rainy season but can also occur in the dry season, in the transition between the dry and rainy seasons, or the transition between the rainy and dry seasons. Fruiting shows greater intensity during the rainy season. *Rhizophora harrisonii* and *R. racemosa* exhibit phenological patterns like those of *R. mangle*. In Amapá State, *R. harrisonii* presented continuous flowering and fruiting throughout the year, but flowering is more intense during the transition from the dry to the rainy season and fruiting is more intense during the rainy season (Fernandes 1999). Similar results were recorded for *R. harrisonii* and *R. racemosa* in Pará State, but the flowering peak occurred in the rainy season (Gardunho 2009).

While the reproductive phenophases of mangrove species are more intense during the dry or rainy seasons, significant correlations with precipitation may be weak or absent (Fernandes 1999; Matni 2007; Bernini and Rezende 2010; Nadia et al. 2012). Most studies have related flowering and fruiting with precipitation and air temperature, but the intensity of solar radiation, photoperiod, relative humidity, evapotranspiration, and interstitial salinity also play important roles in the reproductive phenology of mangrove species (Nadia et al. 2012; Alvarenga 2015; Cardoso et al. 2015; Rodrigues 2015; Lage-Pinto et al. 2021).

Studies conducted at different latitudes show different results. For example, Alvarenga (2015) showed that variations in interstitial salinity promote marked seasonality in reproductive phenophases in a mangrove in Paraná State, while Matni (2007) did not find an effect of salinity on the flowering and fruiting of the mangrove species in Pará State. Bernini and Rezende (2010), who studied a mangrove in Rio de Janeiro State, and Lima (2012), who evaluated a mangrove in Paraná State, demonstrated that the flowering of *L. racemosa* and *R. mangle* mangroves was positively correlated with mean air temperature, but Nadia et al. (2012) did not find such a correlation in a mangrove in Pernambuco State. The flowering of *A. schaueriana* was significantly correlated with photoperiod in Pernambuco State (Nadia et al. 2012), but no correlation was recorded in Paraná State (Lima 2012). The phenological records of Adaime (1985) showed the seasonal influence of low temperatures on flower and fruit formation in the three mangrove species that are typical of the Cananéia region (São Paulo State). On the other hand, Fernandes (1999) found no significant relationship between flowering and abiotic factors and suggested that endogenous factors are responsible for stimulating the formation of mangrove species in Pará State.

Duke (1990) found latitudinal trends in the flowering and fructification of *Avicennia marina* Vierh in Australian mangroves, where the duration of each phenophase increased with higher latitudes. The air temperature was the main factor responsible for stimulating the reproductive cycle, playing a fundamental role in flower formation. Van der Stocken et al. (2017) found clear latitudinal patterns in the release of the fruits and propagules of 47 species of mangroves in the northern and southern hemispheres, with significant positive correlations with precipitation. The authors observed that the propagules/fruits fall from the trees during most of the year without pronounced production peaks in the equatorial zone, but at higher latitudes, the release of propagules is variable and significantly correlated with air temperature.

In general, as mangrove species exhibited a lot of variation in phenological patterns (Tables 4.1, 4.2, 4.3 and 4.4), it is not possible to establish a latitudinal pattern related to abiotic variables. However, phenological data on the species are insufficient to establish a general pattern for the Brazilian coast.

Nadia et al. (2012) showed that biotic factors also shape phenology, as they observed that mangrove species that share the same pollinators present distinct flowering strategies, but the fruiting pattern is similar among species with the same dispersal syndrome. These authors also found that precipitation, air temperature, and day length play an important role in the formation of flowers and fruits. Thus, the reproductive phenological patterns of mangrove species represent a complex response to abiotic and biotic factors. However, the results of the studies remain unclear, so further research on the correlations with these factors is needed to better understand the phenological patterns of Brazilian mangrove species since the climatic variables vary widely along the coast.

Another biotic factor that has been related to reproductive phenology is the forest structure. Silva and Fernandes (2011) evaluated the influence of structural characteristics (height, density, and basal area) on the reproductive phenology of *A. germinans* in Pará State. The authors concluded that the structural attributes do not affect the reproductive phenology of this species, because the phenological events were synchronized, in both quantity and periodicity, in forests with distinct structural characteristics.

Regarding the timing of reproductive phenological phases, some studies have indicated flowering synchrony in populations of *A. germinans* (Fernandes 1999; Rodrigues 2005; Matni 2007; Silva and Fernandes 2011; Lage-Pinto et al. 2021), whereas asynchrony in fruit production is more common in populations of this species (Fernandes 1999; Matni 2007). In *A. schaueriana*, high intraspecific synchrony was observed in flower and fruit production (Nadia et al. 2012), but the reproductive phenophases show lack of synchrony (Fernandes 1999; Matni 2007) or low synchrony in *L. racemosa* (Nadia et al. 2012). In *R. mangle*, synchrony has not been analyzed, because this species exhibits continuous production throughout the year (Nadia et al. 2012) or has been assessed at the peak of flower or fruit production (Rodrigues 2005; Matni 2007), which indicated a lack of synchrony between populations.

4.4 Final Remarks

Direct observation methods have been used mostly to evaluate the reproductive phenology of mangrove species in Brazil. The number of individuals sampled and the frequency of observations of reproductive phenophases have been adequate in most studies, but long-term studies (more than three years) are scarce.

Further studies on reproductive phenology are needed, mainly for *A. germinans* and *A. schaueriana*. There was no record of studies to evaluate the effects of climate change on the reproductive phenology of mangrove species. Therefore, long-term studies are needed and could be made possible through collaboration among scientists who simultaneously collect data along the Brazilian coast. The information may contribute to understanding the responses of mangrove species to climate changes and establishing the reproductive phenological patterns of Brazilian mangrove forests. In addition, the inclusion of a synchrony index would facilitate data comparison.

References

Adaime RR (1985) Produção do bosque de mangue da Gamboa Nóbrega (Cananéia, 25° Lat. S. Brasil). PhD dissertation, Universidade de São Paulo

Adame MF, Virdis B, Lovelock CE (2010) Effect of geomorphological setting and rainfall on nutrient exchange in mangroves during tidal inundation. Mar Freshw Res 61(10):1197–1206

Alongi DM (2002) Present state and future of the world's mangrove forests. Environ Conserv 29: 331–349

Alongi DM (2008) Mangrove forests: Resilience, protection from tsunamis, and responses to global climate change. Estuar Coast Shelf Sci 76:1–13

Alvarenga AMSB (2015) Dinâmica de crescimento das espécies arbóreas de um manguezal no sul do Brasil. Master thesis, Universidade Federal do Paraná

Araújo LDA, Quirino ZGM, Machado IC (2011) Fenologia reprodutiva, biologia floral e polinização de *Allamanda blanchetii*, uma Apocynaceae endêmica da Caatinga. Rev Bras Bot 34(2):211–222

Augspurger CK (1983) Phenology, flowering synchrony, and fruit set of six neotropical shrubs. Biotropica 15:257–267

Ballestrini C, Tezara W, Herrera A (2011) Environmental drivers of leaf phenology in trees of the tropical species *Ficus obtusifolia* Braz. J Plant Physiol 23(2):113–122

Bencke CSC, Morellato LPC (2002) Comparação de dois métodos de avaliação da fenologia de plantas, sua interpretação e representação. Revista Brasil Bot 25(3):269–275

Bernini E, Rezende CE (2010) Litterfall in a mangrove in Southeast Brazil. Panam JAS 5(4): 508–519

Borchert R, Calle Z, Strahler AH, Baertschi A, Magill RE, Broadhead JS, Kamau J, Njoroge J, Muthuri C (2015) Insolation and photoperiodic control of tree development near the equator. New Phytol 205:7–13

Bullock SH, Solis-Magallanes JA (1990) Phenology of canopy trees of a tropical deciduous forest in México. Biotropica 22(1):22–35

Cardoso CS, Chaves FO, Soares MLG (2015) Variação espaço-temporal na produção de propágulos de espécies de mangue no Sudeste do Brasil. Bol Mus Para Emílio Goeldi Cienc Nat 10(3):491–503

Christensen B (1978) Biomass and primary production of *Rhizophora apiculata* Bl. in a mangrove in southern Thailand. Aquat Bot 4:43–52

Cintrón-Molero G, Schaeffer-Novelli Y (1981) Los manglares de la costa brasileña: revisión preliminar de la literatura. Abuchahla GMO (elect rev ed) [S.I] Informe Técnico preparado para la Oficina Regional de Ciéncia y Tecnologia para America Latina y el Caribe de UNESCO y la Universidad Federal de Santa Catarina, 47 pp

Cintrón-Molero G, Schaeffer-Novelli Y (1992) Ecology and management New World mangroves. In: Seeliger U (ed) Coastal plant communities of Latin America. Academic, San Diego, pp 233–258

Cleland EE, Allen JM, Crimmins TM, Dunne JA, Pau S, Travers SE, Zavaleta ES, Wolkovich EM (2012) Phenological tracking enables positive species responses to climate change. Ecology 93(8):1765–1771

Couralet C, Van den Bulcke J, Ngoma LM, Van Acker J, Beeckman H (2013) Phenology in functional groups of central African rainforest trees. J Trop For Sci 25(3):361–374

D'Eça-Neves FF, Morellato LPC (2004) Métodos de amostragem e avaliação utilizados em estudos fenológicos de florestas tropicais. Acta Bot Bras 18(1):99–108

Donato DC, Kauffman JB, Murdiyarso D, Kurnianto S, Stidham M, Kanninen M (2011) Mangroves among the most carbon-rich forests in the tropics. Nat Geosci 4(5):293–297

Duke NC (1990) Phenological trends with latitude in the mangrove tree *Avicennia marina*. J Ecol 78:113–133

Ellison JC (2012) Climate change vulnerability assessment and adaptation planning for mangrove systems. World Wildlife Fund, Washington DC

Encinas-Viso F, Revilla TA, Etienne RS (2012) Phenology drives mutualistic network structure and diversity. Ecol Lett 15:198–208

Engel VL, Martins FR (2005) Reproductive phenology of Atlantic Forest tree species in Brazil: an eleven-year study. J Trop Ecol 46(1):1–16

Fernandes MEB (1999) Phenological patterns of *Rhizophora* L., *Avicennia* L. and *Laguncularia* Gaertn. f. in Amazonian mangrove swamps. Hydrobiologia 413:53–62

Fernandes MEB (2016) Os Manguezais da Costa Norte Brasileira. UFPA/CBRAG, Bragança

Fernandes MEB, Virgulino ARC, Nascimento AAM, Rodrigues LFP (2005) Padrões de floração e frutificação em *Laguncularia racemosa* (L.) Gaertn. f.: uma avaliação metodológica. B Lab Hidro 18:33–38

Fournier LA (1974) Un método cuantitativo para la medición de características fenológicas en árboles. Turrialba 24(4):422–423

Fournier LA, Charpantier C (1975) El tamaño de la muestra y la frecuencia de las observaciones en el estudio de las características fenológicas de los árboles tropicales. Turrialba 25(1):45–48

Freitas L, Bolmgren K (2008) Synchrony is more than overlap: measuring phenological synchronization considering time length and intensity. Rev Brasil Bot 31(4):721–724

Garcia LC, Barros FV, Lemos-Filho JP (2009) Fructification phenology as an important tool in the recovery of iron mining areas in Minas Gerais, Brazil. Braz J Biol 69(3):887–893

Gardunho DCL (2009) Atributos estruturais e fenologia dos manguezais da reserva extrativista marinha de Soure, Ilha do Marajó, Pará. Master thesis, Universidade Federal do Pará

Giri C, Ochieng E, Tieszen LL, Zhu Z, Singh A, Loveland T, Masek J, Duke N (2011) Status and distribution of mangrove forests of the world using earth observation satellite data. Glob Ecol Biogeogr 20:154–159

Hamilton SE, Casey D (2016) Creation of a high spatio-temporal resolution global database of continuous mangrove forest cover for the 21st century (CGMFC-21). Glob Ecol Biogeogr 25: 729–738

Lage-Pinto F, Fernando PS, Carletto AFM, Bernini E (2021) Reproductive phenology and influence of abiotic variables for two mangrove species in northeastern Brasil. Rodriguésia 72:e00392020

Liebsch D, Mikich SB (2009) Fenologia reprodutiva de espécies vegetais da Floresta Ombrófila Mista do Paraná, Brasil. Revi Brasil Bot 32(2):375–339

Lieth H (1974) Introduction to phenology and the modeling of seasonality. In: Lieth H (ed) Phenology and seasonality modeling. Springer, Berlin, pp 3–19

Lima CS (2012) Morfologia foliar e fenologia do componente arbóreo de manguezais das Baías de Guaratuba e Antonina, Paraná, Brasil. Master thesis, Universidade Federal do Paraná

Lugo AE (1998) Mangrove forests: a tough system to invade but an easy one to rehabilitate marine. Poll Bull 37(8-12):427–430

Maciel NC, Soffiati-Netto AA (1998) Novos limites para a distribuição geográfica de *Avicennia germinans* (L.) Stern – Avicenniaceae e *Montrichardia arborecens* (L.) Schott – Araceae, no Rio de Janeiro, Brasil. In: Simpósio de Ecossistemas Brasileiros 4, 2–7 April. ACIESP, Águas de Lindóia

Matni AS (2007) Estudo comparativo da fenologia reprodutiva de *Rhizophora mangle* L., *Avicennia germinans* (L.) Stearn e *Laguncularia racemosa* (L.) Gaertn. f., ao longo da península de Ajuruteua, Bragança, Pará. 2007. Master thesis, Universidade Federal do Pará

Mehlig U (2006) Phenology of the red mangrove, *Rhizophora mangle* L., in the Caeté Estuary, Pará, equatorial Brazil. Aquat Bot 84:158–164

Morellato LPC (2008) Fenologia de plantas e os efeitos das mudanças climáticas. In: Buckeridge MS (ed) Biologia & Mudanças Climáticas no Brasil. RiMa, São Carlos, pp 181–191

Morellato LPC, Romera EC, Talora DC, Takahashi A, Bencke CC, Zipparro VB (2000) Phenology of Atlantic Rain Forest trees: a comparative study. Biotropica 32(4b):811–823

Morellato LPC, Camargo MGG, D'Eça Neves FF, Luize BG, Mantovani A, Hudson IL (2010) The influence of sampling method, sample size, and frequency of observations on plant phenological patterns and interpretation in tropical forest trees. In: Hudson IL, Keatley MR (eds) Phenological research: methods for environmental and climate change analysis. Springer, Dordrecht, pp 99–121

Morellato LPC, Camargo MGG, Gressler E (2013) A review of plant phenology in South and Central America. In: Schwartz MD (ed) Phenology: an integrative environmental science. Springer, Dordrecht, pp 91–113

Morellato LPC, Alberton B, Alvarado ST, Borges B, Buisson E, Camargo MGG, Cancian LF, Carstensen DW, Escobar DFE, Leite PTP, Mendoza I, Rocha NMWB, Soares NC, Silva TSF,

Staggemeier VG, Streher AS, Vargas BC, Peres CA (2016) Linking plant phenology to conservation biology. Biol Conserv 195:60–72

Mumby PJ (2006) Connectivity of reef fish between mangroves and coral reefs: algorithms for the design of marine reserves at seascape scales. Biol Conserv 128(2):215–222

Nadia TL, Morellato LPC, Machado IC (2012) Reproductive phenology of a northeast Brazilian mangrove community: environmental and biotic constraints. Flora 207(9):682–692

Newstrom LE, Frankie GW, Baker HG (1994) A new classification for plant phenology based on flowering patterns in lowland tropical rain forest trees at La Selva, Costa Rica. Biotropica 26(2): 141–159

Perry CL, Mendelssohn IA (2009) Ecosystem effects of expanding populations of *Avicennia germinans* in a Louisiana salt marsh. Wetlands 29(1):396–406

Revilla TA, Encinas-Viso F, Loreau M (2015) Robustness of mutualistic networks under phenological change and habitat destruction. Oikos 124:22–32

Rodrigues LFP (2005) Variação anual dos padrões fenológicos de *Avicennia germinans* L. e *Rhizophora mangle* L. no Furo Grande, Bragança-Pará. Bachelor thesis, Universidade Federal do Pará

Rodrigues GA (2015) Aspectos fenológicos e produção de mudas de três espécies nativas do ecossistema manguezal. Master thesis, Universidade Federal do Rio de Janeiro

Rodríguez-Gallego C, Navarro T (2015) Vegetative and reproductive phenological patterns in coastal dunes in S Spain. Anal Jardin Bot Madrid 72(1):1–17

Santos MCFV (1986) Considerações sobre a ocorrência de *Rhizophora harrisonii* Leechamn e *Rhizophora racemosa* G.F.W. Meyer, no litoral do Estado do Maranhão, Brasil. B Lab Hidro 7: 71–91

Silva LL, Fernandes MEB (2011) Relação entre os atributos estruturais das árvores de *Avicennia germinans* (L.) Stearn e sua fenologia reprodutiva. B Lab Hidro 24(2):51–57

Stevenson PR, Quiñones MJ, Ahumada JA (1998) Annual variation in fruiting pattern using different methods in a lowland tropical forest, Tinigua National Park, Colombia. Biotropica 30(1):129–134

Ulsig L, Nichol CJ, Huemmrich KF, Landis DR, Middleton EM, Lyapustin AI, Mammarella I, Levula J, Porcar-Castell A (2017) Detecting inter-annual variations in the phenology of evergreen conifers using long-term MODIS vegetation index time series. Remote Sens 9(1):49

Valiela I, Bowen JL, York JK (2001) Mangrove forests: one of the world's threatened major tropical environments. BioScience 51:807–815

Van der Stocken T, López-Portillo J, Koedam N (2017) Seasonal release of propagules in mangroves – assessment of current data. Aquat Bot 138:92–99

Van Schaik CP, Terborgh JW, Wrigth SJ (1993) The phenology of tropical forests: adaptative significance and consequences for primary consumers. Annu Rev Ecol Syst 24:353–377

Williams RJ, Myers BA, Eamus D, Duff GA (1999) Reproductive phenology of woody species in North Australian Tropical savanna. Biotropica 31(4):626–636

Wolkovich EM, Cook BI, Davies TJ (2014) Progress towards an interdisciplinary science of plant phenology: building predictions across space, time and species diversity. New Phytol 201(4): 1156–1162

Wong PP, Losada IJ, Gattuso J-P, Hinkel J, Khattabi A, McInnes KL, Saito Y, Sallenger A (2014) Coastal systems and low-lying areas. In: Field CB, Barros VR, Dokken DJ, Mach KJ, Mastrandrea MD, Bilir TE, Chatterjee M, Ebi KL, Estrada YO, Genova RC, Girma B, Kissel ES, Levy AN, MacCracken S, Mastrandrea PR, White LL (eds) Climate change 2014: impacts, adaptation, and vulnerability. Part A: Global and sectoral aspects. Contribution of Working Group II to the fifth assessment report of the Intergovernmental Panel on Climate Change. Cambridge University Press, Cambridge, pp 361–409

Zamith LR, Scarano FR (2004) Produção de mudas de espécies das Restingas do município do Rio de Janeiro, RJ, Brasil. Acta Bot Bras 18(1):161–176

Chapter 5
Genetic and Epigenetic Diversity of Mangrove Plants: Markers of Adaptation in a Changing Environment

Catarina F. Lira and Renan Granado

5.1 Introduction

Mangroves have suffered great habitat loss in the last decades, at risk of disappearing in the next one hundred years (Duke et al. 2007; Polidoro et al. 2010). Brazil is one of the countries with the largest mangrove coverage in the world (Polidoro et al. 2010). Yet, they suffer constantly with intense human pressure, such as pollution, urban growth, and overexploitation of their resources, among other threats, leading to excessive habitat loss and fragmentation and smaller populations (Alongi 2002).

The adaptation process in response to environmental changes is intimately related to the natural genetic variability. This variability can be enriched or impoverished by the evolutionary process through generations as well as by ecological processes as population connectivity. Threats, such as habitat loss and fragmentation, decrease the genetic variability due to genetic drift or inbreeding. On the other hand, mutation and migration can improve variability. The population connectivity fosters gene flow, avoiding the erosion of genetic diversity inside populations and the higher genetic differentiation between populations. The balance between loss and gain of genetic variability might be biased by human impacts, which contribute to much faster and much more intense losses than gains in the species' genetic variability (Otto 2018).

The genetic diversity of a species is measured by the DNA polymorphisms of individuals and populations. There are many molecular markers in the DNA to

C. F. Lira (✉)
Laboratório de Biologia Molecular, Diretoria de Pesquisa, Instituto de Pesquisas Jardim Botânico do Rio de Janeiro, Rio de Janeiro, RJ, Brazil

R. Granado
Laboratório de Biologia Molecular, Diretoria de Pesquisa, Instituto de Pesquisas Jardim Botânico do Rio de Janeiro, Rio de Janeiro, RJ, Brazil

Institute for Plant Sciences, University of Cologne, Cologne, Germany

Y. Schaeffer-Novelli et al. (eds.), *Brazilian Mangroves and Salt Marshes*, Brazilian Marine Biodiversity, https://doi.org/10.1007/978-3-031-13486-9_5

measure the polymorphism level of a species. Some markers are neutral (not affected by natural selection), while others are functional or adaptive and can be directly affected by changes in the environment, thus representing adaptations. Neutral markers have been used in population studies of genetic diversity since the 1980s, but many adaptive markers are also currently used, especially after the advent of DNA sequencing.

Besides the genetic diversity, there are other ways to assess DNA variability and adaptation capacity such as epigenetic markers. Epigenetics are all changes in the DNA that are not associated with nucleotide changes, such as methylation and acetylation of cytosines and histones. Histone methylation or acetylation, cytosine methylation, and small RNAs are epigenetic phenomena directly related to gene expression, gene silencing, and environmental sensing (Rapp and Wendel 2005). The epigenetic diversity has an important role in the organism's adaptation capacity (Lira-Medeiros et al. 2010). Epigenetic diversity can contribute to adaptation to environmental changes, phenotypic plasticity, and selective pressures. Nevertheless, studies focusing on wild species' epigenetic variations and adaptation to environmental changes are still incipient in Brazil.

Although there are several studies on Brazilian mangroves, only a handful focuses on plant species, especially regarding plant genetic diversity. Genetic studies are still overlooked in Brazil due to a lack of funding, hindering scientists from addressing genetic and evolutionary questions. This chapter reviews and discusses the advances in the knowledge of genetic and epigenetic diversity of Brazilian mangrove plants in addition to comparing them with studies from other parts of the globe. There is a great potential for addressing important questions about species evolution and adaptation in the Brazilian mangroves. We hope that this chapter will bring more visibility and interest to genetic studies on such disregarded ecosystems.

5.2 The Importance of Genetic Diversity

The evolution of a species is a process based on natural genetic diversity, where the most suitable organisms will survive and persist in the environment. Natural selection is one evolutionary force that acts by selecting favorable variations in the DNA, while genetic drift is a force that reduces genetic diversity by chance. High levels of genetic diversity are expected in all species that are not under the effects of genetic drift through bottlenecks or founder effects. Scientists assess the level of genetic diversity within a species to evaluate how the evolutionary forces are acting on them.

The genetic diversity of a species is measured by the level of polymorphism of molecular markers in the DNA, either among individuals within a population or among natural populations of a species. Population studies can enlighten information on life history, breeding system (selfing or outcrossing), migration (by calculating gene flow and pollen-to-seed ratio), isolation by distance or fragmentation, bottlenecks, and other processes as well as characteristics of a certain species (Hamrick and Godt 1996).

The higher the genetic diversity, the higher the probability of success of the population or ecosystem when facing threats and environmental changes. High levels of genetic diversity within a population indicate higher outcrossing than selfing rates. This is advantageous for the long-term survival of the population. Lower levels of genetic diversity may indicate high selfing rates, restricted gene flow to other populations, or isolation by distance or fragmentation. So, these analyses can unveil ecological characteristics of the species and become key contributors to conservation strategies and management of natural plant populations.

In contrast with the mangrove species of the Eastern hemisphere, there is little knowledge on the genetic diversity of Brazilian mangroves. It is urgent for the conservation of this ecosystem that more mangrove species and populations from the Americas are studied.

5.3 Brazilian Mangrove Evolutionary Theories

The World mangrove ecosystem has evolved and diversified through a vicariance event that separated Atlantic-East Pacific (AEP) and Indo-West Pacific (IWP) mangroves. This theory is supported by the absence of shared species in these two areas within the genera *Avicennia* and *Rhizophora* (Triest 2008). Within the AEP region, there are two hypotheses for the biogeographic distribution of the species: (a) Vicariance with the Central America Isthmus (CAI) acting as a barrier for mangrove plant dispersal and (b) Long-distance dispersal (LDD) through the Atlantic Ocean causing admixture of American and African populations (Dodd et al. 2002). Both theories are supported by studies using neutral molecular markers.

Low divergence was found between East-American and West-African populations of *A. germinans*, indicating trans-Atlantic LDD events (Dodd et al. 2002; Mori et al. 2015; Nettel and Dodd 2007). Moreover, the lack of shared haplotypes of *A. germinans* in East-Pacific and West-Atlantic populations of Central America corroborated the theory that CAI emergence blocked gene flow between these populations around 3 Ma ago (Dodd et al. 2002; Nettel and Dodd 2007; Mori et al. 2015). A similar genetic structure was found for *Rhizophora mangle* and *R. racemosa*, corroborating both hypotheses for *Rhizophora* species (Cerón-Souza et al. 2010). *Laguncularia* is the third genus occurring in the AEP region. This monospecific genus (the only species being *Laguncularia racemosa*) has been better studied than the others, but there are no works including all regions of occurrence to test those hypotheses.

The mangrove species in Brazil show higher genetic diversity levels on the coastal segments I–V (see Chap. 3; equivalent to the northeast and northern Brazilian coast) than in the rest of the country, as expected. The theory behind this fact is that the Quaternary glaciations limited most of the mangroves in the equatorial regions, while range edge populations went extinct. The Brazilian populations of *R. mangle* corroborate this hypothesis (Pil et al. 2011). The low genetic diversity and high genetic drift at the range edge are also observed in IWP mangrove species, like

Sonneratia alba (Wee et al. 2017), *A. marina* (Arnaud-Haond et al. 2006), and *Lumnitzera racemosa* (Su et al. 2006). More genetic studies are needed with AEP species and populations to uncover the evolutionary patterns of Brazilian mangroves.

The decrease in mangroves' genetic diversity from North to South of the Brazilian coast is probably a result of historic fragmentation leading to genetic drift and restricted gene flow (Pil et al. 2011). Morphological differences can be observed in reduced and scarcely distributed stands in the Southeast Region of Brazil, and the relatively lower biomass. Nevertheless, the increase in studies on the genetic structure would help in elucidating the evolutionary history of mangrove plants in the country.

5.4 Methods in Genetic and Evolutionary Studies

Molecular markers are essential tools for biodiversity studies focusing on genetic diversity. These markers were initially based on morphology, proteins (isoenzymes), or large amounts of DNA. With the development of the polymerase chain reaction (PCR) technique, the former molecular markers were rapidly replaced by DNA-based methods. This new technology consists of DNA amplification, allowing the usage of small amounts of tissue and a high-throughput analysis for genetic population studies (Henry 2013). These methods are also more efficient and accurate than the previous methods.

There are several molecular markers based on PCR. The choice of the method depends on the goal of the study and its scientific question (Table 5.1). If the aim of the study is breeding information, codominant markers, which are markers that allow the identification of heterozygotes, are more suitable due to their robustness on genetic data inherited from the parents. If the goal is to analyze the population structure to answer questions about the evolutionary history of the species, all markers can be used, and the choice will depend on the availability of markers, technique knowledge, and experimental design. On the other hand, for fine-scale structure and local studies, fingerprinting dominant markers are more suitable due to the high number of polymorphisms obtained – despite being numerous and often indistinguishable between parents.

Neutral molecular markers are broadly used for measuring population genetic diversity, because they are not affected by natural selection and can be assessed by different methods. The most used neutral markers are random amplified DNA (RAPD), inter simple sequence repeat (ISSR), amplified fragment length polymorphism (AFLP), microsatellite or simple sequence repeat (SSR), and single nucleotide polymorphism (SNP) – most are based on PCR and sequencing.

Microsatellite and SNP are codominant markers that allow us to assess heterozygosity uncovering the alleles of each marker. On the other hand, dominant markers generate fingerprinting – a series of polymorphisms of markers, such as RAPD, AFLP, and ISSR. Another dominant marker is methyl-sensitive amplified

Table 5.1 Most common scientific goals and the correspondent suitable markers used in the studies on the genetic and epigenetic diversity of mangrove plants in Brazil

Aim of the study	Molecular marker type	Molecular marker
Breeding: Outcrossing and selfing rates, gene flow, number of migrants, paternity, seed-to-pollen ratio, and others	Codominant[a] markers	SSR and SNP
Population structure: Gene flow, isolation by distance, admixture, number of migrants, genetic differentiation, genetic diversity, bottlenecks, founder effect, etc.	Both dominant[b] and codominant[a] markers	All markers
Fine-scale structure: Local genetic diversity, local fragmentation/gene flow, bottlenecks, founder effect, etc.	Dominant[b] markers	Fingerprinting markers: RAPD, ISSR, and AFLP

[a]Codominant markers allow the visualization of allelic polymorphism from both parents and the identification of heterozygotes

[b]Dominant markers do not allow identification of heterozygotes. They are fingerprinting techniques, highly polymorphic but not distinguishing alleles

SSR simple sequence repeat, *SNP* single nucleotide polymorphism, *RAPD* random amplified DNA, *ISSR* inter simple sequence repeat, *AFLP* amplified fragment length polymorphism

polymorphism (MSAP), which allows the analysis of methylated, unmethylated, and hemimethylated cytosine along the genome. This marker is linked to epigenetic phenomena and then linked to the species' adaptation to environmental conditions (Lira-Medeiros et al. 2010). MSAP can be used together with AFLP to separate the genetic and the epigenetic diversities of a species, although this combination is still rarely used albeit with the potential to elucidate evolutionary questions.

The dominant markers are very efficient for wild species studies, because they do not need previous knowledge of the DNA sequence nor expensive laboratory procedures. Fingerprinting techniques are low-cost and highly polymorphic, thus amplifying hundreds or thousands of variable markers for each sample, which is excellent for fine-scale local population studies of wild species. The disadvantage of these markers is the dominance, whereas heterozygosity is not directly assessed but inferred in the statistical analysis; then for parenting, crossing, and mapping analysis (and others), these markers are not suitable (Lira-Medeiros et al. 2006).

Codominant markers show all the alleles, from where heterozygosity and allele frequency can be directly calculated. This allows one to determine the preferential crossing system of the analyzed species, based on the Hardy-Weinberg Equilibrium. Nowadays, codominant markers are easily assessed even on wild species with the advance of sequencing and analysis methodologies. Nonetheless, genomics is still incipient within mangrove conservation efforts (Wee et al. 2019).

With the development of new technologies over time, more efficient tools will facilitate molecular markers analysis, which will become a cheaper way for obtaining fundamental information on fitness, population dynamics, and coping mechanisms to environmental changes (Allendorf et al. 2010).

As the DNA-based methods created a turning point some decades ago, the next-generation sequencing (NGS) technologies have recently created another. The NGS platforms have been applied in SNP analysis of numerous species even those whose genome sequence is not available. They have also revolutionized approaches to biology and notoriously increased the speed at which DNA sequences can be acquired, reducing costs drastically (Henry 2013). In addition, NGS may help researchers in answering many biological questions with high accuracies, such as the phylogeography of wild populations, identification of *loci* affecting fitness, adaptive variation, and inbreeding/outbreeding depression, as well as diminishing sequencing efforts and facilitating the development of new molecular markers (Allendorf et al. 2010; Henry 2013).

5.5 A Look into the Studies on Genetic Diversity

Neutral molecular markers can be assessed by polymorphic fingerprinting (dominant markers) or alleles (codominant markers). Independently of the marker, the results obtained in the laboratory are transferred and processed by specific programs according to the desired analysis, based on Euclidean, Bayesian, or multivariate analysis.

When dealing with dominant markers, heterozygosity cannot be directly calculated, so deviation from Hardy-Weinberg equilibrium (HWE) must be either assumed as null, bypassed, or assessed by other means (Lira-Medeiros et al. 2010). Due to the difficulty in correctly evaluating these genetic data, we exemplify below how to carry out a population analysis to assess the genetic diversity of a species based on dominant markers.

The data collected from the molecular analysis are transferred to tables in the numbers 0 and 1, representing the absence and presence of a marker in each locus. The data is then uploaded to a software, such as Hickory® (Holsinger et al. 2002). The software calculates parameters in genetic diversity by a Bayesian analysis, where uncertainties of inbreeding levels (f) and genetic differentiation (G_{ST}) are incorporated and predicted (Holsinger et al. 2002). This type of software is recommended for studies using dominant markers, because it calculates the deviation of HWE, whereas most of the genetic population programs are based on Euclidean analysis and allele frequency, which points to a bias in small and local experimental designs. The most important genetic indexes inferred by these pieces of software are shown in Table 5.2.

Table 5.2 Most important genetic indexes in population studies. The indexes, range value, and significance of obtained values

Indexes	Value	Significance
P	0–100%	0 – No polymorphism found (all samples are very similar, maybe clones)
		100 – All loci are polymorphic (excellent diversity within samples)
H_S	0–1	The genetic diversity within population equivalent to expected heterozygosity for diploids:
		0 – No diversity (all individuals are very similar, maybe clones)
		1 – Maximum value
H_T	0–1	The total genetic diversity of the species:
		0 – No diversity (all individuals are very similar, maybe clones)
		1 – Maximum value
θ_{ST}	0–1	The genetic differentiation among populations:
G_{ST}		0 – No differentiation (diversity within populations is higher than among populations; great gene flow)
F_{ST}		1 – Populations completely differentiated (no gene flow)
f	0–1	Coefficient of inbreeding:
		0 – No selfing, all outcrossing
		1 – All selfing, no outcrossing

P – Percentage of polymorphic loci, H_S – Mean population heterozygosity level, H_T – Species total heterozygosity level, θ_{ST}, G_{ST}, F_{ST} – Between-population differentiation indexes, f – Inbreeding index

5.6 The Current Knowledge Obtained from Genetic and Epigenetic Markers in Mangroves of the Americas and Brazil

Genetic studies can infer evolutionary background on current populations and species. There are many studies on global and local mangrove populations and their structure, dispersal, connectivity, and how past and future events may have impacted or will impact them.

Using a high-resolution eddy-tide-resolving numerical ocean model, the dispersion of mangrove propagules across the global ocean was simulated (Van der Stocken et al. 2019). Generating connectivity matrices between mangrove habitats using ranges of floating periods, this study found high rates of long-coast transport and transoceanic dispersal across Atlantic, Pacific, and Indian Oceans, but did not observe any connectivity between populations on west Africa and east America (Van der Stocken et al. 2019). However, the spatial scale and distribution of mangrove species remain largely unknown especially because of climate change effects.

One study on the American continent collected samples along the east and west Atlantic coasts and the Pacific coast to analyze the effects of the Atlantic Ocean and the Central American Isthmus (CAI) on genetic diversity and population structure of *Avicennia germinans* (Dodd et al. 2002). They found a high differentiation between the Pacific and Atlantic populations, and a lower one among the Atlantic populations, suggesting that the CAI has had an important influence on the population genetics' structure in this species.

Furthermore, another study analyzing not only *A. germinans*, but also *A. schaueriana*, the only two species of the genus that occur on the Brazilian coast, found new evidence of current hybridization between them using microsatellites (Mori et al. 2015). Studying the sympatric zones in Brazil, they found that this hybridization, contrary to what was thought, is symmetric, because some trees had shared and reciprocal haplotypes from both species, so the gene flow does occur bidirectionally.

In addition, previous studies also found evidence of ancient hybridization between *A. germinans* and *A. bicolor*, a species found in the eastern Pacific coast, supporting that these processes may be more common intense than previously believed. In addition to that, they suggest that the discordant molecular variation patterns between high and low mutations markers for both *A. germinans* and *A. schaueriana* may be due to a recent demographic expansion of both species, probably because they responded differently since the last glaciation, explaining the current distribution of the species in South America (*A. schaueriana* is the only that is present up to the southernmost site of the Brazilian coast) (Mori et al. 2015).

Silva et al. (2021), also studying mangrove species along the South America coast, used landscape genomic approaches to investigate the relative contributions of geographic and environmental variables to genetic variation of *A. schaueriana* and *A. germinans*. Using single nucleotide polymorphism (SNP), they concluded that the large extension and major oceanic currents of the Brazilian coast physically limit dispersal of *Avicennia* species, where the South Equatorial Current acts like a barrier to gene flow in *A. schaueriana* and the isolation pattern best explains the genetic differentiation of *A. germinans*. Additionally, for both species, they observed significant correlations between genetic variation and precipitation regimes, tidal variation, solar radiation, and temperature patterns. Studies like this provide not only an understanding of the effects of the environment on microevolutionary processes and population dynamics but also important information to predict future responses for expected global changes on coastal species for this century. Cruz et al. (2019) found divergences between northern and southern populations along the extremes of South America, mostly associated to biogenesis of photosynthetic apparatus, anthocyanin biosynthesis, and osmotic and hypoxia stress responses.

The story behind the distribution of the red mangrove *Rhizophora mangle* is very interesting. This species can be commonly found around the globe and it is very abundant in Brazil. One study analyzed *R. mangle* individuals along the Brazilian coast and concluded that the populations are not composed of a single panmictic one, presenting well-structured groups with low heterozygosity and low genetic variability (Pil et al. 2011). A conceivable hypothesis to explain the differences and genetic structure of the populations, mainly on the coastal segments I–V (see Chap. 3), is associated with climate change that occurred in the last thousands of years (Pil et al. 2011). The low variability and the reduced richness in the populations of the red mangrove on the Brazilian coast may be explained by the propagule dispersion constraint during the last glaciations, which created refugia interrupting the flow between the north and the south of the country. This hypothesis is supported by palynological and genetic evidence (Pil et al. 2011).

Francisco et al. (2018) not only tested the genetic structure pattern in three species of *Rhizophora* (*R. mangle*, *R. harrisonii*, and *R. racemosa*) but also tested for the presence of a hybrid zone. Covering more than 4,900 km of the Atlantic Coast of South America, the results using microsatellites indicated that *R. mangle* is distantly related to individuals of *R. racemosa* and *R. harrisonii*, where *R. mangle* comprises a "pure" species. Nevertheless, the southern populations differed from the population from the north, while *R. racemosa* and *R. harrisonii* formed a different, non-admixed species and a homogeneous cluster, presenting that hybridization and introgression processes play important roles in the genus. In general, samples from the northern populations presented higher genetic variation and higher frequency of alleles, while southern populations are more homogeneous, as also seen in *Avicennia* (Mori et al. 2015).

However, individuals from both regions were clustered together, indicating that there is no abrupt break between them. The likely reason for the north-south difference regarding genetic variation is maintained by superficial ocean currents as previously suggested by Pil et al. (2011) and Mori et al. (2015), since the high-velocity from the north-northeastward North Current facilitates the movement of propagules from south to north, while the low-velocity south-southwestward Brazil Current creates a barrier to the southwestward dispersal of individuals from the northeastern populations.

Using SNP, Mori et al. (2021) used a phylogeographical approach to test hypotheses of *R. mangle*, *R. racemosa*, and *R. harrisonii* individuals from the Atlantic East Pacific region and South Pacific Islands. Besides finding that the American continent plays an important role as a barrier to gene flow within the genus as other authors suggested, they concluded that taxonomic identification based only on current morphological traits is inconsistent and could not define genetic clustering nor the phylogenetic relationships among groups on both locations. For instance, they found that *R. racemosa* from the Atlantic basin is, probably, because of the continent barrier, genetically more similar to *R. mangle* from the same basin than to *R. racemosa* from the Pacific coast. Also, it was found that *R. harrisonii* is likely composed of two independently originated and separately maintained evolutionary lineages in both sides of the American continent and were positioned between *R. racemosa* and *R. mangle* individuals (as expected due to its putative hybrid nature).

At the same study, testing evolutionary models to best explain *Rhizophora* species evolutionary history, they found a divergence between plants from both oceans' basins and that *R. harrisonii* plants were generated by an admixture process that occurs independently in the Pacific and the Atlantic coasts, although sharing similar morphological traits, indicating that *R. harrisonii* morphotype trees should not be classified as a single hybrid species. Rejecting the unity of *R. harrisoni* and consequently proposing that there are more species than the taxa recognized for the regions, the study also reinforced the taxonomic uncertainties regarding the *Rhizophora* genus.

Moreover, founder effects can also reduce genetic diversity and richness, because they result in the loss of alleles and consequently heterozygosity. A study on the white mangrove *Laguncularia racemosa* in a mangrove restoration site at Guanabara Bay, Rio de Janeiro (Granado et al. 2018), found higher levels of genetic diversity in the autochthonous plants, which survived in the area, than in the ones planted during the restoration effort around 2007 (see Chap. 3, Map 13). Using dominant markers, the results suggest that the population of white mangroves suffered a bottleneck event that consequently increased homozygosity levels. On the other hand, a small population located at Araçá Bay, São Paulo State, showed higher levels of genetic diversity compared to Rio de Janeiro and was considered a potential source of donors to improve the genetic diversity of other areas nearby (Schaeffer-Novelli et al. 2018) (see Chap. 3, Map 14).

Another study found epigenetic diversity differentiation following morphological variation when comparing white mangrove individuals living along a salt marsh to the ones living riverside at Sepetiba Bay (Lira-Medeiros et al. 2010). This was the first study to show that epigenetic diversity can be accounted for in wild populations and can be associated with environmental and morphological characteristics. The theory was corroborated due to nonsignificant genetic differentiation between the same areas of Sepetiba Bay, leading to the conclusion that epigenetic diversity has an important role in plant adaptation decoupled of genetic diversity (Lira-Medeiros et al. 2015).

In summary, it becomes evident that mangrove species all around the world represent a very particular group of plants of unique evolutionary history and intriguing adaptive behavior, which made it possible to overcome all major environmental changes during the evolutionary period. However, recent human activities represent a threat to this resilient group of plants, causing fragmentation and consequently genetic isolation between mangrove populations. This fact accentuates the importance of genetic studies as tools to measure genetic characteristics of natural and impacted populations and subsidizing better restoration strategies and conservation management and planning decisions in a world facing such fast-paced climate changes.

5.7 The Future of Genetic and Epigenetic Studies on Mangrove Plants

Few studies are assessing the genetic and epigenetic diversity of AEP mangrove plants in comparison to many studies on IWP mangroves. A lot of questions are still to be answered about the reproductive biology, dispersal capacity, propagule dormancy, and genetic and epigenetic diversity of AEP species, especially in the coastal segments I–II in the northern Brazilian coast (see Chap. 3), where mangrove ecosystems are most prosperous.

The future of genetic studies is being transposed to genomic studies. New technologies of NGS are revolutionizing the way traditional genetic studies are done. But we still lack computer power to analyze all data that can be generated with full-power NGS and, most importantly, we lack conservation strategies that incorporate genetic and genomic data. The NGS data is particularly useful for clarifying species identity, defining potential protected areas, and understanding adaptation (Wee et al. 2019). The epigenetic studies are also migrating to the NGS platform, which brings a challenge in data analysis but also a new perspective on the adaptive and evolutionary potential of populations and species with low levels of genetic diversity.

With climate changes transforming the global environment, it is possible that increasing sea surface temperature (SST) will allow mangrove ecosystems to expand their range edge in the next years, since low SST is one of the most limiting factors for mangrove occurrence (Ximenes et al. 2018; but see Chap. 3). This could represent another chance for mangroves globally, once they have been facing increasing threats that could lead them to extinction within the next hundred years (Polidoro et al. 2010).

References

Allendorf FW, Hohenlohe PA, Luikart G (2010) Genomics and the future of conservation genetics. Nat Rev Genet 11(10):697–709

Alongi DM (2002) Present state and future of the world's mangrove forests. Environ Conserv 29(03):331–349

Arnaud-Haond S, Teixeira S, Massa SI, Billot C, Saenger P, Coupland G, Duarte CM, Serrão EA (2006) Genetic structure at range edge: low diversity and high inbreeding in Southeast Asian mangrove (*Avicennia marina*) populations. Mol Ecol 15(12):3515–3525

Cerón-Souza I, Rivera-Ocasio E, Medina E, Jiménez JA, McMillan WO, Bermingham E (2010) Hybridization and introgression in New World red mangroves, *Rhizophora* (Rhizophoraceae). Am J Bot 97(6):945–957

Cruz MV, Mori GM, Müller CS, Silva CC, Oh DH, Dassanayake M, Zucchi MI, Oliveira RS, Souza AP (2019) Local adaptation of a dominant coastal tree to freshwater availability and solar radiation suggested by genomic and ecophysiological approaches. Sci Rep 9:19936

Dodd RS, Afzal-Rafii Z, Kashani N, Budrick J (2002) Land barriers and open oceans: effects on gene diversity and population structure in *Avicennia germinans* L. (Avicenniaceae). Mol Ecol 11(8):1327–1338

Duke NC, Meynecke J-O, Dittmann S, Ellison AM, Anger K, Berger U, Cannicci S, Diele K, Ewel KC, Field CD, Koedam N, Lee SY, Marchand C, Nordhaus I, Dahdouh-Guebas F (2007) A world without mangroves? Science 317(5834):41–42

Francisco PM, Mori GM, Alves FM, Tambarussi E, de Souza AP (2018) Population genetic structure, introgression, and hybridization in the genus *Rhizophora* along the Brazilian coast. Ecol Evol 8(6):3491–3504

Granado R, Neta LCP, Nunes-Freitas AF, Voloch CM, Lira CF (2018) Assessing genetic diversity after mangrove restoration in Brazil: why is it so important? Diversity 10(2):27

Hamrick JL, Godt MJW (1996) Effects of life history traits on genetic diversity in plant species. Philos Trans R Soc Lond B 351:1291–1298

Henry RJ (2013) Molecular markers in plants. Blackwell Publishing, Danvers

Holsinger KE, Lewis PO, Dey D (2002) A Bayesian approach to inferring population structure from domimant markers. EEB Articles 1:1–14

Lira-Medeiros CF, Cardoso MA, Ferreira PCG (2006) Uso de microssatélites em estudos de biologia da conservação. Floresta e Ambiente 13(2):25–36

Lira-Medeiros CF, Parisod C, Fernandes RA, Mata CS, Cardoso MA, Ferreira PCG (2010) Epigenetic variation in mangrove plants occurring in contrasting natural environment. PLoS One 5(4):e10326

Lira-Medeiros CF, Cardoso MA, Fernandes RA, Ferreira PCG (2015) Analysis of genetic diversity of two mangrove species with morphological alterations in a natural environment. Diversity 7(2):105–117

Mori GM, Zucchi MI, Sampaio I, Souza AP (2015) Species distribution and introgressive hybridization of two *Avicennia* species from the Western Hemisphere unveiled by phylogeographic patterns. BMC Evol Biol 15(1):61

Mori GM, Madeira AG, Cruz MV, Tsuda Y, Takayama K, Matsuki Y, Suyama Y, Iwasaki T, de Souza AP, Zucchi MI, Kajita T (2021) Testing species hypotheses in the mangrove genus Rhizophora from the Western hemisphere and South Pacific islands. Estuar Coast Shelf Sci 248(5):106948

Nettel A, Dodd RS (2007) Drifting propagules and receding swamps: genetic footprints of mangrove recolonization and dispersal along tropical coasts. Evolution 61(4):958–971

Otto SP (2018) Adaptation, speciation and extinction in the Anthropocene. Proc R Soc B 285: 20182047

Pil MW, Boeger M, Muschner V, Pie MR, Ostrensky A, Boeger WA (2011) Postglacial north-south expansion of populations of *Rhizophora mangle* (Rhizophoraceae) along the Brazilian coast revealed by microsatellite analysis. Am J Bot 98(6):1031–1039

Polidoro BA, Carpenter KE, Collins L, Duke NC, Ellison AM, Ellison JC, Farnsworth EJ, Fernando WS, Kathiresan K, Koedam NE, Livingstone SR, Miyagi T, Moore GE, Nam VN, Ong JE, Primavera JH, Salmo SG III, Sanciangco JC, Sukardjo S, Wang Y, Yong JWH (2010) The loss of species: mangrove extinction risk and geographic areas of global concern. PLoS One 5(4): e10095

Rapp RA, Wendel JF (2005) Epigenetics and plant evolution. New Phytol 168(1):81–91

Schaeffer-Novelli Y, Cintrón-Molero G, Reis-Neto AS, Abuchahla GMO, Neta LCP, Lira-Medeiros CF (2018) The mangroves of Araçá Bay through time: an interdisciplinary approach for conservation of spatial diversity at large scale. Ocean Coast Manage 164:60–67

Silva MF, Cruz MV, Vidal Júnior JD, Zucchi MI, Mori GM, Souza AP (2021) Geographic and environmental contributions to genomic divergence in mangrove forests. Biol J Linnean 132(3): 573–589

Su G-H, Huang Y-L, Tan F-X, Ni X-W, Tang T, Shi S-H (2006) Genetic variation in *Lumnitzera racemosa*, a mangrove species from the Indo-West Pacific. Aquat Bot 84(4):341–346

Triest L (2008) Molecular ecology and biogeography of mangrove trees towards conceptual insights on gene flow and barriers: a review. Aquat Bot 89(2):138–154

Van der Stocken T, Carroll D, Menemenlis D, Simard M, Koedam N (2019) Global-scale dispersal and connectivity in mangroves. Proc Natl Acad Sci 116(3):915–922

Wee AKS, Teo JXH, Chua JL, Takayama K, Asakawa T, Meenakshisundaram SH, Onrizal AB, Ardli ER, Sungkaew S, Suleiman M, Tung NX, Salmo SG III, Yllano OB, Saleh MN, Soe KK, Tateishi Y, Watano Y, Tsuda Y, Kajita T, Webb EL (2017) Vicariance and oceanic barriers drive contemporary genetic structure of widespread mangrove species *Sonneratia alba* J. Sm in the Indo-West Pacific. Forests 8(12):1–21

Wee AKS, Mori GM, Lira-Medeiros CF, Núñez-Farfán J, Takayama K, Faulks L, Shi S, Tsuda Y, Suyama Y, Yamamoto T, Iwasaki T, Nagano Y, Wang Z, Watanabe S, Kajita T (2019) The integration and application of genomic information in mangrove conservation. Conserv Biol 33(1):206–209

Ximenes AC, Ponsoni L, Lira CF, Koedam N, Dahdouh-Guebas F (2018) Does Sea Surface Temperature contribute to determining range limits and expansion of mangroves in Eastern South America (Brazil)? Remote Sens 10(11):1787

Chapter 6
Latitudinal Variation in Brazilian Mangroves' Annual Litterfall as Evidence of Maximum Power and Geoecological Manifestation

Gilberto Cintrón-Molero, Clemente Coelho-Jr, Laís M. Paiva, and Yara Schaeffer-Novelli

6.1 Introduction

One of the major paradigms in mangrove ecology is that mangroves are highly productive systems and are considered an azonal formation present over a wide latitudinal span from the tropics to well over the subtopics, reaching latitudes close to or slightly beyond 34° (Tomlinson 1986). Mangroves visibly do best between latitudes 25°N and S where they manifest their highest productivity in terms of biomass and litterfall (Cintrón and Schaeffer-Novelli 1983; Saenger and Holmes 1991). Although in the New World the species diversity of mangroves is lower than in Asia, mangrove productivity is similar and systems under similar conditions reach similar levels of development (Saenger and Snedaker 1993).

Mangrove aboveground biomass is correlated to litterfall production in the sense that larger trees can support a larger canopy and potential for producing higher litterfall levels. Observations show that productivity is linked to hydrology with riverine systems attaining the highest levels of structural development followed by

G. Cintrón-Molero (✉) · L. M. Paiva
Instituto BiomaBrasil, São Paulo, SP, Brazil

C. Coelho-Jr
Instituto BiomaBrasil, Recife, PE, Brazil

Instituto de Ciências Biológicas, Universidade de Pernambuco, Recife, PE, Brazil

Y. Schaeffer-Novelli
Instituto BiomaBrasil, Recife, PE, Brazil

Instituto Oceanográfico, Universidade de São Paulo, São Paulo, SP, Brazil

IUCN SSC Mangrove Specialist Group, São Paulo, SP, Brazil

Y. Schaeffer-Novelli et al. (eds.), *Brazilian Mangroves and Salt Marshes*, Brazilian Marine Biodiversity, https://doi.org/10.1007/978-3-031-13486-9_6

fringes and, finally, basins (Lugo and Snedaker 1974). However, on an aerial basis, fringes are limited to edges whereas basins can dominate a landscape.

Twilley (1995) presented summaries of biomass and litterfall accumulation and productivity of mangroves with latitude worldwide. These data show a trend for biomass accumulation to/decline with latitude, as would be expected from climatic constraints, but, surprisingly, also show high litterfall production even near latitudinal extremes. In this context, the Brazilian coastline provides an attractive opportunity to examine latitudinal patterning and ecosystem functions as geoecological manifestations, since it extends from Equatorial to Subtropical latitudes (04°26′12″N to 33°45′07″S), an approximately 11 thousand km latitudinal span (IBGE 2016). Cintrón and Schaeffer-Novelli (1981) reported to Brazil that mangroves reach their latitudinal limit along this coast, occupying habitats below the Tropic of Capricorn (23°30′S to 28°30′S). Interestingly, solar energy decreases, and the planetary heat balance shifts from surplus incoming energy to energy deficits at about 35°S to 40°S where radiative deficits prevail (INMET: http://www.inmet.gov.br), which drive productivity patterns.

6.2 Brazilian Mangrove Forests' Annual Litterfall

This chapter shows assembled and analyzed data from published and unpublished studies on Brazilian mangrove forests' annual litterfall along the coast between latitudes 04°N and 27°S. The criteria for data inclusion took into consideration Proctor's (1984) considerations on data comparison issues. The requirements for inclusion were: (1) all litter components had to be included in the study, not merely the leaves; (2) all study sites should be georeferenced or ascertainably situated deriving from other sources; (3) all litterfall data had to represent at least a one-year collection, regardless of the season of commencement or termination of sampling; and (4) all collecting and weighing methods are well accepted in the literature as long as sufficient replicate traps were used to take variability into account.

Northern and southern litterfall values (all components of the litter expressed as g m^2 day^{-1}) were treated as equivalent for mixed and monospecific mangroves communities of the various species (Saenger and Snedaker 1993).

6.2.1 Annual Litterfall

The obtained linear regression shows to be somewhat skewed (skewness = 0.1379), with several point accumulations of high tropical and subtropical values contributing to the skewing (Fig. 6.1). The assembled data ($N = 94$) are presented in Tables 6.1 and 6.2.

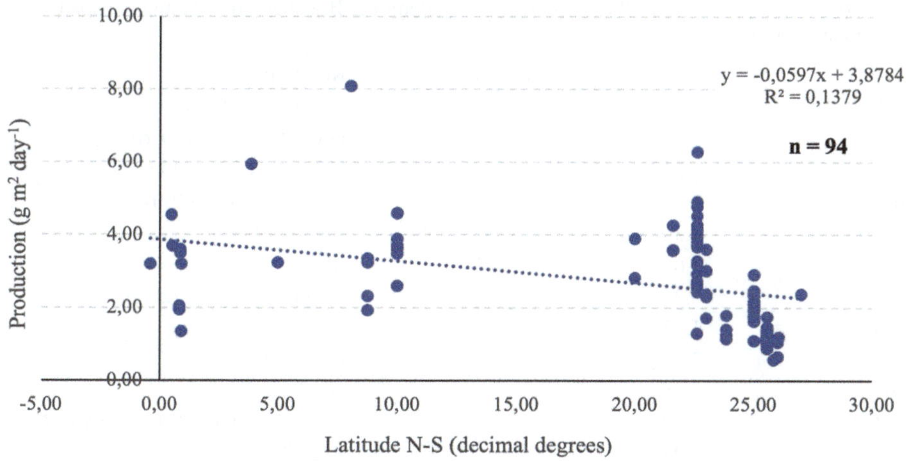

Fig. 6.1 Scatter diagram of mangrove litterfall along the Brazilian coastline. Latitudinal degrees were transformed to latitudinal decimal degrees. Litterfall and latitudinal location fed a best-fit linear regression ($y = ax + b$)

The litterfall values ranged from the highest rate of 8.08 g m^2 day^{-1} in a *Laguncularia racemosa* stand in Olinda (08°S) (Paiva and Coelho-Jr unpublished) to the lowest of 0.65 g m^2 day^{-1} in Babitonga Bay (26°S) (Cunha 2001; Almeida 2004) (see Chap. 3, Maps 8 and 16, respectively). The latitudinal variation on its own was responsible for ca. 14% of the variation in productivity, with a lesser trend in higher latitudes.

6.3 On the Causes for Litterfall Productivity

When there are no site-specific growth constraints, litterfall production is proportional to solar insolation, which, in turn, reflects the generally increased structural complexity of mangrove communities under optimal growing conditions in the tropics.

Pool et al. (1977) presented data on tree height and latitude of mangroves in the western hemisphere. Their data, although not statistically significant, showed a similar trend for the relationship between biomass and latitude. Nevertheless, data on Brazilian litterfall show surprisingly high productivity even near the limits of distribution, as it has been observed in a few other studies, such as Twilley's (1995) and López-Medellín and Ezcurra's (2012) in Mexico.

Litterfall values strongly reflect latitude influences through insolation, temperature, and freshwater availability. Within equatorial latitudes, ample solar radiation in terms of day length duration and solar angle, combined with warm temperatures and suitable site factors, favors high productivity that prevails to about 10°S (Saenger

Table 6.1 Ninety-four litterfall data (g m^2 day^{-1}) along the Brazilian coast latitudinal span with respective references

Latitude	Production (g m^2 day^{-1})	References
(04°N)	3,20	Fernandes (1997)
(00°)	2,05	Gonçalves et al. (2006)
(00°)	1,95	Gonçalves et al. (2006)
(00°)	2,03	Fernandes et al. (2007)
(00°)	3,50	Mehlig (2001)
(00°)	3,60	Mehlig (2001)
(00°)	3,20	Nascimento et al. (2006)
(00°)	1,35	Farias et al. (2006)
(00°)	3,70	Reise (2003)
(00°)	4,55	Nordhaus (2004)
(03°S)	5,94	Travassos et al. (2012) *unpublished*
(05°S)	3,25	Rêgo (1999)
(08°S)	3,24	Longo (2009)
(08°S)	2,32	Longo (2009)
(08°S)	3,35	Longo (2009)
(08°S)	1,94	Longo (2009)
(08°S)	8,08	Paiva and Coelho-Jr (unpublished)
(10°S)	3,49	Menezes (2010)
(10°S)	3,89	Menezes (2010)
(10°S)	4,59	Menezes (2010)
(10°S)	4,58	Menezes (2010)
(10°S)	3,71	Santos (2013)
(10°S)	3,62	Santos (2013)
(14°S)	2,60	Santos (2009)
(20°S)	3,90	Carmo et al. (1998)
(20°S)	2,83	Carmo et al. (1998)
(21°S)	3,58	Bernini and Rezende (2010)
(21°S)	4,26	Bernini and Rezende (2010)
(21°S)	3,59	Bernini and Rezende (2010)
(22°S)	3,81	Chaves (2007)
(22°S)	3,67	Chaves (2007)
(22°S)	3,92	Chaves (2007)
(22°S)	2,93	Chaves (2007)
(22°S)	2,47	Chaves (2007)
(22°S)	4,90	Chaves (2007)
(22°S)	4,11	Chaves (2007)
(22°S)	4,52	Chaves (2007)
(22°S)	4,77	Chaves (2007)
(22°S)	4,22	Chaves (2007)
(22°S)	3,29	Chaves (2007)
(22°S)	4,00	Chaves (2007)
(22°S)	2,55	Chaves (2007)

(continued)

Table 6.1 (continued)

Latitude	Production (g m^2 day^{-1})	References
(22°S)	3,70	Chaves (2007)
(22°S)	6,27	Chaves (2007)
(22°S)	2,68	Chaves (2007)
(22°S)	2,44	Chaves (2007)
(22°S)	4,27	Chaves (2007)
(22°S)	3,73	Chaves (2007)
(22°S)	1,29	Chaves (2007)
(22°S)	2,63	Chaves (2007)
(22°S)	3,95	Chaves (2007)
(22°S)	2,74	Chaves (2007)
(22°S)	3,18	Chaves (2007)
(23°S)	2,31	Machado (2014)
(23°S)	3,02	Machado (2014)
(23°S)	3,60	Machado (2014)
(23°S)	3,02	Machado (2014)
(23°S)	2,33	Machado (2014)
(23°S)	1,72	Machado (2014)
(23°S)	2,38	Silva et al. (1998)
(23°S)	1,79	Lamparelli (1995)
(23°S)	1,40	Lamparelli (1995)
(23°S)	1,25	Lamparelli (1995)
(23°S)	1,14	Lamparelli (1995)
(25°S)	2,34	Valadares (2015)
(25°S)	1,09	Ponte et al. (1990)
(25°S)	2,08	Adaime (1985)
(25°S)	1,67	Adaime (1985)
(25°S)	2,47	Menezes (1994)
(25°S)	1,77	Menezes (1994)
(25°S)	2,11	Menezes (1994)
(25°S)	2,91	Almeida (2004)
(25°S)	2,17	Almeida (2004)
(25°S)	2,05	Almeida (2004)
(25°S)	1,93	Almeida (2004)
(25°S)	1,62	Almeida (2004)
(25°S)	1,12	Larcher (2014)
(25°S)	0,57	Larcher (2014)
(25°S)	0,97	Sessegolo (1997)
(25°S)	1,75	Sessegolo (1997)
(25°S)	1,39	Sessegolo (1997)
(25°S)	1,26	Sessegolo (1997)
(25°S)	0,97	Sessegolo (1997)
(25°S)	0,87	Sessegolo (1997)

(continued)

Table 6.1 (continued)

Latitude	Production (g m^2 day^{-1})	References
(25°S)	1,36	Sessegolo (1997)
(25°S)	1,19	Sessegolo (1997)
(25°S)	1,35	Sessegolo (1997)
(25°S)	1,47	Sessegolo (1997)
(25°S)	1,05	Cunha (2001)
(25°S)	0,65	Cunha (2001)
(26°S)	1,19	Silva (2001)
(27°S)	2,37	Panitz (1986)

and Snedaker 1993; Simrad et al. 2019). However, what is interesting in these data is the peak extending south of about 23°30′S in a region where sun radiative deficits become more pronounced and evident. It seems that this reflects the propensity of mangroves to persist by maximizing the use of available subsidiary energies through autocatalytic growth that reinvests high-quality energy into compartments that reinforce further production, thus generating maximum power. Following Boltzmann (1886), Lotka (1922) proposed the Principle of Maximum Power, which has wide application in all thermodynamic systems. In Ecology, Odum and Pinkerton (1955) associated this phenomenon with autocatalysis. This Principle suggests that evolutionary selection favors systems that can maximize energy flow by using all available site energies and reinvesting captured energy into ways favorable to persistence close to the edge of species' or system's tolerances.

At the ecosystem level, Lotka's Maximum Power Principle (MPP) means selection for increasingly tolerant species or beyond into higher-level designs that combine other systems such as salt marshes. We suggest that the MPP is manifested in mangroves as Maximum Ascendency (sensu Ulanowicz 1980). The Brazilian litterfall data provides empirical evidence for the MPP and associated maximization functions, for example, Goal Functions (i.e., ascendency) described in the literature by Odum (1969) and Jørgesen (1992), among others. The MPP is increasingly recognized as an ecological law of thermodynamics (Odum and Pinkerton 1955; Jørgensen 1992).

In the case of Brazil, the assertiveness of mangroves is combined with coastal geomorphology, tidal regime, and local climate (see Chap. 3) to create extremely favorable conditions that allow mangroves to persist poised at the edge of chaos where environmental conditions preclude further occupation of coastal habitats.

In Southeast Brazil, the coastal escarpment reaches about latitude 23°S; beyond that point, the coastline becomes more indented (or higher fractal dimension) within a region increasingly influenced by the South Atlantic Convergence Zone convective activity (SACZ); at 20°–25°S. The SACZ conveys enhanced precipitation and increasing moisture that, combined with a favorable microtidal (less than 2 m) regime and geomorphology, creates favorable habitats very close to the latitudinal limit. Notably, a microtidal to mesotidal pulsing regime is considered optimal for salt marsh development (Odum et al. 1995) as is for barrier island formation (Davies 1973; Hayes 1975). Thus, the 20°–28°S structural setting of the Brazilian coast

Table 6.2 Ninety-four litterfall data (g m^2 day^{-1}) along the Brazilian coast latitudinal span with respective references

Latitude	Production (g m^2 day^{-1})	References	Latitude	Production (g m^2 day^{-1})	Reference	Latitude	Production (g m^2 day^{-1})	References
(04°N)	3,20	Fernandes (1997)	(22°S)	3,92	Chaves (2007)	(23°S)	1,25	Lamparelli (1995)
(00°)	2,05	Gonçalves et al. (2006)	(22°S)	2,93	Chaves (2007)	(23°S)	1,14	Lamparelli (1995)
(00°)	1,95	Gonçalves et al. (2006)	(22°S)	2,47	Chaves (2007)	(25°S)	2,34	Valadares (2015)
(00°)	2,03	Fernandes et al. (2007)	(22°S)	4,90	Chaves (2007)	(25°S)	1,09	Ponte et al. (1990)
(00°)	3,50	Mehlig (2001)	(22°S)	4,11	Chaves (2007)	(25°S)	2,08	Adaime (1985)
(00°)	3,60	Mehlig (2001)	(22°S)	4,52	Chaves (2007)	(25°S)	1,67	Adaime (1985)
(00°)	3,20	Nascimento et al. (2006)	(22°S)	4,77	Chaves (2007)	(25°S)	2,47	Menezes (1994)
(00°)	1,35	Farias et al. (2006)	(22°S)	4,22	Chaves (2007)	(25°S)	1,77	Menezes (1994)
(00°)	3,70	Reise (2003)	(22°S)	3,29	Chaves (2007)	(25°S)	2,11	Menezes (1994)
(00°)	4,55	Nordhaus (2004)	(22°S)	4,00	Chaves (2007)	(25°S)	2,91	Almeida (2004)
(03°S)	5,94	Travassos et al. (2012)[a]	(22°S)	2,55	Chaves (2007)	(25°S)	2,17	Almeida (2004)
(05°S)	3,25	Rêgo (1999)	(22°S)	3,70	Chaves (2007)	(25°S)	2,05	Almeida (2004)
(08°S)	3,24	Longo (2009)	(22°S)	6,27	Chaves (2007)	(25°S)	1,93	Almeida (2004)

(continued)

Table 6.2 (continued)

Latitude	Production (g m^2 day^{-1})	References	Latitude	Production (g m^2 day^{-1})	Reference	Latitude	Production (g m^2 day^{-1})	References
(08°S)	2,32	Longo (2009)	(22°S)	2,68	Chaves (2007)	(25°S)	1,62	Almeida (2004)
(08°S)	3,35	Longo (2009)	(22°S)	2,44	Chaves (2007)	(25°S)	1,12	Larcher (2014)
(08°S)	1,94	Longo (2009)	(22°S)	4,27	Chaves (2007)	(25°S)	0,57	Larcher (2014)
(08°S)	8,08	Paiva and Coelho-Jr (unpublished)[a]	(22°S)	3,73	Chaves (2007)	(25°S)	0,97	Sessegolo (1997)
(10°S)	3,49	Menezes (2010)	(22°S)	1,29	Chaves (2007)	(25°S)	1,75	Sessegolo (1997)
(10°S)	3,89	Menezes (2010)	(22°S)	2,63	Chaves (2007)	(25°S)	1,39	Sessegolo (1997)
(10°S)	4,59	Menezes (2010)	(22°S)	3,95	Chaves (2007)	(25°S)	1,26	Sessegolo (1997)
(10°S)	4,58	Menezes (2010)	(22°S)	2,74	Chaves (2007)	(25°S)	0,97	Sessegolo (1997)
(10°S)	3,71	Santos (2013)	(22°S)	3,18	Chaves (2007)	(25°S)	0,87	Sessegolo (1997)
(10°S)	3,62	Santos (2013)	(23°S)	2,31	Machado (2014)	(25°S)	1,36	Sessegolo (1997)
(14°S)	2,60	Santos (2009)	(23°S)	3,02	Machado (2014)	(25°S)	1,19	Sessegolo (1997)
(20°S)	3,90	Carmo et al. (1998)	(23°S)	3,60	Machado (2014)	(25°S)	1,35	Sessegolo (1997)
(20°S)	2,83	Carmo et al. (1998)	(23°S)	3,02	Machado (2014)	(25°S)	1,47	Sessegolo (1997)

(21°S)	3,58	Bernini and Rezende (2010)	(23°S)	2,33	Machado (2014)	(25°S)	1,05	Cunha (2001)
(21°S)	4,26	Bernini and Rezende (2010)	(23°S)	1,72	Machado (2014)	(25°S)	0,65	Cunha (2001)
(21°S)	3,59	Bernini and Rezende (2010)	(23°S)	2,38	Silva et al. (1998)	(26°S)	1,19	Silva (2001)
(22°S)	3,81	Chaves (2007)	(23°S)	1,79	Lamparelli (1995)	(27°S)	2,37	Panitz (1986)
(22°S)	3,67	Chaves (2007)	(23°S)	1,40	Lamparelli (1995)			

[a]Unpublished

shapes a mesoscale physiographic/climatic complex that harbors the largest estuarine systems of Southeast Brazil (Angulo et al. 2009). This geomorphic template subsidizes mangrove occupation, where they can act as an endogenous living force capable of modifying geomorphic processes and buffering exogenic (climate-driven limitations). As a result, mangrove productivity instead of decreasing gradually ends abruptly at Laguna, Santa Catarina State, near 28°30′S (see Chap. 3, Map 16), contradicting/contrasting with what is assumed as a pattern for Australia, New Zealand, and Japan (Tomlinson 1986; Friess 2018). We suggest that mangroves here can be considered as "surfing at the edge of chaos," and this is manifested in fractal dynamics on the ground, such as persistent change (or pulsing) at local scales, rather than a steady state.

An appearance of steady state may be induced by constant recruitment. Pulsing at a local scale can induce persistent changes in vegetation patterns consistent with pulsing driven by slight external changes in climate and internal (local scale) changes. A steady state at the landscape level is an emergent function of local scale pulsing. The prevalence of pulsing in nature led Odum et al. (1995) to propose it as a paradigm of nature's organization and suggest that pulsing can be interpreted as oscillations at the edge of chaos.

Mangroves frequently express a high within-region diversity of structural patterns, and an equally high diversity of functional roles (Lugo and Snedaker 1974; Pool et al. 1977). Based on knowledge of the dataset here analyzed, structural characteristics (i.e., high values for height, biomass, and litterfall) indicate optimum habitats, particularly concerning reduced salinity regimes, optimal climatic conditions, site-specific fertility, as also observed where human input of domestic sewage nutrients influences the high values of productivity at Olinda (PE) and Guanabara Bay (RJ) (see Chap. 3, Maps 8 and 13, respectively). The slope of the present regression (0.138) is not significantly different from the slope in Saenger and Snedaker's (1993) regression (0.201, Eq. 5). The high value at Fernando de Noronha (see Chap. 1) (see Chap. 3, 18) can be explained by rich volcanic soil and low salinity values. In contrast, the low-value sites are mainly characterized by natural stress situations, for example, aridity and poor fertility (Cintrón et al. 1978). Although the correspondence is not precise, these general regional trends are consistent with the more specific conclusions drawn by Pool et al. (1977) and Saenger and Snedaker (1993).

Finally, south of 28°30′S, sediment transport by wave action, the microtidal tidal regime (mean range only 0.5 m) and freshwater runoff from a humid climate become too adverse for mangrove establishment. Particularly, a barrier system of Rio Grande do Sul State acts as a dam to freshwater input, which combined with restricted tidal influence favors freshening of the choked lagoon and the emergence of a "cut-off" filter or "brick wall" that impedes mangrove and salt marsh development further south. This barrier system is one of the longest in South America and one of the longest in the world. Salt marshes do reappear at the southern end near Cassino Beach (see Chap. 1) where the inlet allows seawater intrusion into the lower lagoon. The reduction in tidal influence reflects the influence of an amphidromic point off the coast of Rio Grande do Sul that reduces the main lunar component of the tide (Schwiderski 1980) (see Chap. 3, 17).

6.4 Final Remarks

The assembled data on Brazilian litterfall and corresponding analyses fully support Saenger and Snedaker's (1993) results that within the global mangrove community, the indices of organic production are highest at the lower latitudes. Nevertheless, mangrove communities below the Tropic of Capricorn latitudes are capable of larger litterfall rates relative to their biomass than more tropical ones. We suggest that this reflects the propensity of mangroves to persist by maximizing the use of available subsidiary energies combined by a combination of unusually favorable biotic and abiotic factors south of 23°30′S.

The Brazilian coast provides an attractive opportunity to examine latitudinal patterning and ecosystem functions as geoecological manifestations due to its extension. Also, its range makes it possible to study mangroves at the edge of chaos near their latitudinal limit. There is a great potential of addressing important questions about mangrove species' evolution and adaptation, especially in a country with such extensive coastal length. We hope that this chapter will bring more visibility and interest to combined studies on mangrove structure and genetics.

Acknowledgments The present analysis could only have been carried out with the support by Instituto BiomaBrasil, Brazilian National Council for Scientific and Technological Development (CNPq), the logistical support of the Secretaria de Ciência, Tecnologia e Inovação de Pernambuco granted to Espaço Ciência and the Instituto Chico Mendes de Conservação da Biodiversidade (ICMBio), as the access authorization to the National Marine Park of Fernando de Noronha (PARNAMAR Fernando de Noronha).

References

Adaime RR (1985) Produção do bosque de mangue da Gamboa Nóbrega (Cananéia, 25° Lat.S – Brasil). PhD dissertation, Universidade de São Paulo, 305 pp

Almeida R (2004) Ecologia de manguezais: dinâmica da serapilheira e funcionamento do ecossistema, Ilha do Cardoso, Cananéia, São Paulo, Brasil. PhD dissertation, Universidade de São Paulo

Angulo RJ, Lessa GC, Souza MC (2009) The Holocene Barrier Systems of Paranaguá and Northern Santa Catarina Coast, Southern Brazil. In: S Dillenburg, Hesp (eds) Geology and geomorphology of Holocene Coastal Barriers of Brazil. Springer, Berlin/Heidelberg, 135–176 pp

Bernini E, Rezende CE (2010) Litterfall in a mangrove in Southeast Brazil. Pan-Am J Aquat Sie 5: 508–519

Boltzmann LE (1886) Der zweite Hauptsatz der mechanischen Wärmetheorie. Gerold, Vienna

Carmo TMS, Almeida R, Oliveira AR (1998) Consequences of the peeling of *Rhizophora mangle* over the mangrove litterfall. In: Anales del VII Congreso Latinoamericano de Botánica y del XIV Congreso Mexicano de Botánica. Asociación Latinoamericana de Botánica/RLB/Sociedad Botánica de Mexico, Mexico

Chaves FO (2007) Caracterização e relações ambientais da produção de serapilheira em florestas de mangue da baía de Guanabara, Rio de Janeiro – Brasil. PhD dissertation, Universidade de São Paulo, 331 pp

Cintrón G, Schaeffer-Novelli Y (1981) Los manglares de la costa brasileña: revisión preliminar de la literatura. Abuchahla GMO (elect rev ed). Informe Técnico preparado para la Oficina Regional de Ciência y Tecnologia para América Latina y el Caribe de UNESCO y la Universidad Federal de Santa Catarina, 47 pp

Cintrón G, Schaeffer-Novelli Y (1983) Introducción a la ecología del manglar. Oficina Regional de Ciencia y Tecnología de la Unesco para América Latina y el Caribe, Montevideo, 109 pp

Cintrón G, Lugo AE, Pool DJ, Morris G (1978) Mangroves of arid environments in Puerto Rico and adjacent islands. Biotropica 10:110–121

Cunha SR (2001) Estrutura e produção das comunidades de macroprodutores dos manguezais da Baía de Babitonga, Santa Catarina. PhD dissertation, Universidade de Rio Grande, 172 pp

Davies JL (1973) Geographical variation in coastal development. Hafner Publishing Co, New York, 204 pp

Farias ASC, Fernandes MEB, Reise A (2006) Comparação da produção de serapilheira de dois bosques de com diferentes padrões estruturais na península Bragantina, Bragança, Pará. B Mus Paraens Em Goeldi 2(3):61–68

Fernandes MEB (1997) The ecology and productivity of mangroves in the Amazon region, Brazil. PhD dissertation, University of York, 213 pp

Fernandes MEB, Nascimento AAM, Carvalho ML (2007) Estimativa da produção anual de serapilheira dos bosques de mangue no Furo Grande, Bragança-Pará. Rev Árvore 31:949–958

Friess D (2018) Where the tallest mangroves are. Nat Geosci 12:4–5

Gonçalves ASC, Fernandes MEB, Carvalho ML (2006) Variação anual da produção de serapilheira em bosques de mangue no Furo Grande, Bragança, Pará. B Mus Paraens Em Goeldi 2:69–76

Hayes MO (1975) Morphology of sand accumulation in estuaries: an introduction to the symposium. In: Cronin LE (ed) Estuarine research. Vol 1, Chemistry, biology, and the estuarine system. Academic, London, pp 3–22

Instituto Brasileiro de Geografia e Estatística – IBGE (2016) Caracterização do território: posição e extensão. In: Anuário Estatístico do Brasil 1. IBGE, Rio de Janeiro, 457 pp

Jørgesen SE (1992) Integration of ecosystem theories: a pattern, 1st edn. Springer Science, Dordrecht, 383 pp

Lamparelli CC (1995) Dinâmica da serapilheira em manguezais de Bertioga, Região Sudeste do Brasil. PhD dissertation, Universidade de São Paulo, 139 pp

Larcher L, Boeger MRT, Nogueira G, Reissmann CB (2014) Produção de serapilheira em dois manguezais do estado do Paraná, Brasil. Acta Biol Catari 1(1):53–64

Longo AFP (2009) Produtividade primária em bosque de franja e bacia no manguezal do Rio Ariquindá (Tamandaré – Pernambuco, Brasil). Master thesis, Universidade Federal de Pernambuco, 70 pp

López-Medellín X, Ezcurra E (2012) The productivity of mangroves in northewestern Mexico: a meta-analysis of current data. J Coast Conserv 16:399–403

Lotka AJ (1922) Natural selection as a physical principle. Proc Natl Acad Sci U S A 8(6):151–154

Lugo AE, Snedaker SC (1974) The ecology of mangroves. Annu Rev Ecol Syst 5:39–64

Machado MRO (2014) Produção de serapilheira em florestas de mangue de Guaratiba – RJ. Master thesis, Universidade Federal do Rio de Janeiro

Mehlig U (2001) Aspects of tree primary production in an equatorial mangrove forest in Brazil. PhD dissertation, Universität Bremen, 151 pp

Menezes GV (1994) Produção e decomposição em bosques de mangue da Ilha do Cardoso, SP. Master thesis, Universidade de São Paulo, 116 pp

Menezes LCS (2010) Estrutura e produção de serapilheira de floresta de mangue na região estuarina-lagunar do baixo São Francisco sergipano. Master thesis, Universidade Federal de Sergipe, 84 pp

Nascimento RESA, Mehlig U, Menezes MPM (2006) Produção de serapilheira em um fragmento de bosque de terra firme e um manguezal vizinhos na península de Ajuruteua, Bragança, Pará. B Mus Paraens Em Goeldi 2:55–60

Nordhaus I (2004) Feeding ecology of the semi-terrestrial crab *Ucides cordatus* (Decapoda: Brachyura) in a mangrove forest in Northern Brazil. PhD dissertation, Universität Bremen

Odum EP (1969) The strategy of ecosystem development. Science 164:262–270

Odum HT, Pinkerton RC (1955) Time's speed regulator: the optimum efficiency for maximum power output in physical and biological systems. Am Sci 43(2):331–343

Odum WE, Odum EP, Odum HT (1995) Nature's pulsing paradigm. Estuaries 8(4):547–555

Paiva LM, Coelho-Jr C (unpublished) Variação anual da produção de serapilheira do manguezal Chico Science (Olinda, Pernambuco). Bachelor thesis, Universidade de Pernambuco, 62 pp

Panitz CMN (1986) Produção e decomposição de serapilheira no mangue do rio Itacorubi, Ilha de Santa Catarina, Florianópolis, Brasil (27°35'S – 48°31'W). PhD dissertation, Universidade Federal de São Carlos, 601 pp

Ponte ACE, Fonseca IAZ, Claro SMCA (1990) Produção de serapilheira em bosque impactado por petróleo. In: II Simpósio de Ecossistemas da Costa Sul e Sudeste Brasileira: Estrutura, Função e Manejo. ACIESP, Águas de Lindóia, pp 241–253

Pool DJ, Snedaker SC, Lugo AE (1977) Structure of mangrove forests in Florida, Puerto Rico, Mexico and Costa Rica. Biotropica 9:195–212

Proctor J (1984) Tropical forest litterfall. II. The data set. In: Chadwick AC, Sutton SL (eds) Tropical rain forest: ecology and management. Blackwell Scientific Publications, Oxford, pp 83–113

Rêgo RDP (1999) Dinâmica interna do nitrogênio no ecossistema manguezal localizado no estuário Potengi, Natal – RN. Master thesis, Universidade Federal de São Carlos, 64 pp

Reise A (2003) Estimates of biomass and productivity in Fringe Mangroves of North-Brazil. PhD dissertation, Universität Bremen, 196 pp

Saenger P, Holmes N (1991) Physiological, temperature tolerance and behavioral differences between tropical and temperate organisms. In: Connell DW, Hawker DW (eds) Pollution in tropical systems. CRC Press, Boca Raton, pp 69–95

Saenger P, Snedaker SC (1993) Pantropical trends in mangrove above-ground biomass and annual litterfall. Oecologia 96(3):293–299

Santos LL (2009) Estrutura e serapilheira em um manguezal de Ilhéus, Bahia, Brasil. Master thesis, Universidade Estadual de Santa Cruz

Santos TO (2013) Florística, Estrutura Fitossociológica e Produção de Serapilheira do manguezal do rio São Francisco. Master thesis, Universidade Federal de Sergipe, 69 pp

Schwiderski EW (1980) On charting global ocean tides. Rev Geophys Space Phys 18(1):243–268

Sessegolo GC (1997) Estrutura e produção de serapilheira do manguezal do Rio Baguaçu, Baía de Paranaguá, PR. Masther thesis, Universidade Federal do Paraná, 109 pp

Silva MCM (2001) Diagnóstico ambiental do manguezal da baía da Babitonga, Santa Catarina, através do uso de indicadores ecológicos, parâmetros foliares e produtividade de serapilheira. Florianópolis. Master thesis, Universidade Federal de Santa Catarina, 112 pp

Silva CAR, Lacerda LD, Ovalle AR, Rezende CE (1998) The dynamics of heavy metals through litterfall and decomposition in a red mangrove forest. Mangrove Salt Marshes 2(3):149–157

Simard M, Fatoyinbo L, Smetanka C, Rivera-Monroy VH, Castañeda-Moya E, Thomas N, Van der Stocken T (2019) Mangrove canopy height globally related to precipitation, temperature and cyclone frequency. Nat Geosci 12:40–45

Tomlinson PB (1986) The botany of mangroves. Cambridge University Press, Cambridge, 419 pp

Travassos PEPF, Coelho-Jr C, Severi W (2012 – unpublished data) Diagnóstico Ambiental do Manguezal da Baía do Sueste, Fernando de Noronha – PE. Final Report/Process CNPq No. 577369/2008-3, 102 pp

Twilley RR (1995) Properties of mangrove ecosystems related to the energy signature of coastal environments. In: Hall CAS (ed) Maximum power: the ideas and applications of Odum HT. University Press of Colorado, Denver, pp 43–62

Ulanowicz RE (1980) A hypothesis on the development of natural communities. J Theor Biol 85(2): 223–245

Valadares RG (2015) Dinâmica da Produção, Estoque e Decomposição da Serapilheira em um bosque de mangue em São Sebastião, SP. Master dissertation,, Universidade de Santa Cecília, 83 pp

Part III
Mangrove and Salt Marsh Associates

Chapter 7
The Microbiota of Brazilian Mangroves

Ana Júlia F. C. Oliveira, Aline B. Pinto, Bruna B. Zampieri, and Vanessa C. Andrade

7.1 Introduction

Mangrove sediments are complex systems affected by the interaction among geological, hydrological, physicochemical, and biological factors (Köster and Meyer-Reil 2001) that are intensively colonized by microorganisms such as bacteria, archaeon, fungi, cyanobacteria, algae, virus, and protozoa. These microorganisms play an important role in the cycles of matter and energy in sediments (Danovaro et al. 2000; Moreno et al. 2006).

As in the water column, high metabolism leads to an increase in the biomass of microbial producers, consumers, and decomposers, forming a microbial food chain called "microbial loop" (Azam et al. 1983), responsible for the transfer organic matter and energy to the higher trophic levels and nutrient regeneration, contributing to the formation of humic substances and oxygen, which will be consumed through aerobic metabolism, consequently exerting influence on the carbon transference of the food chain of mangrove sediments (Danovaro et al. 2000).

The microbial contribution to the system depends on the density, diversity, and activity of the present microorganisms, which in turn are influenced by biotic and abiotic factors. Among them, densities and species of prey and predator, sediment composition, organic and inorganic matter concentrations, local hydrodynamics, granulometry, and temperature. On the other hand, microorganisms influence the sediments where they occur through their metabolic activity (Peroni and Rossi 1986; Deming and Baross 1993). The main questions in the microbial ecology of

A. J. F. C. Oliveira (✉) · A. B. Pinto
Instituto de Biociências, Universidade Estadual Paulista, São Vicente, SP, Brazil
e-mail: ajuliaf@clp.unesp.br

B. B. Zampieri · V. C. Andrade
Departamento de Bioquímica e Microbiologia, Instituto de Biociências, Universidade Estadual Paulista, Rio Claro, SP, Brazil

Y. Schaeffer-Novelli et al. (eds.), *Brazilian Mangroves and Salt Marshes*, Brazilian Marine Biodiversity, https://doi.org/10.1007/978-3-031-13486-9_7

mangroves are how the microbiological communities are adapted to the conditions of this environment and what are the main factors controlling the diversity, distribution, and activity of these organisms (Moreno et al. 2006).

The importance of bacterial, cyanobacterial, and fungal species is being widened since the products of their metabolism have been the subject of bioprospecting studies and many species have been isolated from mangrove sediments due to their capacity to biodegrade certain compounds and produce enzymes and molecules of commercial interest (Demain 2007; van Dommelen et al. 2016; Luna and Townsend 2007).

This chapter will cover the knowledge on the Brazilian mangrove microbial loop, including studies on the diversity of bacteria, archea, fungi, and viruses in mangrove sediments.

7.2 Microbial Loop

Microorganisms play a major role in the maintenance of marine ecosystems. Nevertheless, for many years, they have received little attention from researchers. Aquatic bacteria have great importance not only in the decomposition of organic matter and in the remineralization of inorganic nutrients (Pomeroy et al. 2007), but also in adding energy to higher levels. In some ecosystems, such as mangroves, microbial processes greatly contribute to productivity, regenerating nutrients and exporting them to adjacent ecosystems (Hemminga et al. 1994; Alongi 1996; Holguin et al. 2001; Sjöling et al. 2005).

Until the end of the 1970s, researchers believed that the role of microorganisms, in the traditional food web, was only as a food source to planktonic organisms, and discarded them in studies of carbon flux dynamics (Fenchel 2008; Andrade 2015). The idea was of a straight flow, where microorganisms were just the base of the food web, giving no direct input of energy and carbon to the higher levels. But, with the development of new techniques and study methods (e.g., estimate growth rates), and the improvement of the existing ones, the idea of a straight flow became outdated (Pomeroy 1974; Steele 1974; Azam et al. 1983; Azam 1998; Fenchel 2008). Herewith, scientists discovered that the counts of microorganisms were underestimated, corresponding to only about 10% of the actual number of organisms in nature (Krambeck et al. 1981; Azam et al. 1983; Andrade 2015).

However, what is the *Microbial loop*? In the early 1980s, Azam et al. (1983) coined the term to describe the result of several ecological interactions between microorganisms involved and the pathways of nutrients (Fig. 7.1). Figure 7.2 displays examples of microorganisms commonly found in Brazilian mangrove sediments.

Most of the dissolved organic matter (DOM) in the marine environment is controlled by microorganisms, mainly by heterotrophic bacteria, cyanobacteria, and heterotrophic nanoflagellates (Pomeroy 1974; Anderson and Ducklow 2001). The cyanobacteria are a significant portion of the autotrophic phytoplanktonic

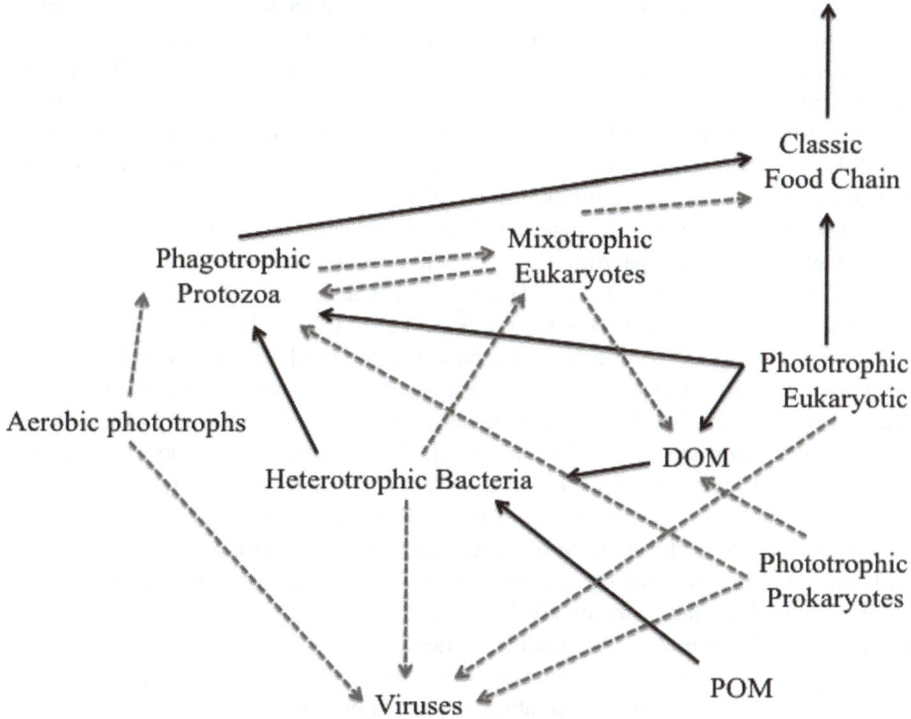

Fig. 7.1 Overview of general microbial loop pathways that also operate in mangrove sediments. Black arrows: Microbial Loop described by Azam et al. (1983). Gray dotted arrows: subsequent additions to the microbial loop. *POM* particulate organic matter. *DOM* dissolved organic matter. (Adapted from Fenchel 2008)

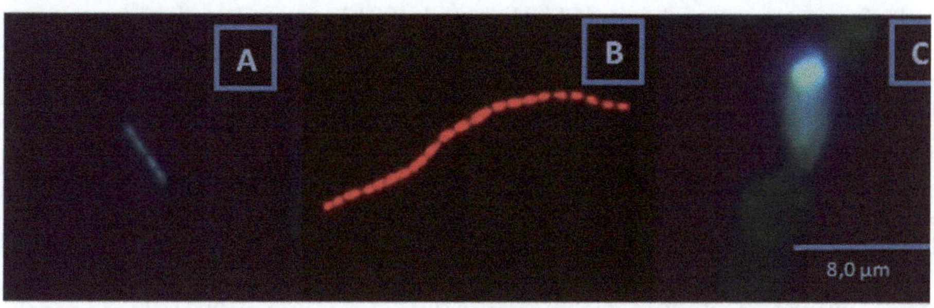

Fig. 7.2 Heterotrophic bacteria, cyanobacteria, and nanoflagellates from Brazilian mangrove sediment samples. (**a**) Heterotrophic bacteria colored with fluorochrome DAPI (4′,6-diamidino-2-phenylindole), (**b**) Self-fluorescence of cyanobacteria chain, and (**c**) Heterotrophic nanoflagellate with DAPI (4′,6-diamidino-2-phenylindole)

biomass, being responsible for a significant part of the primary production in aquatic ecosystems, with densities of about 10^5 cells mL^{-1} (Oliveira 2000; Pomeroy et al. 2007; Pereira 2014). The heterotrophic bacteria are responsible for about 30% of the primary production in coastal waters, with average densities of 10^6 cells mL^{-1} (Pereira 2014). They also have enzymes that develop important activities (e.g., cellulolytic, pectinolytic, amylolytic, and proteolytic), which allow the decomposition of mangrove matter (Matondkar et al. 1981).

Nanoflagellates present average densities of 10^3 cells mL^{-1} and are great predators of autotrophic and heterotrophic picoplankton, developing an essential ecological role in the control of the bacterial community (Caron et al. 1991; González 1993; Fernandes and Mesquita 1999; Oliveira 2000; Pereira 2011; Pereira 2014; Andrade 2015). In the early 1990s, viruses were added to the model of the Microbial Loop, acting in the control of bacterial density (Proctor and Fuhrman 1990; Oliveira 2000). They can change the distribution of particles and other organisms in any trophic level (Proctor and Fuhrman 1990; Bratbak et al. 1992, 1994; Thingstad et al. 1993; Fenchel 2008). According to Bratbak et al. (1994), viral densities vary from 2×10^6 to 50×10^6 Virus Particle-Like (VPL) mL^{-1}, in coastal and estuarine ecosystems, much higher than in offshore and oceanic systems (0.2×10^6–2×10^6 VPL mL^{-1}). Also, the grazing rates of bacteria over viruses are similar to protozoan grazing rates, but since viruses are highly specific, their effects are different and sustain the high diversity of bacteria (Fenchel 2008).

Focusing on the mangrove ecosystem, the mineralization of organic matter in sediments, especially nitrogen fixation, is an important role of microorganisms (Sjöling et al. 2005). The mangrove sediment is rich in organic matter, receiving abundant light and water, resulting in very high biomass production rates (Ghizelini et al. 2012). According to Alongi (2002), the microbial biomass in mangrove sediments is never greater than 1.2% of the total mass of detritus, and 91% of this biomass is composed of bacteria and fungi.

Unfortunately, according to Ghizelini et al. (2012), many Brazilian mangroves are contaminated, which can modify their hydrology, sediment composition, and nutrient dynamics (Lee et al. 2006). There are only a few studies about the microbial loop in Brazil, especially in the sediment of coastal and mangrove areas. Most studies were carried out in the Southeast coast, relating microbial densities to some anthropogenic impacts (e.g., sewage discharge and oil spills), population structure, grazing (e.g., nanoflagellates, viruses), and abiotic factors (Mesquita and Fernandes 1996; Fernandes and Mesquita 1999; Koch and Wolff 2002; Oliveira et al. 2007; Gomes et al. 2010; Souza et al. 2010; Pereira 2011; Moraes et al. 2014; Pereira 2014; Andrade 2015; Quintana et al. 2015; Santana et al. 2021; Tavares et al. 2021; Bezerra et al. 2022).

In recent years, with the development of pyrosequencing and new-generation sequencing techniques, in Brazil, the focus has been on the study of microbial diversity, as can be seen in the next session.

7.3 Microorganisms in Brazilian Mangrove Sediments

Mangroves are known to be rich in microbial diversity, which is fundamental to their homeostasis. According to Alongi (2002), about 91% of the mangrove microbial biomass corresponds to bacteria and fungi involved in nutrient transformation, biogeochemical cycles, and, for some specific groups such as cyanobacteria, in photosynthesis. Besides, they present a high biotechnological potential for being used in bioremediation and the production of certain substances.

These microorganisms not only contribute to high productivity but are also responsible for most of the carbon flow in the sediments, since much of the organic matter retained there undergoes degradation or chemical modification by the microorganisms (Kristensen et al. 1998).

Thus, due to the great importance of microorganisms mangrove ecosystems, knowing their diversity and the processes they carry out can contribute to the advancement of knowledge about the functioning of ecosystems, as well as to the development of protection measures and more effective public policies. In the next sections, we will explore a little more about the diversity of bacteria, archaea, fungi, and viruses in the Brazilian mangroves.

7.3.1 Bacterial and Archea Diversity in Brazilian Mangrove Sediments

Heterotrophic bacteria are 1–2-µm-sized species in the domain Bacteria and live by assimilating dissolved organic compounds from water or by degrading nonliving detrital organic matter (Fig. 7.3). Species in the domain Archaea are also widely present in both sea and freshwater habitats (Sherr and Sherr 2009); however, Archaea studies in Brazilian mangroves are rare.

Bacteria in mangrove sediments, including cyanobacteria, play an important role in the biogeochemical cycles of these coastal ecosystems. These organisms are directly involved in the transformation of nutrients, photosynthesis, nitrogen fixation, methanogenesis, phosphate solubility, sulfate reduction, and production of other substances, including antibiotics and enzymes (Santos et al. 2010). The analysis of prokaryotic communities is crucial to understand ecosystem functioning (Fernandes et al. 2010).

However, for many years, the development of knowledge about microbial diversity has been linked and limited to cultivation methods. The development of molecular techniques capable of identifying noncultivable organisms has considerably increased knowledge of biodiversity in general.

Investigations on marine sediment diversity based on molecular methods began in 1996 (Gray and Herwig 1996), followed by several studies in this field (Urakawa et al. 1999; Shao et al. 2004; Mussmann et al. 2005). Nevertheless, studies on mangrove sediments started with Liang et al. (2006), investigating bacterial diversity

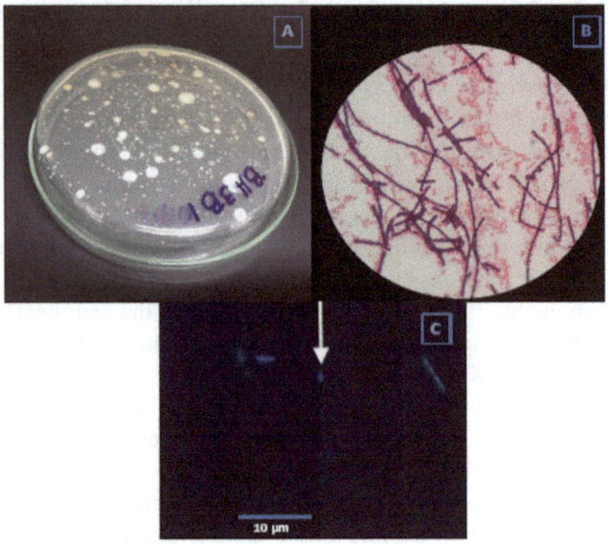

Fig. 7.3 Heterotrophic bacteria isolated from Brazilian mangrove sediments. (**a**) culture media; (**b**) gram; (**c**) epifluorescence microscopy

in subtropical mangroves showing the presence of many gene sequences derived from unknown taxa.

Numerous studies have demonstrated the uniqueness of coastal sediments concerning the composition of their microbiota (Gray and Herwig 1996; Bano et al. 1997; Ravenschlag et al. 1999; Urakawa et al. 1999; Todorov et al. 2000; Polymenakou et al. 2005; Gomes et al. 2010; Santos et al. 2011). However, there is still a gap in the knowledge of microbial biodiversity in mangrove sediments. In Brazil, the studies focus on hydrocarbon-contaminated environments and are mostly conducted in Southeast Brazil.

Dias et al. (2010) conducted a study on pristine mangrove sediments and demonstrated the predominance of the classes Alphaproteobacteria, Gammaproteobacteria, and Acidobacteria. Through molecular techniques (DGGE – Denaturing Gradient Gel Electrophoresis), the authors were able to recover a great diversity of microorganisms as well as to correlate this diversity with environmental factors.

Gomes et al. (2008) conducted a study on urban mangroves under strong anthropogenic pressure and demonstrated the dominance of Alteromonadales, Burkhholderiales, Pseudomonadales, Rhodobacteriales, and Rodocyclales. Members of these groups are involved in hydrocarbon degradation, thus justifying their predominance in human-impacted environments. Gomes et al. (2010) also demonstrated a limited diversity of Betaproteobacteria and Actinobacteria in mangrove sediments from impacted areas in Rio de Janeiro state.

Peixoto et al. (2011) demonstrated that the composition of the microbial community of mangrove sediments reflects the spatial variation of the present pollutants and that the microbial communities are heterogeneously distributed. In this study, Total Petroleum Hydrocarbon (TPH) levels were significantly associated with the presence of bacteria Betaproteobacteria, whereas anthracene and Polycyclic Aromatic Hydrocarbons (PAHs) levels were strongly associated with the presence of Actinobacteria.

Rocha et al. (2016), in a study conducted in the mangroves of Barra Grande, Ceará State, showed a high diversity of Gammaproteobacteria besides the record of three distinct community structures, which evidences the presence of different microhabitats within the same mangrove. Differences in the structure of microbial communities may be associated with the specificity of each site since mangroves are influenced by two distinct environments (marine and terrestrial).

The table below shows studies carried out in different Brazilian mangroves to investigate bacterial diversity in mangroves and estuarine systems using culture-independent methods (Table 7.1).

Brazil still needs many studies that cover all the mangroves in the different regions of the country, in order to fill the gaps in knowledge about the diversity of the microbiota in this ecosystem. Despite the scarcity of studies, some studies have already documented regions with highly diverse bacterial populations and with a very specific microbiome (Rocha et al. 2016; Zampieri et al. 2020). However, we still need to understand how the different types of contaminants influence the diversity and composition of the community, which groups resist these adverse conditions, and which groups have their population reduced.

7.3.2 Fungi Diversity in Brazilian Mangrove Sediments

Fungi are nonmotile eukaryotes, whose cell walls are usually made of polysaccharide chitin presented as microfibrillar bundles, for example, cellulose. Other glycans such as mannan, galactosan, and chitosan may replace chitin in some fungal cell walls (Madigan et al. 2008). These organisms are absorptive heterotrophs, and secrete exoenzymes into the environment, later absorbing the digested nutrients. Most are saprophytes that decompose dead organic matter, but some are parasites of plants and animals.

Fungi are informally divided into unicellular yeasts and filamentous molds based on their overall appearance. Dimorphic fungi have both mold and yeast life cycle stages (Fig. 7.4). Fungal life cycles are usually complex, involving both sexual and asexual forms of reproduction. They occur in diverse habitats; some are aquatic, but most inhabit terrestrial habitats, in soil, or dead plant matter, playing a crucial role in the mineralization of organic carbon.

Mangrove forests are among the most productive ecosystems in the world and are characterized by intense carbon processing (Kristensen et al. 1998). Fungi are thought to play an important role in organic matter decomposition pathways in this ecosystem (Jones 2000). Their diversity has been widely studied, especially on mangrove trees and decomposing litter (Ananda and Sridhar 2004).

Mangrove trees are interesting study objects for mycologists, because the bases of their trunks and the aerating roots are permanently or intermittently submerged, whereas the upper parts of roots and trunks are rarely reached by the saltwater. Thus,

Table 7.1 List of studies using culture-independent methods to access bacteria diversity in mangroves and estuarine systems

Study area	State	Method	Sample	References
Mangrove	RJ	Sequencing of 16S rRNA gene and T-RFLP	Sediment	Brito et al. (2006)
Mangrove	CE	PCR analysis of 16S rRNA gene and DGGE	Water	Sousa et al. (2006)
Estuary of Guanabara Bay	RJ	16S rRNA PCR, DGGE, and clones library	Water	Vieira et al. (2007)
Mangrove at Guanabara Bay	RJ	16S rRNA PCR, DGGE and clones library	Sediment	Gomes et al. (2008)
Mangrove at Guanabara Bay	RJ	16S rRNA PCR, DGGE and clones library	Sediment	Gomes et al. (2010)
Mangrove of Cardoso Island	SP	Denaturing gradient gel electrophoresis (DGGE) and analysis of 166 sequences from a clone library	Sediment	Dias et al. (2010)
Mangrove of Todos os Santos Bay	BA	Denaturing gradient gel electrophoresis (DGGE)	Sediment	Peixoto et al. (2011)
Mangrove of Cananéia and Bertioga	SP	Pyrosequencing of 16S rRNA gene	Sediment	Andreote et al. (2012)
São Sebastião Channel and Búzios Island	SP, RJ	16S library and pyrosequencing	Water and sediment	Carlos et al. (2013)
Cardoso Island	SP	T-RFLP and pyrosequencing of 16S rRNA gene	Sediment	Mendes and Tsai (2014)
Barra Grande mangrove	CE	T-RFLP	Sediment	Rocha et al. (2016)
Camamu Bay	BA	MiSeq sequencing	Water	Affe et al. (2018)
São Sebastião Channel and Santos Bay	SP	MiSeq sequencing	Sediment	Zampieri et al. (2020)

Modified from Bruce et al. (2012). Brazilian states: Ceará (CE), Bahia (BA), Rio de Janeiro (RJ), São Paulo (SP). See Chap. 3, Maps 5, 11, 13, and 14, respectively

terrestrial fungi and lichens occupy the upper half of the trees and marine species occupy the lower. At the interface, there is an overlap between marine and terrestrial fungi (Kohlmeyer 1969).

According to Sridhar (2013), the first monograph on marine mycology "A Treatise on Fungi in Oceans and Estuaries" was published by Johnson and Sparrow in 1961. In 1975, Hughes summarized the studies on marine mycology up to that year. Kohlmeyer and Kohlmeyer (1979) listed 209 filamentous fungi in their taxonomic and ecological monograph (Marine Mycology: The Higher Fungi), subsequently elevating them to 321 species (Kohlmeyer and Volkmann-Kohlmeyer

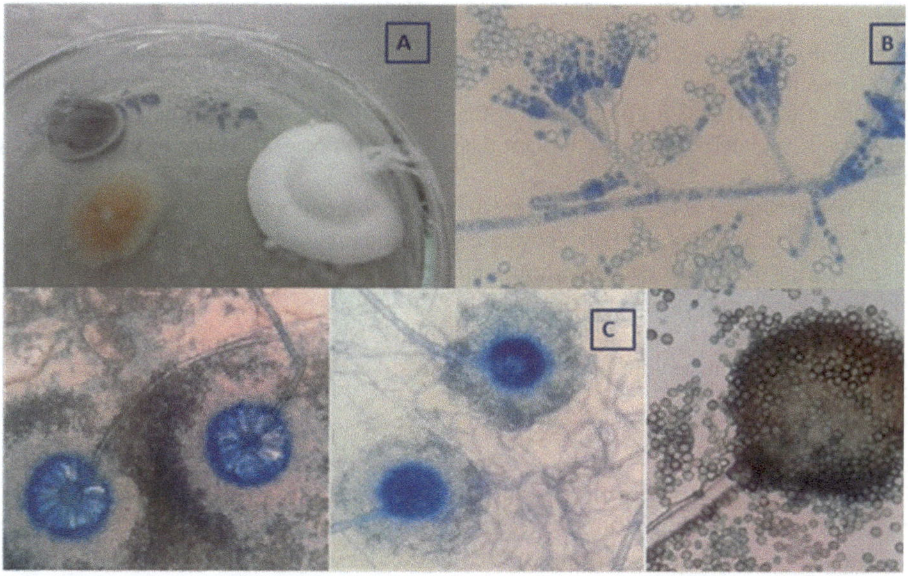

Fig. 7.4 Fungi isolated from Brazilian mangrove sediments. (**a**) Culture showing colonies of *Aspergillus* (gray and orange) and *Penicillum* (white); (**b**) *Penicilum* conidia; and (**c**) *Aspergillus* conidia

1991), until Hyde and Pointing (2000) reached the mark of 444 species. The literature review by Schmit and Shearer (2003) reveals the occurrence of 625 fungal species from different mangrove habitats. About 200 species have been considered as obligate marine fungi, whereas 131 species that occur in mangrove sediment or peat are not restricted to the marine environment.

In Brazil, there is little knowledge of mangrove fungi in comparison to what is known about bacteria. The molecular/culture-independent methods that have been widely used to study bacteria have not yet been reported on the same scale for fungi. Some studies, like Soares et al. (1997), found ascomycetes, basidiomycetes, zygomycetes, and mitosporic in the mangroves. More often, studies try to find a biotechnological application to fungi, focusing on species such as *Cyclothyrium* spp., *Penicillium* spp., and *Psilocybe* spp., isolated from the Santos' estuarine sediment and showing the ability to degrade phenanthrene and pyrene. A study by Gomes et al. (2011) isolated and identified filamentous fungi from mangrove sediments of Pernambuco, in Northeast Brazil. They found *Penicillium* and *Aspergillus* as the dominant genera, followed by *Trichoderma*, *Fusarium*, *Phoma*, *Talaromyces*, *Cladosporium*, *Eupenicillium*, *Gongronella*, *Microsphaeropsis*, *Mucor*, *Stibella*, and *Thielavia*.

Sebastianes et al. (2013) showed through molecular identification that the fungal community associated with some mangroves in São Paulo State is composed of at least 34 different genera, the most frequent being *Diaporthe, Colletotrichum,*

Fusarium, *Trichoderma*, and *Xylaria*. The results indicated that the mangrove fungal community possesses a vast diversity and richness of endophytic fungi. The data also revealed a large reservoir of fungal genetic diversity inhabiting these Brazilian mangrove forests and highlighted substantial differences between the fungal communities associated with distinct plant tissues, plant species, level of human impact, and seasonality.

A few Brazilian studies have recently focused on endophytic fungi (Costa et al. 2012; Sebastianes et al. 2013). These are microorganisms that invade the tissues of plants during their life cycle without causing disease symptoms, and many show biotechnological purposes, producing a high diversity of substances with potential medical, agricultural, and industrial applications. All the above-mentioned studies revealed a large reservoir of fungal genetic diversity inhabiting these Brazilian mangrove forests.

Ghizelini et al. (2012) emphasized the existence of few publications about fungal communities in mangrove sediments. There are many reasons for this, including the fact that short fragments of ribosomal DNA, which are frequently used to describe bacterial diversity, are not suitable to adequately describe fungal diversity. This fact shows that there is a real need to refine and establish specific molecular markers for fungal taxonomy. This is the reason why even today there is a certain difficulty in accessing fungal communities in mangroves, despite the progress that has already been made. In addition to that, the diversity of environments in which Brazilian mangroves can be found hinders generalization. Each of them possesses specificities and different degrees of conservation. This way, much effort is still needed to better characterize fungal diversity in this unique and extensive area, such as the study by Ghizelini et al. (2019) on the effects of a massive oil spill that affected most of the northeastern Brazilian coast on mangrove sediment fungal diversity.

7.3.3 Virus Diversity in Brazilian Mangrove Sediments

A virus is a small parasite that cannot reproduce by itself. Once it infects a susceptible cell, however, a virus can adjust the cell machinery to produce more viruses. Most viruses have either RNA or DNA as genetic material. The nucleic acid may be single- or double-stranded. The entire infectious virus particle, called a virion, consists of nucleic acid and an outer protein shell. The simplest viruses contain only enough RNA or DNA to encode four protein types. The most complex can encode 100–200 proteins (Lodish et al. 2000).

Viruses infect numerous microorganisms including, predominantly, bacteria (bacteriophages or phages) but also archaea, protists, and fungi. They are the most abundant and ubiquitous biological entities on Earth and are important drivers of ecosystem functioning. Little is known, however, about the vast majority of these viruses that live off microorganisms (Allen and Abedon 2014). Studies show that viruses present incredible abundance. The estimated overall abundance of marine viruses in the world's oceans is on the order of 10^{30} (Suttle 2005, 2007), a value that exceeds by ten times the abundance of prokaryotes (Suttle 2005).

Viruses are no longer viewed as static but instead as dynamic players within the microbial ecology of the oceans. Through cell lysis, viruses return photosynthesis-derived carbon to the ocean. Some studies show that viruses may be responsible for 50% of all bacterial mortality (Fuhrman and Noble 1995). The particles also have a function in the control of bacterial blooms and in altering the genetic composition of their host. More research is required to fully understand the processes occurring in the microbial loop. However, it is clear that viruses play a central role in this loop, and that many questions remain to be answered.

According to Weitz and Wilhelm (2012), the study of viruses in the oceans has been key to many discoveries, like the finding that viruses have evolved novel lineages of key photosynthetic genes, as seen in cyanophages infecting *Synechococcus* and *Prochlorococcus* (Mann et al. 2003). Considering so much new information about viruses, the identification of many novel viral families defies much of the conventional knowledge about the viral life history, for example, the discovery of "giant" algal-infecting viruses (Derelle et al. 2008), and the discovery of viruses that exploit other viruses (La Scola et al. 2008).

Although no specific studies show the role of viruses in mangroves areas, especially in Brazil, one can infer that the same findings for the marine environment would take place in the mangrove environment. Thus, there are still gaps in the knowledge on virus diversity in mangroves all over the world.

7.4 Final Remarks

The field of marine microbial ecology is just at the beginning of its specific technological evolution. Such advances in technologies include direct-imaging methods, sequencing technologies, and bioinformatics that have revealed a previously unknown world of viral diversity in natural environments. However, microbial activities in the mangrove ecosystem are highly complex, especially the microbial loop itself. For this reason, there is a need for more studies about the structure of microbial communities, their function and impact on productivity, potentially changing paradigms in the management and conservation strategies of mangroves in the country.

References

Affe HMJ, Rigonato J, Nunes JMC, Menezes M (2018) Metagenomic analysis of cyanobacteria in an oligotrophic tropical estuary, South Atlantic. Front Microbiol 9:1393

Allen H, Abedon S (2014) Virus ecology and disturbances: impact of environmental disruption on the viruses of microorganisms. Front Microbiol 5:700

Alongi D (1996) The dynamics of benthic nutrient pools and fluxes in tropical mangrove forests. J Mar Res 54(1):123–148

Alongi D (2002) Present state and future of the world's mangrove forests. Environ Conserv 29(3): 331–349

Ananda K, Sridhar K (2004) Diversity of filamentous fungi on decomposing leaf and woody litter of mangrove forests of southwest coast of India. Curr Sci India 87:1431–1437

Anderson T, Ducklow H (2001) Microbial loop carbon cycling in ocean environments studied using a simple steady-state model. Aquat Microb Ecol 26:37–49

Andrade V (2015) Estrutura e Funcionamento da Alça Microbiana nos Sedimentos da Baía do Araçá. Master thesis, Universidade Estadual Paulista

Andreote F, Jimenez D, Chaves D, Dias A, Luvizotto D et al (2012) The microbiome of Brazilian mangrove sediments as revealed by metagenomics. PLoS One 7(6):e38600

Azam F (1998) Microbial control of oceanic carbon flux: the plot thickens. Science 280(5364): 694–696

Azam F, Fenchel T, Field J, Gray J, Meyer-Reil L, Thingstad F (1983) The ecological role of water-column microbes in the Sea. Mar Ecol Prog Ser 10:257–263

Bano N, Nisa M, Khan N, Saleem M, Harrison P, Ahmed S, Azam F (1997) Significance of bacteria in the flux of organic matter in the tidal creeks of the mangrove ecosystem of the Indus River Delta, Pakistan. Mar Ecol Prog Ser 157:1–12

Bezerra WM, Tavares TCL, Nogueira VLR, Normando LRO, Bomfim TA, Angelim AL, Melo VMM (2022) Bacteriome associated with *Rhizophora mangle* sediments within Brazil semi-arid mangroves. Arq Ciên Mar 55:34–51

Bratbak G, Heldal M, Thingstad T, Riemann B, Haslund O (1992) Incorporation of viruses into the budget of microbial C-transfer A first approach. Mar Ecol Prog Ser 83:273–280

Bratbak G, Thingstad F, Heldal M (1994) Viruses and the microbial loop. Microb Ecol 28(2): 209–221

Brito E, Guyoneaud R, Goñi-Urriza M, Ranchou-Peyruse A, Verbaere A et al (2006) Characterization of hydrocarbonoclastic bacterial communities from mangrove sediments in Guanabara Bay, Brazil. Res Microbiol 157:752–762

Bruce T, De Castro A, Kruger R, Thompson C, Thompson F (2012) Microbial diversity of Brazilian biomes. In: Nelson KE, Jones-Nelson B (eds) Genomics applications for the developing world. Springer, Berlin, pp 217–247

Carlos C, Torres T, Ottoboni L (2013) Bacterial communities and species-specific associations with the mucus of Brazilian coral species. Sci Rep 3:1624

Caron D, Lim E, Miceli G, Waterbury J, Valois F (1991) Grazing and utilization of chroococcoid cyanobacteria and heterotrophic bacteria by protozoa in laboratory cultures and a coastal plankton community. Mar Ecol Prog Ser 76:205–217

Costa I, Maia L, Cavalcanti M (2012) Diversity of leaf endophytic fungi in mangrove plants of Northeast Brazil. Braz J Microbiol 43(3):1165–1173

Danovaro R, Marrale D, Dell'Anno A, Croce ND, Tselepides A, Fabiano M (2000) Bacterial response to seasonal changes in labile organic matter composition on the continental shelf and bathyal sediments of the Cretan Sea. Prog Oceanogr 46(2–4):345–366

Demain A (2007) The business of biotechnology. Ind Biotechnol 3:269–283

Deming J, Baross J (1993) The early diagenesis of organic matter: bacterial activity. In: Engel MH, Macko SA (eds) Organic geochemistry. Plenum Press, New York, pp 119–144

Derelle E, Ferraz C, Escande M, Eychenie S, Cooke R, Piganeau G, Desdevises Y, Bellec L, Moreau H, Grimsley N (2008) Life-cycle and genome of OtV5, a large DNA virus of the pelagic marine unicellular green alga *Ostreococcus tauri*. PLoS One 5:2250–2251

Dias A, Andreote F, Rigonato J, Fiore M, Melo I, Araújo W (2010) The bacterial diversity in Brazilian non-disturbed mangrove sediment. Antonie van Leeuw J Microb 98:541–551

Fenchel T (2008) The microbial loop – 25 years later. J Exp Mar Biol Ecol 366:99–103

Fernandes A, Mesquita H (1999) An experimental study of nanoflagellate bacterivory. Rev Microbiol 30:365–372

Fernandes S, Bonin P, Michotey V, Lokabharathi P (2010) Denitrification: an important pathway for nitrous oxide production in tropical mangrove sediments (Goa, India). J Environ Qual 39: 1507–1516

Fuhrman J, Noble P (1995) Viruses and Protists cause similar bacterial mortality in Coastal Sea waters. Limnol Oceanogr 7:1236–1242

Ghizelini AM, Mendonça-Hagler L, Macrae A (2012) Microbial diversity. Braz J Microbiol 43 (4):1242–1254

Ghizelini AM, Martins KG, Gießelmann UC, Santoro E, Pasqualette L, Mendonça-Hagler LCS, Rosado AS, Macrae A (2019) Fungal communities in oil contaminated mangrove sediments – who is in the mud? J Mar Pollut Bull 139:181–188

Gomes NCM, Borges LR, Paranhos R, Pinto FN, Mendonça-Hagler LCS, Smalla K (2008) Exploring the diversity of bacterial communities in sediments of urban mangrove forests. FEMS Microbiol Ecol 66:96–109

Gomes NCM, Flocco C, Costa R, Junca H, Vilchez R, Pieper D, Krögerrecklenfort E, Paranhos R, Mendonça-Hagler LCS, Smalla K (2010) Mangrove microniches determine the structural and functional diversity of enriched petroleum hydrocarbon-degrading consortia. FEMS Microbiol Ecol 74(2):276–290

Gomes NCM, Cleary DFR, Calado R, Costa R (2011) Mangrove bacterial richness. Integr Biol 4(4):419–423

González J (1993) Grazing by marine nanoflagellates on virus and virus-sized particles: ingestion and digestion. Mar Ecol Prog Ser 94:1–10

Gray J, Herwig R (1996) Phylogenetic analysis of the bacterial communities in marine sediments. Appl Environ Microbiol 62:4049–4059

Hemminga M, Slim F, Kazungu J, Ganssen G, Nieuwenhuze J, Kruyt N (1994) Carbon outwelling from a mangrove forest with adjacent seagrass beds and coral reefs (Gazi Bay, Kenya). Mar Ecol Prog Ser 106:291–301

Holguin G, Vazquez P, Bashan Y (2001) The role of sediment microorganisms in the productivity, conservation, and rehabilitation of mangrove ecosystems; an overview. Biol Fertil Soils 33:265–278

Hughes G (1975) Studies of fungi in oceans and estuaries since 1961. I. Lignicolous, caulicolous and foliicolous species. Oceanogr Mar Biol 13:69–180

Hyde K, Pointing S (2000) Marine mycology – a practical approach. Fungal Diversity Press, Hong Kong

Jones E (2000) Marine fungi: some factors influencing biodiversity. Fungal Divers 4:53–73

Koch V, Wolff M (2002) Energy budget and ecological role of mangrove epibenthos in the Caeté estuary, North Brazil. Mar Ecol Prog Ser 228:119–130

Kohlmeyer J (1969) Ecological notes on fungi in mangrove forests. Trans Br Mycol Soc 53:237–250

Kohlmeyer J, Kohlmeyer E (1979) Marine mycology: the higher fungi, 1st edn. Academic, New York. 690 pp

Kohlmeyer J, Volkmann-Kohlmeyer B (1991) Illustrated key to the filamentous higher marine fungi. Bot Mar 34:1–61

Köster M, Meyer-Reil L (2001) Characterization of carbon and microbial biomass pools in shallow water coastal sediments of the southern Baltic Sea (Nordrügensche Bodden). Mar Ecol Prog Ser 214:25–41

Krambeck C, Krambeck H, Overbeck J (1981) Microcomputer-assisted biomass determination of plankton bacteria on scanning electron micrographs. Appl Environ Microbiol 42:142–149

Kristensen E, Jensen M, Banta G (1998) Transformation and transport of inorganic nitrogen in sediments of a southeast Asian mangrove forest. Aquat Microb Ecol 15(2):165–175

La Scola B, Desnues C, Pagnier I, Robert C, Barrassi L, Fournous G, Merchant M, Suza-Monti M, Forterre P, Koonin E, Raoult D (2008) The virophage as a unique parasite of the giant mimivirus. Nature 455(7209):100–165

Lee S, Dunn R, Young R, Connolly R, Dale P, Dehayr R, Lemckert C, McKinnon S, Powell B, Teasdale P, Welsh D (2006) Impact of urbanization on coastal wetland structure and function. Austral Ecol 31:149–163

Liang J, Chen Y, Lan C, Tam N, Zan Q, Huang L (2006) Recovery of novel bacterial diversity from mangrove sediment. Mar Biol 150:739–747

Lodish H, Berk A, Zipursky S, Matsudaira P, Baltimore D, Darnell J (2000) Molecular cell biology, 4th edn. WH Freeman, New York

Luna B, Townsend M (2007) Tipranavir: the first nonpeptidic protease inhibitor for the treatment of protease resistance. Clin Ther 11:2309–2318

Madigan MT, Martinko JM, Dunlap PV, Clark DP (2008) Brock biology of microorganisms, 12nd edn. Pearson Benjamin Cummings, San Francisco

Mann N, Cook A, Millard A, Bailey S, Clokie M (2003) Marine ecosystems: bacterial photosynthesis genes in a virus. Nature 424:6950

Matondkar S, Mahtani S, Mavinkurve S (1981) Studies on mangrove swamps of Goa. I. Heterotrophic bacterial flora from mangrove swamps. Mahasagar B Natl Inst Ocean 14:325–327

Mendes L, Tsai S (2014) Variations of bacterial community structure and composition in mangrove sediment at different depths in southeastern Brazil. Diversity 6:827–843

Mesquita H, Fernandes A (1996) Variação de curta escala temporal de bactérias, plcofltoplâncton e nanoheterótrofos na região de Ubatuba – SP, Brasil. Rev Bras Oceanogr 44(1):47–56

Moraes P, Franco D, Pellizari V, Yukio P, Sumida G (2014) Effect of plankton-derived organicmatter on themicrobial community of coastal marine sediments. J Exp Mar Biol Ecol 461:257–266

Moreno M, Ferrero T, Granelli V, Marin V, Albertelli G, Fabiano M (2006) Across shore variability and trophodynamic features of meiofauna in a microtidal beach of the NW Mediterranean. Estuar Coast Shelf Sci 66:357–367

Mussmann M, Ishii K, Rabus R, Amann R (2005) Diversity and vertical distribution of cultured and uncultured Deltaproteobacteria in an intertidal mud flat of the Wadden Sea. Environ Microbiol 7:405–418

Oliveira A (2000) Regeneração de nutrientes por alguns componentes da alça microbiana. PhD Dissertation, Universidade de São Paulo

Oliveira A, Hollnagel H, Mesquita H, Fontes R (2007) Physical, chemical and microbiological characterization of the intertidal sediments of Perequê Beach, Guarujá (SP), Brazil. Mar Pollut Bull 54:921–927

Peixoto R, Chaer GM, Carmo FL, Araújo FV, Paes JE, Volpon A, Santiago GA, Rosado AS (2011) Bacterial communities reflect the spatial variation in pollutant levels in Brazilian mangrove sediment. Antonie van Leeuw J Microb 99(2):341–354

Pereira C (2011) Densidade de Bactérias Heterotróficas, Cianobactérias e Nanoflagelados Heterótrofos em dois Estuários do Estado de São Paulo, Universidade Estadual Paulista, São Vicente, 29 pp

Pereira C (2014) Efeitos da contaminação fecal sobre a dinâmica e a estrutura da alça microbiana em areias de praias do litoral norte do Estado de São Paulo. Master thesis 79 pp, Universidade Estadual Paulista

Peroni C, Rossi G (1986) Determination of microbial activity in marine sediments by resazurin reduction. J Chem Ecol 2:205–209

Polymenakou P, Bertilsson S, Tselepides A, Stephanou E (2005) Bacterial community composition in different sediments from the Eastern Mediterranean Sea: a comparison of four 16S ribosomal DNA clone libraries. Microb Ecol 50:447–462

Pomeroy L (1974) The ocean's food web, a changing paradigm. Bioscience 24:499–504

Pomeroy L, Williams P, Azam F, Hobbie J (2007) The microbial loop. Oceanography 20(2):28–33

Proctor L, Fuhrman J (1990) Viral mortality of marine bacteria and cyanobacteria. Nature 343:60–62

Quintana C, de Moraes P, Yoshinaga M, Wakeham S, Sumida P (2015) Microbial biomass response to different quantities and sources of organic matter in Brazilian coastal sediments. Mar Ecol 36(3):766–779

Ravenschlag K, Sahm K, Pernthaler J, Amann R (1999) High bacterial diversity in permanently cold marine sediments. Appl Environ Microbiol 65:3982–3989

Rocha L, Colares G, Nogueira V, Paes F, Melo V (2016) Distinct habitats select particular bacterial communities in mangrove sediments. Int J Med Microbiol 2016:3435809

Santana CO, Spealman P, Melo VMM, Gresham D, Jesus TB, Chinalia FZ (2021) Effects of tidal influence on the structure and function of prokaryotic communities in the sediments of a pristine Brazilian mangrove. Biogeosciences 18:2259–2273

Santos H, Carmo F, Paes J, Rosado A, Peixoto R (2010) Bioremediation of mangroves impacted by petroleum. Water Air Soil Pollut 216:329–350

Santos H, Cury J, Carmo F, Santos A, Tiedje J, van Elsas J (2011) Mangrove bacterial diversity and the impact of oil contamination revealed by pyrosequencing: bacterial proxies for oil pollution. PLoS One 6:1–8

Schmit J, Shearer C (2003) A checklist of mangrove-associated fungi, their geographical distribution and known host plants. Mycotaxon 53:423–477

Sebastianes F, Romão-Dumaresq A, Lacava P, Harakava R, Azevedo J, Melo I, Pizzirane-Kleiner A (2013) Species diversity of culturable endophytic fungi from Brazilian mangrove forests. Curr Genet 59:153–166

Shao P, Chen Y, Zhou H, Qu L, Ma Y, Li H, Jiao N (2004) Phylogenetic diversity of Archae in prawn farm sediment. Mar Biol 149:133–142

Sherr E, Sherr B (2009) Encyclopedia of microbiology, 3rd edn. Academic, San Diego. 4600 pp

Sjöling S, Mohammed S, Lyimo T, Kyaruzi J (2005) Benthic bacterial diversity and nutrient processes in mangroves: impact of deforestation. Estuar Coast Shelf Sci 63:397–406

Soares CAG, Maury M, Pagnocca FC, Araújo FV, Mendonça-Hagler LC, Hagler AN (1997) Ascomycetous yeasts from tropical intertidal dark mud of southeast Brazilian estuaries. J Gen Appl Microbiol 43:265–272

Sousa OV, Macrae A, Menezes FGR, Gomes NCM, Vieira RHSF, Mendonça-Hagler LCS (2006) The impact of shrimp farming effluent on bacterial communities in mangrove waters, Ceara, Brazil. Mar Pollut Bull 52(12):1725–1734

Souza V, Laut L, Silva F, Figueiredo J, Vital H, Frazão E (2010) Benthic foraminifera and bacterial activity as a proxy for environmental characterization in Potengi Estuary, Rio Grande do Norte, Brazil. Anu Inst Geocienc UFRJ 33(1):20–34

Sridhar K (2013) Mangrove fungal diversity of west coast of India. In: Bhatt JR, Ramakrishna M, Remadevi OK, Nilaratna BP, Venkataraman K (eds) Mangroves of India: their biology and uses. Zoological Survey of India, Kolkata, pp 161–182

Steele JH (1974) The structure of marine ecosystems. Harvard University Press, Cambridge, 128 pp

Suttle C (2005) Viruses in the sea. Nature 437:356–361

Suttle C (2007) Marine viruses – major players in the global ecosystem. Nat Rev Microbiol 5(10): 801–812

Tavares TCL, Bezerra WM, Normando LRO, Rosado AS, Melo VMM (2021) Brazilian semi-arid mangroves-associated microbiome as pools of richness and complexity in a changing world. Front Microbiol 12:715991

Thingstad T, Heldal M, Bratbak G, Dundas I (1993) Are viruses important partners in pelagic fend webs? Trends Ecol Evol 8(6):209–213

Todorov J, Chistoserdov A, Aller J (2000) Molecular analysis of microbial communities in mobile deltaic muds of Southeastern Papua New Guinea. FEMS Microbiol Ecol 33:147–155

Urakawa H, Kita-Tsukamoto K, Ohwada K (1999) Microbial diversity in marine sediments from Sagami Bay and Tokyo Bay, Japan, as determined by 16S rRNA gene analysis. Microbiology 145:3305–3315

van Dommelen S, van der Meel R, van Solinge W, Coimbra M, Vader P, Schiffelers R (2016) Cetuximab treatment alters the content of extracellular vesicles released from tumor cells. Nanomedicine (UK) 11(8):881–890

Vieira R, Clementino M, Cardoso A, Oliveira D, Albano R et al (2007) Archaeal communities in a tropical estuarine ecosystem: Guanabara Bay, Brazil. Microb Ecol 54:460–468

Weitz J, Wilhelm S (2012) Ocean viruses and their effects on microbial communities and biogeochemical cycles. F1000 Biol Rep 4:17

Zampieri B, Andrade V, Chinellato R, Garcia A, Oliveira M, Brucha G, Oliveira A (2020) Heavy metal concentrations in Brazilian port areas and their relationships with microorganisms: can pollution in these areas change the microbial community? Environ Monit Assess 192:512

Chapter 8
Mangrove Macroalgal Communities

Nair S. Yokoya, Franciane M. Pellizzari, Rafael de Felício, Lorene Armstrong, Hosana M. Debonsi, Silvia Maria P. B. Guimarães, and Mutue T. Fujii

8.1 Introduction

Despite the high variation in abiotic parameters in estuaries, mangroves harbor a great diversity of organisms, from primary producers to consumers in high trophic levels. This ecosystem's macroalgal community shows a characteristic low species diversity (Oliveira 1984), growing on pneumatophores, seedlings, saplings, roots, and in the lower trunks of mangrove trees, including *Rhizophora mangle* L., *Avicennia schaueriana* Stapf & Leechmann, and *Laguncularia racemosa* (L.) Gaertner (Cordeiro-Marino et al. 1992).

Mangrove macroalgae are essential components of coastal ecosystems, providing several ecological services, such as nutrient retention, habitat provision, breeding and spawning grounds, and food source to diverse invertebrates and larvae of marine and terrestrial organisms. These will then serve as food for juvenile stages of fishes

N. S. Yokoya (✉) · S. M. P. B. Guimarães · M. T. Fujii
Centro de Conservação da Biodiversidade, Instituto de Pesquisas Ambientais, São Paulo, SP, Brazil

F. M. Pellizzari
Laboratório de Ficologia e Qualidade de Água Marinha, Universidade Federal do Paraná, campus Paranaguá, Paranaguá, PR, Brazil

R. de Felício
Laboratório Nacional de Biociências, Centro Nacional de Pesquisa em Energia e Materiais, Campinas, SP, Brazil

L. Armstrong
Departamento de Ciências Farmacêuticas, Universidade Estadual de Ponta Grossa, Ponta Grossa, PR, Brazil

H. M. Debonsi
Faculdade de Ciências Farmacêuticas de Ribeirão Preto, Universidade de São Paulo, Ribeirão Preto, SP, Brazil

Y. Schaeffer-Novelli et al. (eds.), *Brazilian Mangroves and Salt Marshes*, Brazilian Marine Biodiversity, https://doi.org/10.1007/978-3-031-13486-9_8

and invertebrates of commercial importance. Therefore, mangrove macroalgae have a primordial role in the maintenance of life and contribute significantly to the conservation of coastal and marine biodiversity.

8.2 Species Composition

The macroalgal community growing on the several microhabitats in the mangrove trees (Figs. 8.1–8.4) is dominated by a few genera, mainly *Bostrychia* Mont., *Caloglossa* (Harv.) G. Martens, and *Catenella* Grev. This assemblage is widely known as "Bostrychietum" (sensu Post 1968). Although composed of a few red algae, the delimitation of these species within each genus is quite complicated due to the wide morphological plasticity.

In Brazil, before Fontes (2012), Kano (2015), and Sena (2016), the identification of the mangrove macroalgae was primarily based on morphological features. The identification of specimens based solely on morphological characters may result in either underestimation or overestimation of diversity, since the phenotypic plasticity

Figs. 8.1–8.4 Mangrove macroalgal community. **8.1–8.2**: Macroalgal assemblage growing on the pneumatophores, rhizosphere, and on the trunks of mangrove trees at Rio Escuro, Ubatuba, São Paulo State (see Chap. 1 for location) (Bostrychietum *sensu* Post 1968). **8.3–8.4**: Details of the Bostrychietum (scale = 5 mm)

Table 8.1 Macroalgal species reported to the Brazilian mangroves

RHODOPHYTA
Ceramiales
Delesseriaceae
Caloglossa apomeiotica J.A. West and Zuccarello
C. confusa Krayesky, J.A. West and Kamiya
C. kamiyana Freshwater, Cath.E. Miller and Frankovich[a]
C. leprieurii (Mont.) G. Martens
C. rotundata Kamiya
Rhodomelaceae
Bostrychia binderi Harvey
B. calliptera (Mont.) Mont.
B. kelanensis Grunow ex E. Post
B. montagnei Harv.
B. moritziana (Sond. ex Kütz.) J. Agardh
B. pilulifera Mont.
B. radicans (Mont.) Mont.
B. tenella (J.V. Lamour.) J. Agardh
Dawsoniocolax bostrychiae (A.B.Joly & Yam.-Tomita) A.B.Joly & Yam.-Tomita
Gigartinales
Caulacanthaceae
Catenella caespitosa (Wither.) L.M. Irvine
CHLOROPHYTA
Ulotrichales
Gayraliaceae
Gayralia brasiliensis Pellizzari, M.C. Oliveira and N.S. Yokoya
G. oxysperma (Kütz.) K.L. Vinogr. ex Scagel et al.
Bryopsidales
Udoteaceae
Boodleopsis pusilla (Collins) W.R. Taylor, A.B. Joly and Bernat.
B. vaucherioidea Calderón-Saenz and Schnetter
Cladophorales
Boodleaceae
Cladophoropsis membranacea (C. Agardh) Børgesen
Cladophoraceae
Pseudorhizoclonium africanum (Kütz.) Boedeker
Rhizoclonium riparium (Roth) Kütz. ex Harv.

Sources: Paula et al. (1989), Fujii et al. (1990), Fontes (2012), Pellizzari et al. (2013), Sena (2016), Fontes et al. (2016), Kano et al. (2017), Wynne (2017), and Freshwater et al. (2021)

[a]*Caloglossa kamiyana* from Brazilian mangroves was previously identified as *C. ogasawaraensis* Okamura

of some taxa may lead to misidentifications (Krayesky et al. 2012). Sena (2016) and Kano et al. (2017) have used molecular approaches based on the 5' region of the cox1 gene that encodes cytochrome c oxidase subunit 1 (COI-5P), as proposed by Saunders (2005); the universal plastid amplicon (UPA), as proposed for DNA barcoding in algae (Presting 2006; Sherwood and Presting 2007), and the gene encoding for the large subunit of RuBisCO (rbcL), which was tested as a DNA barcoding tool (Saunders and Kucera 2010).

The current estimation of the most common macroalgae in Brazilian mangroves points to 22 species (Table 8.1), with fifteen Rhodophyta and seven Chlorophyta

(Hadlich and Bouzon 1985; Paula et al. 1989; Fujii et al. 1990; Fontes 2012; Pellizzari et al. 2013; Fontes et al. 2016; Sena 2016; Kano et al. 2017; Wynne 2017; Freshwater et al. 2021). In some of these studies, the authors revealed that many taxa are genetically more diverse than what the morphology suggests. Species among the genus *Caloglossa* are specifically challenging to identify because of their widespread phenotypic plasticity (Kamiya et al. 1999, 2016; Krayesky et al. 2011, 2012; Kano et al. 2017; Wynne 2017; Freshwater et al. 2021). Some red algae have a more specific occurrence in Brazilian mangroves such as *Murrayella periclados* (C. Agardh) F. Schmitz, *Polyshiphonia subtilissima* Montagne, *P. tepida*, and *P. howei* (Cunha et al. 1999, Cutrim et al. 2004, Fontes et al. 2016, Farraboti 2018). In Brazil, the two last species are currently named *Vertebrata foetidissima* (Cocks ex Bornet) Díaz-Tapia and Maggs, and *Wilsonosiphonia fujiae* D. Bustamante, Won, and T.O. Cho, respectively. Few green algae had also occasional occurrences, as *Caulerpa fastigiata* Montagne (Fontes et al. 2007), *Cladophora vagabunda* (L.) Hoek, *Ulva flexuosa* Wulfen and *Chaetomorpha* sp. (Machado and Nassar 2007).

The most diverse genus is *Bostrychia*, comprising eight species widely distributed in mangroves (Menezes et al. 2015). The species of *Bostrychia* present morphological differences that allow their identification (Figs. 8.5–8.10). The thalli of *Bostrychia* are polysiphonous, typical of Rhodomelaceae, and the species distinction is based on the branching pattern, presence or absence of cortication, degree of polysiphonous branches, and the type of haptera (King and Puttock 1989).

Two types of haptera are described in *Bostrychia* species: Cladohapteron and peripherohapteron. The first is originated from pericentral and axial cells, like a first main branching with negative geotropism to attach to the substrate. Cladohapteron is found in *B. kelanensis*, *B. moritziana, B. pilulifera*, and *B. radicans*. On the other hand, peripherohapteron is originated from pericentral and cortical cells in the ventral side of the stoloniferous branches, and it can be found in *B. binderi*, *B. calliptera, B. montagnei,* and *B. tenella.*

Molecular and phylogenetic data have provided advances for the knowledge of the taxonomy of the genus *Bostrychia*. An important example is the *B. tenella* complex, in which *B. binderi* was synonymized to *B. tenella*, since the morphological characters to distinguish both species overlapped (King et al. 1988; King and Puttock 1989). However, *B. binderi* was resurrected from *B. tenella* species complex based on molecular and morphological data of specimens collected worldwide (Zuccarello et al. 2015).

Dawsoniocolax bostrychiae is a Rhodomelacean adelphoparasite originally described growing on *Bostrychia radicans* (Figs. 8.11–8.12), and later also on *B. moritziana*. The thallus is small shaped with whitish warts. Guimarães (1993) described reproductive structures of female and male gametophytes as tetrasporophytes.

Among the red algal species growing on mangroves, the only one not belonging to Ceramiales is *Catenella caespitosa* (Caulacanthaceae, Gigartinales). This species is easily recognized by its cartilaginous, segmented sausage-shaped thallus (Figs. 8.13–8.14).

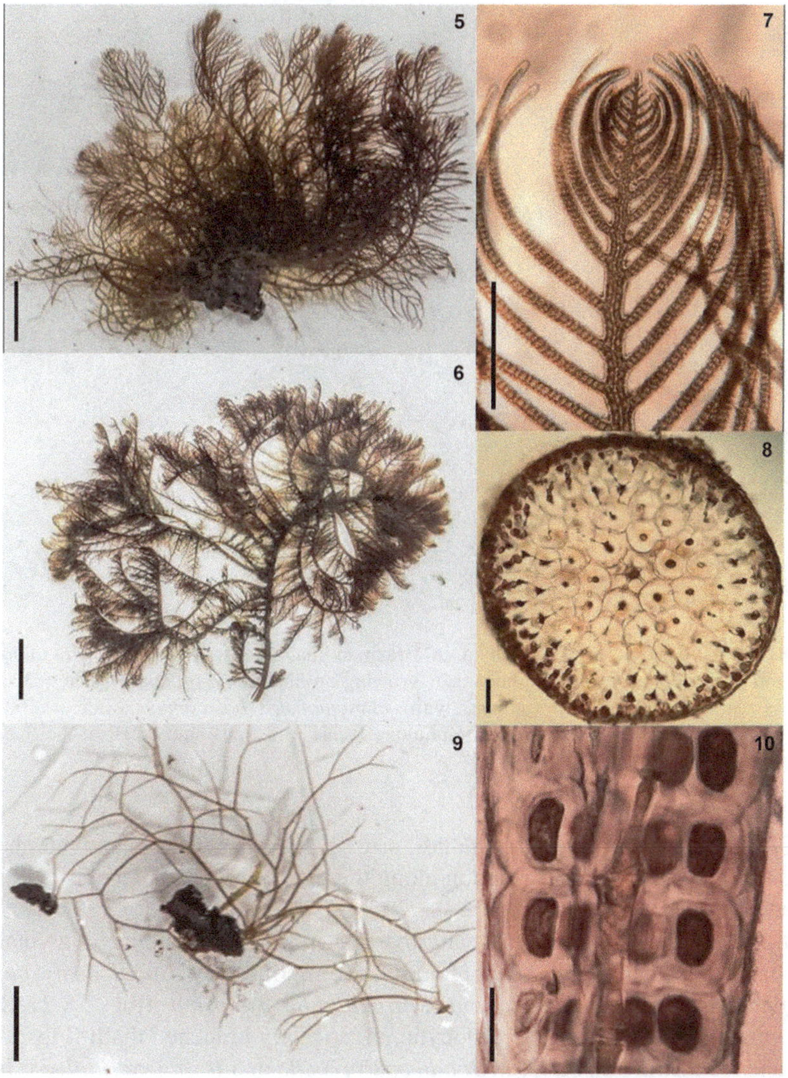

Figs. 8.5–8.10 Species of *Bostrychia* (Ceramiales, Rhodophyta) in Brazilian mangroves. **8.5**: *B. radicans* (scale = 5 mm). **8.6**: *B. montagnei*, gross morphology (scale = 5 mm). **8.7**. *B. calliptera*, detail of the thallus (scale = 100 μm). **8.8**: *B. montagnei*, cross section of the thallus showing dense cortication (scale = 25 μm). **8.9–8.10**: *B. kelanensis*, gross morphology (scale = 1 mm), and longitudinal section of the ecorticated thallus (scale = 50 μm)

In Brazil, the genus *Caloglossa* is represented by five species: *C. apomeiotica*, *C. confusa*, *C. leprieurii*, *C. kamiyana*, and *C. rotundata* (Figs. 8.15–8.20). Except for *C. kamiyana*, the other four species are morphologically similar, and the diagnostic characteristics are difficult to distinguish. Nevertheless, these taxa are genetically different (Kano et al. 2017). The main morphological characteristics used to

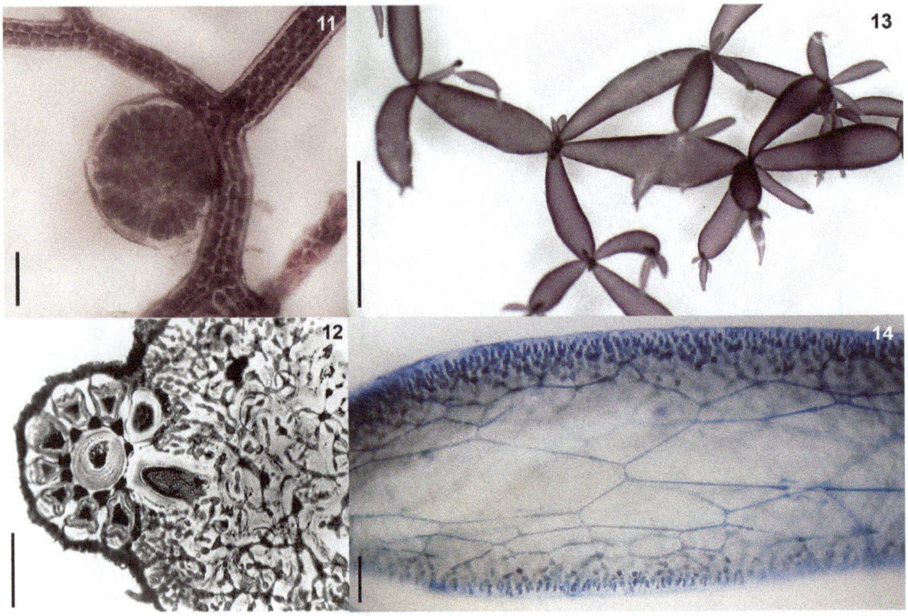

Figs. 8.11–8.14 Species of Rhodophyta in Brazilian mangroves. **8.11–8.12**: *Dawsoniocolax bostrychiae*, a Rhodomelacean adelphoparasite growing on *Bostrychia radicans* (scale = 300 μm); and cross section of *Bostrychia radicans* with *Dawsoniocolax bostrychiae* (scale = 100 μm). **8.13–8.14**: *Catenella caespitosa*, gross morphology (scale = 5 mm); and detail of the thallus in cross section (scale = 200 μm)

identify *Caloglossa* species are thallus size, blade shape, type and number of rhizoids per cell and their distribution along the thalli, and characteristic of vein on the blade.

Considering the Chlorophyta, mangrove green algae have filamentous thalli as found in Bryopsidales and Cladophorales, exemplified by *Boodleopsis pusilla*, *Cladophoropsis membranacea*, and *Rhizoclonium riparium* (Figs. 8.21–8.25). *Boodleopsis* is characterized by cenocytic, divaricately branched thalli (Figs. 8.21–8.22), and the distinction between *Boodleopsis pusilla* and *B. vaucherioidea* is based on the dichotomy degree. The first species presents regularly constricted thalli, and the angle of dichotomy is smaller than that presented by *B. vaucherioidea*, which has divaricated branching with an angle of 90–140°, without constrictions.

Among the monostromatic green algae described in Brazilian mangroves, two taxa were recognized based on the differences in life history, thallus ontogeny, and type of swarmer liberation: *Monostroma* sp. (zoospores germinate into a laminar blade) and *Ulvaria oxysperma* (Kützing) Bliding (zoospore germination gives rise to a saccate stage, and later forming laminar blade) (Cordeiro-Marino et al. 1993; Braga 1997). However, *U. oxysperma* is currently regarded as a synonym of *Gayralia oxysperma* (Kützing) K.L.Vinogradova ex Scagel et al. (Guiry and Guiry 2018).

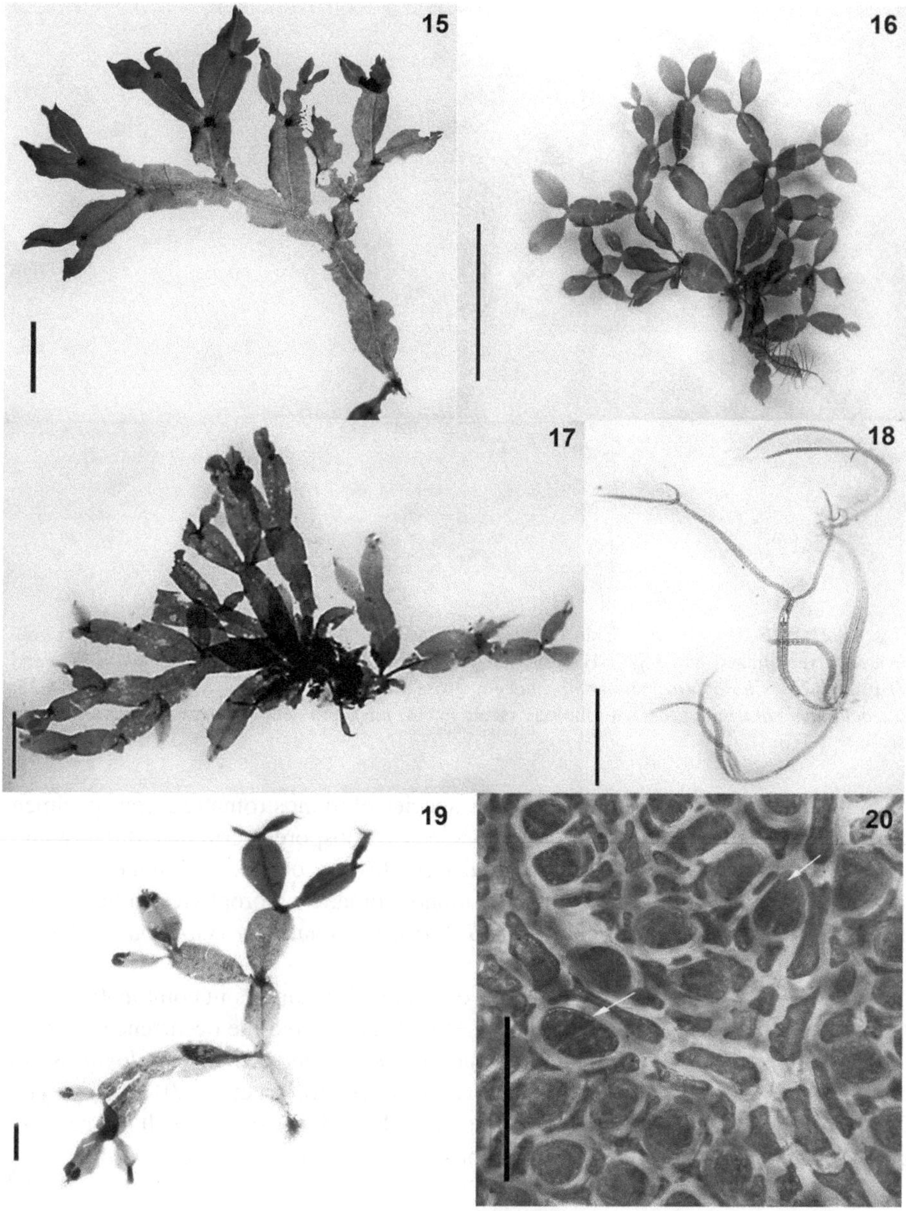

Figs. 8.15–8.20 Species of *Caloglossa* in Brazilian mangroves: **8.15–8.16** *C. apomeiotica*, gross morphology (scale = 10 mm) and detail of the bisporangia in surface view (arrows, scale = 100 μm). **8.17** *C. confusa* (scale = 1 mm). **8.18** *C. leprieurii* (scale = 5 mm). **8.19** *C. kamiyana* (scale = 10 mm). **8.20** *C. rotundata* (scale = 10 mm). (Modified from Kano et al. 2017)

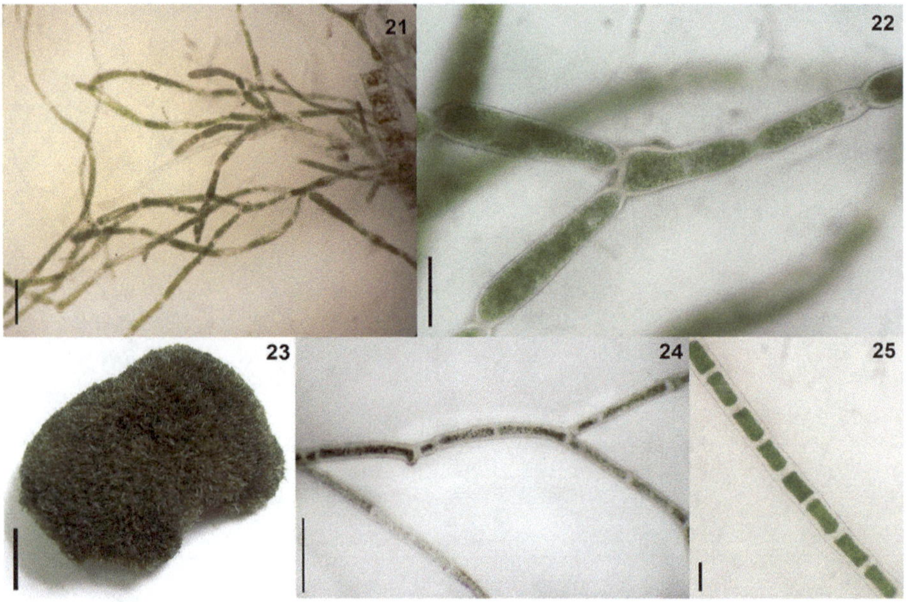

Figs. 8.21–8.25 Species of Chlorophyta in Brazilian mangroves. **8.21–8.22**: Cenocytic, divaricately branched and regularly constricted thalli in *Boodleopsis pusilla* (scales = 200 and 50 μm). **8.23**: *Cladophoropsis membranacea*, gross morphology (scale = 1 mm). **8.24–8.25**: *Rhizoclonium riparium*, gross morphology (scale = 500 μm), and detail of the unisseriate filament (scale = 100 μm)

Pellizzari et al. (2013) identified a new species of monostromatic green specimen based on molecular markers. In this new species, zoospores germinate directly into leafy monostromatic blades, and asexual reproduction occurs by biflagellate zoospores. Hence, the status of mangrove monostromatic chlorophytes indicates the occurrence of two species of *Gayralia*: *G. brasiliensis* and *G. oxysperma* (Pellizzari et al. 2013) (Figs. 8.26 and 8.27).

Therefore, taxonomical studies based on molecular markers in combination with morphological and developmental analyses have elucidated the occurrence of cryptic species in some genera of mangrove macroalgae, as reported to *Caloglossa* (Kano et al. 2017; Freshwater et al. 2021), and *Gayralia* (Pellizzari et al. 2013). However, further taxonomical studies are required to better knowledge of the small filamentous green algae belonging to the orders Bryopsidales and Cladophorales.

8.3 Diversity of Algal Propagules

Macroalgae are sessile organisms and can be dispersed by propagules (e.g., spores, gametes, and zygotes) or by free-floating seaweeds in different dispersal ranges. Algal propagules are spread in the marine environment and their richness in

Figs. 8.26 and 8.27 Monostromatic green algae in Brazilian mangroves. **8.26**: General aspect of *Gayralia brasiliensis*. **8.27**: Initial developmental stage of *G. brasiliensis* growing on net cultivation. Scales = 1 mm

comparison to the *in situ* populations can be distinct, and this phenomenon was studied mainly in the coastal habitats (Zechman and Mathieson 1985).

Temporal and spatial variations on the composition of macroalgal propagules in the water column are influenced by many factors, including, according to Hoffmann (1987): (a) Periodic variations in reproductive activity (related to reproductive seasonality and endogenous rhythms); (b) propagule production, which is associated with life strategies (e.g., opportunistic algae produce a larger number of propagules); and (c) environmental factors, which play an important role on algal reproduction, and spore release, settlement, and germination.

The only study on macroalgal propagules in mangroves was carried out by Cordeiro-Marino et al. (1990) who evaluated the composition of macroalgal propagules in Perequê River mangrove, Cardoso Island, Southeast Brazil (see Chap. 3, Map 14). Seawater samples (500 mL) were collected in three sites along a salinity gradient over 1 year, and aliquots of 100 mL were filtered in fiberglass membrane, and cultured under controlled conditions of temperature, photoperiod, and irradiance in the laboratory, with culture medium at a salinity of 25. After 2 weeks, plantlets growing on the membranes were isolated and cultured for several months until their taxonomic identification. The diversity of algal propagules was composed of 35 taxa, including 14 green algae, 16 red algae, and 5 brown algae (Table 8.2). However, within the total of 35 taxa, only eight species (26.7%) effectively colonized and grew successfully in the studied mangrove (Braga et al. 1990a, b; Yokoya et al. 1999). Propagule composition varied among the collecting sites and seasons (Table 8.2). The higher diversity of algal propagules was observed in the spring (southern hemisphere) with the presence of Ulvales (*Ulva* spp.), Ceramiales (*Callithamnion* and *Polysiphonia*), Gelidiales (*Gelidium*), and Ectocarpales (*Bachelotia* and *Feldmannia*). Except for *Gelidium* sp., these species are considered opportunistic, as they produce a large number of propagules of high dispersal capacity. These results evidenced that although the propagule flora is more diversified, only a few species can survive, attach, and grow as *in situ* populations in Brazilian mangroves.

Table 8.2 Composition of macroalgal propagules in seawater sampled in three different sites (S1, S2 and S3) in the mangrove of Cardoso Island, São Paulo State, Southeast Brazil. (After Cordeiro-Marino et al. 1990)

	February			June			October			February		
Taxon	S1	S2	S3	S1	S2	S3	S1	S2	S3	S1	S2	S3
CHLOROPHYTA												
1. ***Boodleopsis pusilla***	+	+	+	−	−	−	+	+	+	−	−	−
2. Chaetophoraceae	+	+	+	+	+	+	−	+	+	−	−	−
3. ***Cladophoropsis membranacea***	−	+	−	−	−	−	+	−	+	−	−	−
4. ***Gayralia brasiliensis***	−	−	−	−	−	−	+	−	−	−	−	−
5. ***Rhizoclonium tortuosum***	−	−	−	+	−	+	−	−	−	−	−	−
6. *Rhizoclonium* sp.	−	+	−	−	−	−	−	−	+	−	−	−
7. *Ulva clathrata*	+	+	−	+	−	−	−	+	+	+	−	−
8. *U. flexuosa*	−	−	−	−	+	−	+	+	+	−	+	−
9. *U. lactuca*	−	−	−	−	−	−	+	−	+	−	−	−
10. *U. linza*	−	−	−	−	−	−	−	−	+	−	−	−
11. *U. micrococca*	−	+	−	+	−	+	+	−	+	−	−	−
12. *Ulva* sp. 1	−	−	−	−	−	−	−	+	−	−	−	−
13. *Ulva* sp. 2	−	−	−	−	−	−	−	−	+	+	−	−
14. *Ulva* sp. 3	+	−	+	+	+	+	+	+	+	+	+	+
RHODOPHYTA												
15. *Acrochaetium* sp.	−	−	−	+	−	−	−	−	+	−	−	−
16. *Antithamnionella breviramosa*	−	−	−	−	−	−	−	+	−	−	−	−
17. ***Bostrychia calliptera***	−	−	−	−	−	+	−	−	−	−	−	−
18. ***B. moritziana***	+	−	−	−	+	−	−	−	−	−	+	−
19. ***B. radicans***	+	+	+	+	+	+	−	+	+	−	+	+
20. *Callithamnion* sp.	−	−	−	−	+	−	+	+	+	−	−	−
21. ***Caloglossa leprieurii***	−	−	+	−	−	−	−	−	+	−	−	−
22. ***C. kamiyana***	+	+	−	−	−	−	−	−	−	−	−	−
23. *Centroceras clavulatum*	−	−	−	−	−	−	+	−	−	−	−	−
24. *Ceramium codii*	−	+	−	−	−	−	+	−	−	−	−	−
25. *Erythrotrichia carnea*	−	−	−	+	−	−	−	−	−	−	−	−
26. *Herposiphonia secunda*	−	+	−	+	−	−	−	−	−	−	−	−
27. *Heterosiphonia crispella*	−	−	+	−	−	−	−	−	−	−	−	−
28. *Gelidium pusillum*	−	−	−	−	+	−	−	−	−	−	−	−
29. *Gelidium* sp.	−	−	−	+	+	−	+	+	+	−	−	−
30. *Polysiphonia subtilissima*	−	−	−	−	−	−	+	+	+	−	−	−
OCHOROPHYTA/PHAEOPHYCEAE												
31. *Bachelotia antillarum*	+	+	+	+	+	+	+	+	+	+	−	−
32. *Feldmannia irregularis*	−	−	−	+	−	−	+	−	+	−	−	−

(continued)

Table 8.2 (continued)

	February			June			October			February		
Taxon	S1	S2	S3	S1	S2	S3	S1	S2	S3	S1	S2	S3
33. *Feldmannia* sp.	–	–	–	+	–	–	+	–	–	–	–	–
34. *Sphacelaria rigidula*	–	–	–	–	–	–	+	–	–	–	–	–
35. *Sphacelaria* sp.	–	–	–	–	–	+	–	–	–	–	–	–
TOTAL	**8**	**12**	**7**	**13**	**9**	**7**	**16**	**12**	**19**	**4**	**4**	**2**

+ presence; – absence
S1: downstream site (salinity of 22–32); S2: intermediate site (salinity of 15–32); S3: upstream site (salinity of 15–32)
Species written in bold were recorded as *in situ* populations in the Cardoso Island mangroves (Braga et al. 1990a, b; Yokoya al. 1999)

Currently, the available molecular tools, such as DNA barcoding, have allowed a more accurate identification at the species level (e.g., Pellizzari et al. 2013; Kano et al. 2017), and consequently, the propagule-originating species, thus facilitating studies on species richness and dispersion of macroalgal propagules. Therefore, further research can contribute to understand the propagule dispersal in mangroves and the influence of seaweed communities from nearby rocky shores.

8.4 Ecology

Ecological studies on Brazilian mangroves were addressed mainly to evaluate the distribution, seasonality, biomass variation, zonation, colonization, and succession in macroalgal communities. However, these studies are restricted to certain regions, and there are gaps in the knowledge on macroalgal communities from mangroves along the Brazilian coast.

Bostrychia radicans and *B. calliptera* have been reported as the predominant species considering the percentage cover or biomass in the majority of Brazilian mangrove macroalgal communities (Eston et al. 1991; Cunha et al. 1999; Cutrim et al. 2004; Machado and Nassar 2007; Fontes et al. 2016), except for the green alga *Cladophoropsis membranacea*, which presented the highest biomass in Vila Velha and Suape mangroves, Pernambuco State (Fontes et al. 2007) (see Chap. 3, Map 8). *B. radicans* presented higher percentage cover and predominated in all sites and substrata (pneumatophores, plantlets, etc.) of mangroves of Cardoso Island, São Paulo State (Eston et al. 1991) (see Chap. 3, Map 14). Besides, macroalgal communities of Picinguaba and Fazenda rivers, in Ubatuba, São Paulo State, are predominated by *B. radicans* with the highest biomass values followed by *B. calliptera* and *B. moritziana*, and no clear spatial and temporal patterns in the distribution of the species were observed probably due to the similar environmental conditions of both rivers associated with the lack of a dry season throughout the year (Machado and Nassar 2007). Similarly, *B. radicans* also showed the highest biomass and frequency percentages in the mangrove of Boa Viagem Beach, Maranhão State (Fontes et al. 2016) (see Chap. 3, Map 3). However, *B. calliptera* predominated with higher biomass values in Parná-Açu and Tauá-Mirim mangroves, in Maranhão State (Cutrim et al. 2004), and Babitonga Bay mangrove, Santa Catarina State (Cunha et al. 1999) (see Chap. 3, Map 16).

There are no clear patterns in the recruitment and vertical distribution of macroalgae in the southeastern Brazilian mangroves, and this could be related to the absence of competition for bare space during colonization processes, as reported by Eston et al. (1992). Besides, the range of vertical distribution of macroalgae on trunks of *R. mangle, L. racemosa*, and *A. schaueriana* seems to be associated with tidal levels; however, zonation pattern was not observed, since *Rhizoclonium* spp., *Bostrychia calliptera*, and *B. radicans* occurred over the vertical range of distribution (Yokoya et al. 1999). Wilkinson (1980) suggested that a low number of species might reduce interspecific competition, resulting in a less well-defined zonation in the estuarine algal community. On the other hand, a clear zonation pattern in macroalgal communities was observed in mangroves of Puerto Rico (Almodovar and Pagan 1971) and Japan (Tanaka and Chihara 1987).

Temporal variations on species composition in mangroves of Cardoso Island are related to temperature, tidal levels, and tolerance of each species to withstand emersion, whereas spatial variations are related to salinity and light (Yokoya et al. 1999). Macroalgal distribution and biomass on trunks and roots of mangrove trees in Babitonga Bay were influenced by salinity gradients and flooding frequency (Cunha and Costa 2002). Salinity gradients also influenced species composition of macroalgal communities in Itapanhaú River mangroves, Bertioga, São Paulo State (Farraboti 2018).

Following Dawes (1998), the stages of the successional process consider pioneer species, intermediate, and stable forms (the final mature/climax stage). Pioneer species are usually annual, showing rapid growth and high reproduction rates, high productivity, short life histories, and simple morphology. All these characteristics allow them to colonize barren and disturbed areas, such as numerous species of Ulvales. In contrast, perennial species have long life histories, complex morphology, slow growth, and spend resources producing chemical and structural grazing defenses. However, the macroalgal colonization pattern on artificial substrates in the southeastern Brazilian mangroves showed that some perennial species may present early settlement, with *Bostrychia radicans* settling throughout the colonization process (Eston et al. 1992). Therefore, these authors suggested that mangrove macroalgal colonization patterns illustrate a successional series in which pioneer communities are also the final ones, and perennial species settled on available substrates as early colonists. Also, only species already observed on natural substrates settled on PVC tubes and wood canes placed as artificial substrates (Eston et al. 1992). Pellizzari et al. (2007) observed recruitment of *Gayralia brasiliensis* in artificial substrates (polypropylene nets) placed to evaluate the potential of its cultivation in estuaries (Fig. 8.28). Besides, no other propagules were attached in artificial or natural substrates during this experiment, suggesting an "exclusion hypothesis" of recruitment and successional patterns on mangroves, probably

Fig. 8.28 Cultivation of *Gayralia brasiliensis* in polypropylene nets in the estuarine complex of Paranaguá Bay, Paraná State, South Brazil (Pellizzari et al. 2007)

associated with competition and the production of chemical compounds against biofouling and/or herbivory (Karsten et al. 2000; Cassolato et al. 2008).

Little information on the herbivory and fauna associated with mangrove macroalgae is available. Macrofauna diversity associated with the Bostrychietum community of pneumatophores in Araçá Bay, São Paulo State (see Chap. 3, Map 14), comprised mostly of omnivorous amphipods and desiccation-resistant detritivorous species, and emersion time of Bostrychietum and the period of the year affect the community structure, for both seaweeds and the associated fauna (García et al. 2016). Besides, epiphytic macroalgae were considered determinants for the fauna, since the volume of sediment retained among stems and algal biomass influenced their distribution (Lopes 2011).

8.5 Cultivation

The world market for seaweeds including species from mangrove-estuarine complexes is increasing due to a demand by cosmetic, pharmaceutical, and food industries. These sorts of experiments indicated that some mangrove algae species recruit and grow well even when seeded on artificial collectors, suggesting that some species are suitable for management and cultivation (Pellizzari et al. 2007). This may be relevant as an alternative livelihood of fishermen communities facing the present decline in fisheries. Asian countries, such as South Korea and Japan, already cultivate monostromatic green algae in the brackish water of inner bays and estuaries (Ohno and Largo 1998).

Among the species of Brazilian mangrove macroalgae, ecophysiological responses and life history of species belonging to Ulotrichales have been well investigated due to cultivation purposes. Cordeiro-Marino et al. (1993) and Braga (1997) described the reproduction and thallus ontogeny of mangrove monostromatic green algae, aiming to elucidate their taxonomy.

Pellizzari et al. (2006) described the *Gayralia* sp. phenology comprising recruitment and percentage cover in these populations, establishing databases for cultivation in Paranaguá Bay, while Pellizzari et al. (2007) reported details on cultivation strategies of the edible *Gayralia*. Pellizzari et al. (2008) studied in detail the life history, thallus ontogeny, and the effects of temperature, irradiance, and salinity on the growth of two species of *Gayralia*. Later these authors published the new species, *Gayralia brasiliensis*, based on morphology, ontogeny, and phylogenetic position (Pellizzari et al. 2013). The authors reported that *Gayralia* sp. grew better at 18–22 °C and that recruitment occurred year-round in Paranaguá Bay, with higher recruitment rates during autumn (see Chap. 3, Map 15).

Based on recruitment and succession studies, Pellizzari et al. (2007) underwent *Gayralia* pilot cultivation using floating polypropylene nets in shallow subtidal areas or set in fixed systems in the intertidal zone. The best locations for natural recruitment (by asexual zooids) and frond growth were in the outer estuary, surrounding the mangrove fringes, where mother-fronds usually grow attached to stems and roots. Net biomass production was 458 ± 157 g m^{-2} (wet weight) only 45 days after zooid settlement. Although higher growth rates have been observed in nets in floating rafts, the fixed systems showed more stable production with few disturbances from epiphytes, fouling, and grazing, mainly during the winter. Peña-Salamanca (2008) studied the spatial-temporal dynamics of algal biomass associated with mangrove roots in Buenaventura Bay, in the Pacific Coast of Colombia, observing that the average biomass per species at the mouth of the estuary was significantly higher than in the inner estuary (annual average of 30.7 ± 10.8 and 13.8 ± 4.1 g m^{-2}, respectively). These results could be related to the level of tidal flood and the vertical distribution on the root; both seem to be the most limiting factors for algal biomass.

8.6 Chemical Diversity and Biotechnological Application

Marine environments comprise a large variety of organisms with adaptations to extreme conditions or wide environmental variation, and the search for new medicinal or cosmetic active compounds has been proven promising. Marine Natural Products (MNPs) have taken a progressively important position as drugs or as lead structures for bioinspired chemicals (Maciel et al. 2018). The review of Blunt et al. (2016) reported the extraordinary number of 1378 new compounds in 456 papers for 2014, which were isolated from marine microorganisms, phytoplankton, seaweeds (including mangrove macroalgae), sponges, and cnidarians.

Most MNPs often demonstrate remarkable pharmacological potential such as anticancer, antiviral, anti-inflammatory, antimalarial, antioxidant, antifungal, and antibacterial properties. Besides, functional foods and nutraceuticals containing antioxidant peptides isolated from the marine environment have become a topic of attention for pharmaceutical products and the healthy food industry (Maciel et al. 2018).

8.6.1 Mangrove Macroalgae

In Brazil, studies on the chemical diversity of mangrove macroalgae are focused on a few genera, *Bostrychia* and *Gayralia*, mainly from São Paulo and Paraná states, and need to be extended to macroalgae from other regions of Brazil as well as to other taxonomic groups.

Earlier chemical studies on the genus *Bostrychia* were focused on the yield of primary metabolism compounds. Among these compounds, we could exemplify sulfated polysaccharides from *B. montagnei*, which showed antiviral effects against *Herpes simplex*, and demonstrated anticoagulant activity (Noseda et al. 1999; Duarte et al. 2002, 2010). However, the most extensively studied primary metabolites in *Bostrychia* spp. are the polyols (low molecular weight carbohydrates) with multiple functions for cell osmoregulation, and for protecting and stabilizing organelle systems, protein synthesis, and enzymatic functions (Zuccarello and West 2011). Studies on polyol in *Bostrychia* spp. collected in São Paulo mangroves showed that *B. binderi, B. tenella*, and *B. moritziana* produced sorbitol and dulcitol, *B. calliptera* and *B. montagne* produced sorbitol, dulcitol, and traces of digeneaside, and *B. radicans* showed only sorbitol (Karsten et al. 1992; Zuccarello and West 2011).

Further studies aiming to discover new metabolites with different structures led to the isolation of two aromatic compounds from *Bostrychia tenella*, the sulfate metabolite potassium 4-(hydroxymethyl)-benzenosulfonate and the compound 1-methoxyphenethyl alcohol, the latter described previously as a synthetic product (de Felício et al. 2008) (Fig. 8.29). De Oliveira et al. (2012) discovered two new amides (*N*,4-dihydroxy-*N*-(2′-hydroxyethyl)-benzamide and *N*,4-dihydroxy-*N*-(2′-‑hydroxyethyl)-benzeneacetamide) in *B. radicans* (Fig. 8.29).

Volatile compounds were identified from the nonpolar fractions of marine algae *Bostrychia radicans* and *B. tenella*, both growing on the rocky shore at Dura Beach, in Ubatuba, São Paulo State, and from *B. radicans* and *B. calliptera*, both collected in the mangrove at Escuro River, also in Ubatuba. A solid-phase microextraction (SPME) and GC-MS analyses showed the presence of aldehydes in all samples but in higher percentages in mangrove species as well as the alcohols class. Moreover, fatty acids and hydrocarbons could also be observed in the fractions (de Oliveira et al. 2009).

Mangrove macroalgae have been exposed to high solar radiation and produce mycosporine-like amino acids (MAAs), which have antioxidant and photoprotective properties. The MAA contents were quantified in several *Bostrychia* species, including *B. calliptera* and *B. radicans* from São Paulo mangroves, and the MAA profiles evidenced different chemotypes and can be used as taxonomic markers (Orfanoudaki et al. 2020a). Moreover, phytochemical profiles of several samples of *Bostrychia calliptera* from different countries, including samples collected in Cardoso Island mangroves, revealed that MAAs and brominated and sulfated phenols are suitable chemotaxonomic markers since the three different chemotypes were the same lineages evidenced by DNA sequence data (Orfanoudaki et al. 2020b).

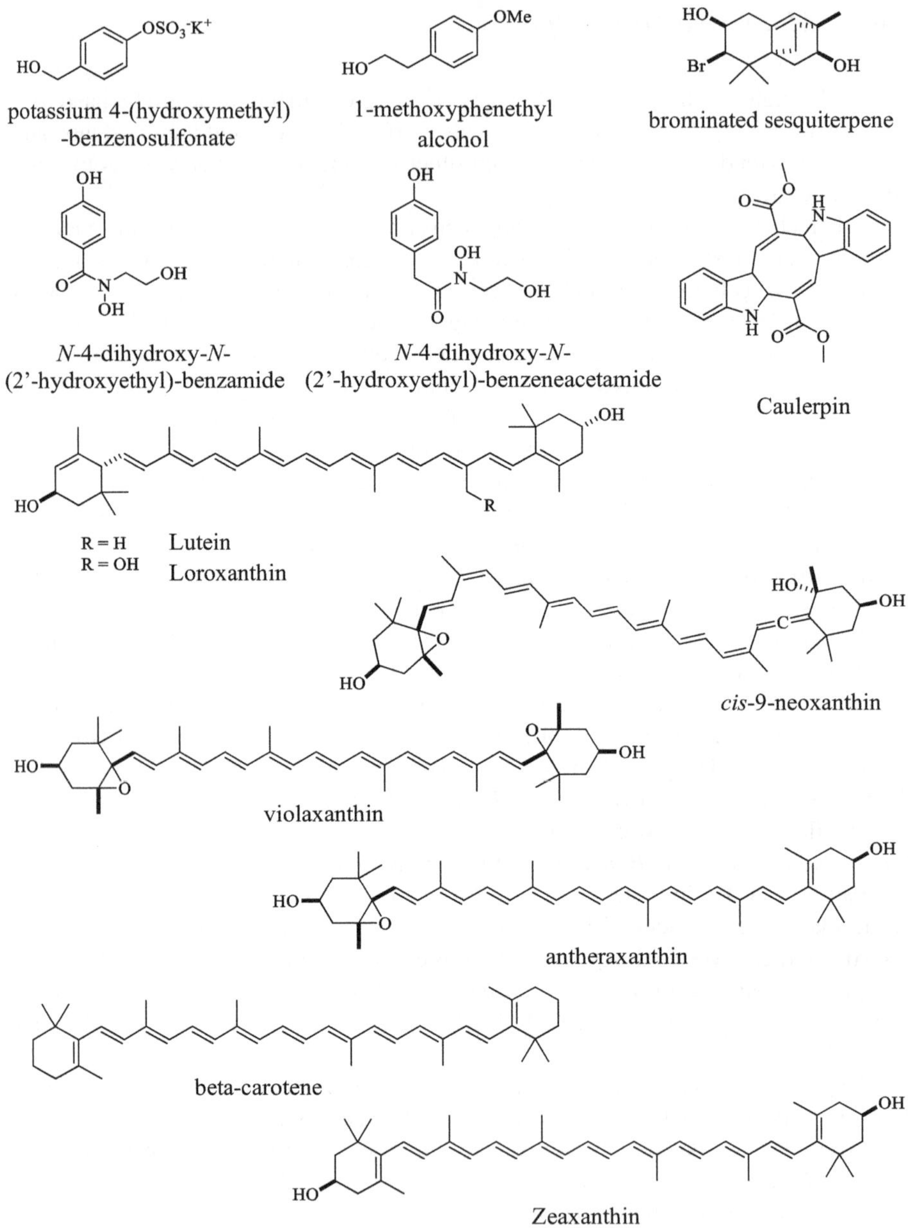

Fig. 8.29 Examples of metabolites found in macroalgae from Brazilian mangroves algae

A homogeneous sulfated heterorhamnan carbohydrate was obtained from the green seaweed *Gayralia brasiliensis* (previously identified as *G. oxysperma*) collected from mangroves in the Paraná State (Cassolato et al. 2008). The backbone is constituted by 3- and 2- linked rhamnosyl units, the latter being 50% substituted at C-3 by side chains containing 2-sulfated glucuronic and galacturonic acids and xylosyl units. The 3- and 2- linked rhamnosyl units are unsulfated (20%), disulfated (16%), and mostly monosulfated at C-2 (27%) and C-4 (37%). The branched and sulfated heterorhamnan of *G. brasiliensis* had high and specific activity against the *Herpes simplex* virus (Cassolato et al. 2008). Besides, antioxidant activities based on the 2,2-diphenyl-1-picrylhydrazyl (DPPH) radical scavenging method, by quantification of phenolic and carotenoid contents, were reported to *G. brasiliensis* from Paraná mangroves (Bernardi et al. 2016).

Although the chemical studies on Brazilian mangrove macroalgae are limited to *Bostrychia* and *Gayralia*, other classes of compounds were reported in other genera. As an example, a brominated sesquiterpene called (2S,3S,4aR,6R,7S)-6-bromo-2,5,5-trimethyl-3,4,5,6,7,8-hexahydro-2H- 2,4a-ethanonaphthalene-3,7-diol was isolated from the alga *Caloglossa leprieurii* from Xisha Island in the South China Sea (Fraga 2003).

Yoshii et al. (2004) investigated the carotenoid compositions of Cladophorales from Japan. The HPLC analysis showed that pigments of *Cladophoropsis membranacea* and *C. vaucheriiformis* (Areschoug) Papenfuss were classified as lutein type, while *C. fasciculatus* (Kjellman) Wille was included in loroxanthin type, which was considered as an inconsistency once it was expected for them to have the same precursor type. *Rhizoclonium grande* Børgesen was also included as loroxanthin type. Carotenoids 9-*cis*-neoxanthin, violaxanthin, antheraxanthin, lutein, and β-carotene were detected in all mentioned species and a small amount of zeaxanthin was detected only in *Cladophoropsis vaucheriiforms*. The authors concluded that the ancestral state for the Cladophorales was the loroxanthin type, based on the distribution pattern of the carotenoid types (Fig. 8.29).

Aqueous extract of *Rhizoclonium hieroglyphicum* (C.Agardh) Kützing from the Nan River in northern Thailand showed strong antioxidant properties with high free-radical scavenging capacity probably due to its large amount of sulfated polysaccharides and polyphenolic compounds. In addition, the authors of the study have evidenced a gelling ability of this extract that could be useful in nutritional, pharmaceutical, and cosmetic products (Mungmai et al. 2014).

Ethanolic extracts of *Cladophoropsis* sp. from the Persian Gulf showed cytotoxic effects on three tumor cell lines and, therefore, may be good candidates for further obtaining novel anticancer compounds. Moreover, stronger cytotoxic effects on estrogen-negative breast cancer cell line (MDA-MB-231(ER−)) in comparison to estrogen-positive cells (MCF-7 and T-47D) suggest that the extract of *Cladophoropsis* sp. may have an estrogen receptor/progesterone receptor-independent mechanism acting on cellular growth inhibition (Erfani et al. 2015).

Based on the information presently described, Brazilian mangrove species belonging to *Caloglossa* and Cladophorales have the potential to be a source of compounds with biological activities, and further studies are required to characterize their chemical profile.

8.6.2 Algicolous Fungi

Fungi colonize the internal tissues of different marine organisms, such as sponges, jellyfishes, and algae, being predominantly responsible to produce bioactive secondary metabolites that cannot be found in terrestrial organisms. Their relationship to the hosts is usually described as a symbiotic association of mutualism (Blunt et al. 2016; Maciel et al. 2018).

Many compounds biosynthesized by marine algae and associated endophytic fungi have been studied in the search of finding new natural alternatives to neutralizing the damage caused by solar radiation, for example, mycosporine produced by fungi and mycosporine-like amino acids found in algae, cyanobacteria, and invertebrates (Pallela et al. 2010).

It is important to mention that fungal chemical studies showed variations in the number of new compounds, with the figures of 318 in 2014 (Blunt et al. 2016), 371 in 2015 (Blunt et al. 2017), and 328 in 2016 (Blunt et al. 2018). Moreover, an increasing noteworthy number of compounds was observed in new substances isolated from endophytic fungi from mangroves, reaching 142 in 2016 (Blunt et al. 2018). These elevated numbers justify further studies employing endophytic fungi (Blunt et al. 2017, 2018).

In Brazil, information on fungi associated with mangrove macroalgae is limited to *Bostrychia* species. Endophytic fungi isolated from *Bostrychia radicans* and *B. tenella* were proven to be potential sources of bioactive secondary metabolites, acting as antimicrobial, cytotoxic, and exhibiting photoprotective properties (Erbert et al. 2012; de Felício et al. 2015; Pavão et al. 2016; Maciel et al. 2018).

The endophytic fungus *Phomopsis longicolla* isolated from *Bostrychia radicans* yielded three major compounds (Fig. 8.30): 18-deoxycytochalasin H, mycophenolic acid, and dicerandrol C; the latter showed antibacterial activity against *Staphylococcus aureus* (ATCC6538) and *S. saprophyticus* (ATCC 15305), with minimum inhibitory concentrations of 1 and 2 μg mL^{-1} (1.33 and 2.66 μm, respectively) (Erbert et al. 2012). Further studies on *P. longicolla* from *B. radicans* led to the isolation of another known structure called phomoxanthone A (PhoA), focusing on its cytotoxicity, genotoxicity, and mutagenicity in healthy lymphocytes and promyelocytic leukemia HL60 cells. Cells were treated with PhoA at concentrations from 0.01 to 100.0 μg mL^{-1}, not showing cytotoxicity, genotoxicity, or mutagenicity in lymphocytes at any tested concentration (Pavão et al. 2016). Otherwise, PhoA was highly cytotoxic, genotoxic, and mutagenic to HL60 cells. The obtained results highlight the selectivity of PhoA, which had a distinct performance on the cancer cell line, whereas no damage has been observed on healthy lymphocytes. The findings of this work suggest this compound has potential as a lead drug against cancer (Pavão et al. 2016).

Ten endophytic fungi strains were isolated from *Bostrychia tenella*, and they are identified as *Acremonium sp.*, *A. implicatum*, *Eurotium sp.*, *Nigrospora oryzae*, *Penicillium decaturense*, *P. waksmanii*, *Phomopsis sp.*, *Trichoderma atroviride*, *Xylaria sp.*, and Xylariaceae (de Felício et al. 2015). Furthermore, this is the first

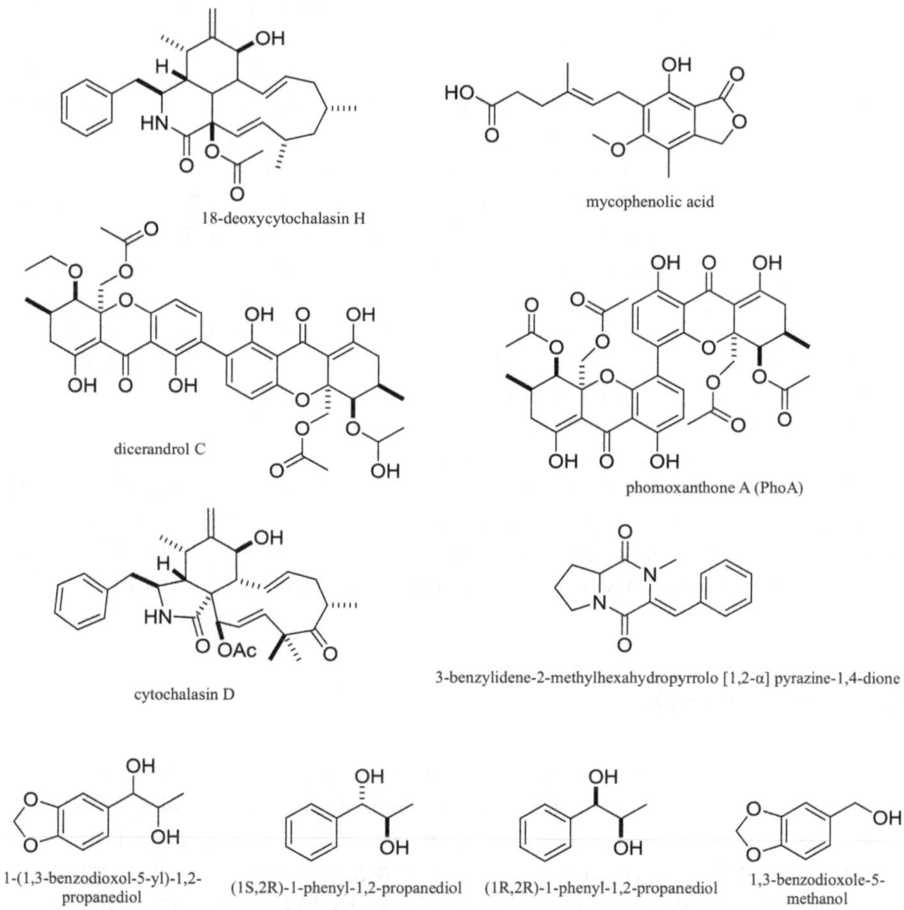

Fig. 8.30 Examples of metabolites isolated from algicolous fungi from Brazilian mangroves

description of *Acremonium implicatum* as a marine endophyte and *Xylaria* sp., *Trichoderma atroviride* and *Nigrospora oryzae* as marine seaweed endophytes. Their crude extracts and organic fractions obtained from mycelia after fungal cultivation were evaluated for diverse activities: Cytotoxic activity using SF-295 (glioblastoma), HL-60 (leukemia), and HCT-8 (human colon carcinoma) cell lines; antifungal activity against *Cladosporium cladosporioides* and *C. sphaerospermum* phytopathogenic fungi; and antibacterial potential against *Staphylococcus aureus* Rosenbach 1884 (ATCC 6538) and *Klebsiella pneumoniae* (Schroeter 1886) Trevisan 1887 (ATCC 13883) (de Felício et al. 2015). The strains of *Penicillium* are highlighted by their positive results in a diversity of assays, exhibiting efficient cell growth inhibition (80–100%) in all tumor cell lines, fungal growth inhibition of *Cladosporium*, and antibacterial activity against *S. aureus* and *K. pneumoniae*.

Besides, the other five strains were active in at least one test (de Felício et al. 2015). In addition, this work led to the isolation of a compound called cytochalasin D, which had its first report from a marine seaweed endophyte *Xylaria* sp. This compound was described at the first time as a metabolite from terrestrial fungi with antitumor and antibiotic properties (de Felício et al. 2015).

Maciel et al. (2018) evaluated the photoprotective activity of compounds isolated from the endophytic fungus *Annulohypoxylon stygium,* one of the strains isolated from the red alga *Bostrychia radicans.* This work yielded the isolation of five compounds: Two bioactive metabolites, the novel substance 3-benzylidene-2-methylhexahydropyrrolo [1,2-α] pyrazine-1,4-dione, which was previously described as a synthetic product in the literature, and the known compound 1-(1,3-benzodioxol-5-yl)-1,2-propanediol; two diastereomers (1*S*,2*R*)-1-phenyl-1,2-propanediol and (1*R*,2*R*)-1-phenyl-1,2-propanediol; and the 1,3-benzodioxole-5-methanol. The compounds were evaluated through the UVA/UVB absorption range, photostability, and phototoxicity using the 3T3 NRU phototoxicity test (OECD TG 432). Metabolites 3-benzylidene-2-methylhexahydropyrrolo [1,2-α] pyrazine-1,4-dione and 1-(1,3-benzodioxol-5-yl)-1,2-propanediol demonstrated expressive UVB absorption, photostability, and no phototoxicity (Fig. 8.30). This research allowed the discovery of a new photoprotective class of natural products.

8.7 Final Remarks

This chapter outlined the current state of knowledge on Brazilian mangrove macroalgae, and the later approaches in the elucidation of taxonomic problems in some genera, physiological and ecological processes, as well as chemical diversity. Taxonomical studies based on molecular markers in combination with morphological and developmental data have elucidated the occurrence of cryptic species in some genera, such as *Caloglossa* and *Gayralia.* The higher diversity of algal propagules is an interesting approach to understanding seaweed adaptation, distribution, and survival in mangrove ecosystems. Moreover, the strategies to survive in these conditions could explain the large number of compounds biosynthesized by mangrove algae and associated endophytic fungi, which are new natural products with potential to be considered as bioactive compounds for targeting pharmacological and cosmetic purposes. Studies on chemical diversity are focused on *Bostrychia* and *Gayralia* from the southern and southeastern regions, and need to be extended to mangroves from other regions of Brazil as well as to other taxonomic groups.

Finally, this chapter evidenced that the mangrove macroalgal commnities are well studied in certain regions, and there are gaps, which need to be filled in order to understand the macroalgal communities from mangroves distributed along the Brazilian coast.

References

Almodovar LR, Pagan FA (1971) Notes on a mangrove lagoon and mangrove channels at la Parguera, Puerto Rico. Nova Hedwigia 21:241–253

Bernardi J, Vasconcelos ERTPP, Lhullier C, Gerber T, Neto PC, Pellizzari FM (2016) Preliminary data of antioxidant activity of green seaweeds (Ulvophyceae) from the Southwestern Atlantic and Antarctic Maritime islands. Hidrobiologica 26:233–239

Blunt JW, Copp BR, Keyzers RA, Munro MH, Prinsep MR (2016) Marine natural products. Nat Prod Rep 33:382–431

Blunt JW, Copp BR, Keyzers RA, Munro MH, Prinsep MR (2017) Marine natural products. Nat Prod Rep 34:235–294

Blunt JW, Copp BR, Keyzers RA, Munro MH, Prinsep MR (2018) Marine natural products. Nat Prod Rep 35:8–53

Braga MRA (1997) Recruitment of two species of monostromatic blade-like chlorophytes, *Monostroma* sp. and *Ulvaria oxysperma* (Ulvales, Chlorophyta), in São Paulo state, Brazil. Phycol Res 45:153–161

Braga MRA, Fujii MT, Yokoya NS, Eston VR, Plastino EM, Cordeiro-Marino M (1990a) Macroalgae reproductive patterns in mangroves of Ilha do Cardoso, SP, Brazil. In: Anais do II Simposio Sobre Ecossistemas da Costa Sul e Sudeste Brasileira. Estrutura, função e manejo, vol 1. ACIESP, São Paulo, pp 314–326

Braga MRA, Fujii MT, Cordeiro-Marino M (1990b) Monostromatic green algae (Ulvales, Chlorophyta) of São Paulo and Paraná states (Brazil): distribution, growth, and reproduction. Rev Bras Bot 20:197–203

Cassolato JEF, Noseda MD, Pujol CA et al (2008) Chemical structure and antiviral activity of the sulfated heterorhamnan isolated from the green seaweed *Gayralia oxysperma*. Carbohydr Res 343:3085–3095

Cordeiro-Marino M, Yokoya NS Guimarães SMPB et al (1990) Aporte de propágulos de algas bentônicas a manguezais da Ilha do Cardoso, SP, Brasil. In: Annals of V Congreso Latinoamericano de Botánica, Havana

Cordeiro-Marino M, Braga MRA, Eston VR, Fujii MT, Yokoya NS (1992) Mangrove macroalgal communities of Latin America: the state of art and perspectives. In: Seeliger U (ed) Coastal plant communities of Latin America. Academic, New York, pp 51–64

Cordeiro-Marino M, Braga MRA, Fujii MT, Guimarães SMPB, Mitsugui EM (1993) Monostromatic green algae from Espírito Santo State, Brazil: life-history, growth and reproduction in culture. Rev Bras Biol 53:285–293

Cunha SR, Costa CS (2002) Gradientes de salinidade e frequência de alagamento como determinantes da distribuição e biomassa de macroalgas associadas a troncos de manguezais na Baía de Babitonga, SC. Braz J Aquat Sci Technol 6:93–102

Cunha SR, Nascimento J, Lima GB, Zacharjasiewicz G, Crestani DEV, Mafra LL Jr, Pazeto FD, Sant'Anna F, CSB C (1999) Distribuição e biomassa de macroalgas em um manguezal da Baía da Babitonga, SC: Resultados preliminares. Notas Téc Facimar 3:1–15

Cutrim MVJ, Silva EF, Azevedo ACG (2004) Distribuição vertical das macroalgas aderidas em rizóforos de *Rhizophora mangle* Linnaeus nos manguezais de Parna-Açu e Tauá-Mirim (Ilha de São Luís/MA-Brasil). Boletim do Laboratório de Hidrobiologia 17:9–18

Dawes CJ (1998) Marine Botany, 2nd edn. Wiley, New York

de Felício R, Debonsi HM, Yokoya NS (2008) Potassium 4-(hydroxymethil)-benzenosulfonate: a novel metabolite isolated from the marine red alga *Bostrychia tenella* (Rhodomelaceae, Ceramiales). Quím Nova 3:837–839

de Felício R, Pavão G, de Oliveira ALL, Erbert C, Conti R, Pupo MT, Furtado NAJC, Ferreira EG, Costa-Lotufo LV, Young MCM, Yokoya NS, Debonsi HM (2015) Antibacterial, antifungal and cytotoxic activities exhibited by endophytic fungi from the Brazilian marine red alga *Bostrychia tenella* (Ceramiales). Rev Bras Farmacog 25:641–650 de Oliveira ALL, Silva DB, Turatti ICC, Yokoya NS, Debonsi H (2009) Volatile constituents of Brazilian *Bostrychia* species (Rhodomelaceae) from mangrove and rocky shore. Biochem Syst Ecol 37:761–765

de Oliveira ALL, Silva DB, Lopes NP, Debonsi H, Yokoya NS (2012) Chemical constituents from red algae *Bostrychia radicans* (Rhodomelaceae): new amides and phenolic compounds. Quím Nova 35:2186–2188

Duarte MER, Noseda MD, Cardoso MA, Tulio S, Cerezo AS (2002) The structure of a galactan sulfate from the red seaweed *Bostrychia montagnei*. Carbohydr Res 337:1137–1144

Duarte MER, Noseda DG, Noseda MD, Pujol CA, Damonte EB (2010) Inhibitory effect of sulfated galactans from the marine alga *Bostrychia montagnei* on *Herpes simplex* virus replication in vitro. Phytomedicine 8:53–58

Erbert C, Lopes AA, Yokoya NS, Furtado NAJC, Conti R, Pupo MT, Lopes JLC, Debonsi HM (2012) Antibacterial compound from the endophytic fungus *Phomopsis longicolla* isolated from the tropical red seaweed *Bostrychia radicans*. Bot Mar 55:435–440

Erfani N, Nazemosadat Z, Moein M (2015) Cytotoxic activity of ten algae from the Persian Gulf and Oman Sea on human breast cancer cell lines; MDA-MB-231, MCF-7, and T-47D. Pharmacol Res 7:133–137

Eston VR, Yokoya NS, Fujii MT, Braga MRA, Plastino EM, Cordeiro-Marino M (1991) Mangrove macroalgae in Southeastern Brazil: spatial and temporal patterns. Rev Bras Biol 51:829–837

Eston VR, Braga MRA, Cordeiro-Marino M, Fujii MT, Yokoya NS (1992) Macroalgal colonization patterns on artificial substrates inside Southeastern Brazilian mangroves. Aquat Bot 42:315–325

Farraboti E (2018) Ocorrência de macroalgas bentônicas em dois bosques de manguezal em um estuário subtropical. Master thesis, Universidade Santa Cecília

Fontes KAA (2012) Taxonomia e distribuição geográfica do gênero *Bostrychia montagnei* (Ceramiales, Rhodophyta) nos manguezais do litoral brasileiro. Master thesis, Universidade Federal Rural de Pernambuco

Fontes KAA, Pereira SMB, Zickel CS (2007) Macroalgas do "Bostrychietum" aderido em pneumatóforos de duas áreas de manguezal do Estado de Pernambuco, Brasil. Iheringia 62: 31–38

Fontes KAA, Lisboa AT, Castro RS (2016) Macroalgas aderidas em pneumatóforos de *Avicennia germinans* (L.) Stearn na praia de Boa Viagem, São José de Ribamar–Maranhão. Acta Tecnol 11:33–45

Fraga BM (2003) Natural sesquiterpenoids. Nat Prod Rep 20:392–413

Freshwater DW, Miller CE, Frankovich TA, Wynne MJ (2021) DNA sequence analyses reveal two new species of *Caloglossa* (Delesseriaceae, Rhodophyta) from the skin of West Indian Manatees. J Mar Sci Eng 9:163. https://doi.org/10.3390/jmse9020163

Fujii MT, Yokoya NS, Cordeiro-Marino M (1990) *Stictosiphonia kelanensis* (Grunow ex Post) King & Puttock (Rhodomelaceae, Rhodophyta): a new record from Atlantic mangroves. Hoehnea 17:93–97

García AF, Bueno M, Leite FPP (2016) The Bostrychietum community of pneumatophores in Araçá Bay: an analysis of the diversity of macrofauna. J Mar Biol Assoc UK 96:1617–1624

Guimarães SMPB (1993) Morphology and systematics of the red algal parasite *Dawsoniocolax bostrychiae* (Choreocolacaceae, Rhodophyta). Phycologia 32:251–258

Guiry MD, Guiry GM (2018) Algaebase. World-wide electronic publication, National University of Ireland, Galway. http://www.algaebase.org. Searched on 17 Apr 2018

Hadlich RM, Bouzon ZL (1985) Contribuição ao levantamento taxonômico dasalgas marinhas bentônicas do mangue do Itacorubi, Florianópolis, Santa Catarina, Brasil. II – Rhodophyta. Ínsula 15:89–116

Hoffmann AJ (1987) The arrival of seaweed propagules at the shores: a review. Bot Mar 30:151–165

Kamiya M, Tanaka J, King RJ, West JA, Zuccarello GC, Kawai H (1999) Reproductive and genetic distinction between broad and narrow entities of *Caloglossa continua* (Delesseriaceae, Rhodophyta). Phycologia 38:356–367

Kamiya M, West JÁ, Karsten U, Ganesan EK (2016) Molecular and morphological delineation of *Caloglossa beccarii* and related species (Delesseriaceae, Rhodophyta). Phycologia 55(6): 640–649

Kano CH (2015) Delesseriaceae (Ceramiales, Rhodophyta) no sudeste brasileiro baseado em morfologia, DNA barcode e distribuição geográfica. Master thesis, Instituto de Botânica da Secretaria de Estado do Meio Ambiente

Kano CH, Sena FS, Cassano V, Fujii MT (2017) *Caloglossa* species diversity (Delesseriaceae, Rhodophyta) based on morphology and DNA data with emphasis on southeastern Brazil. Braz J Bot 40:551–564

Karsten U, West JA, Zuccarello G (1992) Polyol content of *Bostrychia* and *Stictosiphonia* (Rhodomelaceae, Rhodophyta) from field and culture. Bot Mar 35:11–19

Karsten U, Sawall T, West JA, Wiencke C (2000) Ultraviolet sunscreen compounds in epiphytic red algae from mangroves. Hydrobiologia 432:159–171

King RJ, Puttock CF (1989) Morphology and taxonomy of *Bostrychia* and *Stictosiphonia* (Rhodomelaceae/Rhodophyta). Aust Syst Bot 2:1–73

King RJ, Puttock CF, Vickery RS (1988) A taxonomic study on the *Bostrychia tenella* complex (Rhodomelaceae, Rhodophyta). Phycologia 27:10–19

Krayesky DM, Norris J, West JA, Fredericq S (2011) The *Caloglossa leprieurii*-complex (Delesseriaceae, Rhodphyta) in the Americas: the elucidation of overlooked species based on molecular and morphological evidence. Cryptogam Algol 32:37–62

Krayesky DM, Norris JN, West JA, Kamiya M, Viguerie M, Wysor B, Fredericq S (2012) Two new species of *Caloglossa* (Delesseriaceae, Rhodophyta) from the Americas, *C. confusa* and *C. fluviatilis*. Phycologia 51:513–530

Lopes OL (2011) Fauna associada às macroalgas epífitas no manguezal do Rio das Garças, Guaratuba, Paraná. Master thesis, Universidade Federal do Paraná

Machado GEM, Nassar CA (2007) Assembléia de macroalgas de dois manguezais do Núcleo Picinguaba – Parque Estadual da Serra do Mar, São Paulo, Brasil. Rodriguésia 58:835–846

Maciel OMC, Tavares RSN, Caluz DRE, Gaspar LR, Debonsi HM (2018) Photoprotective potencial of metabolites isolated from algae-associated fungi *Annulohypoxylon stygium*. J Photochem Photobiol B 178:316–322

Menezes M, Bicudo CEM, Moura CWN, Alves AM, Santos AA, Pedrini AG, Araújo A, Tucci A, Fajar A, Malone C, Kano CH, Sant'Anna CL, Branco CZ, Odebrecht C, Peres CK, Neuhaus EB, Eskinazi-Leça E, Aquino E, Nauer F, Silva WJ (2015) Update of the Brazilian floristic list of Algae and Cyanobacteria. Rodriguésia 66:1047–1062

Mungmai L, Jiranusornkul S, Peerapornpisal Y, Sirithunyalug B, Leelapornpisid P (2014) Extraction, characterization and biological activities of extracts from freshwater macroalga [*Rhizoclonium hieroglyphicum* (C. Agardh) Kützing] cultivated in Northern Thailand. Chiang Mai J Sci 41:14–26

Noseda MD, Tulio S, Duarte MER (1999) Polysaccharides from the red seaweed *Bostrychia montagnei*: chemical characterization. J Appl Phycol 11:35–40

Ohno M, Largo DL (1998) The seaweed resources of Japan. In: Critchley AT, Ohno M (eds) Seaweed resources of the world. Japan International Cooperation Agency, Yokosuka, pp 1–14

Oliveira EC (1984) Brazilian mangrove vegetation with special emphasis on the seaweeds. In: Por FD, Dor I (eds) Hydrobiology of the Mangal. W Junk Publishers, The Hague, pp 55–66

Orfanoudaki M, Hartmann A, Kamiya M, West J, Ganzera M (2020a) Chemotaxonomic study of *Bostrychia* spp. (Ceramiales, Rhodophyta) based on their mycosporine-like amino acid content. Molecules 25:3273

Orfanoudaki M, Hartmann A, Ngoc HN, Gelbrich T, West J, Karsten U, Ganzera M (2020b) Mycosporine-like amino acids, brominated and sulphated phenols: suitable chemotaxonomic markers for the reassessment of classification of *Bostrychia calliptera* (Ceramiales, Rhodophyta). Phytochemistry 174(112344)

Pallela R, Na-Young Y, Kim S-K (2010) Anti-photoaging and photoprotective compounds derived from marine organisms. Mar Drugs 8:1189–1202

Paula EJ, Ugadim Y, Kanagawa AI (1989) Macroalgas de Manguezais da Ilha de Maracá – Estado do Amapá, Brasil. Ínsula 19:95–114

Pavão GB, Venâncio VP, de Oliveira AL, Hernandes LC, Almeida MR, Antunes LM, Debonsi HM (2016) Differential genotoxicity and cytotocity of phomoxanthone A isolated from the fungus *Phomopsis longicolla* in HL60 cells and peripheral blood lymphocytes. Toxicol in Vitro 37: 211–217

Pellizzari FM, Oliveira EC, Yokoya NS (2006) Coverage and recruitment of the edible green macroalga *Gayralia* sp. (Monostromataceae) in Paranaguá Bay, Southern Brazil. J Coast Res 39 (SI):157–159

Pellizzari FM, Absher TM, Yokoya NS, Oliveira E (2007) Cultivation of the edible green seaweed *Gayralia* (Chlorophyta) in Southern Brazil. J Appl Phycol 19:63–69

Pellizzari FM, Oliveira EC, Yokoya NS (2008) Life-history, thallus ontogeny, and the effects of temperature, irradiance and salinity on the growth of the edible green seaweed *Gayralia* spp. (Chlorophyta) from Southern Brazil. J Appl Phycol 20:75–82

Pellizzari FM, Oliveira MC, Medeiros AS, Yokoya NS, Oliveira E (2013) Morphology, ontogeny, and phylogenetic position of *Gayralia brasiliensis* sp. nov. (Ulotrichales, Chlorophyta) from the southern coast of Brazil. Bot Mar 56:197–205

Peña-Salamanca EJ (2008) Dinámica espacial y temporal de la biomasa algal asociada a las raíces de mangle en la Bahía de Buenaventura, Costa Pacífica de Colombia. Bol Invest Mar Cost 37: 55–70

Post E (1968) Zur Verbreitungs-Okologie des Bostrychietum. Hydrobiologia 31:241–316

Presting GG (2006) Identification of conserved regions in the plastid genome: implications for DNA barcoding and biological function. Can J Bot 84:1434–1443

Saunders GW (2005) Applying DNA barcoding to red macroalgae: a preliminary appraisal holds promise for future applications. Philos Trans R Soc Biol Sci 360:1879–1888

Saunders GW, Kucera H (2010) An evaluation of *rbc*L, tufA, UPA, LSU and ITS as DNA barcode markers for the marine green macroalgae. Cryptogam Algol 31:487–528

Sena FS (2016) Diversidade de espécies de macroalgas associadas no manguezal da Ilha Barnabé, Baixada Santista, SP, Brasil, com base em "DNA Barcode". Master thesis, Universidade de São Paulo

Sherwood AR, Presting GG (2007) Universal primers amplify a 23S rDNA plastid marker in eukaryotic and cyanobacteria. J Phycol 43:605–608

Tanaka J, Chihara M (1987) Species composition and vertical distribution of macroalgae in brackish waters of Japanese mangrove forests. Bull Nat Sci Mus 13:141–150

Wilkinson M (1980) Estuarine benthic algae and their environment: a review. In: Price JH, Irvine EG, Farnham WF (eds) The shore environment, vol 2. Academic, Oxford, pp 425–486

Wynne MJ (2017) A checklist of benthic marine algae of the tropical and subtropical Western Atlantic: fourth revision. Nova Hedwig Beih 145:1–202

Yokoya NS, Plastino EM, Braga MRA, Fujii MT, Cordeiro-Marino M, Eston VR, Harari J (1999) Temporal and spatial variations in the structure of macroalgal communities associated with mangrove trees of Ilha do Cardoso, São Paulo state, Brazil. Rev Bras Bot 22:195–204

Yoshii Y, Hanyuda T, Wakana I, Miyagi K, Arai S, Ueda K, Inouye I (2004) Carotenoid compositions of *Cladophora* balls (*Aegagropila linnaei*) and some members of the Cladophorales (Ulvophyceae, Chlorophyta): Their taxonomic and evolutionary implication. J Phycol 40:1170–1177

Zechman FW, Mathieson AC (1985) The distribution of seaweed propagules in estuarine, coastal and offshore waters of New Hampshire, USA. Bot Mar 28:283–294

Zuccarello GC, West JA (2011) Insights into the evolution and speciation in the red alga *Bostrychia*: 15 years of research. Algae 26:21–32

Zuccarello GC, Muangmai N, Preuss M, Sanchez LB, Loiseaux de Goër S, West JA (2015) The *Bostrychia tenella* species complex: morphospecies and genetic cryptic species with resurrection of *B. binderi*. Phycologia 54:261–270

Chapter 9
Macro- and Meiofaunal Communities in Brazilian Mangroves and Salt Marshes

Helio H. Checon, Guilherme N. Corte, Beatriz P. Cunha, Yasmina M. L. Shah Esmaeili, Gustavo Fonseca, and Antônia Cecília Z. Amaral

9.1 Benthic Environment

Mangrove forests are coastal habitats commonly characterized by a sedimentary environment composed mainly of muddy sediments. Organisms inhabiting mangroves are subjected to pressing conditions, many of which vary on a short time scale, from a few hours (semidiurnal tides) to a few days (moon cycle). Tidal cycles, anoxic sediment, and salinity fluctuation due to brackish estuarine waters are among the conditions shared with other coastal environments. What makes mangroves and salt marshes unique is the presence of vegetation, which is responsible for changing sedimentary dynamics and providing resources and refuge for a wide range of macro- and meiofaunal organisms. Due to these factors, the presence of mangroves enhances the diversity of benthic species in coastal ecosystems, especially for mollusks and crustaceans (Nagelkerken et al. 2008; Gorman and Turra 2016), despite the harsh and variable physical environment. As mangroves provide a variety of microhabitats, many organisms inhabiting mangroves are also found on adjacent ecosystems, such as estuaries, beaches, and rocky shores. The high productivity of mangroves represents an important source of organic matter for the resident fauna, but also for neighboring communities when exported to nearby coastal ecosystems.

H. H. Checon (✉) · G. N. Corte
Instituto Oceanográfico, Universidade de São Paulo, São Paulo, SP, Brazil

Departamento de Biologia Animal, Instituto de Biologia, Universidade Estadual de Campinas, Campinas, SP, Brazil

B. P. Cunha · Y. M. L. Shah Esmaeili · A. C. Z. Amaral
Departamento de Biologia Animal, Instituto de Biologia, Universidade Estadual de Campinas, Campinas, SP, Brazil

G. Fonseca
Instituto do Mar, Universidade Federal de São Paulo, Santos, SP, Brazil

Y. Schaeffer-Novelli et al. (eds.), *Brazilian Mangroves and Salt Marshes*, Brazilian Marine Biodiversity, https://doi.org/10.1007/978-3-031-13486-9_9

Aside from the sedimentary environment, mangroves also house a very diverse fauna that lives attached to mangrove aerial roots or is associated with macroalgae that grow on mangrove roots and pneumatophores. The most illustrative macroalgae association known as *Bostrychietum* (see Chap. 8) is commonly seen in Brazilian mangroves growing on pneumatophores and trunks of trees, such as those of the genera *Avicennia* and *Laguncularia.* The presence of these macroalgae enhances habitat complexity and resource availability, providing refuge and food to the resident fauna and to the settling larvae of species that otherwise would not be able to recruit in mangroves. Tree structures are not only important for macroalgae establishment, but also as a substrate for encrusting macrofauna, mainly suspension feeders such as bivalves.

Brazil has the second-largest mangrove area in the world, extending from subtropical to tropical climates (Spalding 2010). Aside from the expected latitudinal variations, the tidal regime varies greatly from the south (microtidal) to the north coast (macrotidal) of the country, and vegetation changes gradually to the dominance of *Spartina alterniflora* Loisel toward the southern coast (see Chaps. 1 and 3). Tidal variation can deeply affect mangrove benthic fauna, mainly by changing activity and zonation patterns between and within species (Nishida et al. 2006; Wunderlich and Pinheiro 2012). Hence, the patterns of benthic community structure (i.e., abundance, dominant taxon, diversity) may vary greatly along the coast, as already demonstrated in the Brazilian estuaries (Lana and Bernardino, 2018). In the following sections, we present an overview of the composition of benthic fauna in mangroves and salt marshes of Brazil, highlighting the most representative taxa. It is important to mention that this chapter reviews the studies carried out in Brazil but is a nonexhaustive review of the main findings. Thus, at the end of each section, we provide additional references to those looking to further delve into the benthic biodiversity of Brazilian mangroves.

9.2 Benthic Fauna

Like other benthic environments, such as sandy beaches, coastal bays, and seagrass habitats, a great diversity of taxa can be found in mangroves. The faunistic composition ranges from mobile to sessile, detritivores to predators, and epifaunal to infaunal species. Many species have behavioral or physiological adaptations to cope with the varying tides and salinity and the overall oxygen-depleted sediments due to the environmental conditions generated by the location within estuaries and the high decomposition rate within the sediment. An important difference from other intertidal soft-bottom habitats (e.g., sandy beaches) is that, due to the presence of vegetation, epifaunal organisms are more common. As the mangrove and saltmarsh biotic structures enhance the environmental suitability for many species, these species are also important for the maintenance of the ecosystem by participating in the cycling of nutrients and organic matter, enhancing the mineralization, and oxygenating the sediment via bioturbation. Below we present the most representative groups of mangrove invertebrate fauna, including examples of species commonly found on the Brazilian coast.

9.2.1 Mangrove Crabs

Mangrove crabs are likely the most prominent faunistic group in mangrove ecosystems (see also Chap. 10). They are conspicuous organisms that can be seen climbing trees, hiding among the mangrove roots, and burrowing in the muddy sediment. Their assemblage may attain a very high diversity in mangroves, being composed of taxa from many different families. The most abundant mangrove crab superfamilies along the Brazilian coast are Ocypodoidea (e.g., Ocypodidae, Ucididae) and Grapsoidea (e.g., Grapsidae, Sesarmidae, Varunidae), like patterns registered in mangroves around the globe (Branco 1990; Oshiro et al. 1998; Colpo et al. 2011; Sousa et al. 2015). Colpo et al. (2011) suggest that the forest's successional stage is related to the dominance of one superfamily over another, with Ocypodoidea dominating pioneer stages, and Grapsoidea, the more mature forests.

The Ucididae family is represented mainly by one species, the swamp ghost-crab *Ucides cordatus* (Fig. 9.1a), popularly known as "caranguejo-uçá" in Brazil. This is likely the most common and well-known benthic species in Brazilian mangroves,

Fig. 9.1 Brazilian mangrove crabs. (**a**) *Ucides cordatus*; (**b**) *Goniopsis cruentata* (Photos: Clemente Coelho-Jr)

Fig. 9.4 (**a**) *Uca* spp.; (**b**) *Cardisoma guanhumi*. (Photos: C. Coelho-Jr)

occurring abundantly and continuously in mangroves along the entire coast, albeit varying in density at local scales (Schories et al. 2003; Sandrini-Neto and Lana 2011). This species inhabits burrows mainly in the inner area of mangroves, where it feeds primarily on leaves and plant detritus (Nordhaus and Wolff 2007). Its feeding rate is so high that *U. cordatus* is considered to be responsible for the removal and processing of leaf litter in Brazilian mangroves (Schories et al. 2003; Nordhaus et al. 2009). It is also the most important mangrove species in terms of economic exploitation (see Sect. 9.4). This is the most studied species in Brazilian mangroves, with research focusing on the population structure, ecological responses, and fisheries in addition to impact assessments (Diele et al. 2005; Duarte et al. 2016; Banci et al. 2017; Pinheiro et al. 2018).

Fiddler crabs (Ocypodidae), popularly known as "chama-marés" in Brazil, also occur in great abundance and are a widely studied faunal component in Brazilian mangroves, with many studies that focus on population, behavior, and community ecology (e.g., Costa and Negreiros-Fransozo 2001; Koch et al. 2005; Bezerra et al. 2006; Machado et al. 2013; Thurman et al. 2013; Checon and Costa 2017). There are currently ten fiddler crab species recognized in Brazilian mangroves: *Leptuca cumulanta*, *L. leptodactyla*, *L. uruguayensis*, *Minuca burgersi*, *M. mordax*, *M. rapax*, *M. thayeri*, *M. vocator*, and *Uca maracoani* (see Fig. 9.4a) in addition to one endemic species (*Minuca victoriana*). Due to this diversity, fiddler crabs are found in different zones along the mangrove, including the unvegetated edge to the innermost areas (Checon and Costa 2017). Most of these species occur along the entire Brazilian coast, except for *L. uruguayensis*, which is restricted to the mangroves of southern and southeastern areas, and *M. victoriana,* whose distribution is restricted to the southeastern coast, but with a trend of northward and southward expansion noted recently (Castiglioni et al. 2010; Thurman et al. 2013). The ecology and behavior of fiddler crabs can be further explored in works such as Thurman et al. (2017), De Grande et al. (2018), Vianna et al. (2020), and Arakaki et al. (2020).

The blue land crab *Cardisoma guanhumi* (Gecarcinidae), popularly known as "guaiamum" in Brazil, has a wide distribution along the coast. Like *U. cordatus*, that is a species that is commonly exploited for economic purposes (Sect. 9.4). However, while *U. cordatus* is mostly found in the intertidal and fringe areas of mangroves, *C. guanhumi* occurs in the apicum zone, the dry upper area of mangroves (Oliveira-Neto et al. 2008). Occurrence in this zone makes this species more susceptible to suppression of mangroves, which has resulted in population decline, especially in the more urbanized southeastern coast (Amaral and Jablonski 2005). It has been found to adapt to terrestrial conditions outside of mangroves in some areas; however, this adaptation comes with a cost, and individuals inhabiting the apicum zones of mangroves can excavate more suitable burrows than individuals adapted to the terrestrial environment (Oliveira-Neto et al. 2008). Further information can be found in studies such as Cavalcanti et al. (2009), Firmo et al. (2012), and Silva et al. (2014).

The red mangrove-root crab *Goniopsis cruentata* (Gecarcinidae), popularly known as "maria-mulata" or "aratu-vermelho" in Brazil, is another species with continuous distribution along the coast (Fig. 9.1b). This species is a nonburrowing crab that inhabits crevices and holes inside logs, as well as empty burrows left by other species. It has a crucial role in the food web, being one of the most prominent omnivores in Brazilian mangroves (Koch and Nordhaus 2010). As an active and generalist predator, it feeds on smaller crabs, but may also consume carrion, mangrove propagules, and detritus (Koch and Nordhaus 2010; Andrade et al. 2012). Gut content studies indicate that plant detritus is likely the main source of food for this species (Lima-Gomes et al. 2011). Further research on this species can be found in studies such as Lira et al. (2013), Davanso et al. (2013) and Hirose et al. (2015).

The Varunidae crab *Neohelice granulata* (formerly known as *Chasmagnathus granulatus*), popularly known as "catanhão" in Brazil, is also very abundant, but its distribution is restricted to the southern and southeastern coasts, occurring in mangroves and salt marshes, where it is the main grazer of *Spartina alterniflora* (Freitas et al. 2015). The species is widely studied in terms of behavior, grazing pressure, influence on sedimentation, and pollution impacts (Spivak 2010; Freitas et al. 2015). According to Spivak (2010), it is the sixth most studied crab species in the world, due to its abundance in South Brazil and Argentina.

Other numerically important crab species inhabiting Brazilian mangroves are the Xanthidae *Eurythium limosum,* the Grapsidae *Sesarma rectum, Pachygrapsus gracilis,* and, especially, the tree crab *Aratus pisonii* (mangrove-tree crab) (Fig. 9.3), commonly found climbing mangrove trees or walking between pneumatophores. Most of these species are not restricted to Brazilian mangroves and can be

Fig. 9.3 *Aratus pisonii* H. Milne Edwards and *Melampus coffea* L. (Photo: Clemente Coelho-Jr)

found distributed along the Western Atlantic Coast. Blue crabs (Portunidae) occur in mangroves at low tide levels in estuarine areas, represented mainly by the omnivore crabs from the genus *Callinectes,* mainly *C. sapidus* and *C. danae,* both popularly known as "siria zul" in Brazil.

9.2.2 Other Crustaceans

Aside from brachyurans, peracarid crustaceans are also commonly found in mangrove forests. These organisms may occur under different conditions and habitats. Contrary to the brachyuran, which are found more commonly in the sediment, peracarid crustaceans, especially amphipods, are more abundant in association with macroalgae mats. For instance, amphipods and tanaids were reported as the most common organisms associated with the *Bostrychietum* on pneumatophores in a Southeast Brazilian mangrove (García et al. 2015). Amphipods and tanaids can also be found on decomposing mangrove leaves, especially those of *Rhizophora mangle* (Oliveira et al. 2012).

Shrimps are also commonly found in Brazilian mangroves. Species of importance as fishery resources are the pink shrimps *Farfantepenaeus brasiliensis* and *F. paulensis* (popularly known as camarão-rosa in Brazil), occupy mangrove estuaries and adjacent bay areas, although the latter species is much more present in the estuary (Costa et al. 2008). The seabob shrimp *Xiphopenaeus kroyeri* (popularly known as *camarão-sete-barbas* in Brazil*)* is also registered in bay areas surrounded by mangroves (Grabowski et al. 2016). Despite the occurrence of shrimp species, populational and ecological studies in Brazilian mangroves of this group are scarce, and most are directed toward the characterization and effects of shrimp farming, as discussed in Sect. 9.5.

Depending on sedimentary conditions and availability of gastropod shells, hermit crabs of the families Paguridae (*Pagurus criniticornis*) and Diogenidae (*Clibanarius antillensis*, *C. sclopetarius, C. vittatus*, and *Calcinus tibicen*) may also occur in the mangroves (Amaral et al. 2010; Sousa et al. 2015). Finally, although not commonly surveyed in Brazilian mangroves, barnacles (Cirripedia) can be frequently found attached to the trunks and pneumatophores of mangrove trees. Farrapeira (2006) reported five species attached to mangrove trunks and wood debris (*Amphibalanus amphitrite*, *A. eburneus*, *A. improvisus*, *Chthamalus proteus*, and *Euraphia rhizophorae*), albeit noting the degraded condition of the study area in Recife (Pernambuco State, PE) (see Chap. 3, Map 8) could have limited the occurrence of barnacles since they are pollutant-sensitive animals. Another study in an estuarine area on the Island of Itamaracá (PE) registered six species attached to mangrove trees (*Amphibalanus amphitrite*, *A. improvisus*, *A. reticulatus*, *Chtalamus proteus*, *Euraphia rhizophorae*, and *Microeuraphia rhizophorae*) (Farrapeira 2008).

9.2.3 Mollusks

Mollusks are the second most common epifaunal component in mangroves (Nagelkerken et al. 2008). In some mangroves, mollusks reach high densities, being very abundant elements of the epifauna (mainly Gastropoda, oysters, and mussels) and infauna (most Bivalvia, except oysters, and mussels).

Bivalvia is likely the most known mollusk group inhabiting mangroves, mainly due to the occurrence of economically important species. Infaunal bivalves such as the heart clam *Anomalocardia brasiliana (flexuosa)* (popularly known as "berbigão" or "vôngole" in Brazil) and *Tagelus plebeius* (popularly known as "unha-de-velho" in Brazil) are examples of infaunal bivalves commonly found in mangroves, albeit also common on adjacent ecosystems and captured for human consumption (Nishida et al. 2006). However, epifaunal bivalves are also very common in mangroves.

The most prominent examples of epifaunal bivalves are mangrove oysters, belonging to the genera *Crassostrea*, especially *C. rhizophorae* and *C. gasar* (Fig. 9.2). These filter-feeding oyster species live attached to the trunks and the

Fig. 9.2 *Crassostrea rhizophorae* Guilding. (Photo: Clemente Coelho-Jr)

rhizosphere, fragments of rocks or other hard substrates, and may form dense aggregations in the lower intertidal area within the mangrove. Other abundant sessile bivalves inhabiting mangroves are the mussels *Mytella charruana* and *M. guyanensis* (popularly known as "sururu" in Brazil). Contrary to *Crassostrea* sp., these species mainly attach to the superficial mud sediment, forming mussel beds, although sometimes they are found attached to prop roots and pneumatophores.

Mytella spp. mussel beds increase substratum complexity, and an associated fauna may occur in areas where they are found (Nishida and Leonel 1995). Both the mangrove oysters and mussels are economically important, being collected by artisanal collectors, as well as cultivated in aquaculture systems (Nishida et al. 2006). Due to their filter-feeding habit, these species are studied as indicators of anthropic interference, based on the bioaccumulation of toxic compounds in their tissues (Molisani et al. 2004; Vaisman et al. 2005).

Shipworms (Bivalvia, Teredinae) are bivalves popularly known in Brazil as "teredo" or "turu." They can be found along the entire Brazilian coast, and they are known for their unusual life form, which allows them to bore and digest wood, forming extensive galleries inside mangrove logs. The number of species along the Brazilian coast varies among locations. In a mangrove in Paraná, South Brazil, ten species were recorded (*Bankia bagidaensis, B. fimbriatula, B. gouldi, B. rochi, Lyrodus floridanus, Nausitora fusticula, Neoteredo reynei, Teredo bartschi, T. mindanensis, T. navalis*) (Müller and Lana 1986), the highest number among all studies in the country. Six species were found in a mangrove at Dura Beach (Ubatuba, São Paulo), Southeast Brazil, in which *Nausitora fausticula* was the most abundant, and the species *Bankia* sp. and *Neoteredo reynei* were limited because of low salinity (Lopes and Narchi 1993). In contrast, in Northern Brazilian mangroves, Beasley et al. (2005) found four shipworm species, whereas Filho et al. (2013) recorded only *N. reynei* inhabiting mangrove logs (Filho et al. 2013). Although this may suggest a latitudinal pattern in shipworm richness, other factors such as salinity, zonation, and tree density are likely to influence differences between mangroves.

Gastropods are also commonly found in mangroves. These mollusks usually occupy the sedimentary environment, as epifauna, but may also be found climbing on the branches of mangrove trees. Some species, such as the pulmonated gastropod *Melampus coffea* (coffee bean snail) (Fig. 9.3), exhibit this climbing behavior during the high tide, returning to the sediment to forage during low tides (Maia and Tanaka 2007). The Littorinidae genera *Littorina* and *Littoraria* may also be abundantly found climbing on mangrove trees. Moreover, a survey by Queiroz and Dias (2014) on a hypersaline mangrove in northeast Brazil found that gastropods were the most abundant organisms inhabiting algal fronds of *Gracillaria* spp., especially the fat dove snail *Parvanachis obesa* (Collumbellidae) and the Virgin Nerite *Neritina virginea,* popularly known as "aruá-do-mangue" in Brazil (Neritidae). Additional findings regarding mollusks in Brazilian mangroves can be found in studies such as Lima et al. (2017), Muniz et al. (2019), and Saad et al. (2019).

9.2.4 Polychaetes

Polychaetes are one of the most numerically abundant organisms inhabiting mangrove sediments. The number of studies on these organisms in Brazilian mangroves, however, is limited in comparison to crustaceans and mollusks, as the worldwide trend (Nagelkerken et al. 2008). Contrary to the observed positive correlation between mollusk and crustacean abundance and the presence of mangroves, polychaetes in Brazilian mangroves and salt marshes do not seem to follow that pattern (Netto and Lana 1997; Pagliosa and Lana 2005). There is a decrease in density in mangrove areas, which may be attributed to changes in sedimentary conditions caused by mangrove plants (Checon et al. 2017). Nonetheless, studies on Brazilian mangroves reveal a high diversity of polychaete species (Lana et al. 1997; Netto and Gallucci 2003; Pagliosa and Lana 2005; Checon et al. 2017).

The polychaete *Capitella capitata* is part of a species complex likely composed of many cryptic species (Silva et al. 2017), and one of the most common polychaetes of Brazilian mangroves (Lana et al. 1997; Netto and Gallucci 2003; Checon et al. 2017). This species complex is widely recognized as a biological indicator of organic enrichment and is associated with very fine sediments (Pearson and Rosenberg 1978). Given the anthropic impacts on Brazilian mangroves, its occurrence may not only be linked to organic contamination but also be an effect of the naturally high content of organic matter of mangroves. Other capitellids, such as *Heteromastus filiformis, H. similis*, and *Mediomastus californiensis*, are found in much lower frequency and abundance compared to *C. capitata* (Pagliosa and Lana 2005; Checon et al. 2017).

Spionid polychaetes are also fairly common, especially the *Polydora* genus, which can be found in the sediment and boring oyster shells (Sabry et al. 2013). Nereididae polychaetes are rather dominant in mangroves and salt marshes, with *Laeonereis culveri* and *Nereis oligohalina* being the most reported species (Netto and Lana 1997; Pagliosa and Lana 2005). The Nereidid polychaetes *Namalycastis* sp. and *Namanereis* sp. are frequently associated with shipworm galleries, with the former being called "mother of turu" (Aviz et al. 2009), and decomposing mangrove leaves (Oliveira et al. 2012). Other reported species in Brazilian mangroves and salt marshes are *Glycinde multidens* (Goniadidae), *Isolda pulchella* (Ampharetidae), *Sigambra grubei* (Pilargidae), and *Typosyllis* sp. (Syllidae) (Netto and Lana 1997; Netto and Gallucci 2003; Pagliosa and Lana 2005; Checon et al. 2017). Additional research on polychaetes in Brazilian mangroves can be found in studies such as Melo et al. (2013), Silva-Camacho et al. (2017), and Tavares-Cutrim et al. (2018).

9.2.5 Other Groups

The macrofaunal groups reviewed in the previous items are the most recognized and likely most abundant organisms found in mangrove forests in Brazil. However, other faunal taxa may occur in these environments, especially in the sublittoral. Studies evaluating these groups in Brazilian mangroves are generally lacking, which compromises the recognition of the true biodiversity of mangroves, and the ecological role that these other groups may play in the mangrove ecosystem.

Echinoderms may be found in the shallow sublittoral of coastal ecosystems, including mangroves. For instance, the holothuroidean *Holothuria dakarensis* was recently registered for the first time in the Southwestern Atlantic, in a hypersaline mangrove in Northeast Brazil (Prata et al. 2014). Ophiuroideans can also be found associated with mangrove areas on the Brazilian coast, for example, *Amphipholis squamata, Ophiactis savignyi*, and *Ophiocoma echinate* (Lima et al. 2011).

Other groups, such as sponges, ascidians, and Actinairian anemones, can be found in the lower intertidal area of mangroves, particularly attached to the vegetation (Rosa-Filho and Farrapeira 1998).

9.2.6 Meiofauna

Meiofaunal organisms are numerically dominant in mangrove sediments (Netto and Gallucci 2003), even in the most organically enriched anoxic and sulfidic sediments, where macrofauna becomes rare (Zeppilli et al. 2018). The meiobenthic fauna plays an important ecological role in mangroves since it takes part in litter degradation (Somerfield et al. 1998) and is a significant component of the diet of commercially relevant fish and crustacean species (Coull et al. 1995; Dittel et al. 1997). For the most abundant meiobenthic taxa, the nematodes, it has been indicated that, at the global scale, richness is higher closer to the equator (Brustolin et al. 2018), a pattern distinct from the other benthic taxa, which may have evolved independently of mangrove tree richness.

At a local scale, the meiobenthos community structure shows high density and dominance. In Brazil, meiofauna studies are restricted to a few mangroves, in which nematodes accounted for 70–90% of the total sampled meiofauna (Gomes et al. 2002; Netto and Gallucci 2003; Paula et al. 2006; Santos et al. 2009; Venekey et al. 2019). Total nematode genus richness is similar among studies, with ca. 80 genera (Gerlach 1958; Netto and Gallucci 2003; Pinto et al. 2013), except for some mangroves from the Amazonian coast with less than 40 genera (Venekey et al. 2019). In this area, the low diversity of nematofauna together with low meiofauna density and richness is probably related to the high local hydrodynamics, since the large Amazon River discharges cause significant fluctuations in salinity (Rosa Filho and Aviz 2013, Venekey et al. 2019). There is so far no evidence for exclusively mangrove-specific nematode taxa (Nagelkerken et al. 2008), but there are some genera commonly found in mangrove areas (Fonseca and Netto 2015). The families Chromadoridae and Linhomoidae are the most relevant in terms of abundance and richness. *Terschellingia* (Linhomoidae), for example, accounting for more than half of the nematode abundance in Curuçá, Pará State (PA), North Brazil (Venekey and Melo 2016) (see Chap. 3, Map 2), also numerically dominated the nematode fauna in Santa Catarina (South Brazil), together with *Haliplectus* (Haliplectidae) and

Anoplostoma (Anoplostomatidae) (Netto and Gallucci 2003). In Itamaracá (PE), in Northeast Brazil, *Terschellingia* was found in all sampled microhabitats, and it was the most abundant in sandy mud, sponge sediments, and microhabitats composed of pneumatophore, algae, and sediments (Pinto et al. 2013). Nematodes in mangroves are mainly structured by the sediment characteristics and detritus biomass (Netto and Gallucci 2003; Pinto et al. 2013). Species richness is positively correlated to the amount of detritus and silt percentage (Netto and Gallucci 2003).

Harpacticoid copepods are also frequently found in high numbers in mangroves (Gomes et al. 2002). Like nematodes, there is little evidence of specialized mangrove copepod taxa; however, some species of the Darcythompsoniidae family can only be found on mangrove leaf litter (Por 1984; Somerfield et al. 1998). One study in Cananéia, state of São Paulo, registered 14 copepod species in the mangroves (Por et al. 1984) (see Chap. 3, Map 14). Copepods are also found in association with mangrove macroalgae (Garcia et al. 2015).

The mangrove phytal meiofauna is often dominated by acari (Nagelkerken et al. 2008, Gallucci et al. 2020) and is mostly found on turf growing on prop roots and pneumatophores (Chatterjee 2015). Halacarids were the second most abundant taxa to be found in a state of Santa Catarina mangrove studied by Netto and Gallucci (2003). The species *Rhombognathus abirus* and *R. picinguabensis* were described in the mangrove at Fazenda Beach, Ubatuba, state of São Paulo (Pepato and Silveira 2015).

Kinorhynchs are often reported as one of the rarest meiofaunal taxa in ecological studies, representing <1% of the total abundance (e.g., Netto and Gallucci 2003). Yet, in the Itamaracá (PE) mangroves, kinorhynchs were found in densities equal to, or even higher than, those observed for copepods (Gomes et al. 2002; Santos et al. 2009). Platyhelminthes and other soft-bodied taxa, such as gnasthostomulids and gastrotrichs, may be just as abundant as nematodes and copepods in mangroves; however, the techniques of meiofauna sampling often damage these flatworms, leading to underestimation of their abundance (Alongi 1987; Fonseca et al. 2010).

9.3 Ecological Role

The mangrove benthic fauna is associated with important ecological processes, especially crabs, widely known as important ecosystem engineers (Kristensen 2008). Crab burrowing and colonization cause significant bioturbation and detritus consumption, both contributing to sediment oxidation, increase in microbial activity, nutrient cycling, and diverse biochemical pathways (Koch and Wolff 2002; Kristensen 2008). Bioturbation by crabs also changes the environment's physical properties, such as surface topography and the distribution of sediment (Lee 2008).

Ucides cordatus is a key species for the homeostasis of Brazilian mangrove ecosystems. Along with the tree crab *Aratus pisonii*, they are the main leaf consumers in the country's mangroves, with *U. cordatus* being a detritus feeder, whereas *A. pisonii* is a fresh leaf consumer (Koch and Nordhaus 2010). This leaf consumption plays a major role in the cycling of nutrients that were not exported by the tides to adjacent areas.

The presence of vegetation directly influences the occurrence of crabs, but the crabs' activity also in turn impacts tree diversity. Propagule predation by the red crab *Goniopsis cruentata* is suggested to directly influence the composition of mangrove stands, as it predates preferentially on propagules of *Laguncularia racemosa* and *Avicennia schaueriana*, hampering the recruitment by these species and favoring the establishment of *Rhizophora mangle* (Ferreira et al. 2013).

Mangrove crabs and other benthic fauna can also provide habitat for other organisms by enhancing environmental complexity. Galleries created by *U. cordatus* serve as a home for other crustaceans in south Brazil, especially for the tanaid *Sinolobus stanfordi* (Kassuga and Masunari 2008). Tunnels created in mangrove logs by the shipworm *Neoteredo reynei* promote an enhancement of habitat complexity and create a habitat where a diverse microbenthic community thrives. Aviz et al. (2009) found that these abandoned galleries are densely colonized by decapod crustaceans, polychaetes, amphipods, and gastropods.

Crassostrea sp. oyster banks form reef-like formations that enable the occurrence of other organisms. Fishes, such as economically relevant *Centropomus* sp. (or "robalo" in Brazil) and *Atherinella brasiliensis* (or "peixe-rei" in Brazil), were observed in a mangrove oyster farm in Ceará State, Northeast Brazil, during high tide (Freitas et al. 2006). It is known that oysters also enhance the occurrence of other benthic groups such as crustaceans, mollusks, and polychaetes (Minchinton and Ross 1999; McAfee et al. 2016). However, studies with this associated fauna in Brazilian mangroves are still lacking.

9.4 Species of Social-Economic Relevance

Many coastal communities depend directly or indirectly on subsistence and commercial fishing, crab catching, and mollusk gathering in mangrove areas. These species are mainly exploited as a food source in regional cuisine, but also for medicinal, religious, and artistic purposes.

By far the most prominent benthic organisms in this context are brachyuran crabs, especially the mangrove crab *Ucides cordatus*, the blue land crab *Cardisoma guanhumi* (Fig. 9.4b), the mangrove root crab *Goniopsis cruentata*, and the blue swimming crabs (*Callinectes* spp.). These species represent the most important economically explored macrobenthic species in Brazilian mangroves, as shown by local surveys in coastal communities (Glaser 2003; Santos et al. 2017). A deeper analysis of the social importance and the biological aspect of economically important brachyuran crabs can be found in Chap. 10 of this book.

Bivalves are the second most exploited macrobenthic group for commercial purposes. The most prominent examples are Mytillidae mussels, the mangrove oyster *Crassostrea rhizophorae,* the stout razor clam *Tagelus plebeius*, and shipworm species. The clam *Anomalocardia brasiliana* (Fig. 9.5), one of the most

Fig. 9.5 *Anomalocardia brasiliana* Gmelin. (Photo: Guilherme Moraes de Oliveira Abuchahla)

consumed bivalves in Brazil, inhabits the mangrove borders, although it presents a higher abundance in estuarine areas rather than in the mangroves.

The importance of mangrove macrobenthic exploitation has been declining in the past years, mainly due to aquaculture expansion, overexploitation (stock depletion), and urbanization (habitat loss), and industrialization (contamination and pollution) of coastal cities. Nonetheless, many human communities, especially on the Northern and Northeastern coasts, still heavily rely on these resources for their subsistence. Communities around the Caeté River estuary (PA) (see Chap. 3, Map 2) have a strong dependence on the capture of mangrove species, especially *U. cordatus*, with up to 80% of the population relying on such activity (Glaser 2003).

Mangrove fauna also has cultural relevance in some regions. Especially in North and Northeast Brazil (although not restricted to), animal-derived material (e.g., extracts from tissues and organs, powdered components, ointments, crab fat, and animal infusions) is used as alternative medicine and in religious rituals or as souvenirs. Some examples include the fat of *U. cordatus* and *C. guanhumi*, used to treat hemorrhages and aid cicatrization, respectively (Costa-Neto and Marquez 2000); the flesh and shell of *C. rhizophorae* are used for the treatment of an array of unrelated illnesses (e.g., osteoporosis, stomach pain, flu, and cancer) (Alves and Rosa 2006); the flesh and bile of *G. cruentata* are used to treat venereal diseases (Alves and Rosa 2006). Other examples include the clam *A. brasiliana*, fiddler crabs, the mussel *M. charruana*, and the tree crab *A. pisonii*, which are used either in popular medicine practices, as well as to produce ornaments and artisanal goods (Alves and Rosa 2006; Alves et al. 2007). On the Northeast, the neogastropod *N. virginea* is commonly captured to make ornaments, such as necklaces, due to the unique patterns of their shells. The preservation of such traditions is important from a cultural standpoint; however, it should be noted that this may add pressure on populations, especially considering that most of these species are also commercialized in the food industry.

Mangroves are not only important for macrobenthic species of economic relevance. Many economically important fish species use mangroves both as a refuge from predators and as feeding ground, preying upon benthic species. That is one of the reasons why mangroves are considered nursing habitats for juvenile fishes. Some examples in Brazil are the sea catfish *Sciades herzbergii* (popularly known as "bagre"), a generalist feeder in mangroves, preying upon crustaceans, polychaetes, and mollusks (Giarizzo and Saint-Paul 2008); the banded puffer *Colomesus psittacus* (popularly known as "baiacu" in Brazil), a specialist feeder, preying mainly upon barnacles and brachyuran crabs (Krumme et al. 2007); *Eucinomostus* spp. (popularly known as "carapicu" in Brazil), a transitory inhabitant during juvenile stages, which enters mangrove areas to prey on polychaetes (Chaves and Otto 1999); and the flatfishes *Citharichthys* sp., *Etropus* sp., and *Achirus lineatus*, which feed on a diversity of brachyuran crabs, mollusks, polychaetes, and gammarid amphipods (Chaves and Serenato 1998). Conclusively, mangrove benthic species are a key element in the maintenance of fish stocks, important for ecological and socioeconomical equilibria.

9.5 Anthropic Impact on Mangrove Benthic Fauna

Worldwide, coastal areas are under increasing threat from anthropic activities. Mangroves are no exception, and loss by deforestation for urban, aquaculture, and industrial activities is a major concern for this ecosystem's conservation. In Brazil, estimates point to a considerable loss of the total mangrove area in the country just over the last three decades (Ferreira and Lacerda 2016). Furthermore, aside from the obvious impacts on the benthic fauna through the loss of complexity due to mangrove suppression, anthropic activities developed within and surrounding the estuaries affect the fauna directly.

As a consequence of the use of benthic species for commercial purposes, overexploitation of economically important species represents a strong pressure on the Brazilian mangrove fauna. Currently, there are regulations to control the exploitation of species such as *Ucides cordatus* and *Cardisoma guanhumi* during breeding periods. Nonetheless, that does not inhibit the effects of overexploitation on these species. *Ucides cordatus,* alongside *Neohelice granulata,* are considered as nearly threatened (NT, per the IUCN criteria), due to the decline in the population of *U. cordatus* and the limited range of *N. granulata* (Pinheiro and Boos 2016). *Cardissoma guanhumi* populations are rapidly declining because of exploitation and loss of *apicum* zones along the Brazilian coast (Firmo et al. 2012), and the species is considered critically endangered (CR, per IUCN criteria), with an estimated decrease of 88% in the population throughout three generations (Pinheiro and

Boos 2016). Populations of *Goniopsis cruentata* have also seen historical shrinking, reported by local communities in Pernambuco State, which is attributed to, aside from overexploitation, increased tourism, aquiculture, and incoming pesticides from upland crops (Maciel and Alves 2009). Overexploitation of *G. cruentata* also increased in recent years due to a concurrent decrease of *U. cordatus* populations (Botelho et al. 2004). This species, however, is still considered as least concern (LC, per IUCN criteria) (Pinheiro and Boos 2016). It is important to notice, however, that there are sustainable examples of fauna capture on the Brazilian coast (Diele et al. 2010) and that not every instance results in threats to local fauna (see Chap. 10). For fiddler crabs, most species have wide distributions along the coasts and are considered not threatened, except for *U. uruguayensis* and *U. victoriana*, classified as nearly threatened (NT) due to the limited range (Pinheiro and Boos 2016).

Aquaculture, especially shrimp farming, had the largest contribution to worldwide mangrove loss during the last century (Valiela et al. 2001). The implementation of aquaculture in northeast Brazil is perceived by local communities as the cause for the decrease in abundance of *U. cordatus*, which in turn hinders the livelihood of these communities who rely on the species (Santos et al. 2017). More systemic impacts have also been registered, with impacts of shrimp farming at Todos os Santos Bay, state of Bahia (see Chap. 3, Map 11), affecting the abundance, richness, and diversity of macrobenthic fauna as well as the chemistry of the water column (Hatje et al. 2016; Ribeiro et al. 2016), with an increase in the concentration of metals and organic nutrients (Ribeiro et al. 2016). The history of implementation of shrimp farming in Brazil and its impacts can be found in studies such as Guimarães et al. (2010), Queiróz et al. (2013), Tenório et al. (2015), Ferreira and Lacerda (2016), and Bernardino et al. (2018).

Climate change is another major concern for the conservation of mangrove ecosystems and their benthic fauna. Contrary to other coastal ecosystems, such as sandy beaches, predictions of impacts vary between regions, with studies indicating expansion in some areas, and contraction and even disappearance in others. Recent predictions for the Brazilian coast indicate that mangroves from the Northern region are likely to expand, whereas in more encroached areas, such as the Southeast coast, rising sea level is likely to cause habitat loss and consequently affect local biodiversity (Godoy and Lacerda 2015), as smaller mangroves are more likely to be affected by tide and wave energy effects. This loss in biodiversity would have a strong socio-ecological impact on this coastal sector.

The presence of solid waste and pollutants is another problem in Brazilian mangroves that impact benthic macrofauna. Mangroves are susceptible to this type of pollution due to the retention potential of the pneumatophores of *Avicennia schaueriana* and *Laguncularia racemosa* and leaves of *Spartina alterniflora*. In this sense, mangroves end up as sinks of plastic waste and debris that arrive in the estuaries (Ivar do Sul and Costa 2014; Vermeiren et al. 2016). Plastic wastes (e.g., bags, food wrapping) are the bulk component of this problem on the Southern and Southeastern coasts, and they derive mostly due to urban activities and the illegal occupation of the coastline (Cordeiro and Costa 2010; Vieira et al. 2013). Yet the impacts of this pollution on the benthic fauna are little studied in Brazil (but see

Gorman and Turra 2016). It is known that fiddler crabs (*Minuca rapax*) may ingest microplastic pellets that in turn accumulate in their internal organs, causing potentially harmful effects (Brennecke et al. 2015). The same likely occurs for other species, given the evidence of microplastic ingestion by many benthic species worldwide (Wright et al. 2013).

Mangroves are ecosystems prone to the retention of organic and inorganic contaminants. Studies in Guanabara Bay, state of Rio de Janeiro (RJ), show that mangroves retain heavy metals that would otherwise end up in coastal waters (Machado et al. 2002) (see Chap. 3, Map 13). Other studies show heavy metal accumulation in Brazilian mangroves (e.g., Meyer et al. 1998; Fonseca et al. 2013), which results in negative effects on the resident fauna. They were found to accumulate in the tissues of *Ucides cordatus* and *Callinectes danae* subjected to inorganic pollution in the Santos estuary, state of São Paulo, resulting in physiological alteration in oxygen consumption (Harris and Santos 2000). Malformations have been noted in *U. cordatus* inhabiting mangroves in heavily urbanized areas (Pinheiro and Toledo 2010). Heavy metals above the limit of safety for human consumption have been found in the tissue of *G. cruentata* in impacted mangroves (Vedolin et al. 2020). In the mangroves near Sepetiba Bay (RJ), a heavily industrialized area, species such as *Crassostrea gigas, Anomalocardia brasiliana, Callinectes danae*, and *Cardisoma guanhumi* show higher heavy metal levels than permitted by law for human consumption (Molisani et al. 2004). This contamination tends to be higher in filter-feeding organisms, such as bivalves, so much that *Crassostrea rhizophorae* has been chosen as an indicator species to monitor impact, especially due to its tendency to accumulate mercury (Vaisman et al. 2005).

Polycyclic aromatic hydrocarbon concentrations reach levels that might cause adverse effects on aquatic organisms in Northeast Brazil (Cavalcante et al. 2009). In the southeast coast, contamination by organochlorines decreased the density of *Neohelice granulate* once the pollutant accumulated in their eggs (de Sousa et al. 2008).

The above-mentioned examples are a sample of Brazilian studies on anthropogenic impacts on mangroves; however, more studies further discuss those and other aspects of the interaction between humans and the mangrove fauna (e.g., Faraco and Lana 2003; Soares-Gomes et al. 2010; França et al. 2012; Pinheiro et al. 2012; Pagliosa et al. 2016; Bernardino et al. 2020). The extensive literature available on the subject in Brazil shows the extent and relevance that anthropic activities have on these ecosystems and shows how the benthic fauna are under pressure and can be used to monitor changes and the status of mangroves.

9.6 Final Remarks

The biodiversity of Brazilian mangroves' benthic fauna is widely recognized. However, there are important regional gaps in the knowledge of this diversity, limiting the potential for better understanding and improved conservation strategies (Amaral and Jablonski 2005).

The growing human pressures on mangrove ecosystems along the Brazilian coast demonstrate the need to carry on further studies on their description and evaluation of their extent. The effects on the benthic fauna are mostly negative and there is no evidence of recovery. However, many gaps still need to be resolved regarding anthropic impacts on mangroves in Brazil, such as studies that assess the consequences of habitat fragmentation, as well as the implications of contamination at the population and community level. Considering that many of these species are ecologically and economically relevant, the comprehension of the negative effects is paramount to developing adequate conservation and management practices. Without such studies, it is hard to evaluate the exact condition of benthic fauna in Brazilian mangroves and to subsidize new management strategies based on the acquired knowledge.

Ferreira and Lacerda (2016) recently suggested measures for the protection of Brazilian mangroves, such as the integration of native communities in policymaking, sustainable aquaculture practices, and the planting of mangrove trees to lessen the speed of habitat loss. The extent to which these actions will be carried out and the outcomes on the recovery or protection of benthic fauna are still to be unfolded and studied.

Acknowledgments The authors would like to thank Clemente Coelho Júnior and Guilherme Moraes de Oliveira Abuchahla for providing images to this chapter. The authors acknowledge the Fundação de Amparo à Pesquisa do Estado de São Paulo (Fapesp) for financial support to HHC (2018/22036-0) and GNC (grant 2017/17071-9), linked to the FAPESP Global Climate Change Research Program thematic project (grant 2015/03804-9), to BPC (2017/21986-2) and YMLSE (2018/05099-9); and the Conselho Nacional de Desenvolvimento Científico e Tecnológico for financial support to ACZA (301551/2019-7).

References

Alongi DM (1987) Intertidal zonation and seasonality of meiobenthos in tropical mangrove estuaries. Mar Biol 95:447–458

Alves RRN, Rosa IL (2006) From cnidarians to mammals: the use of animals as remedies in fishing communities in NE Brazil. J Ethnopharmacol 107:259–276

Alves RRN, Rosa IL, Santana GG (2007) The role of animal-derived remedies as complementary medicines in Brazil. Bioscience 57:949–955

Amaral ACZ, Jablonski S (2005) Conservation of marine and coastal biodiversity in Brazil. Conserv Biol 19:625–631

Amaral ACZ, Migotto A, Turra A, Schaeffer-Novelli Y (2010) Araçá: biodiversidade, impactos e ameaças. Biota Neotropica 10:219–264

Andrade EB, Lima-Júnior TB, Leite-Júnior JMA, Leite JRSA (2012) Predation by native fish and feeding by crab species on *Leptodactylus macrosternum* Miranda-Ribeiro, 1926 (Anura: Leptodactylidae) in northeastern, Brazil. Herpetol Notes 5:173–175

Arakaki JY, De Grande FR, Arvigo AL, Pardo JCF, Fogo BR, Sanches FHC, Miyai CA, Marochi MZ, Costa TM (2020) Battle of the borders: is a range-extending fiddler crab affecting the spatial niche of a congener species? J Exp Mar Biol Ecol 532:151455

Aviz D, de Mello CF, da Silva PF (2009) Macrofauna associada às galerias de *Neoteredo reynei* (Bartsch, 1920) (Mollusca: Bivalvia) em troncos de *Rhizophora mangle* Linnaeus durante o período menos chuvoso, em manguezal de São Caetano de Odivelas, Pará (costa norte do Brasil). Bol Mus Para Emilio Goeldi Cienc Nat 4:47–55

Banci KRS, Mori GM, Oliveira MA, Paganelli FL, Pereira MR, Pinheiro MAA (2017) Mar Pollut Bull 116:440–447

Beasley CR, Fernandes CM, Gomes CP, Brito BA, Santos SML, Tagliaro CH (2005) Molluscan diversity and abundance among coastal habitats in Northern Brazil. Ecotropica 11:9–20

Bernardino A, Gomes LEO, Hadlich HL, Andrades R, Correa LB (2018) Mangrove clearing impacts on macrofaunal assemblages and benthic food webs in a tropical estuary. Mar Pollut Bull 126:228–235

Bernardino AF, Sanders CJ, Bissoli LB, Gomes LEO, Kauffman JB, Ferreira TO (2020) Land use impacts on benthic bioturbation potential and carbon burial in Brazilian mangrove ecosystems. Limnol Oceanogr 65:2366–2376

Bezerra LEA, Dais CB, Santana GZ, Matthews-Cascon H (2006) Spatial distribution of fiddler crabs in a tropical mangrove of Northeast Brazil. Sci Mar 70:759–766

Botelho RO, Andrade CER, Santos MCF (2004) Estudo da População de Aratu-do-Mangue, *Goniopsis cruentata* (Latraille, 1803) (Crustacea, Decapoda, Grapsidae) no Estuário do Rio Camaragibe (Alagoas - Brasil). Bol Tec Cient CEPENE 12:91–98

Branco JO (1990) Aspectos ecológicos dos Brachyura (Crustacea: Decapoda) no manguezal do Itacorubi, SC – Brasil. Rev Bras Zool 7:165–179

Brennecke D, Ferreira EC, Costa TMM, Appel D, Gama BAP, Lenz M (2015) Ingested microplastics (>100 μm) are translocated to organs of the tropical fiddler crab *Uca rapax*. Mar Pollut Bull 96:491–495

Brustolin MC, Nagelkerken I, Fonseca G (2018) Large-scale distribution patterns of mangrove nematodes: a global meta-analysis. Ecol Evol 8:4734–4742

Castiglioni DS, Almeida AO, Bezerra LEA (2010) More common than reported: range extension, size–frequency and sex-ratio of *Uca* (*Minuca*) *victoriana* (Crustacea: Ocypodidae) in tropical mangroves, Brazil. Mar Biodivers Rec 3:1–8

Cavalcante RM, Sousa FW, Nascimento RF, Silveira ER, Freire GSS (2009) The impact of urbanization on tropical mangroves (Fortaleza, Brazil): evidence from PAH distribution in sediments. J Environ Manag 91:328–335

Cavalcanti BALP, Silva DB, Silva LMB, Rodrigues GG (2009) Socioeconomic aspects of the production chain of species Cardisoma guanhumi in the northeast coast of Brazil. J Environ Anal Prog 4:65–70

Chatterjee T (2015) A checklist of halacarid and pontarachnid mites (Acari: Halacaridae and Pontarachnidae) associated with mangroves. Ecol Mont 2:334–341

Chaves PTC, Otto G (1999) The mangrove as a temporary habitat for fish: the *Eucinostomus* Species at Guaratuba Bay, Brazil (25° 52′S;48° 39′W). Braz Arch Biol Technol 42:61–68

Chaves PTC, Serenato A (1998) Diversidade de dietas na assembléia de linguados (Teleostei, Pleuronectiformes) do manguezal da Baía de Guaratuba, Paraná, Brasil. Rev Bras Oceanogr 46(1):61–68

Checon HH, Costa TM (2017) Fiddler crab (Crustacea: Ocypodidae) distribution and the relationship between habitat occupancy and mouth appendages. Mar Biol Res 3:618–629

Checon HH, Corte GN, Silva CF, Schaeffer-Novelli Y, Amaral ACZ (2017) Mangrove vegetation decreases density but does not affect species richness and trophic structure of intertidal polychaete assemblages. Hydrobiology 795:169–179

Colpo KD, Chacur MM, Guimarães FJ, Negreiros-Fransozo ML (2011) Subtropical Brazilian mangroves as a refuge of crab (Decapoda: Brachyura) diversity. Biodivers Conserv 20:3239–3250

Cordeiro CAMM, Costa TM (2010) Evaluation of solid residues removed from a mangrove swamp in the São Vicente Estuary, SP, Brazil. Mar Pollut Bull 60:1762–1767

Costa TM, Negreiros-Fransozo ML (2001) Morphological adaptation of the second maxilliped in semiterrestrial crabs of the genus *Uca* from a subtropical Brazilian mangrove. Nauplius 9:123–131

Costa RC, Lopes M, Castilho AL, Fransozo A, Simões SM (2008) Abundance and distribution of juvenile pink shrimps *Farfantepenaeus* spp. in a mangrove estuary and adjacent bay on the northern shore of the São Paulo State, southeastern Brazil. Invertebr Reprod Dev 52:51–58

Costa-Neto EM, Marques JGW (2000) Faunistic resources used as medicines by artisanal fishermen from Siribinha Beach, State of Bahia, Brazil. J Ethnobiol 20:93–109

Coull BC, Greenwood JG, Fielder DR, Coull BA (1995) Subtropical Australian juvenile fish eat meiofauna: experiments with winter whiting Sillago maculata and observations on other species. Mar Ecol Prog Ser 125:13–19

Davanso MB, Moreira LB, Pimentel MF, Costa-Lotufo LV, Abessa DMS (2013) Biomarkers in mangrove root crab *Goniopsis cruentata* for evaluating quality of tropical estuaries. Mar Environ Res 91:80–88

De Grande FR, Granado P, Sanches FHC, Costa TM (2018) Organic matter affects fiddler crab distribution? Results from field and laboratorial trials. Estuar Coast Shelf Sci 212:138–145

De Sousa AS, Torres JPM, Meire RO, Neves RC, Couri MS, Serejo CS (2008) Organochlorine pesticides (OCs) and polychlorinated biphenyls (PCBs) in sediments and crabs (*Chasmagnathus granulata*) from mangroves of Guanabara Bay, Rio de Janeiro State, Brazil. Chemosphere 73:186–192

Diele K, Koch V, Saint-Paul U (2005) Population structure, catch composition and CPUE of the artisanally harvested mangrove crab *Ucides cordatus* (Ocypodidae) in the Caeté estuary, North Brazil: indications for overfishing? Aquat Living Resour 18:169–178

Diele K, Araújo ARR, Glaser M, Salzmann U (2010) Artisanal fishery of the mangrove crab *Ucides cordatus* (Ucididae) and first steps toward a successful co-Management in Bragança, North Brazil. In: Saint-Paul U, Schneider H (eds) Mangrove dynamics and management in Northeastern Brazil. Springer, Berlin/Heidelberg, pp 287–297

Dittel AI, Epifanio CE, Cifuentes LA, Kirchman DL (1997) Carbon and nitrogen sources for shrimp postlarvae fed natural diets from a tropical mangrove system. Estuar Coast Shelf Sci 45:629–637

Duarte LFA, Souza CA, Nobre CR, Pereira CDS, Pinheiro MAA (2016) Multi-level biological responses in *Ucides cordatus* (Linnaeus, 1763) (Brachyura, Ucididae) as indicators of conservation status in mangrove areas from the western Atlantic. Ecotoxicol Environ Saf 133:176–187

Faraco LFD, Lana PC (2003) Response of polychaetes to oil spills in natural and defaunated subtropical mangrove sediments from Paranaguá bay (SE Brazil). Hydrobiologia 496:321–328

Farrapeira CMR (2006) Barnacles of the estuarine region of Recife, Pernambuco. Trop Oceanogr 34:100–119

Farrapeira CMR (2008) Cirripedia Balanomorpha del estuario del Río Paripe (Isla de Itamaracá Pernambuco Brasil). Biota Neotrop 8(3):31–39

Ferreira AC, Lacerda LD (2016) Degradation and conservation of Brazilian mangroves, status and perspectives. Ocean Coast Manag 125:38–46

Ferreira AC, Ganade G, Freire FAM, Attayde JL (2013) Propagule predation in a Neotropical mangrove: the role of the Grapsid crab *Goniopsis cruentata*. Hydrobiologia 707:135–146

Filho CS, Tagliaro CH, Beasley CR (2013) Seasonal abundance of the shipworm *Neoteredo reynei* (Bivalvia, Teredinidae) in mangrove driftwood from a northern Brazilian beach. Iheringia 98: 17–23

Firmo AMS, Tognella MMP, Silva SR, Barboza RRRD, Alves RRN (2012) Capture and commercialization of blue land crabs ("guaiamum") *Cardisoma guanhumi* (Latreille, 1825) along the coast of Bahia State, Brazil: an ethnoecological approach. J Ethnobiol Ethnomed 8:12

Fonseca G, Netto SA (2015) Macroecological patterns of estuarine nematodes. Estuar Coasts 38: 612–619

Fonseca VG, Carvalho GR, Sung W, Johnson HF, Power DM, Neill SP, Packer M, Blaxter ML, Lambshead JD, Thomas WK, Creer S (2010) Second-generation environmental sequencing unmasks marine metazoan biodiversity. Nat Commun 1:98

Fonseca EF, Baptista Neto JA, Silva CG (2013) Heavy metal accumulation in mangrove sediments surrounding a large waste reservoir of a local metallurgic plant, Sepetiba Bay, SE Brazil. Environ Earth Sci 70:643–650

França MC, Francisquini MI, Cohen MCL, Pessenda LCR, Rossetti DF, Guimarães JTF, Smith CB (2012) The last mangroves of Marajó Island—Eastern Amazon: impact of climate and/or relative sea-level changes. Rev Palaeobot Palynol 187:50–65

Freitas LEL, Feitosa CV, Araújo ME (2006) Mangrove oyster (*Crassostrea rhizophorae*) (Guilding, 1928) farming areas as artificial reefs for fish: a case study in the State of Ceará, Brazil. Braz J Oceanogr 54:31–39

Freitas RF, Schrack EC, Sieg RD, Silliman BR, Costa CSB (2015) Grazing scar characteristics impact degree of fungal facilitation in *Spartina alterniflora* leaves in a South American Salt Marsh. Braz Arch Biol Technol 58:103–108

Gallucci F, Christofoletti RA, Fonseca G, Dias GM (2020) The effects of habitat heterogeneity at distinct spatial scales on hard-bottom associated communities. Diversity 12:1–13

Garcia A, Bueno M, Leite FPP (2015) The Bostrychietum community of pneumatophores in Araçá Bay: an analysis of the diversity of macrofauna. J Mar Biol Assoc UK 96:1–8

Gerlach SA (1958) Die mangroveregion tropischer Küsten als Les-bensraum. Zeitschrift für Morphologie und ökologie der Tierre 46:636–730

Giarrizzo T, Saint-Paul U (2008) Ontogenetic and seasonal shifts in the diet of the pemecou sea catfish *Sciades herzbergii* (Siluriformes: Ariidae), from a macrotidal mangrove creek in the Curuçá estuary, Northern Brazil. Rev Biol Trop 56:861–873

Glaser M (2003) Interrelations between mangrove ecosystem, local economy and social sustainability in Caeté Estuary, North Brazil. Wetl Ecol Manag 11:265–272

Godoy MDP, Lacerda LD (2015) Mangroves' response to climate change: a review of recent findings on mangrove extension and distribution. An Acad Bras Ciênc 87:651–667

Gomes CAA, dos Santos PJP, Alves TNC, Rosa Filho JS, Sousa Santos LP (2002) Variação temporal da meiofauna em área de manguezal em Itamaracá – Pernambuco. Atlântica RG 24(2): 89–96

Gorman D, Turra A (2016) The role of mangrove revegetation as a means of restoring macrofaunal communities along degraded coasts. Sci Total Environ 566-567:223–229

Grabowski RC, Simões SM, Castilho AL (2016) Population structure, sex ratio and growth of the seabob shrimp *Xiphopenaeus kroyeri* from coastal waters of southern Brazil. Zookeys 457:253–269

Guimarães AS, Travassos P, Souza-Filho PWM, Gonçalves FD, Costa F (2010) Impacts of aquaculture on mangrove áreas in the northern Pernambuco Coast (Brazil) using remote sensing and geographic information system. Aquac Res 41:828–838

Harris RR, Santos MCF (2000) Heavy metal contamination and physiological variability in the Brazilian mangrove crabs *Ucides cordatus* and *Callinectes danae* (Crustacea: Decapoda). Mar Biol 137:691–703

Hatje V, Souza MM, Ribeiro LF, Eça GF, Barros F (2016) Detection of environmental impacts of shrimp farming through multiple lines of evidence. Environ Pollut 219:672–684

Hirose GL, Souza LS, Silva SLR, Alves DFR, Negreiros-Fransozo ML (2015) Population structure of the red mangrove crab, Goniopsis cruentata (Decapoda: Grapsidae) under different fishery impacts: implications for resource management. Rev Bras Trop 63:443–457

Ivar do Sul JA, Costa MF (2014) The present and future of microplastic pollution in the marine environment. Environ Pollut 185:352–364

Kassuga A, Masunari S (2008) Carcinofauna acompanhante de galerias de *Ucides cordatus* (Linnaeus) (Crustacea Ocypodidae) no manguezal do rio Pinheiros, Cabaraquara, Baía de Guaratuba, PR. Acta Biol Par 37:255–258

Koch V, Nordhaus I (2010) Feeding ecology and ecological role of north brazilian mangrove crabs. In: Saint-Paul U, Schneider H (eds) Mangrove dynamics and management in North Brazil, Ecological studies (analysis and synthesis), vol 211. Springer Verlag, Heidelberg, pp 265–273

Koch V, Wolff M (2002) Energy budget and ecological role of mangrove epibenthos in the Caeté estuary, North Brazil. Mar Ecol Prog Ser 228:119–130

Koch V, Wolff M, Diele K (2005) Comparative population dynamics of four fiddler crabs from a north Brazilian mangrove ecosystem. Mar Ecol Prog Ser 291:177–188

Kristensen E (2008) Mangrove crabs as ecosystem engineers; with emphasis on sediment processes. J Sea Res 59:30–43

Krumme U, Keuthen H, Saint-Paul U, Villwock W (2007) Contribution to the feeding ecology of the banded puffer fish *Colomesus psittacus* (Tetraodontidae) in north Brazilian mangrove creeks. Braz J Biol 67:383–392

Lana PC, Bernardino A (2018) Brazilian Estuaries: a benthic perspective. Springer, Cham

Lana PC, Couto ECG, Almeida MVO (1997) Polychaete distribution and abundance in intertidal flats of Paranaguá Bay (SE Brazil). Bull Mar Sci 60:433–442

Lee SY (2008) Mangrove macrobenthos: assemblages, services, and linkages. J Sea Res 59:16–29

Lima MLF, Correia MD, Sovierzoski H, Manso CLC (2011) New records of Ophiuroidea (Echinodermata) from the shallow waters of Maceió, state of Alagoas, Brazil. Mar Biodivers Rec 4:e97

Lima SFB, Lucena RA, Santos GM, Souza JW, Christoffersen ML, Guimarães CR, Oliveira GS (2017) Inventory of mollusks from the estuary of the Paraíba River in northeastern Brazil. Biota Neotrop 17:e20160239

Lima-Gomes RC, Cobo VJ, Fransozo A (2011) Feeding behaviour and ecosystem role of the red mangrove crab *Goniopsis cruentata* (Latreille, 1803) (Decapoda, Grapsoidea) in a subtropical estuary on the Brazilian coast. Crustaceana 84:735–747

Lira JJPR, Calado TCS, Araújo MSLC (2013) Condition factor of Goniopsis cruentata (Crustacea, Brachyura, Grapsidae) from Mundaú/Manguaba estuarine complex, Alagoas, Brazil. Iheringia 102:285–291

Lopes SGBC, Narchi W (1993) Levantamento e distribuição das espécies de Teredinidae (Mollusca – Bivalvia) no manguezal da Praia Dura, Ubatuba, São Paulo. Bol Inst Oceanogr 41:29–38

Machado W, Silva-Filho EV, Oliveira RR, Lacerda LD (2002) Trace metal retention in mangrove ecosystems in Guanabara Bay, SE Brazil. Mar Pollut Bull 44:1277–1280

Machado GB, Gusmão-Júnior JB, Costa TM (2013) Burrow morphology of *Uca uruguayensis* and *Uca leptodactylus* (Decapoda: Ocypodidae) from a subtropical mangrove forest in the Western Atlantic. Integr Zool 8:307–314

Maciel DC, Alves AGC (2009) Conhecimentos e práticas locais relacionados ao aratu *Goniopsis cruentata* (Latreille, 1803) em Barra de Sirinhaém, litoral sul de Pernambuco, Brasil. Biota Neotrop 9:29–36

Maia RC, Tanaka MO (2007) Avaliação de efeitos de espécies de mangue na distribuição de *Melampus coffeus* (Gastropoda, Ellobiidae) no Ceará, Nordeste do Brasil. Iheringia 97:379–382

McAfee D, Cole VJ, Bishop MJ (2016) Latitudinal gradients in ecosystem engineering by oysters vary across habitats. Ecology 97:929–939

Melo KR, Tagliaro CH, Beasley CR (2013) Seasonal changes in the subtidal benthic macrofauna of a mangrove coast in northern Brazil. J Coast Res 65:87–92

Meyer U, Hagen W, Medeiros C (1998) Mercury in a northeastern Brazilian mangrove area, a case study: potential of the mangrove oyster *Crassostrea rhizophorae* as bioindicator for mercury. Mar Biol 131:113–121

Minchinton TE, Ross PM (1999) Oysters as habitat for limpets in a temperate mangrove forest. Aust J Ecol 24:157–170

Molisani MM, Marins RV, Paraquetti HHM, Bidone ED, Lacerda LD (2004) Environmental changes in Sepetiba Bay, SE Brazil. Reg Environ Chang 4:17–27

Müller ACP, Lana PC (1986) Teredinidae (Mollusca, Bivalvia) do Litoral do Paraná. Nerítica 1:27–48

Muniz APA, Duarte RCS, Mota ELS, Lima JO, Dias TLP (2019) Mollusks associated with the *Avicennia schaueriana* (Magnoliophyta: Acanthaceae) roots of a hypersaline estuary in north-eastern Brazil. Acta Limnol Bras 31:e16

Nagelkerken I, Blaber SJM, Bouillon S, Green P, Haywood M, Kirton LG, Meynecke J-O, Pawlik J, Penrose HM, Sasekumar A, Somerfield PJ (2008) The habitat func-tion of mangroves for terrestrial and marine fauna: a review. Aquat Bot 89:155–185

Netto SA, Gallucci F (2003) Meiofauna and macrofauna communities in a mangrove from the Island of Santa Catarina, South Brazil. Hydrobiologia 505:159–170

Netto SA, Lana PC (1997) Intertidal zonation of benthic macrofauna in a subtropical salt marsh and nearby unvegetated flat (SE, Brazil). Hydrobiologia 353:171–180

Nishida AK, Leonel RMV (1995) Occurrence, population dynamics and habitat characterization of *Mytella guyanensis* in the Paraíba do Norte River estuary. Bolm Inst Oceanogr 43:41–49

Nishida AK, Nordi N, Alves RNN (2006) The lunar-tide cycle viewed by crustacean and mollusc gatherers in the state of Paraíba, Northeast Brazil and their influence in collection attitudes. J Ethnobiol Ethnomed 2:28

Nordhaus I, Wolff M (2007) Feeding ecology of the mangrove crab *Ucides cordatus* (Ocypodidae): food choice, food quality and assimilation efficiency. Mar Biol 151:1665–1681

Nordhaus I, Diele K, Wolff M (2009) Activity patterns, feeding and burrowing behaviour of the crab *Ucides cordatus* (Ucididae) in a high intertidal mangrove forest in North Brazil. J Exp Mar Biol Ecol 374:104–112

Oliveira AB, Rizzo AE, Conceição E, Couto G (2012) Benthic macrofauna associated with decomposition of leaves in a mangrove forest in Ilhéus, State of Bahia, Brazil. J Mar Biol Assoc U K 92:1479–1487

Oliveira-Neto JF, Pie MR, Chammas MA, Ostrensky A, Boeger WA (2008) Phylogeography of the blue land crab, *Cardisoma guanhumi* (Decapoda: Gecarcinidae) along the Brazilian coast. J Mar Biol Assoc UK 88(7):1417–1423

Oshiro LMY, Silva R, Silva ZS (1998) Composição da fauna de braquiúros (Crustacea, Decapoda) dos manguezais da Baía de Sepetiba-RJ. Nauplius 6:31–40

Pagliosa PR, Lana PC (2005) Impact of plant cover removal on macrobenthic community structure of a subtropical salt marsh. Bull Mar Sci 77:1–18

Pagliosa PR, Oortman MS, Rovai AS, Soriano-Sierra EJ (2016) Is mangrove planting insufficient for benthic macrofaunal recovery when environmental stress is persistent? Ecol Eng 95:290–301

Paula JHC, Rosa Filho JS, Souza ALB, Aviz D (2006) A meiofauna como indicadora de impactos da carcinicultura no estuário de Curuçá (PA). Boletim Laboratório Hidrobiologia 19:61–72

Pearson TH, Rosenberg R (1978) Macrobenthic succession in relation to organic enrichment and pollution of the marine environment. Oceanogr Mar Biol Ann Rev 16:229–311

Pepato AR, Silveira PSA (2015) Two new species of *Rhombognathus* (Halacaridae, Trombidiformes) from a Mangrove in the northern littoral zone of São Paulo State (Brazil). Zootaxa 3905:500–510

Pinheiro MAA, Boos H (2016) Livro vermelho dos crustáceos do Brasil. Avaliação 2010-2014. Sociedade Brasileira de Carcinologia, Porto Alegre

Pinheiro MAA, Toledo TR (2010) Malformation in the crab *Ucides cordatus* (Linnaeus, 1763) (Crustacea, Brachyura, Ocypodidae), in São Vicente (SP), Brazil

Pinheiro MAA, Gandara e Silva PP, LFA D, Almeida AA, Zanotto FP (2012) Accumulation of six metals in the mangrove crab *Ucides cordatus* (Crustacea: Ucididae) and its food source, the red mangrove *Rhizophora mangle* (Angiosperma: Rhizophoraceae). Ecotoxicol Environ Saf 81: 114–121

Pinheiro MAA, Souza MR, Santos LCM, Fontes RCF (2018) Density, abundance and extractive potential of the mangrove crab, Ucides cordatus (Linnaeus, 1763) (Brachyura, Ocypodidae): subsidies for fishery management. An Acad Bras Cienc 90:1381–1395

Pinto TK, Austen MCV, Warwick RM, Somerfield PJ, Esteves AM, Castro FJV, Fonseca-Genevois VG, Santos PJP (2013) Nematode diversity in different microhabitats in a mangrove region. Mar Ecol 34:257–268

Por FD (1984) Notes on the benthic copepods of the mangal ecosystem. In: Por FD, Dor I (eds) Hydrobiology of the Mangal – the ecosystem of the mangrove forests. Developments in hydrobiology, vol 20. W. Junk, The Hague, pp 67–70

Por FD, Prado-Por MS, Oliveira EC (1984) The mangal of the estuary and lagoon system of Cananeia (Brazil). In Por FD, Dor I(eds.) Hydrobiology of the Mangal—The Ecosystem of the MangroveForests. Dr. W. Junk Publishers, TheHague, pp. 211–228

Prata J, Dias TLP, Christoffersen ML (2014) Occurrence of Holothuria (Holothuria) dakarensis (Holothuroidea: Echinodermata) in the South-Western Atlantic, with notes on distribution and ecology. Mar Biodivers Rec 7:e9

Queiroz RNM, Dias TLP (2014) Molluscs associated with the macroalgae of the genus Gracilaria (Rhodophyta): importance of algal fronds as microhabitat in a hypersaline mangrove in northeastern Brazil. Braz J Biol 74:52–63

Queiroz L, Rossi S, Meireles J, Coelho C (2013) Shrimp aquaculture in the federal state of Ceará, 1970-2012: trends after mangrove forest privatization in Brazil. Ocean Coast Manag 73:54–62

Ribeiro LF, Eça GF, Barros F, Hatje V (2016) Impacts of shrimp farming cultivation cycles on microbenthic assemblages and chemistry of sediments. Environ Pollut 211:307–315

Rosa Filho JS, Aviz D (2013) Macrobenthic communities of an Amazonian estuary (Guajará Bay, Brazil): temporal and spatial changes. J Coast Res 65:123–128

Rosa-Filho JS, Farrapeira-Assunção CM (1998) Zonação vertical da macrofauna bentônica dos andares supra litoral e médio litoral dos manguezais da região da Ilha de Itamaracá (Pernambuco, Brazil). Cad Ômega 4:59–68

Saad LO, Cunha CM, Colpo KD (2019) How mollusk assemblages respond to different urbanization levels: characterization of the malacofauna in subtropical Brazilian mangroves. Mar Biodivers 49:989–999

Sabry RC, Gesteira TCV, Magalhães ARM, Barraco MA, Guertler C, Ferreira LP, Vianna RP, Silva PM (2013) Parasitological survey of mangrove oyster, *Crassostrea rhizophorae*, in the Pacoti River Estuary, Ceará State, Brazil. J Invertebr Pathol 112:24–32

Sandrini-Neto L, Lana PC (2011) Distribution patterns of the crab *Ucides cordatus* (Brachyura, Ucididae) at different spatial scales in subtropical mangroves of Paranaguá Bay (southern Brazil). Helgol Mar Res 66:258

Santos PJP, Botter-Carvalho ML, Nascimento-Júnior AB, Marinho RGC, Carvalho PVVC, Valença APMC (2009) Response of estuarine meiofauna assemblage to effects of fertilizer enrichment used in the sugar cane monoculture Pernambuco, Brazil. Braz J Oceanogr 57(1): 43–55

Santos LCM, Gasalla MA, Dahdouh-Guebas F, Bitencourt MD (2017) Socio-ecological assessment for environmental planning in coastal fishery areas: a case study in Brazilian mangroves. Ocean Coast Manag 138:60–69

Schories D, Barletta-Bergan A, Barletta M, Krumme U, Mehlig U, Rademaker V (2003) The keystone role of leaf-removing crabs in mangrove forests of North Brazil. Wetl Ecol Manag 11: 243–255

Silva CC, Schwamborn R, Lins Oliveira JE (2014) Population biology and color patterns of the blue land crab, *Cardisoma guanhumi* (Latreille 1828) (Crustacea: Gecarcinidae) in the Northeastern Brazil. Braz J Biol 74:949–958

Silva CF, Seixas VC, Barroso R, Di Domenico M, Amaral ACZ, Paiva PC (2017) Demystifying the *Capitella capitata* complex (Annelida, Capitellidae) diversity by morphological and molecular data along the Brazilian coast. PLoS One 12:e0177760

Silva-Camacho DS, Gomes RS, Santos JNS, Araújo FG (2017) Distribution of benthic fauna in sediment grains and prop roots of a mangrove channel in South-Eastern Brazil. J Mar Biol Assoc UK 97:377–385

Soares-Gomes A, Neves RL, Aucélio R, Van der Ven PH, Pitombo FB, Mendes CLT, Ziolli RL (2010) Changes and variations of polycyclic aromatic hydrocarbon concentrations in fish, barnacles and crabs following an oil spill in a mangrove of Guanabara Bay, Southeast Brazil. Mar Pollut Bull 60:1359–1363

Somerfield PJ, Gee JM, Aryuthaka C (1998) Meiofaunal communities in a Malaysian mangrove forest. J Mar Biol Assoc UK 78:717–732

Sousa DB, Santos NB, Oliveira VM, Carvalho-Neta RNF, Almeida ZS (2015) Carcinofauna bêntica estuarina de dois manguezais da costa amazônica maranhense, Brasil. Iheringia Sér Zool 105:339–347

Spalding M (2010) World atlas of mangroves. Routledge, London

Spivak ED (2010) The crab *Neohelice* (=*Chasmagnathus*) *granulata*: an emergent animal model from emergent countries. Helgol Mar Res 64(3):149–154

Tavares-Cutrim AS, Santos Souza LK, Passos Ribeiro L, Oliveira VM, Almeida ZS (2018) Estructura de la comunidad de poliquetos en un manglar de la costa norte de Brasil. Acta Biol Colomb 23:286–294

Tenório GS, Souza-Filho PWM, Ramos EMLS, Alves PJO (2015) Mangrove shrimp farm mapping and productivity on the Brazilian Amazon coast: environmental and economic reasons for coastal conservation. Ocean Coast Manag 104:65–77

Thurman C, Faria SC, McNamara JC (2013) The distribution of fiddler crabs (*Uca*) along the coast of Brazil: implications for biogeography of the western Atlantic Ocean. Mar Biodivers Rec 6:1–21

Thurman CL, Faria SC, McNamara JC (2017) Geographical variation in osmoregulatory abilities among populations of ten species of fiddler crabs from the Atlantic coast of Brazil: a macrophysiological analysis. J Exp Mar Biol Ecol 497:243–253

Vaisman AG, Marins RV, Lacerda LD (2005) Characterization of the mangrove oyster, *Crassostrea rhizophorae*, as a biomonitor for mercury in tropical estuarine systems, Northeast Brazil. Bull Environ Contam Toxicol 74:582–588

Valiela I, Bowen JL, York JK (2001) Mangrove Forests: One of the World's Threatened Major Tropical Environments: at least 35% of the area of mangrove forests has been lost in the past two decades, losses that exceed those for tropical rain forests and coral reefs, two other well-known threatened environments. Bioscience 51:807–815

Vedolin MC, Trevizani TH, Angeli JLF, Petti AMV, Figuera RLC (2020) Assessment of metal concentration in *Goniopsis cruentata* (Latreille, 1803) (Decapoda, Grapsidae) from two Brazilian mangroves under different anthropogenic influences. Reg Stud Mar Sci 36:101305

Venekey V, Melo TPG (2016) Nematodes as indicators of shrimp farm impact on an amazonian estuary (Curuçá, Pará, Brazil). Braz J Oceanogr 64:75–88

Venekey V, Melo TPG, Rosa Filho JS (2019) Effects of seasonal fluctuation of amazon river discharge on the spatial and temporal changes of meiofauna and nematodes in the amazonian coast. Estuar Coast Shelf Sci 227:106330

Vermeiren P, Muñoz CC, Ikejima K (2016) Sources and sinks of plastic debris in estuaries: a conceptual model integrating biological, physical, and chemical distribution mechanisms. Mar Pollut Bull 113:7–16

Vianna BS, Miyai CA, Augusto A, Costa TM (2020) Effects of temperature increase in the biology and behavior of fiddler crabs. Physiol Behav 215:112765

Vieira BP, Dias D, Nakamura EM, Arai TI, Hanazaki N (2013) Is there temporal variation on solid wastes stranding in mangroves? A case study in Ratones mangrove, Florianópolis, Brazil. Biotemas 26:79–86

Wright SL, Thompson RC, Galloway TS (2013) The physical impacts of microplastics on marine organisms: a review. Environ Pollut 178:483–492

Wunderlich AC, Pinheiro MAA (2012) Mangrove habitat partitioning by Ucides cordatus (Ucididae): effects of the degree of tidal flooding and tree-species composition during its life cycle. Helgol Mar Res 67:279–289

Zeppilli D, Leduc D, Fontanier C, Fontaneto D, Fuchs S, Gooday AJ, Goineau A, Ingels J, Ivanenko VN, Kristensen RM, Neves RC, Sanchez N, Sandulli R, Sarrazin J, Sørensen MV, Tasiemski A, Vanreusel A, Autret M, Bourdonnay L, Claireaux M, Coquillé V, De Wever L, Rachel D, Marchant J, Toomey L, Fernandes D (2018) Characteristics of meiofauna in extreme marine ecosystems: a review. Mar Biodivers 48(1):35–37

Chapter 10
Brachyuran Crabs of Brazilian Mangrove and Salt Marsh Ecosystems

Anders J. Schmidt and Karen Diele

10.1 Introduction

Brachyuran crabs (referred to in this chapter as "crabs" for the sake of simplification) have significant cultural and socioeconomic importance around the world. In Brazil, the oldest records of human harvesting mangrove and marsh crabs date back to at least 6000 years BP (Corrêa et al. 1978). Today, thousands of coastal inhabitants still rely on crab harvesting for their livelihood, and the appreciation for this natural resource appears in gastronomy, folk festivals, popular music, and literature (e.g., Diele et al. 2010a). Within the realm of science, the number of articles focusing on Brazilian crabs has grown considerably in recent decades, from studies on individual species to crab communities.

The aim of this chapter is to present an overview of the crab biodiversity of Brazilian mangroves and salt marshes, building on the general information on the benthic biodiversity presented in Chap. 9 and including information on their geographic and zonal distribution, as well as on the most conspicuous biological features of the different species and their ecological and socioeconomic roles. The chapter will be particularly useful to people unfamiliar with this faunal group but working in Brazilian mangrove and salt marsh ecosystems, and for experts from other parts of the world aiming to work in Brazil.

A. J. Schmidt (✉)
Centro de Formação em Desenvolvimento Territorial, Universidade Federal do Sul da Bahia, Campus Paulo Freire, Teixeira de Freitas, BA, Brazil

K. Diele
School of Applied Sciences, Edinburgh Napier University, Edinburgh, UK

Centre for Conservation and Restoration Science, Edinburgh Napier University, Edinburgh, UK

Y. Schaeffer-Novelli et al. (eds.), *Brazilian Mangroves and Salt Marshes*, Brazilian Marine Biodiversity, https://doi.org/10.1007/978-3-031-13486-9_10

10.2 Criteria for Species Compilation

We categorized as mangrove crabs those species that spend at least one stage of their life cycle in this ecosystem, associated with vegetation, decaying wood, leaf litter, or sediment, between the lower and the higher levels of perigean spring tides. This tidal range includes not only areas dominated by the mangrove trees *Avicennia germinans*, *A. schaueriana*, *Laguncularia racemosa*, and *Rhizophora mangle*, but also the adjacent subtidal and the upper supratidal zones, which contain non-mangrove species sensu stricto (see Lana and Bernardino 2018). This chapter will take those marginal zones into consideration, even though they are mainly populated by salt marsh plants (e.g., *Spartina* spp.) or terrestrial vegetation (e.g., *Hibiscus pernambucensis* Arruda), since they are closely connected to mangroves, that is, hosting several crab species that depend on the mangrove forest (Schmidt et al. 2013).

The categorization of salt marsh crabs comprises species that occur in this ecosystem in latitudes where mangrove vegetation is no longer present, such is the case of the south of the Santa Catarina and the entire Rio Grande do Sul states (see Chap. 3, Maps 16 and 17). There, large flats of halophile herbaceous plants, mainly *Spartina* spp. and *Scirpus maritimus*, are the dominant vegetation. Given the ecological connectivity of these marshes to adjacent shallow water meadows, often occupied by the marine spermatophyte *Ruppia maritima* (Bemvenuti 1987), crab species occurring in the subtidal zone were also included in this compilation.

This chapter focuses on the most abundant and conspicuous mangrove and salt marsh crab species, while exotic species, cryptic commensal species, and occasional occurrences of species typical of other habitats (e.g., rocks, pillars, or other hard substrates introduced by humans) are not considered here. The species were categorized in terms of maximum carapace width: small (<40 mm); medium (between 40 and 60 mm); and large (> 60 mm) (Diele et al. 2010b). Besides, we cite the most notable characteristics of each species' adult stage regarding their behavior and ecology.

10.3 Species Richness and Geographical Distribution

A total of 36 Brazilian crab species matched the above-mentioned criteria; 30 of them occur exclusively in mangroves, five in both mangroves and salt marshes, and one in salt marshes only (Abele 1992; Coelho-Filho and Coelho 1996; Melo 1996; Bemvenuti 1998; Coelho et al. 2008; Marochi and Masunari 2011; Melo 2008; Almeida et al. 2010; Diele et al. 2010b; Colpo et al. 2011; Boos et al. 2012; Negromonte et al. 2012; Thurman et al. 2013).

Eight families of Brachyura are present in mangroves: Ocypodidae (10 species), Sesarmidae (7 species), Panopeidae (6 species), Portunidade (6 species), Grapsidae (3 species), Varunidae (2 species), and Ucididae and Gecarcinidae (each with one species) (Table 10.1). Compared to mangroves, salt marshes have considerably lower biodiversity, including six species from four families of Brachyura: Portunidae (2 species), Varunidae (2 species), Sesarmidae and Ocypodidae (each with one species) (Table 10.1).

Species richness increases from the Equator, in the extreme northern region of Brazil (Amapá State: 20 crab species) (see Chap. 3, Map 1), to the south, toward the Tropic of Capricorn – peaking in the state of Rio de Janeiro, with 33 species (Fig. 10.1). This peak may be explained by the fact that more temperate species reach their northernmost occurrence in Rio de Janeiro (e.g., *Neohelice granulata* and *Uca uruguayensis*), coexisting with species of warmer climates (e.g., *Leptuca cumulanta* and *Panopeus lacustris*). Species that are less tolerant to low temperatures gradually decrease toward the southern coast of Brazil, resulting in a reduction in crab species richness between the states of Rio de Janeiro and Santa Catarina (4 species less in the latter compared to the former state) (Fig. 10.1). With mangroves lacking in southern Brazilian estuaries, these areas become dominated by salt marshes, characterized by considerably lower crab species richness, with only 6 species (Table 10.1 and Fig. 10.1).

10.4 Zonation

In mangrove and salt marsh ecosystems, there is a co-occurrence of soft-bottom horizontal zonation from the channel margin to the dryland, with a hard-bottom vertical zonation extending from the roots to the top of the plants (Schmidt et al. 2013; Schaeffer-Novelli et al. 2016). These zones are generated principally by gradients of seawater flooding and, in the case of horizontal zonation, by freshwater drainage from the upland. Flooding contributes to the appearance of secondary gradients of sediment grain size, organic matter, salinity, and pH of interstitial water, among others (e.g., Sandrini-Neto and Lana 2011; Schmidt 2012). The vegetation zones resulting from these gradients decisively influence, in turn, temperature gradients, shifting from lower temperatures in shaded arboreal zones to higher temperatures at the herbaceous marginal zones (e.g., Schmidt 2012; Schmidt et al. 2013).

Crabs associate with different ecological zones according to their tolerance to submersion, desiccation, temperature, as well as in response to morphological adaptations, such as pericardial sacks that store water for their gills and pereopods adapted for swimming; physiological adaptations, such as osmotic regulation and respiration: and behavioral adaptations, such as the ability to burrow and search for appropriate microhabitats (e.g., Warner 1969; Principe et al. 2018).

Crab distribution is also influenced by substrate, porewater salinity, food availability, presence of predators, and intra- as well as interspecific competition (Schmidt 2012; Schmidt et al. 2013; Thurman et al. 2013; Grande et al. 2018). The large number of environmental gradients, biotic and abiotic factors that influence the local distribution of species, allied to geographic distributions, makes it

Table 10.1 List of species (and families) of brachyuran crabs in Brazilian mangroves and salt marshes and their range along the 17 coastal Brazilian states (see Fig. 10.1)

Species	Family	Range N-S	Ecosystem	
			mangrove	Salt marsh
Cardisoma guanhumi	Gecarcinidae	PA-SC	x	
Goniopsis cruentata	Grapsidae	AP-SC	x	
Pachygrapsus gracilis	Grapsidae	AP-RS	x	
Pachygrapsus transversus	Grapsidae	AP-RS	x	
Leptuca cumulanta	Ocypodidae	AP-RJ	x	
Leptuca leptodactyla	Ocypodidae	MA-SC	x	
Leptuca uruguayensis	Ocypodidae	RJ-RS	x	x
Minuca burgersi	Ocypodidae	PA-SC	x	
Minuca mordax	Ocypodidae	AP-SC	x	
Minuca rapax	Ocypodidae	AP-SC	x	
Minuca thayeri	Ocypodidae	AP-SC	x	
Minuca victoriana	Ocypodidae	CE-SP	x	
Minuca vocator	Ocypodidae	AP-SC	x	
Uca maracoani	Ocypodidae	SP-SC	x	
Acantholobulus bermudensis	Panopeidae	PA-SC	x	
Eurytium limosum	Panopeidae	PA-SC	x	
Panopeus americanus	Panopeidae	PA-SC	x	
Panopeus austrobesus	Panopeidae	RJ-RS	x	
Panopeus lacustris	Panopeidae	AP-RJ	x	
Panopeus occidentalis	Panopeidae	MA-RS	x	
Callinectes bocourti	Portunidae	AP-RS	x	
Callinectes danae	Portunidae	AP-RS	x	x
Callinectes exasperatus	Portunidae	AP-SC	x	
Callinectes marginatus	Portunidae	PA-PR	x	
Callinectes ornatus	Portunidae	AP-RS	x	
Callinectes sapidus	Portunidae	AL-RS	x	x
Armases angustipes	Sesarmidae	PA-SC	x	
Armases benedicti	Sesarmidae	AP-SC	x	
Armases rubripes	Sesarmidae	PA-RS	x	x
Aratus pisonii	Sesarmidae	AP-SC	x	
Sesarma crassipes	Sesarmidae	AP-PE	x	
Sesarma curacaoense	Sesarmidae	AP-ES	x	
Sesarma rectum	Sesarmidae	AP-SC	x	
Ucides cordatus	Ucididae	AP-SC	x	
Cyrtograpsus angulatus	Varunidae	RJ-RS		x
Neohelice granulata	Varunidae	RJ-RS	x	x

Coastal Brazilian states, from North (N) to South (S): Amapá (AP); Pará (PA); Maranhão (MA); Piauí (PI); Ceará (CE); Rio Grande do Norte (RN); Paraíba (PB); Pernambuco (PE); Alagoas (AL); Sergipe (SE); Bahia (BA); Espírito Santo (ES); Rio de Janeiro (RJ); São Paulo (SP); Paraná (PR); Santa Catarina (SC); and Rio Grande do Sul (RS). See Chap. 3, Maps: 1-AP, 2-PA, 3-MA, 4-PI, 5-CE, 6-RN, 7-PB, 8-PE, 9-AL, 10-SE, 11-BA, 12-ES, 13-RJ, 14-SP, 15-PR, 16-SC, and 17-RS

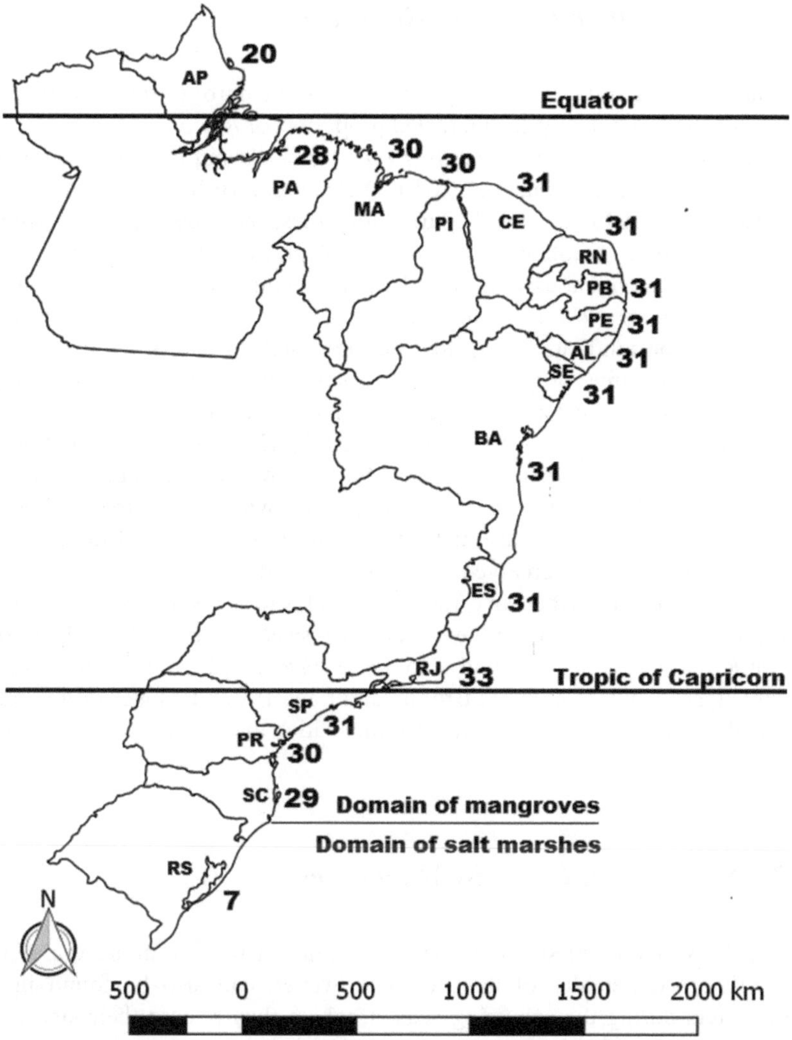

Fig. 10.1 Number of brachyuran species in Brazilian mangroves and salt marshes. Coastal Brazilian states, from North to South: Amapá (AP); Pará (PA); Maranhão (MA); Piauí (PI); Ceará (CE); Rio Grande do Norte (RN); Paraíba (PB); Pernambuco (PE); Alagoas (AL); Sergipe (SE); Bahia (BA); Espírito Santo (ES); Rio de Janeiro (RJ); São Paulo (SP); Paraná (PR); Santa Catarina (SC); and Rio Grande do Sul (RS). See Chap. 3, Maps: 1-AP, 2-PA, 3-MA, 4-PI, 5-CE, 6-RN, 7-PB, 8-PE, 9-AL, 10-SE, 11-BA, 12-ES, 13-RJ, 14-SP, 15-PR, 16-SC, and 17-RS

difficult to establish a standard pattern of crab zonation in Brazilian mangroves and salt marshes. In the following section, we present some recurrent features of crab zonation on hard and soft bottoms; however, it must be noted that the species cited here do not always occur simultaneously in the same geographic areas along the Brazilian coast (Table 10.1).

10.4.1 Hard Bottom Crabs in Mangroves

In mid- and low-intertidal zones, especially at the mangrove fringe, at the trees' lower bottom, between the oyster-incrusted prop roots of *Rhizophora mangle*, one is likely to find small grapsid crabs (*Pachygrapsus gracilis* and *P. transversus*) (Warner 1969; Melo 1996; Almeida et al. 2010; Diele et al. 2010b). These crabs feed on algae and small invertebrates, mostly during low tides, and show typical fast-moving grapsid behavior (Christofoletti et al. 2010). Another, even more agile grapsid crab, *Goniopsis cruentata,* inhabits the intertidal zone, being most abundant at the seaward mangrove fringe (Warner 1969; Diele et al. 2010b). This medium-sized crab wanders around on mudflats during low tide in search of diverse edible resources such as mangrove tree propagules, algae, small invertebrates, and sediment-containing microorganisms (von Hagen 1977; Diele et al. 2010b; Lima-Gomes et al. 2011; Ferreira et al. 2013; Wellens et al. 2015). *G. cruentata* does not burrow, exclusively taking advantage of other crabs' burrows when threatened. At high tide, when water covers the substrate, this crab climbs backward up into trees, where they move slowly or remain stationary; they will jump into the water if threatened but return to the tree when the perceived danger has passed. Climbing trees is more of a strategy to avoid predators than an effort to search for food resources in *G. cruentata* (von Hagen 1977). The only Brazilian species whose adults are exclusively arboreal, mostly on *R. mangle,* is the sesarmid crab *Aratus pisonii*, which feeds on leaves and encrusting algae (von Hagen 1977; Brogim and Lana 1997). Females of this species migrate to the seaward mangrove fringe for incubation and larval release (e.g., Leme and Fransozo 1998).

10.4.2 Soft Bottom Crabs in Mangroves

In Brazil, the soft bottom subtidal zone is dominated by swimming crabs of the genus *Callinectes*. Females of this genus, however, can also be found in deep offshore waters during the spawning season, where they release their larvae (e.g., Branco and Masunari 2000). Besides *Callinectes danae* and *C. sapidus*, one of the most intimately mangrove-linked swimming crab species is *C. exasperatus*. These crabs visit the inner forest during high tide, where they predate on crustaceans, mollusks, and fish, or act as scavengers (Carvalho and Couto 2010). At low tide, *C. exasperatus* tends to stay in tidal pools near the prop roots of *R. mangle* or return to the subtidal zone.

The soft-bottom intertidal zone of mangroves is the preferred habitat of burrowing crabs. The burrows are used as shelters against predators and competitors, help to avoid adverse environmental conditions (e.g., Araújo and Calado 2011), and, in many cases, are the feeding ground for previously stored food (e.g., Nordhaus et al. 2006). The crabs also retrieve to the burrows for molting (Alves and Nishida 2002) and mating (e.g., Crane 1975; Christy 2007). Besides, burrows also serve as recruitment sites (Schmidt and Diele 2009; Schmidt 2012).

Burrows by Panopeidae crabs are rather cryptic, as they are small and predominantly situated between the oyster-encrusted rhizophores of *R. mangle* or underneath pieces of deadwood. The latter are the habitat of the small *Acantholobulus bermudensis*, *Panopeus americanus*, *P. austrobesus*, and *P. occidentalis*, all frequent in the low-intertidal zone. From the low-intertidal, advancing to the mid-intertidal zone, medium-sized *Panopeus lacustris* and smaller *Eurytium limosum* (Coelho-Filho and Coelho 1996; Diele et al. 2010b) can be found. Most panopeid crabs leave their burrows at high tide to feed on encrusting oysters and barnacles using their strong claws to break the resistant outer shells (Whitefleet-Smith and Hardin 2014).

The small fiddler crabs (Ocypodidae Family) are independent of consolidated substrates. The most striking feature of these crabs is the large, asymmetrical chelipeds of the males used for what is called "claw-waving display" (Crane 1975). This display consists basically of a rhythmic elevation and lowering of the larger claw; the movements are complex and vary according to species and context (Christy and Salmon 1984). In most Brazilian fiddler crabs, the claw-waving display, together with specific sound-making, is used by the males to defend their burrows and attract females for mating (Crane 1975; Christy 2007). The females wander on the mud surface looking for males and usually use the male's burrow for breeding (Christy 2007). When not involved in reproduction, fiddler crabs spend their time feeding on nutrient-rich microorganisms such as bacteria, fungi, protozoa, and microalgae, by removing them from the sediment using their buccal appendages (Diele and Koch 2010a).

As the substrate is fundamental for feeding and burrowing, the characteristics of the sediment, together with water salinity and the shore level, will decisively influence the tidal spatial distribution of fiddler crabs (Koch et al. 2005; Thurman et al. 2013). There is a predominance of *Uca maracoani*, the largest Brazilian fiddler crab in body size, in the low-intertidal zone where it burrows in the unvegetated mudflat adjacent to the mangrove forest (Koch et al. 2005; Diele et al. 2010b; Thurman et al. 2013). In muddy sediments between mangrove trees, chimney-building fiddler crabs predominate. Chimneys are sedimentary protector walls surrounding the entire edge of the burrow opening (Crane 1975; Gusmão-Júnior et al. 2012). These chimney-like structures are built by *Minuca vocator* (from the low- to mid-intertidal), *M. thayeri* (mid-intertidal), and *Leptuca cumulanta* (from the mid to the upper-intertidal) (Warner 1969; Koch et al. 2005; Diele et al. 2010b; Thurman et al. 2013). This sequence of fiddler crabs in the intertidal zone is not always complete and rigorous, and the presence or absence of species will depend also on the salinity in a given part of the estuary and the geographic distribution of each crab species (Thurman et al. 2013).

Although fiddler crabs dominate in terms of abundance, for example, 45 burrows/m^2 in *M. vocator* (Colpo and Negreiros-Fransozo 2016), the large burrows of the ucidid *Ucides cordatus*, whose carapace width reaches up to 9 cm (Pinheiro et al. 2005; Diele and Koch 2010b), are the most conspicuous structures in the intertidal zone of a typical Brazilian mangrove. Despite its large size, *U. cordatus* is not easily seen, since the crabs are rather cryptic, spending up to 85% of the time inside their burrows (Nordhaus et al. 2009). When they emerge, they mostly stay right near the burrow entrance, only occasionally moving within a radius of about 1 m to collect plant material or mud (Nordhaus et al. 2009). The burrows are often sealed with mud and remain closed for numerous days when the crabs are vulnerable during molting periods (Alves and Nishida 2002; Piou et al. 2007).

The cryptic routine behavior of *U. cordatus* changes drastically during the popularly known as "andadas," in which a synchronized mass mate-searching occurs. This event is closely linked to lunar and storm cycles (Diele 2000; Diele et al. 2005; Diele and Koch 2010a; Schmidt et al. 2012). During the *andadas*, many mostly male *U. cordatus* individuals leave their burrows, exploring other burrows and fighting between each other while searching for females. Copulations are rarely seen on the mud surface, suggesting that they mate inside burrows. When females fertilize their eggs with stored sperm and extrude the egg masses, they often climb mangrove trees during high tide (Schmidt et al. 2008a). As it follows, they return to their burrows to incubate the eggs, chasing away any crab that might come near. Most females release their larvae from their burrows, but in some locations, spawning migrations toward channel margins have been reported (Góes et al. 2000; Schmidt 2006). Larvae are exported to offshore waters with ebb tidal currents and reinvade the estuary after 3–4 weeks (Diele 2000). Larval settlement is triggered by conspecific odors (Diele and Simith 2007; Simith et al. 2013, 2017) and recruitment occurs inside the burrows of conspecifics (Schmidt and Diele 2009; Schmidt 2012), mainly located in the upper intertidal zone and lower supratidal, with sandy-muddy substrates, most typically dominated by *L. racemosa* trees (Schmidt et al. 2009, 2013; Wunderlich and Pinheiro 2013).

The upper-intertidal zone is also the habitat of small sesarmid crabs (e.g., Araújo et al. 2014) that feed on plant material (Brogim and Lana 1997), some burrowing, others nonburrowing. Although some species of this family inhabit low- to mid-intertidal zones, for example, *Armases benedicti* and *A. rubripes* (Diele et al. 2010b; Lima et al. 2006), sesarmids such as *Armases angustipes*, *Sesarma curacaoense*, and *S. rectum* are more conspicuously found in the upper-intertidal underneath *L. racemosa* trees (e.g., Araújo et al. 2014). They also enter the salt flats in the supratidal zone, which, in some places in Brazil, are known as *apicuns*.

In the lower supratidal, there is still some shading by smaller mangrove trees and this zone often functions as a nursery for juvenile *U. cordatus* crabs (Schmidt et al. 2009, 2013). However, the most abundant species in this zone are small fiddler crabs, which can reach high population densities, for example, 240 ind./m^2 of *Leptuca leptodactyla* (Masunari 2006). In localities with saltier interstitial water and sandier sediment, *L. leptodactyla* and *Minuca rapax* occur, while *Leptuca uruguayensis* is more prominent in muddy sediment (Masunari 2006; Thurman et al. 2010). Males of all three species build semicircular sediment structures above the burrow entrance, called hoods or semidomes, related to mate searching (Masunari 2012; Ribeiro et al. 2016; Carvalho et al. 2018). Non-hood-building species are also found in the supratidal zone influenced by upland freshwater, as the species *Minuca mordax*, *M. burgersi*, and *M. victoriana* (Warner 1969; Masunari 2006; Almeida et al. 2010; Thurman et al. 2010).

Fiddler crabs and sesarmid crabs are found throughout the supratidal zone until the transition to the landward vegetation, which is dominated by plant species such as *Acrostichum aureum*, *Hisbiscus pernambucensis*, *Dalbergia ecastophyllum*, and *Aechmea blanchetiana*, a bromeliad in which *Armases angustipes* and *A. rupripes* seek shelter (Melo 1996; Fischer et al. 1997). This zone is also the recruitment area for the largest Brazilian mangrove crab, the blue land crab *Cardisoma guanhumi* (Schmidt et al. 2008b, 2013). This gecarcinid crab has a carapace width of up to 11 cm (e.g., Hurtado et al. 2015) and their burrows, which often reach down to the groundwater, can be found in the adjacent landward forests known as "matas de restinga" in Brazil. These crabs harvest leaves and fruits, and occasionally carrion, which they consume in their burrows (Herreid II 1963; Firmo et al. 2012). They only abandon their burrows for longer times during their reproductive period, when males and females emerge in great numbers to mate. The egg-bearing females migrate to the edge of the tidal water to release their larvae (Gifford 1962; Firmo et al. 2012), occasionally even climbing trees for that purpose. These mass movements of *C. guanhumi* are also called *andadas* and are influenced by the same geophysical factors as *U. cordatus*.

10.4.3 Hard Bottom Crabs in Salt Marshes

In salt marshes, hard bottom crabs are less common compared to mangrove forests. This is likely due to the lower structural complexity of the short and thin plants. Between stems of *Spartina* spp., the small herbivorous sesarmids *Armases rubripes* (Bemvenuti 1998) compete for space and food with the medium-sized varunid *Neohelice granulata* (Capítoli et al. 1977). Despite being a typical soft-bottom crab, *N. granulata* frequently climbs into the marsh canopy and grazes on the leaf blades of *S. alterniflora* (Freitas et al. 2014).

10.4.4 Soft Bottom Crabs in Salt Marshes

Salt marshes have a soft bottom horizontal zonation that is quite variable, both spatially and temporally. In Brazil, this ecosystem occurs predominantly in localities with low amplitude tides (micro-tides), and meteorological conditions can strongly influence both frequency and duration of flooding (Costa and Davy 1992). The portunid *Callinectes sapidus*, and, to a lesser extent, *C. danae,* can be found in the subtidal zone, predominantly in meadows of *Rupia maritima* adjacent to salt marshes (Weber and Levy 2000). The juveniles of these swimming crabs spend the summer feeding on algae and polychaetes, whilst larger specimens prey upon mollusks and crustaceans, including the varunid *Cyrtograpsus angulatus* (Kapusta and Bemvenuti 1998). *C. angulatus*, another predator of benthic macrofauna, also occurs in the subtidal zone of salt marshes in its juvenile phase, and small individuals can even be found under rocks in the intertidal zone (Boschi 1964; Bemvenuti 1998).

In the intertidal salt marsh zone, especially among large meadows of *Spartina* spp., generalist feeders such as the varunid *Neohelice granulata* prevail, inhabiting semi-permanent burrows between stems and roots (Boschi 1964; Bemvenuti 1998). The herbivore and deposit-feeder *Armases rubripes* hides in the vegetation in the intertidal zone (Bemvenuti 1998), although sometimes digging burrows (Fischer et al. 1997). A typical burrowing crab and deposit feeder is the ocypodid crab *Uca uruguayensis*, abundant on more compact substrates in the supratidal zone of salt marshes (Boschi 1964; Ribeiro et al. 2016).

10.5 Ecological Role

Most Brazilian mangrove and salt marsh crabs are burrowers, and their digging can affect physicochemical sediment properties, such as changes in its reduction state. For example, in a mangrove in Northeast Brazil, more oxidizing conditions were found in sediments bioturbated in areas of higher densities of *U. cordatus* (Araújo et al. 2012). In contrast, in North Brazil, at half of the previously mentioned burrow density, the same effect could be observed at the immediate vicinity of the burrow walls (Pülmanns et al. 2014), but not further away (Pülmanns et al. 2016). The increased sediment-air interface at the walls allows aerobic infauna to thrive and increases aerobic microbial decomposition during low tide. This may result in increased sediment-generated CO_2 being released near burrows (e.g., Leopold et al. 2013; Pülmanns et al. 2014).

Crab burrows can also increase sediment drainage, which facilitates the removal of excess salts from the sediment, as observed in complex galleries of sesarmid crabs that dominate in mangroves in the Indo-West Pacific (e.g., Stieglitz et al. 2000). However, in Brazil, the most abundant crab species (ucidids and ocypodids) build relatively simple, nonbranched burrows (e.g., Araújo and Calado 2011; Machado et al. 2013), mostly with single openings, and this might explain why their burrows do not significantly increase sediment desalination (Pülmanns et al. 2016; Pestana et al. 2017).

The forest floor of Brazilian mangroves is often free of mangrove leaves mostly due to the feeding of *U. cordatus*, the key-litter feeder in the region (Schories et al. 2003). In the state of Pará, North Brazil, these crabs consume up to 81% of the litter and propagule production (Nordhaus et al. 2006). A large part of the consumption is returned to the sediment as feces, a growth medium for microorganisms, as evidenced by the much higher microbial biomass in *U. cordatus* feces compared to sediment (Nordhaus 2004). The species fragments litter into small pieces through sloppy feeding, and detritus-feeding crabs further grind these to even smaller particles. The consecutive increased surface detritus area promotes microbial degradation and benefits microbe-feeding fiddler crabs. Hence, the litter- and detritus-feeding activities of mangrove crabs accelerate nutrient cycling in this ecosystem (Nordhaus and Wolff 2007). Crabs can also influence biogeochemical cycles by transporting organic matter together with sediment in the process of excavation and maintenance of burrows, as reported with *Leptuca leptodactyla* by Natálio et al. (2017).

Besides the importance of crabs for nutrient cycling and sediment properties, they can influence the distribution of mangrove trees through selective propagule feeding. In Brazil, *U. cordatus* and *G. cruentata* are the most important propagule predators (Nordhaus et al. 2006; Ferreira et al. 2013; Wellens et al. 2015), and most studies have shown that the crabs prefer the fleshier, more nutritious, and palatable *Avicennia* propagules over *Rhizophora* and *Laguncularia* (McKee 1995; Souza and Sampaio 2011). In Panama, this differential predation seems to be one key factor preventing the formation of *Avicennia* stands in the low intertidal, where the density of *U. cordatus* and *G. cruentata* is highest (Sousa and Mitchell 1999). Future research should focus on generating a better understanding of the ecological role of any given mangrove crab species in different environmental contexts (Pülmanns et al. 2016).

Crabs are also important for the trophic processes in salt marshes, but the most abundant species in these ecosystems have a less specialized diet and feed on more than one trophic level (Bemvenuti and Colling 2010). As a secondary consumer, the abundant *Neohelice granulata* can prey upon other invertebrates, but it mainly acts as a primary consumer feeding upon leaves and detritus of *Spartina* spp. (Bemvenuti 1987; D'Incao et al. 1990). The selective herbivory by *N. granulata* on *Spartina alterniflora* at the mid-marsh promotes the dominance of the plant *Scirpus maritimus* in this zone and restricts the occurrence of the former to the low-marsh zone (Costa et al. 2003). The salt marsh benthic diversity is also affected by the bioturbation caused by individuals of *N. granulata*. For example, the intense sediment disturbance caused by their burrowing activities negatively affects surface meiofauna, reducing population densities of copepods and ostracods (Rosa and Bemvenuti 2005).

10.6 Fisheries and Socioeconomic Relevance

Edible crab species sustain the livelihoods of coastal human communities and are important economic resources in Brazil. *U. cordatus*, popularly known as *caranguejo-uçá* or simply *caranguejo*, is the main commercial resource harvested from almost all Brazilian mangroves. The harvesting techniques are artisanal and vary from place to place. One of the most traditional capture techniques consists of simply reaching deep with the arm into the crab burrow and pulling out the crab. As some burrows are up to 2 m deep, the crab harvesters also use auxiliary techniques to catch the crabs (Diele et al. 2010a). They often use a hook made of just a simple rebar with curved end or a wooden stick with an affixed rebar hook (Schmidt 2006; Schmidt et al. 2009; Diele et al. 2010a). Clogging the entrance of the burrow with mud (locally called "tapamento") induces the crab to climb up to uncover it, which makes them more accessible to capture (Nascimento et al. 2016).

These active capturing techniques have in recent years been replaced by techniques involving traps. The most widely used trap is called "redinha" (Portuguese for "small net"), which is simply a tangled net of polypropylene threads (obtained from old grain sacks) affixed to two thin pieces of mangrove branches (broken from the mangrove trees) at the entrance of the burrow. When the crab emerges from its burrow, it becomes entangled and can easily be caught (Nascimento et al. 2012, 2016 Duarte et al. 2014). A less-used trap is called "lacinho" (Portuguese for "small lasso"), a technique that consists of fixing a noose of thin nylon line at the exit to the burrow. When the crab tries to leave the burrow, it will be "lassoed" by the line.

The mangrove crab *U. cordatus* is also captured using baited traps in some locations (Carvalho and Igarashi 2009). The trap consists of a cylinder (usually a PVC tube, plastic bottle, or can) with one end permanently sealed, and the other having a trap door tensed with rubber bands but held open by a wooden lever lightly attached to a metal clip (the trigger), which pierces the cylinder wall. The clip is baited inside the cylinder (usually with leaves of *R. mangle*) and the trap is placed onto a burrow opening (as if it was an extension of the tunnel). Upon entering the cylinder and trying to remove the bait, the crab will release the wooden lever and the rubber bands will shut the trap door.

That baited trap is called "ratoeira" (Portuguese for "mouse trap"), as it was originally designed to capture rats and mice, and is commonly used to capture *Cardisoma guanhumi*, the blue land crab, locally known as "guanhamum" or "guaiamum." The bait used to capture the *guaiamum* mainly consists of aromatic fruits such as lemons, pineapples, and jackfruits. Another type of baited trap used to capture this crab is a wooden box with a levered door and roof held open by a trigger, known as "arapuca." When the crabs enter and touch the bait, the trap door falls by its weight (Botelho and Santos 2005; Firmo et al. 2012).

Another widely harvested crab in Brazilian mangroves, despite its smaller size, is *Goniopsis cruentata*, popularly known as "aratu." This crab is captured using a variety of techniques. One of the most rudimentary methods is to simply throw handfuls of mud at the rapidly scurrying crabs to temporarily immobilize them. Another technique is almost like fishing; the bait (e.g., mangrove leaves and other crustaceans such as *Aratus pisonii*) is tied to a line attached to a bamboo pole, which is then held by the fisher sitting on tree roots, dangling the fishing rod over the mud where the *aratus* are foraging. The crabs will run to the bait, grab it, and hold onto it so that they can easily be collected (Botelho and Santos 2005; Maciel and Alves 2009). The capture of *aratus* can also involve baited funnel traps made of wooden splints tied together. The crabs will enter the funnel trap to eat the bait but are then unable to leave (Botelho and Santos 2005). These traditional capture methods have been rapidly replaced in many parts of Brazil by nocturnal catch using lanterns. The strong, sudden light momentarily paralyzes the animal, which can then be easily captured.

Swimming crabs are popularly known in Brazil as "siris." Different species of *Callinectes* are also captured in the mangrove forest with the aid of a rebar hook, as during low tide, many individuals remain in pools in the intertidal zone. Nevertheless, the subtidal zone is the best place to capture other species of swimming crabs.

As they are scavengers, swimming crabs are attracted to bait such as meat, chicken, or fish. The bait can be tied in witch-hat nets, trotlines, or funnel traps placed on the sediment (Botelho and Santos 2005; Severino-Rodrigues et al. 2009; Magalhães et al. 2011). Nonbaited gears, such as trawling nets, are also employed in shallow waters, mostly in regions near salt marshes (Maier and Neto 2009).

Commercialization varies according to species. The highly valued *guaiamum* (*C. guanhumi*) is held in tanks and fed with fruits and grains to fatten up, and later be sold bigger and alive. *U. cordatus* crabs are also sold alive, but without being fattened. They are usually transported in large quantities alive on trucks, and many individuals die during transport and are discarded (Legat et al. 2006). In some places, the crabs are processed for their meat by cooking and removing their exoskeletons before being sold. However, meat processing is more common in *aratus* (*G. cruentata*), never sold alive, and in *siris* (*Callinectes* spp.), rarely sold alive. Meat processing is typically undertaken by women and younger members of the family (Nascimento et al. 2017). The sold meat is often called "siri catado" in northeast Brazil.

The different forms of harvesting, processing, and trading Brazilian mangrove and salt marsh crabs have in common the low costs of the equipment used, which makes these activities viable to populations of very low income, providing alternatives to unemployment – a historic problem in Brazil (Diele et al. 2010a). That solution is not, however, absent of drawbacks, since crab harvesting is often an unhealthy occupation (e.g., Glaser and Diele 2004), and the economic gains are severely reduced by the strong presence of middlemen within the production chain (e.g., Nascimento et al. 2017). That added to sparse public investments in these traditional (and often isolated) coastal communities, limits access to education, health services, and basic sanitation.

10.7 Threats, Management, and Conservation

Overfishing, diseases, habitat destruction, and pollution are the main factors threatening mangrove and salt marsh crab populations in Brazil. Given that many thousands of people harvest crabs along the coast, it might be expected that overharvesting would threaten many species. However, scientifically robust stock assessment studies are still sparse to date and therefore, for most harvesting areas, it is unknown whether the targeted crab populations are overfished. Nonetheless, as a precautionary measure, *U. cordatus*, *C. guanhumi*, and *C. sapidus* were even included in the "National List of Aquatic Invertebrate Species and Fish Overexploited or Threatened with Overexploitation" (Pinheiro and Rodrigues 2011) in 2004, but this list was replaced by the Ordinance 445/2014. Furthermore, different states have their regulations, for example, São Paulo State Decree No. 63,853 established in 2018.

Despite intense harvesting, in the few localities where exploited *U. cordatus* populations have been monitored for several years, the stocks appeared to be sustained, for example, Bragança, Pará State (Diele et al. 2005) and Canavieiras and Caravelas, Bahia State (Schmidt 2014a, b) (see Chap. 3, Maps 2 and 11, respectively). That does not mean, however, that the species is not suffering from other threats. Successive mass die-offs caused by a fungal disease impacted numerous populations from 1997 to 2005 throughout almost all of Northeast Brazil (Schmidt et al. 2008c; Orélis-Ribeiro et al. 2010). This region has also suffered from significant habitat loss, especially in recruitment zones of *U. cordatus* and *C. guanhumi* in the *apicuns*, due to environmental degradation caused by shrimp farming and salt extraction (Schmidt et al. 2013; Lima and Bailey 2015; Schaeffer-Novelli et al. 2016).

Shrimp ponds and artificial salt pans also impact the spawning migrations of *C. guanhumi* females from upland areas to the mangrove forest as well as the migration of recruits from the supratidal mangrove zone to the landward forests (Schmidt et al. 2013; Schaeffer-Novelli et al. 2016). In upland areas, where urban and tourism infrastructure growth is most pronounced, the species competes with humans for habitat. Because human activities, including catching small juveniles, greatly impact *C. guanhumi* populations, this species is currently officially regarded as critically endangered according to Ordinance 445/2014, as mentioned above.

Swimming crabs are also not free from threats, even in sublittoral areas without human occupation. Throughout Brazil, *Callinectes* spp., including young individuals, are the most frequent bycatch of shrimp trawling (e.g., Ruas et al. 2017). Additionally, as they live essentially submerged their entire lives cycles, swimming crabs are continuously exposed to pollutants such as heavy metals (e.g., Lacerda et al. 2012), although crabs that are not submerged most of their lifecycles, that is, *C. guanhumi, G. cruentata*, and *U. cordatus*, also suffer from those impacts (e.g., Horta et al. 2011; Pinheiro et al. 2012; Carneiro et al. 2018).

Considering so many threats, there is much to be done to guarantee the conservation of mangrove and salt marsh crab populations in Brazil. There is an urgent need for reliable stock assessments from a wide range of localities in the country to help evaluate current fishing pressures, as well as studies in populations' statuses to inform communities and regulators. Monitoring artisanal harvesting of crabs presents a considerable challenge as landings occur in several small sites scattered along the coast, and the exact number of involved fishers is unknown. Nevertheless, population parameters of some crab species have been continuously monitored in some locations in Northeast and North Brazil through combined efforts of governmental organs, universities, and NGOs, for example, São João da Ponta (PA) and Canavieiras and Caravelas (BA) (Schmidt 2014a, b).

The harvesting of *U. cordatus*, *C. guanhumi,* and *Callinectes* spp. (but not *G. cruentata*) has been subject to governmental regulation in terms of minimum capture sizes as well as protection of egg-bearing females and reproductive events. Safeguarding the reproduction, namely, the *andadas* (synchronous mass-mating events) of *U. cordatus* and *C. guanhumi*, remains a challenge to date, due to critical knowledge gaps regarding their temporal occurrence.

To better understand the *andada* phenomenon, a network of researchers (the REMAR network)[1] has been studying the synchrony of reproductive activities with geophysical cycles across multiple areas of Brazil since 2013. Since 2017, this research is supported by a supplementary participatory citizen-science approach where the public contributes with observational *andada* data via purpose-built Android and iOS smartphone applications *Remar_Cidadão*, freely downloadable on Google Play Store (ICMBio 2017, 2018) or Apple App Store. With the inclusion of both conventional scientists and citizen-scientists' inputs and the support of environmental governmental agencies, the REMAR network is predicting with high accuracy the occurrence of andadas in most of the Brazilian coast, thus helping to define adequate temporal bans to avoid existing social and economic conflicts. Bans based on REMAR forecasts have already been established for the North and Northeast regions from 2020 to 2024 by governmental agencies, a procedure likely to be continued in the longer-term future (Schmidt et al. 2012).

A currently controversial topic is the prohibition of certain harvesting techniques. For example, the use of *redinha* traps to capture *U. cordatus* is now prohibited by law, even though this technique is still widely used throughout the country. The respective legislation was implemented by Government agencies without sound scientific knowledge regarding the degree of damage caused (e.g., ghost-fishing and pollution by plastic waste). It is also currently unknown whether these commonly used traps are better or worse than any of the other techniques. Locally agreed regulations regarding harvesting techniques have recently been established by official participatory agreements and appear promising, especially within extractive reserves designed to promote the sustainable use of natural resources (see Chap. 16). Such agreements have a great potential to regulate fishing activities by empowering traditional populations.

10.8 Final Remarks

Brazilian mangroves and salt marshes are the habitat of diverse ecologically and economically important brachyuran crab species. Degradation and pollution threaten these habitats and their associated species, including the many important ecosystem services they provide. Across the country, coastal poor communities depend on harvesting mangrove and salt marsh crabs for their livelihoods, yet neither the number of people extracting crabs is known to date, nor the sustainability of the respective activity. This hampers the conservation of the targeted species, and their harvesting management, putting the future of the associated human livelihoods at risk. Research efforts need to be increased to fill these knowledge gaps to inform stakeholders. The recent participatory and transdisciplinary approaches implemented in some locations appear promising and show great potential to be geographically expanded to help generate robust data on a national scale.

[1] http://crabdata.napier.ac.uk/public/

References

Abele LG (1992) A review of the grapsid crab genus *Sesarma* (Crustacea: Decapoda: Grapsidae) in America, with the description of a new genus. Smithsonian Institution Press, Washington DC

Almeida AO, Souza GBG, Boehs G, Bezerra LEA (2010) Shallow-water anomuran and brachyuran crabs (Crustacea: Decapoda) from southern Bahia, Brazil. Lat Am J Aquat Res 38(3):329–376

Alves RRN, Nishida AK (2002) A Ecdise do Caranguejo-uçá, *Ucides cordatus* L. (Decapoda, Brachyura) na Visão dos Caranguejeiros. Interciência 27(3):110–117

Araújo MSLC, Calado T (2011) Burrows architecture of the crab *Ucides cordatus* (LINNAEUS, 1763) (Crustacea, Decapoda, Ucididae) in a mangrove swamp of Brazil. Tropical Oceanography

Araújo MSLC, Castiglioni DS, Coelho PA (2012) Width-weight relationship and condition factor of *Ucides cordatus* (Crustacea, Decapoda, Ucididae) at tropical mangroves of Northeast Brazil. Iheringia 102(3):277–284

Araújo MSLC, Tenório D, Castiglioni D (2014) Diversidade e distribuição dos Crustacea Brachyura dos manguezais dos rios Ariquindá e Mamucabas, litoral sul de Pernambuco, Brasil. Rev Gestão Costeira Integrada 14:483–499

Bemvenuti CE (1987) Predation effects on a benthic community in estuarine soft sediments. Atlantica 9(1):5–32

Bemvenuti CE (1998) Invertebrados bentônicos. In: Seeliger U, Odebrecht C, Castello JPO (eds) Ecossistemas Costeiro e Marinho do Extremo Sul do Brasil. Ecoscientia, Rio Grande, pp 46–50

Bemvenuti CE, Colling LA (2010) Relações tróficas na comunidade bentônica da região estuarina da Lagoa dos Patos, RS, Brasil. Cad Ecol Aquat 5(1):1–8

Boos H, Buckup GB, Buckup L, Araujo PB, Magalhães C, Almerão MP, Santos RA, Mantelatto FL (2012) Checklist of the Crustacea from the state of Santa Catarina, Brazil. Checklist 8(6): 1020–1046

Boschi EE (1964) Los Crustáceos Decápodos Brachyura del Litoral Bonaerense. Bol Inst Biol Mar 6:1–99

Botelho ERO, Santos MCF (2005) A cata de crustáceos e moluscos no manguezal do Rio Camaragibe – Estado de Alagoas: aspectos socioambiental e técnico-econômico. Bol Técn Cient CEPENE 13(2):77–96

Branco J, Masunari S (2000) Reprodutive ecology of the blue crab, *Callinectes danae* Smith, 1869 in the Conceição Lagoon system, Santa Catarina Isle, Brazil. Rev Bras Biol 60:17–27

Brogim RA, Lana PC (1997) Espectro alimentar de *Aratus pisonii*, *Chasmagnathus granulata* e *Sesarma rectum* (Decapoda; Grapsidae) em um manguezal da Baía de Paranaguá. Iheringia 83: 35–43

Capítoli RR, Bemvenuti CE, Gianuca NM (1977) Ocorrência e observações bio-ecológicas do caranguejo *Metasesarma rubripes*. Atlantica 2(1):50–62

Carneiro L, Silva D, Reis L, Oliveira DAF, Maciel LC, Garcia KS, Soares SAR, Queiroz AFS (2018) Distribuição de elementos traço em tecidos de *Goniopsis cruentata* (Latreille, 1803) capturados nos manguezais do sul da Bahia – Brasil e avaliação do potencial de risco no consumo. Quím Nova 41:959–968

Carvalho FL, Couto ECG (2010) Dieta do siri *Callinectes exasperatus* (Decapoda,Portunidae) no estuário do rio Cachoeira, Ilhéus, Bahia. UNICiências 14(2):325–334

Carvalho H, Igarashi M (2009) A utilização do forjo na captura do caranguejo uçá (*Ucides cordatus*) na comunidade de Tapebas em Fortaleza – CE. Biotemas 22:69

Carvalho R, Pardo J, Costa T (2018) Construction and structure of the semidomes of the fiddler crab *Minuca rapax* (Brachyura: Ocypodidae) in southern Brazil. J Crustac Biol 38:241–244

Christofoletti R, Murakami V, Oliveira DN, Barreto R, Flores A (2010) Foraging by the omnivorous crab *Pachygrapsus transversus* affects the structure of assemblages on sub-tropical rocky shores. Mar Ecol Prog Ser 420:125–135

Christy JH (2007) Predation and the reproductive behavior of fiddler crabs (Genus *Uca*). In: Duffy JE, Thiel M (eds) Evolutionary ecology of social and sexual systems: crustaceans as model organisms. Oxford Scholarship Online, Oxford, pp 211–232

Christy JH, Salmon M (1984) Ecology and evolution of mating Systems of Fiddler Crabs (Genus *Uca*). Biol Rev 59:483–509

Coelho PA, Almeida AO, Bezerra LEA (2008) Checklist of the marine and estuarine Brachyura (Crustacea: Decapoda) of northern and northeastern Brazil. Zootaxa 1956:1–58

Coelho-Filho P, Coelho P (1996) Sinopse dos Crustáceos Decápodos Brasileiros (Família Xanthidae). Tropical Oceanography

Colpo KD, Negreiros-Fransozo ML (2016) Sampling technique affects the population structure assessments of fiddler crab *Minuca vocator* (Herbst, 1804) (Ocypodidae: Gelasiminae). Nauplius 24:e2016015

Colpo KD, Chacur MM, Guimarães FJ, Negreiros-Fransozo ML (2011) Subtropical Brazilian mangroves as a refuge of crab (Decapoda: Brachyura) diversity. Biodivers Conserv 20:3239–3250

Corrêa MMG, Zwink W, Brum INS (1978) Ocorrências de crustáceos no sambaqui Zé Espinho. In: Kneip LM (ed) Coletores e pescadores pré-históricos de Guaratiba, Rio de Janeiro. UFRJ/UFF, Niterói, pp 217–227

Costa CSB, Davy AJ (1992) Coastal salt marsh communities of Latin America. In: Seeliger U (ed) Coastal plant communities of Latin America. Academic, Cambridge, pp 179–199

Costa CSB, Marangoni J, Azevedo A (2003) Plant zonation in irregularly flooded salt marshes: relative importance of stress tolerance and biological interactions. J Ecol 91:951–965

Crane J (1975) Fiddler crabs of the world, Ocypodidae: genus *Uca*. Princeton University Press, Princeton

D'Incao F, Silva KG, Ruffino ML, Braga AC (1990) Hábito alimentar do caranguejo *Chasmagnathus granulata* Dana, 1851 na Barra do Rio Grande, RS (Decapoda, Grapsiade). Atlantica 12(2):85–93

Diele K (2000) Life history and population structure of the exploited mangrove crab *Ucides cordatus cordatus* (L.) (Decapoda: Brachyura) in the Caeté estuary, North Brazil. PhD dissertation, University of Bremen

Diele K, Koch V (2010a) Comparative population dynamics and life histories of north Brazilian mangrove crabs, genera *Uca* and *Ucides* (Ocypodoidea). In: Saint-Paul U, Schneider H (eds) Mangrove dynamics and Management in North Brazil. Ecological studies. Springer-Verlag, Berlin/Heidelberg, pp 275–286

Diele K, Koch V (2010b) Growth and mortality of the exploited mangrove crab *Ucides cordatus* (Ucididae) in N-Brazil. J Exp Mar Biol Ecol 395:171–180

Diele K, Simith DJ (2007) Effects of substrata and conspecific odour on the metamorphosis of mangrove crab megalopae, *Ucides cordatus* (Ocypodidae). J Exp Mar Biol Ecol 348:174–182

Diele K, Koch V, Saint-Paul U (2005) Population structure, catch composition and CPUE of the artisanally harvested mangrove crab *Ucides cordatus* (Ocypodidae) in the Caeté estuary, North Brazil: indications for overfishing? Aquat Living Resour 18:169–178

Diele K, Araújo ARR, Glaser M, Salzmann U (2010a) Artisanal fishery of the mangrove crab *Ucides cordatus* (Ucididae) and first steps toward a successful co-Management in Bragança, North Brazil. In: Saint-Paul U, Schneider H (eds) Mangrove dynamics and Management in North Brazil. Springer-Verlag, Berlin/Heidelberg, pp 287–297

Diele K, Koch V, Abrunhosa FA, Lima JF, Simith DJB (2010b) The brachyuran crab Community of the Caeté Estuary, North Brazil: species richness, zonation and abundance. In: Saint-Paul U, Schneider H (eds) Mangrove dynamics and Management in North Brazil. Springer-Verlag, Berlin/Heidelberg, pp 251–264

Duarte LFDA, Duran RS, Mendonça JT, Pinheiro MAA (2014) Fishery of the Uçá crab *Ucides cordatus* (Linnaeus, 1763) in a mangrove area in Cananéia, state of São Paulo, Brazil: fishery performance, exploitation patterns and factors affecting the catches. Braz J Oceanogr 62:187–199

Ferreira A, Ganade G, Morais Freire F, Attayde J (2013) Propagule predation in a Neotropical mangrove: the role of the Grapsid crab *Goniopsis cruentata*. Hydrobiologia 707:135–146

Firmo AM, Tognella MM, Silva SR, Barboza RRRD, Alves RRN (2012) Capture and commercialization of blue land crabs ("guaiamum") *Cardisoma guanhumi* (Latreille, 1825) along the coast of Bahia State, Brazil: an ethnoecological approach. J Ethnobiol Ethnomed 8:12

Fischer E, Araujo A, Duarte L (1997) Consumption of bromeliad flowers by the crab *Metasesarma rubripes* in a Brazilian Coastal Forest. Crustaceana 70:118–123

Freitas RF, Schrack EC, Sieg RD, Silliman BR, Costa CSB (2014) Grazing scar characteristics impact degree of fungal facilitation in *Spartina alterniflora* leaves in a South American Salt Marsh. Braz Arch Biol Technol 58:103–108

Gifford CA (1962) Some observations on the general biology of the land crab, *Cardisoma guanhumi* (Latreille), in South Florida. Biol Bull 123:207–223

Glaser M, Diele K (2004) Asymmetric outcomes: assessing central aspects of the biological, economic and social sustainability of a mangrove crab fishery, *Ucides cordatus* (Ocypodidae), in North Brazil. Ecol Econ 49:361–373

Góes P, Sampaio FDF, Carmo TMS, Toso GC, Leal MS (2000) Comportamento e período reprodutivo do caranguejo do mangue *Ucides cordatus*. In: Abstracts of the V Simpósio de Ecossistemas Brasileiros: Conservação, ACIESP, Vitória, 11–15 October 2000

Grande FRD, Granado P, Sanches FHC, Costa TM (2018) Organic matter affects fiddler crab distribution? Results from field and laboratorial trials. Estuar Coast Shelf Sci 212:138–145

Gusmão-Júnior JBL, Machado GBO, Costa TM (2012) Burrows with chimneys of the fiddler crab *Uca thayeri*: construction, occurrence, and function. Zool Stud 51(5):598–605

Herreid C III (1963) Observations on the feeding behavior of *Cardisoma guanhumi* (Latreille) in Southern Florida1. Crustaceana 5:176–180

Horta MAP, Ferreira AP, Luzardo AJR, Pinto WJ, Horta M, Faro ARMC (2011) Risk analysis of cadmium intake by fish consumers in a sub-tropical coastal lagoon, Sepetiba Bay – SE, Brazil. Rev Bras Promoção Saúde 24(1):46–53

Hurtado EG, Rodríguez YG, Delgado YG, Consuegra AH (2015) Densidad poblacional de *Cardisoma guanhumi* (Latreille, 1825) Crustacea: Brachyura: Gercarcinidae) en el Parque Nacional Ciénaga de Zapata, Cuba. Boletín del Centro de Investigaciones Biológicas 49(2): 110–124

Instituto Chico Mendes de Conservação da Biodiversidade – ICMBio (2017) Aplicativo será usado em prol da sustentabilidade dos caranguejos. http://www.icmbio.gov.br/portal/images/stories/comunicacao/downloads/icmbioemfoco446.pdf. Accessed 17 Nov 2017

Instituto Chico Mendes de Conservação da Biodiversidade – ICMBio (2018) Aplicativo monitora andadas reprodutivas de caranguejo. http://www.icmbio.gov.br/portal/ultimas-noticias/20-geral/10130-aplicativo-monitora-andadas-reprodutivas-de-caranguejo. Accessed 28 Dec 2018

Kapusta SC, Bemvenuti CE (1998) Atividade nictemeral de alimentação de juvenis de *Callinectes sapidus*, Rathbun, 1895 (Decapoda: Portinidae) numa pradaria de Ruppia marítima L. e num plano não vegetado, numa enseada estuarina da Lagoa dos Patos, RS, Brasil. Nauplius 6:41–52

Koch V, Wolff M, Diele K (2005) Comparative population dynamics of four fiddler crabs (Ocypodidae, genus Uca) from a North Brazilian mangrove ecosystem. Mar Ecol Prog Ser 291:177–188

Lacerda LD, Campos RC, Santelli RE (2012) Metals in water, sediments, and biota of an offshore oil exploration area in the Potiguar Basin, Northeastern Brazil. Environ Monit Assess 185: 4427–4447

Lana PC, Bernardino AF (eds) (2018) Brazilian estuaries: a benthic perspective, vol. 1. Brazilian marine biodiversity. Springer, Cham

Legat JFA, Mota RI, Puchnick-Legat A, Bittencourt C, Santana WS (2006) Considerations about *Ucides cordatus* fishing in the Parnaíba river Delta region, Brazil. J Coastal Res 39:1281–1283

Leme MHA, Fransozo MLN (1998) Reproductive patterns of *Aratus pisonii* (Decapoda: Grapsidae) from an estuarine area of São Paulo Northern Coast, Brazil. Rev Biol Trop 46(3):673–678

Leopold A, Marchand C, Deborde J, Chaduteau C, Allenbach M (2013) Influence of mangrove zonation on CO_2 fluxes at the sediment–air interface (New Caledonia). Geoderma 202-203:62–70

Lima JSG, Bailey C (2015) Shrimp farming as a coastal zone challenge in Sergipe state, Brazil: balancing goals of conservation and social justice. In: Finkl C, Makowski C (eds) Environmental management and governance. Springer Verlag, Heidelberg/Berlin, pp 233–252

Lima GV, Soares MRS, Oshiro LM (2006) Reproductive biology of the sesarmid crab *Armases rubripes* (Decapoda, Brachyura) from an estuarine area of the Sahy River, Sepetiba Bay, Rio de Janeiro, Brazil. Iheringia Ser Zool 96:47–52

Lima-Gomes RD, Cobo V, Fransozo A (2011) Feeding behaviour and ecosystem role of the red mangrove crab *Goniopsis cruentata* (Latreille, 1803) (Decapoda, Grapsoidea) in a subtropical estuary on the Brazilian coast. Crustaceana 84:735–747

Machado GBO, Gusmão-Junior JBL, Costa TM (2013) Burrow morphology of *Uca uruguayensis* and *Uca leptodactylus* (Decapoda: Ocypodidae) from a subtropical mangrove forest in the western Atlantic. Integr Zool 8:307–314

Maciel DC, Alves AGC (2009) Conhecimentos e práticas locais relacionados ao aratu *Goniopsis cruentata* (Latreille, 1803) em Barra de Sirinhaém, litoral sul de Pernambuco, Brasil. Biota Neotropica 9:29–36

Magalhães HFD, Neto EMC, Schiavetti A (2011) Saberes pesqueiros relacionados à coleta de siris e caranguejos (Decapoda: Brachyura) no município de Conde, Estado da Bahia. Biota Neotrop 11:45–54

Maier ELB, Neto PQS (2009) A pesca do siri no estuário da lagoa dos patos – RS. Geografia Ensino Pesquisa Santa Maria 13(2):363–370

Marochi MZ, Masunari S (2011) Os caranguejos Eriphiidae, Menippidae, Panopeidae e Pilumnidae (Crustacea Brachyura) de águas rasas do litoral do Paraná, com chave pictórica de identificação para as espécies. Biota Neotrop 11:21–33

Masunari S (2006) Distribuição e abundância dos caranguejos *Uca* Leach (Crustacea, Decapoda, Ocypodidae) na Baía de Guaratuba, Paraná, Brasil. Rev Bras Zool 23(4):901–914

Masunari S (2012) Hood construction as an indication of the breeding period of the fiddler crab *Uca (Leptuca) leptodactyla* Rathbun, 1898 (Decapoda, Ocypodidae) from Guaratuba Bay, southern Brazil. Crustaceana 85:1153–1169

McKee KL (1995) Mangrove species distribution and propagule predation in Belize: an exception to the dominance-predation hypothesis. Biotropica 27:334

Melo GAS (1996) Manual de identificação dos Brachyura (caranguejos e siris) do litoral brasileiro. Editora Plêiade, São Paulo

Melo GAS (2008) The Brachyura (Decapoda) of Ilha Grande Bay, Rio de Janeiro, Brazil. Nauplius 16(1):1–22

Nascimento DM, Ferreira EN, Bezerra DM, Rocha PD, Alves RR, Mourão JS (2012) Capture techniques use of Caranguejo-uçá crabs (*Ucides cordatus*) in Paraíba state (northeastern Brazil) and its socio-environmental implications. An Acad Bras Cienc 84:1051–1064

Nascimento DM, Alves AGC, Alves RRN, Barboza RRD, Diele K, Mourão JS (2016) An examination of the techniques used to capture mangrove crabs, *Ucides cordatus*, in the Mamanguape River estuary, northeastern Brazil, with implications for management. Ocean Coast Manag 130:50–57

Nascimento DM, Alves RRN, Barboza RRD, Rilke R, Barboza D, Schmidt AJ, Diele K, Mourão JS (2017) Commercial relationships between intermediaries and harvesters of the mangrove crab *Ucides cordatus* (Linnaeus, 1763) in the Mamanguape River estuary, Brazil, and their socio-ecological implications. Ecol Econ 131:44–51

Natálio LF, Pardo JC, Machado GB, Fortuna MD, Gallo DG, Costa TM (2017) Potential effect of fiddler crabs on organic matter distribution: a combined laboratory and field experimental approach. Estuar Coast Shelf Sci 184:158–165

Negromonte AO, Araújo MDSLD, Coelho PA (2012) Decapod crustaceans from a marine tropical mangrove ecosystem on the Southern Western Atlantic, Brazil. Nauplius 20:247–256

Nordhaus I (2004) Feeding ecology of the semi-terrestrial crab *Ucides cordatus cordatus* (Decapoda: Brachyura) in a mangrove forest in northern Brazil. PhD dissertation, University of Bremen

Nordhaus I, Wolff M (2007) Feeding ecology of the mangrove crab *Ucides cordatus* (Ocypodidae): food choice, food quality and assimilation efficiency. Mar Biol 151:1665–1681

Nordhaus I, Wolff M, Diele K (2006) Litter processing and population food intake of the mangrove crab *Ucides cordatus* in a high intertidal forest in northern Brazil. Estuar Coast Shelf Sci 67: 239–250

Nordhaus I, Diele K, Wolff M (2009) Activity patterns, feeding and burrowing behaviour of the crab *Ucides cordatus* (Ucididae) in a high intertidal mangrove forest in North Brazil. J Exp Mar Biol Ecol 374:104–112

Orélis-Ribeiro R, Boeger WA, Vicente VA, Chammas M, Ostrensky A (2010) Fulfilling Koch's postulates confirms the mycotic origin of Lethargic Crab Disease. Antonie van Leeuwenhoek 99:601–608

Pestana DF, Pülmanns N, Nordhaus I, Diele K, Zimmer M (2017) The influence of crab burrows on sediment salinity in a *Rhizophora*-dominated mangrove forest in North Brazil during the dry season. Hydrobiologia 803:295–305

Pinheiro MAA, Rodrigues AMT (2011) Crustáceos sobre-explotados e o plano nacional de gestão dos caranguejos uçá (*Ucides cordatus*), guaiamú (*Cardisoma guanhumi*) e do siri-azul (*Callinectes sapidus*): uma estratégia para evitar que passem ao status de ameaçados de extinção. Revista CEPSUL Biodiversidade Conservação Marinha 2:50–57

Pinheiro MAA, Fiscarelli AG, Hattori GY (2005) Growth of the Mangrove Crab *Ucides cordatus* (Brachyura, Ocypodidae). J Crustacean Biol 25:293–302

Pinheiro MAA, Silva PPGE, Duarte LFDA, Almeida AA, Xanotto FP (2012) Accumulation of six metals in the mangrove crab *Ucides cordatus* (Crustacea: Ucididae) and its food source, the red mangrove *Rhizophora mangle* (Angiosperma: Rhizophoraceae). Ecotoxicol Environ Saf 81: 114–121

Piou C, Berger U, Hildenbrandt H, Grimm V, Diele K, D'Lima C (2007) Simulating cryptic movements of a mangrove crab: recovery phenomena after small scale fishery. Ecol Model 205:110–122

Principe SC, Augusto A, Costa TM (2018) Differential effects of water loss and temperature increase on the physiology of fiddler crabs from distinct habitats. J Therm Biol 73:14–23

Pülmanns N, Diele K, Mehlig U, Nordhaus I (2014) Burrows of the semi-terrestrial crab *Ucides cordatus* enhance CO_2 release in a North Brazilian Mangrove Forest. PLoS One 9(10):e109532

Pülmanns N, Mehlig U, Nordhaus I, Saint-Paul U, Diele K (2016) Mangrove crab *Ucides cordatus* removal does not affect sediment parameters and stipule production in a one year experiment in northern Brazil. PLoS One 11(12):e0167375

Ribeiro PD, Christy JH, Iribarne OO, Núñez JD (2016) Hood-building dynamics and mating mode in the temperate fiddler crab *Uca uruguayensis* Nobili, 1901. J Crustac Biol 36:507–514

Rosa L, Bemvenuti C (2005) Effects of the burrowing crab *Chasmagnathus granulata* (Dana) on meiofauna of estuarine intertidal habitats of Patos Lagoon, Southern Brazil. Braz Arch Biol Technol 48(2):267–274

Ruas VM, Becker C, D'Incao F (2017) Evaluation of the Blue Crab *Callinectes sapidus* Rathbun, 1896 Bycatch in Artisanal Fisheries in Southern Brazil. Braz Arch Biol Technol 60:1

Sandrini-Neto L, Lana PC (2011) Distribution patterns of the crab *Ucides cordatus* (Brachyura, Ucididae) at different spatial scales in subtropical mangroves of Paranaguá Bay (southern Brazil). Helgol Mar Res 66:167–174

Schaeffer-Novelli Y, Sierra EJS, Vale CC, Bernini E, Rovai AS, Pinheiro MAA, Schmidt AJ, Almeida R, Coelho C Jr, Menghini RP, Martinez DI, Abuchahla GMO, Cunha-Lignon M, Charlier-Sarubo S, Shirazawa-Freitas J, Cintrón-Molero G (2016) Climate changes in mangrove forests and salt marshes. Braz J Oceanogr 64(2):37–52

Schmidt AJ (2006) Estudo da dinâmica populacional do caranguejo-uçá, *Ucides cordatus cordatus* (Linnaeus 1763) (Crustacea-Decapoda-Brachyura), e dos efeitos de uma mortalidade em massa desta espécie em manguezais do sul da Bahia. Master thesis, Universidade de São Paulo

Schmidt AJ (2012) Ritmos de acasalamento e habitat de recrutamento do caranguejo-uçá (*Ucides cordatus*), e suas implicações para gestão em manguezais de Caravelas, BA. PhD dissertation, Universidade Federal do Rio Grande

Schmidt AJ (2014a) Relatório da avaliação de estoque de caranguejo-uçá na RESEX do Cassurubá – a report to the Conservation International. Projeto Pesca Mais Sustentável, Caravelas

Schmidt AJ (2014b) Relatório da avaliação de estoque de caranguejo-uçá na RESEX de Canavieiras – A report to the Conservation International. Projeto Pesca Mais Sustentável, Caravelas

Schmidt AJ, Diele K (2009) First field record of mangrove crab *Ucides cordatus* (Crustacea: Decapoda: Ucididae) recruits co-inhabiting burrows of conspecific crabs. Zoologia-Curitiba 26:792–794

Schmidt AJ, Bemvenuti CE, Diele K, Galli OBS, Serafim KB, Jesus JBG, Theil CMI (2008a) Teste de método de monitoramento de migrações reprodutivas e observações sobre o comportamento do caranguejo-uçá, *Ucides cordatus*, em manguezal de Caravelas, BA. In: Abstracts of the V Congresso Brasileiro sobre Crustáceos, Sociedade Brasileira de Carcinologia, Gramado, 10–13 November 2008

Schmidt AJ, Rabelo LB, Galli OBS, Theil CI, Bemvenuti CE, Jesus JBG, Serafim KB (2008b) Distribuição espacial do caranguejo uçá *Ucides cordatus* e do guaiamum *Cardisoma guanhumi*, em uma transição de manguezal para restinga em Caravelas/BA. In: Abstracts of the IV Simpósio Brasileiro de Oceanografia, Instituto Oceanográfico de São Paulo, São Paulo, 8–12 December 2008

Schmidt AJ, Theil CMI, Galli OBS (2008c) Estudos preliminares sobre efeitos de uma mortalidade em massa. Bol Tecn Cient CEPENE Tamandaré PE 16:51–57

Schmidt AJ, Oliveira MA, Souza EP, May M, Araújo SMB (2009) Estudo comparativo da dinâmica populacional de caranguejo-uçá, *Ucides cordatus* (Linnaeus, 1763) (Custacea-decapoda-brachyura), em áreas afetadas e não afetadas por uma mortalidade em massa no Sul da Bahia, Brasil. Bol Tec Cient CEPENE Tamandaré PE 17(1):41–64

Schmidt AJ, Bemvenuti CE, Diele K (2012) Effects of geophysical cycles on the rhythm of mass mate searching of a harvested mangrove crab. Anim Behav 84(2):333–340

Schmidt AJ, Bemvenuti CE, Diele K (2013) Sobre a definição da zona de apicum e a sua importância ecológica para populações de caranguejo-uçá, *Ucides cordatus*. Bol Tec Cient CEPENE Tamandaré PE 19(1):9–25

Schories D, Bergan AB, Barletta M et al (2003) The keystone role of leaf-removing crabs in a mangrove forests of North Brazil. Wetl Ecol Manag 11:243–255

Severino-Rodrigues E, Soares FC, Graça-Lopes R, Souza KH, Canéo VOC (2009) Diversidade e biologia de espécies de Portunidae (Decapoda, Brachyura) no Complexo Estuarino-Lagunar de Iguape, Ilha Comprida e Cananeia, São Paulo, Brasil. Bol Inst Pesca 35(1):47–60

Simith DJB, Abrunhosa FA, Diele K (2013) Chemical induction in mangrove crab megalopae, *Ucides cordatus* (Ucididae): do young recruits emit metamorphosis-triggering odours as do conspecific adults? Estuar Coast Shelf Sci 131:264–270

Simith DJB, Abrunhosa FA, Diele K (2017) Metamorphosis of the edible mangrove crab *Ucides cordatus* (Ucididae) in response to benthic microbial biofilms. J Exp Mar Biol Ecol 492:132–140

Sousa WP, Mitchell BJ (1999) The effect of seed predators on plant distributions: is there a general pattern in mangroves? Oikos 86:55–66. https://doi.org/10.2307/3546569

Souza MMA, Sampaio EVSB (2011) Predation on propagules and seedlings in mature and regenerating mangroves in the coast of Ceará, Brazil. Hydrobiologia 661:179–186

Stieglitz T, Ridd P, Müller P (2000) Passive irrigation and functional morphology of crustacean burrows in tropical mangrove swamp. Hydrobiologia 421(1):69–76

Thurman CL, Hanna J, Bennett C (2010) Ecophenotypic physiology: osmoregulation by fiddler crab (*Uca* spp.) from the northern Caribbean in relation to ecological distribution. Mar Freshw Behav Physiol 43(5):339–356

Thurman CL, Faria SC, McNamara JC (2013) The distribution of fiddler crabs (*Uca*) along the coast of Brazil: implications for biogeography of the western Atlantic Ocean. Mar Biodivers 6:E1

von Hagen HO (1977) The tree climbing crabs of Trinidad. Stud Fauna Curaçao Caribb Islands 54: 25–59

Warner GF (1969) The occurrence and distribution of crabs in a Jamaican mangrove swamp. J Anim Ecol 38:379–389

Weber LI, Levy JA (2000) Genetic population structure of the swimming crab *Callinectes danae* (Crustacea: Decapoda) in southern Brazil. Hydrobiologia 420:203–210

Wellens S, Sandrini-Neto L, González-Wangüemert M, Lana P (2015) Do the crabs *Goniopsis cruentata* and *Ucides cordatus* compete for mangrove propagules? A field-based experimental approach. Hydrobiologia 757:117–128

Whitefleet-Smith LA, Harding JM (2014) Size selectivity by Atlantic mud crabs *Panopeus herbstii* (Milne Edwards) feeding on ivory barnacles *Balanus eburneus* (Gould). J Shellfish Res 33:25–33

Wunderlich AC, Pinheiro MAA (2013) Mangrove habitat partitioning by *Ucides cordatus* (Ucididae): effects of the degree of tidal flooding and tree-species composition during its life cycle. Helgol Mar Res 67:279–289

Chapter 11
Fish Communities of Mangroves and Salt Marshes Along the Brazilian Coast

Teodoro Vaske Júnior and Matheus M. Rotundo

11.1 Introduction

Mangroves provide structural heterogeneity for fish species, expanding shelter areas and providing nutrients to early life stages (Nagelkerken et al. 2000; Laegdsgaard and Johnson 1995, 2001; Dorenbosch et al. 2004). Several marine species, including reef species, depend on the connectivity with habitats offered by mangroves in estuarine environments for one or more life stages (Huxham et al. 2004; Mumby et al. 2004; Rooker et al. 2018; Bastos et al. 2022). Such connectivity is directly associated with fishery production in adjacent marine areas, in part due to the carbon flow and the trophic processes in mangroves (Giarrizzo et al. 2006; Corrêa and Uieda 2007).

For fish, much of the mangrove ecosystem is only available for a short time due to tidal variations. Thus, populations migrate in different estuarine areas where abiotic factors (mainly temperature, salinity, depth, turbidity, and dissolved oxygen) are also responsible for each species' movements (Joyeux et al. 2009; Vilar et al. 2013). Estuaries, salt marshes, and mangroves are habitats where fishes find ideal conditions for mating, breeding, foraging, and feeding, finding shelter from predators, and optimal growth of larvae and juveniles. Fish such as anchovies, croakers, sardines, groupers, and mullets, among others, swim through the mangroves for breeding and feeding. Some fish species spend their entire lives within the mangrove and adjacencies, while others use the area only during part of their life cycle. Several commercial fish depend in some way on the mangrove for their survival, being captured by artisanal fishery at some point of the year (Gasalla and Rossi-Wongtschowski 2004; Siliprandi et al. 2019).

T. Vaske Júnior (✉)
Instituto de Biociências, Universidade Estadual Paulista, São Vicente, SP, Brazil

M. M. Rotundo
Acervo Zoológico, Universidade Santa Cecília, Santos, SP, Brazil

Y. Schaeffer-Novelli et al. (eds.), *Brazilian Mangroves and Salt Marshes*, Brazilian Marine Biodiversity, https://doi.org/10.1007/978-3-031-13486-9_11

Structural complexity and environmental characteristics encourage the use of mangroves by fish, and these factors may influence separately or together. The high turbidity and refuge areas originating from the mangrove prop roots and pneumatophores act as shelter for several species during high tide. The great availability of food, as a function of the intense primary production, also favors the abundance of crustaceans in different stages of development, which, for instance, will serve as a food supply for many planktivore and carnivore fishes.

In addition to the physiological adaptations necessary to inhabit mangroves, fish species bear morphological adaptations that benefit them over transient species (e.g., Gobiidae, Blenniidae, Anablepidae). In general, species well adapted to these habitats are rarely captured by common fishing gears like trawls and traps, because of the ease of hiding and avoidance, as fishing gears are commonly used in deeper waters. In this way, many studies do not truly estimate the real diversity of fish in mangroves. This fact can be corroborated, for example, as in the case of fire in the port area of the Santos and São Vicente estuary (SP-Brazil), where, due to mortality, 41 uncatalogued species were recorded for the region (Rotundo et al. 2020). Nevertheless, when analyzed, it is possible to realize that the ichthyofauna associated with mangroves is relatively similar (at the family level) in different regions of the planet (Krishnamurthy et al. 1984), even when considering regional differences caused by the biogeographic patterns of distribution.

11.2 Review on Fish Diversity in Brazilian Mangroves and Salt Marshes

This is a review on fish communities of Brazilian mangroves regarding species richness, latitudinal and environmental occurrence, and artisanal fisheries. Several fish species inhabit estuaries at least during one stage of their life cycles. According to Andrade-Tubino et al. (2008), the last synthesis of Brazilian estuarine fishes showed 304 species belonging to 83 families in 22 estuaries along the coast from the Amazon estuary (Pará State) to the Arroio-Chuí estuary (Rio Grande do Sul State). The most frequently found families were Sciaenidae, Gobiidae, Serranidae, Ariidae, Haemulidae, Gerreidae, Paralichthyidae, and Syngnathidae, in order of relevance. In the same study, only 11 species were common to all localities. Another study that covered estuarine fishes between 0° and 25°S was performed in four large estuaries, aiming to determine the length-weight relationship of 70 species distributed in 25 families (Joyeux et al. 2009).

More recently, Blaber and Barletta (2016) reviewed the status of estuarine fishes in 22 estuaries in South America (13 of them in Brazil), including comments in sustainable management and conservation. Their research mainly focused on biomass and biodiversity of fishes, nursery habitats, exploitation of living resources (flora and fauna), conservation and recovery of the riparian vegetation, treatment of domestic, industrial, and agricultural effluents, water circulation barriers, and dredging strategies. The conclusions of this study highlight that there is scarce information that deals with an overview of mangrove fish for the entire Brazilian coast, that is, the species can be grouped into habitat characteristics and geographic regions of occurrence. In this way, the present study surveys the diversity of fish species that have been documented in mangroves, estuaries, and salt marshes along the Brazilian coast, which represents the most updated list yet available, including updated taxonomy once some species have recently been renamed.

A total of 37 points were collected from the literature whenever the species were inhabiting waters with expressive variable salinity as observed in estuaries, mangroves, and salt marshes along the Brazilian coast (Fig. 11.1).

According to the consulted literature on fish occurrence and diversity in Brazilian estuaries, mangroves, and salt marshes in these 37 localities, there are to date 604 species of fish in these environments, distributed in 123 families (Andreata et al. 1989, 2002a, b; Andreata and Marca 2013; Araújo and Azevedo (2001); Araújo et al. 2017; Araújo et al. 1998; Azevedo et al. 2007; Azpelicueta et al. 2019; Barbanti et al. 2013; Barletta et al. 2003, 2005; Barreiros et al. 2009; Barros et al. 2011; Basilio et al. 2009; Bolzan 2014; Caires et al. 2019; Carvalho-Filho et al. 2019; Carvalho-Filho et al. 2019; Carvalho et al. 2020; Carvalho-Neta and Castro 2008; Castro et al. 2010; Cattani et al. 2016; Catelani et al. 2014; Chao et al. 2021; Chaves and Bouchereau 2000; Claudino et al. 2015; Contente 2013; Contente et al. 2011; Corrêa and Uieda 2007; Costa and Camara 2012; Dallevo-Gomes et al. 2020; Dias et al. 2011; Favero and Dias 2015; Figueiredo et al. 2021; Garcia and Vieira 1997, 2001; Garcia et al. 2003; Giarrizzo 2007; Giarrizzo et al. 2006; Giglio and Freitas 2013; Guedes et al. 2005; Hostim-Silva et al. 2013; Jimenez et al. 2013; Lamas et al. 2016; Lima 2010; Littmann et al. 2021; Loebmann and Vieira 2005; Lopes et al. 1998; Lourenço 2016; Marceniuk et al. 2016, 2017, 2020; Martins and Vendel 2014; Melo et al. 2016; Neves et al. 2011; Paiva et al. 2009, 2013; Paiva and Araujo 2016; Passos et al. 2012; Pereira and Rocha 2015; Parenti 2019; Petean et al. 2020; Pickett et al. 2020; Pinheiro 2010; Pinheiro-Jr et al. 2017; Ramos et al. 2011; Ramos and Vieira 2001; Reis-Filho et al. 2010; Ribeiro et al. 2018; Rocha and Dias 2015; Rotundo et al. 2020; Santos et al. 2015; Silva et al. 2016; Silva et al. 2018; Silva Jr 2012; Soares-Filho 1996; Souza-Conceição et al. 2013; Tavera et al. 2018; Tavera et al. 2018; Teixeira et al. 2017; Vendel and Chaves 2006; Viana and Frédou 2014; Vieira 2006; Vilar et al. 2011; Xavier et al. 2012) (Table 11.1).

To illustrate the most representative species of mangroves on the Brazilian coast, 90 most common species were selected in order of frequency of occurrence in the four political divisions of the country in the coast (north, northeast, southeast, and south) and the three habitats (Table 11.2), so that they can be easily recognized and compared with other estuarine, mangrove, or salt marsh fishes in the world.

The table summary above shows that over 43% (262) of the recorded species were present in all four coastal political regions. The most widespread and numerous taxonomic families are Carangidae, Characidae, Cichlidae, Sciaenidae, and Serranidae. These families display fishes of all habitats, many of which are considered of economic relevance in the country.

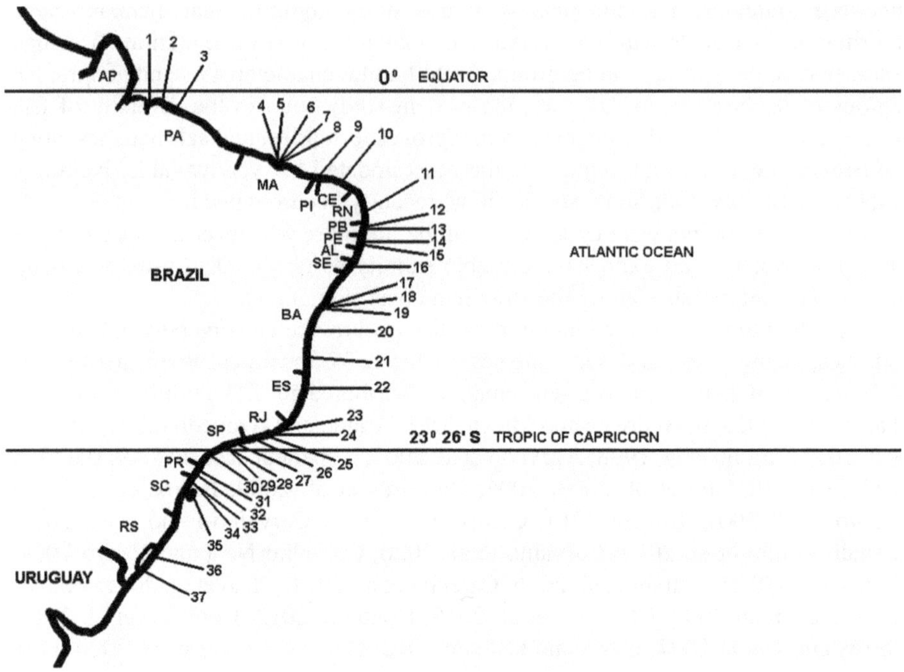

Fig. 11.1 Mangroves, estuaries, and saltmarshes that were considered in the review of fish occurrence and diversity along the Brazilian coast. North region: (1) Curuçá Estuary (PA), (2) Caeté Estuary (PA), and (3) Pará River (PA); see Chap. 3, Maps 1 and 2. Northeast region: (4) Paciência River (MA), (5) Anil River (MA), (6) Cachorros River (MA), (7) Raposa Mangrove (MA), (8) Caranguejo Island (MA), (9) Timonha and Ubatuba Estuary (PI), (10) Jaguariba and Mundaú Estuary (CE), (11) Guaraíras Lagoon (RN), (12) Mamanguape River (PB), (13) Comandatuba Estuary (PB), (14) Goiana and Santa Cruz Estuary (PE), (15) Formoso River (PE), (16) Japaratuba River (SE), (17) Mataripe Estuary (BA), (18) Itaparica Island (BA), (19) Contas River (BA), (20) Itacaré River (BA), and (21) Cassurubá River (BA); see Chap. 3, Maps 3–11 and 18. Southeast region: (22) São Mateus Estuary (ES), (23) Rodrigo de Freitas Lagoon and Guanabara Bay (RJ), (24) Marapendi Lagoon (RJ), (25) Ribeira Bay (RJ), (26) Mambucaba River (RJ), (27) Ubatuba and Araçá Bay (SP), (28) Bertioga Channel (SP), (29) Santos and São Vicente Estuary (SP), and (30) Cananéia Estuary (SP); see Chap. 3, Maps 12, 13, and 14. South region: (31) Paranaguá Estuary (PR), (32) Guaraguaçu and Guaratuba Bay (PR), (33) Saí River (SC), (34) Babitonga Bay (SC), (35) Ratones River (SC), (36) Peixes Lagoon (RS), and (37) Patos Lagoon (RS); see Chap. 3, Maps 15, 16, and 17

The area in the coast with the largest number of exclusive species is the north region (54), all of them either considered fresh- or brackish-water species. Two of the assessed families cannot be found in that region: Narcinidae and Rivulidae. The former represented by the Brazilian electric ray or lesser numbfish *Narcine brasiliensis*, and the latter by three species: the killifish *Anablepsoides cearensis*, found in the lower portion of the Mundaú River estuary (CE); *Atlantirivulus santensis*, found throughout the São Paulo State coast; and *Kryptolebias ocellatus*, commonly found in Northeast Brazil and some areas of Espírito Santo State

Table 11.1 List of fish species found in estuaries, mangroves, and salt marshes along the Brazilian coast in different habitats

Taxon	Habitat	Distribution
Family Acanthuridae		
Acanthurus bahianus Castelnau, 1855	M	N, NE, SE, S
Acanthurus chirurgus (Bloch, 1787)	M	N, NE, SE, S
Acanthurus coeruleus Bloch & Schneider, 1801	M	N, NE, SE
Family Acestrorhynchidae		
Acestrorhynchus grandoculis Menezes & Géry, 1983	F	N
Family Achiridae		
Achirus achirus (Linnaeus, 1758)	FBM	N, NE
Achirus declivis Chabanaud, 1940	FBM	N, NE, SE, S
Achirus lineatus (Linnaeus, 1758)	BM	N, NE, SE, S
Apionichthys dumerili Kaup, 1858	FB	N
Catathyridium garmani (Jordan, 1889)	BM	SE, S
Trinectes microphthalmus (Chabanaud, 1928)	BM	N, NE, SE, S
Trinectes paulistanus (Miranda Ribeiro, 1915)	BM	N, NE, SE, S
Family Albulidae		
Albula goreensis Valenciennes, 1847	M	N, NE
Albula nemoptera (Fowler, 1911)	BM	N, NE, SE
Albula vulpes (Linnaeus, 1758)	M	N, NE, SE, S
Family Anablepidae		
Anableps anableps (Linnaeus, 1758)	FB	N, NE
Anableps microlepis Müller & Troschel, 1844	FB	N, NE
Jenynsia lineata (Jenyns, 1842)	FB	SE, S
Jenynsia multidentata (Jenyns, 1842)	FB	SE, S
Family Anostomidae		
Leporinus copelandii Steindachner, 1875	F	NE, SE
Leporinus fasciatus (Bloch, 1794)	F	N, NE
Leporinus friderici (Bloch, 1794)	F	N, SE, S
Leporinus piau Fowler, 1941	F	NE
Schizodon fasciatus Spix & Agassiz, 1829	F	N
Family Antennariidae		
Antennarius multiocellatus (Valenciennes, 1837)	M	N, NE, SE
Antennarius striatus (Shaw, 1794)	M	N, NE, SE, S
Histrio histrio (Linnaeus, 1758)	M	N, NE, SE, S
Family Apogonidae		
Apogon pseudomaculatus Longley, 1932	M	N, NE, SE, S
Family Apteronotidae		
Apteronotus albifrons (Linnaeus, 1766)	F	N, NE, SE
Sternarchella terminalis (Eigenmann & Allen, 1942)	F	N
Sternarchorhamphus muelleri (Steindachner, 1881)	F	N
Family Arhynchobatidae		
Sympterygia acuta Garman, 1877	M	SE, S
Sympterygia bonapartii Müller & Henle, 1841	M	SE, S

(continued)

Table 11.1 (continued)

Taxon	Habitat	Distribution
Family Ariidae		
Amphiarius phrygiatus (Valenciennes, 1840)	BM	N, NE
Amphiarius rugispinis (Valenciennes, 1840)	BM	N, NE
Aspistor luniscutis (Valenciennes, 1840)	BM	N, NE, SE, S
Aspistor quadriscutis (Valenciennes, 1840)	FBM	N, NE
Bagre bagre (Linnaeus, 1766)	BM	N, NE, SE, S
Bagre filamentosus (Swainson, 1839)	BM	N, NE, SE, S
Cathorops agassizii (Eigenmann & Eigenmann, 1888)	FB	N, NE
Cathorops spixii (Agassiz, 1829)	FBM	N, NE, SE, S
Genidens barbus (Lacepède, 1803)	BM	NE, SE, S
Genidens genidens (Cuvier, 1829)	BM	N, NE, SE, S
Genidens machadoi (Miranda Ribeiro, 1918)	BM	SE, S
Genidens planifrons (Higuchi et al., 1982)	BM	S
Notarius bonillai (Miles, 1945)	F	N, NE
Notarius grandicassis (Valenciennes, 1840)	BM	N, NE, SE
Potamarius grandoculis (Steindachner, 1877)	F	NE, SE
Sciades couma (Valenciennes, 1840)	FBM	N, NE
Sciades herzbergii (Bloch, 1794)	BM	N, NE
Sciades parkeri (Traill, 1832)	BM	N, NE, SE
Sciades passany (Valenciennes, 1840)	BM	N, NE
Sciades proops (Valenciennes, 1840)	FB	N, NE
Family Aspredinidae		
Aspredinichthys filamentosus (Valenciennes, 1840)	FB	N
Aspredinichthys tibicen (Valenciennes, 1840)	FBM	N, NE
Aspredo aspredo (Linnaeus, 1758)	FBM	N
Family Atherinopsidae		
Atherinella blackburni (Schultz, 1949)	BM	NE, SE
Atherinella brasiliensis (Quoy & Gaimard, 1825)	BM	N, NE, SE, S
Membras dissimilis (Carvalho, 1956)	M	SE, S
Odontesthes argentinensis (Valenciennes, 1835)	FBM	SE, S
Odontesthes bonariensis (Valenciennes, 1835)	FBM	NE, SE, S
Odontesthes incisa (Jenyns, 1841)	FBM	S
Family Auchenipteridae		
Ageneiosus inermis (Linnaeus, 1766)	F	N
Ageneiosus ucayalensis Castelnau, 1855	F	N
Pseudauchenipterus nodosus (Bloch, 1794)	FB	N, NE, SE
Trachelyopterus galeatus (Linnaeus, 1766)	F	N, NE
Trachelyopterus striatulus (Steindachner, 1877)	F	SE
Family Balistidae		
Balistes capriscus Gmelin, 1789	M	N, NE, SE, S
Balistes vetula Linnaeus, 1758	M	N, NE, SE

(continued)

Table 11.1 (continued)

Taxon	Habitat	Distribution
Family Batrachoididae		
Amphichthys cryptocentrus (Valenciennes, 1837)	M	N, NE
Batrachoides surinamensis (Bloch & Schneider, 1801)	BM	N, NE
Opsanus beta (Goode & Bean, 1880)[a]	FBM	SE, S
Porichthys plectrodon Jordan & Gilbert, 1882	M	N, NE
Porichthys porosissimus (Cuvier, 1829)	M	SE, S
Thalassophryne maculosa Günther, 1861	M	N, NE
Thalassophryne montevidensis (Berg, 1893)	M	NE, SE, S
Thalassophryne nattereri Steindachner, 1876	M	N, NE, SE
Thalassophryne punctata Steindachner, 1876	M	NE
Family Belonidae		
Strongylura marina (Walbaum, 1792)	FBM	N, NE, SE, S
Strongylura timucu (Walbaum, 1792)	FBM	N, NE, SE, S
Tylosurus acus acus (Lacepède, 1803)	M	N, NE, SE
Tylosurus crocodilus (Péron & Lesueur, 1821)	BM	N, NE
Family Blenniidae		
Entomacrodus nigricans Gill, 1859	M	N, NE
Hypleurochilus fissicornis (Quoy & Gaimard, 1824)	M	SE, S
Lupinoblennius paivai (Pinto, 1958)	FB	NE, SE
Omobranchus punctatus (Valenciennes, 1836)[a]	BM	SE, S
Parablennius marmoreus (Poey, 1876)	M	N, NE, SE, S
Parablennius pilicornis (Cuvier, 1829)	M	SE, S
Scartella cristata (Linnaeus, 1758)	M	N, NE, SE, S
Family Bothidae		
Bothus ocellatus (Agassiz, 1831)	M	N, NE, SE
Bothus robinsi Topp & Hoff, 1972	M	N, NE, SE, S
Family Callichthyidae		
Callichthys callichthys (Linnaeus, 1758)	F	N, NE, SE, S
Corydoras paleatus (Jenyns, 1842)	F	SE, S
Hoplosternum littorale (Hancock, 1828)	F	N, SE, S
Family Carangidae		
Alectis ciliaris (Bloch, 1787)	M	N, NE, SE
Caranx bartholomaei Cuvier, 1833	M	N, NE, SE, S
Caranx crysos (Mitchill, 1815)	BM	N, NE, SE, S
Caranx hippos (Linnaeus, 1766)	FBM	N, NE, SE, S
Caranx latus Agassiz, 1831	FBM	N, NE, SE, S
Caranx lugubris Poey, 1860	M	N, NE, SE, S
Caranx ruber (Bloch, 1793)	M	N, NE, SE, S
Chloroscombrus chrysurus (Linnaeus, 1766)	BM	N, NE, SE, S
Hemicaranx amblyrhynchus (Cuvier, 1833)	M	N, NE, SE, S
Oligoplites palometa (Cuvier, 1832)	FBM	N, NE, SE, S
Oligoplites saliens (Bloch, 1793)	BM	N, NE, SE, S

(continued)

Table 11.1 (continued)

Taxon	Habitat	Distribution
Oligoplites saurus (Bloch & Schneider, 1801)	FBM	N, NE, SE, S
Parona signata (Jenyns, 1841)	M	SE, S
Selar crumenophthalmus (Bloch, 1793)	M	N, NE, SE
Selene brownii (Cuvier, 1816)	BM	N, NE, SE
Selene setapinnis (Mitchill, 1815)	BM	N, NE, SE, S
Selene vomer (Linnaeus, 1758)	BM	N, NE, SE, S
Seriola lalandi Valenciennes, 1833	BM	N, NE, SE, S
Trachinotus carolinus (Linnaeus, 1766)	BM	N, NE, SE, S
Trachinotus cayennensis Cuvier, 1832	BM	N
Trachinotus falcatus (Linnaeus, 1758)	BM	N, NE, SE, S
Trachinotus goodei Jordan & Evermann, 1896	M	N, NE, SE, S
Trachinotus marginatus Cuvier, 1832	M	SE, S
Trachurus lathami Nichols, 1920	M	N, NE, SE, S
Uraspis secunda (Poey, 1860)	M	N, NE, SE, S
Family Carcharinidae		
Carcharhinus limbatus (Valenciennes, 1839)	BM	N, NE, SE, S
Carcharhinus porosus (Ranzani, 1839)	BM	N, NE, SE, S
Rhizoprionodon lalandii (Valenciennes, 1839)	M	N, NE, SE, S
Rhizoprionodon porosus (Poey, 1861)	FBM	N, NE, SE, S
Family Centropomidae		
Centropomus ensiferus Poey, 1860	FBM	N, NE, SE
Centropomus irae Carvalho-Filho et al., 2019	BM	N
Centropomus parallelus Poey, 1860	FBM	N, NE, SE, S
Centropomus pectinatus Poey, 1860	FBM	N, NE, SE
Centropomus undecimalis (Bloch, 1792)	FBM	N, NE, SE, S
Family Chaetodontidae		
Chaetodon ocellatus Bloch, 1787	M	N, NE, SE
Chaetodon striatus Linnaeus, 1758	M	N, NE, SE, S
Family Characidae		
Astyanax bimaculatus (Linnaeus, 1758)	F	N, NE, SE, S
Astyanax eigenmanniorum (Cope, 1894)	F	SE, S
Astyanax fasciatus (Cuvier, 1819)	F	N, NE, SE, S
Astyanax giton Eigenmann, 1908	F	SE
Astyanax janeiroensis Eigenmann, 1908	F	SE
Astyanax lacustris (Lütken, 1875)	F	SE, S
Astyanax taeniatus (Jenyns, 1842)	F	NE, SE
Cheirodon ibicuhiensis Eigenmann, 1915	F	S
Cheirodon interruptus (Jenyns 1842)	F	S
Compsura heterura Eigenmann, 1915	F	NE, SE, S
Diapoma alburnus (Hensel, 1870)	F	S
Hemigrammus guyanensis Géry, 1959	F	N, NE
Hemigrammus marginatus Ellis, 1911	F	N, NE, SE, S

(continued)

Table 11.1 (continued)

Taxon	Habitat	Distribution
Hemigrammus rodwayi Durbin, 1909	F	N, NE
Hyphessobrycon bifasciatus Ellis, 1911	F	SE, S
Hyphessobrycon boulengeri (Eigenmann, 1907)	F	S
Hyphessobrycon igneus Miquelarena et al., 1980	F	SE, S
Hyphessobrycon luetkenii (Boulenger, 1887)	F	SE, S
Hyphessobrycon meridionalis Ringuelet et al., 1978	F	S
Macropsobrycon uruguayanae Eigenmann, 1915	F	S
Mimagoniates inequalis (Eigenmann, 1911)	F	S
Moenkhausia costae (Steindachner, 1907)	F	NE
Oligosarcus hepsetus (Cuvier, 1829)	F	SE, S
Oligosarcus jenynsii (Günther, 1864)	FB	SE, S
Oligosarcus robustus Menezes, 1969	F	S
Phenacogaster calverti (Fowler, 1941)	F	NE, SE
Pseudocorynopoma doriae Perugia, 1891	F	S
Serrapinnus heterodon (Eigenmann, 1915)	F	NE, SE
Serrapinnus piaba (Lütken, 1875)	F	N, NE, SE, S
Triportheus auritus (Valenciennes, 1850)	F	N
Triportheus angulatus (Spix & Agassiz, 1829)	F	N, NE
Family Cichlidae		
Astronotus ocellatus (Agassiz, 1831)	F	N
Australoheros facetus (Jenyns, 1842)	F	SE, S
Cichla ocellaris Bloch & Schneider, 1801	FB	N, NE
Cichla orinocensis Humboldt, 1821	F	N
Cichla pinima Kullander & Ferreira, 2006	F	N
Cichla pleiozona Kullander & Ferreira, 2006	F	N
Cichla temensis Humboldt, 1821	F	N
Cichlasoma bimaculatum (Linnaeus, 1758)	F	N
Cichlasoma orientale Kullander, 1983	F	NE
Cichlasoma portalegrense (Hensel, 1870)	F	S
Coptodon rendalli (Boulenger, 1897)[a]	FB	NE, SE, S
Crenicichla johanna Heckel, 1840	F	N
Crenicichla lacustris (Castelnau, 1855)	F	SE
Crenicichla lepidota Heckel, 1840	F	SE, S
Crenicichla lugubris Heckel, 1840	F	N
Crenicichla menezesi Ploeg 1991	F	NE
Crenicichla reticulata (Heckel, 1840)	F	N, NE
Crenicichla saxatilis (Linnaeus, 1758)	F	N, NE
Geophagus brasiliensis (Quoy & Gaimard, 1824)	FB	SE, S
Geophagus proximus (Castelnau, 1855)	F	N
Geophagus surinamensis (Bloch, 1791)	F	N
Oreochromis niloticus (Linnaeus, 1758)[a]	FB	NE, SE, S

(continued)

Table 11.1 (continued)

Taxon	Habitat	Distribution
Family Clariidae		
Clarias gariepinus (Burchell 1822)[a]	F	SE, S
Family Clinidae		
Ribeiroclinus eigenmanni (Jordan, 1888)	M	SE, S
Family Clupeidae		
Brevoortia aurea (Spix & Agassiz, 1829)	M	SE, S
Brevoortia pectinata (Jenyns, 1842)	M	SE, S
Harengula clupeola (Cuvier, 1829)	M	N, NE, SE, S
Harengula jaguana Poey, 1865	FBM	N, NE, SE, S
Lile piquitinga (Schreiner & Miranda Ribeiro, 1903)	FB	N, NE, SE
Opisthonema oglinum (Lesueur, 1818)	BM	N, NE, SE, S
Platanichthys platana (Regan, 1917)	FB	NE, SE, S
Ramnogaster arcuata (Jenyns, 1842)	M	SE, S
Rhinosardinia amazonica (Steindachner, 1879)	FB	N, NE
Rhinosardinia bahiensis (Steindachner, 1879)	FB	N, NE, SE
Sardinella aurita Valenciennes, 1847	BM	N, NE, SE, S
Sardinella brasiliensis (Steindachner, 1879)	BM	N, NE, SE, S
Family Congridae		
Conger orbignianus Valenciennes, 1837	M	SE, S
Family Crenuchidae		
Characidium bimaculatum Fowler, 1941	F	NE
Characidium rachovii Regan, 1913	F	S
Family Ctenoluciidae		
Boulengerella cuvieri (Spix & Agassiz, 1829)	F	N
Family Curimatidae		
Curimata inornata Vari, 1989	F	N
Cyphocharax gilbert (Quoy & Gaimard, 1824)	F	NE, SE
Cyphocharax saladensis (Meinken, 1933)	F	S
Cyphocharax voga (Hensel, 1870)	F	S
Steindachnerina elegans (Steindachner, 1875)	F	NE
Steindachnerina notonota (Miranda Ribeiro, 1937)	F	NE
Family Cynodontidae		
Rhaphiodon vulpinus Spix & Agassiz, 1829	F	N
Family Cynoglossidae		
Symphurus diomedeanus (Goode & Bean, 1885)	M	N, NE, SE, S
Symphurus jenynsi Evermann & Kendall, 1906	M	SE, S
Symphurus plagusia (Bloch & Schneider, 1801)	BM	N, NE, SE, S
Symphurus tessellatus (Quoy & Gaimard, 1824)	BM	N, NE, SE, S
Symphurus trewavasae Chabanaud, 1948	M	SE, S
Family Dactylopteridae		
Dactylopterus volitans (Linnaeus, 1758)	M	N, NE, SE, S
Dactyloscopus crossotus Starks, 1913	M	N, NE, SE

(continued)

Table 11.1 (continued)

Taxon	Habitat	Distribution
Family Dasyatidae		
Bathytoshia centroura (Mitchill, 1815)	BM	N, NE, SE, S
Dasyatis hypostigma Santos & Carvalho, 2004	BM	SE, S
Fontitrygon geijskesi (Boeseman, 1948)	BM	N
Hypanus berthalutzae Petean et al., 2020	BM	N, NE, SE
Hypanus guttatus (Bloch & Schneider, 1801)	M	N, NE, SE, S
Hypanus marianae (Gomes, Rosa & Gadig, 2000)	BM	NE
Hypanus say (Lesueur, 1817)	BM	N, NE, SE, S
Family Diodontidae		
Chilomycterus antennatus (Cuvier, 1816)	M	N, NE
Chilomycterus antillarum Jordan & Rutter, 1897	M	N, NE
Chilomycterus reticulatus (Linnaeus, 1758)	M	SE
Chilomycterus spinosus (Linnaeus, 1758)	M	N, NE, SE, S
Diodon holocanthus Linnaeus, 1758	M	N, NE, SE, S
Diodon hystrix Linnaeus, 1758	M	N, NE, SE, S
Family Doradidae		
Lithodoras dorsalis (Valenciennes, 1840)	F	N
Family Echeneidae		
Echeneis naucrates Linnaeus, 1758	M	N, NE, SE, S
Phtheirichthys lineatus (Menzies, 1791)	M	N, NE, SE, S
Remora remora (Linnaeus, 1758)	M	N, NE, SE, S
Family Eleotridae		
Butis koilomatodon (Bleeker, 1849)[a]	FBM	NE, SE
Dormitator maculatus (Bloch, 1792)	FBM	N, NE, SE, S
Eleotris pisonis (Gmelin, 1789)	FBM	N, NE, SE, S
Erotelis smaragdus (Valenciennes, 1837)	FBM	N, NE
Guavina guavina (Valenciennes, 1837)	FBM	N, NE, SE, S
Family Elopidae		
Elops smithi McBride et al., 2010	M	N, NE, SE, S
Family Engraulidae		
Anchoa cayorum (Fowler, 1906)	M	N, NE
Anchoa filifera (Fowler, 1915)	BM	N, NE, SE, S
Anchoa hepsetus (Linnaeus, 1758)	BM	N, NE, SE, S
Anchoa januaria (Steindachner, 1879)	M	N, NE, SE, S
Anchoa lyolepis (Evermann & Marsh, 1900)	M	N, NE, SE, S
Anchoa marinii Hildebrand, 194	M	N, NE, SE, S
Anchoa parva (Meek & Hildebrand, 1923)	FBM	N, NE, SE, S
Anchoa spinifer (Valenciennes, 1848)	FBM	N, NE, SE, S
Anchoa tricolor (Spix & Agassiz, 1829)	BM	NE, SE, S
Anchovia clupeoides (Swainson 1839)	M	N, NE, SE, S
Anchovia surinamensis (Bleeker, 1865)	FB	N, NE
Anchoviella brevirostris (Günther, 1868)	BM	N, NE, SE, S

(continued)

Table 11.1 (continued)

Taxon	Habitat	Distribution
Anchoviella guianensis (Eigenmann, 1912)	FB	N
Anchoviella lepidentostole (Fowler, 1911)	FBM	N, NE, SE, S
Cetengraulis edentulus (Cuvier, 1829)	BM	N, NE, SE, S
Engraulis anchoita Hubbs & Marini, 1935	M	N, NE, SE, S
Engraulis eurystole (Swain & Meek, 1884)	M	N, NE
Lycengraulis batesii (Günther, 1868)	FB	N, NE
Lycengraulis grossidens (Spix & Agassiz, 1829)	FBM	N, NE, SE, S
Pterengraulis atherinoides (Linnaeus, 1766)	FB	N, NE
Family Ephippidae		
Chaetodipterus faber (Broussonet, 1782)	FBM	N, NE, SE, S
Family Erythrinidae		
Hoplias malabaricus (Bloch, 1794)	FB	N, NE, SE, S
Family Exocoetidae		
Cheilopogon melanurus (Valenciennes, 1847)	M	N, NE, SE
Hirundichthys affinis (Günther, 1866)	M	N, NE
Parexocoetus brachypterus (Richardson, 1846)	M	N, NE, SE, S
Family Fistularidae		
Fistularia petimba Lacepède, 1803	M	N, NE, SE, S
Fistularia tabacaria Linnaeus, 1758	M	N, NE, SE, S
Family Gagidae		
Urophycis brasiliensis (Kaup, 1858)	M	SE, S
Family Gerreidae		
Diapterus auratus Ranzani, 1842	FBM	N, NE, SE, S
Diapterus rhombeus (Cuvier, 1829)	BM	N, NE, SE, S
Eucinostomus argenteus Baird & Girard, 1855	FBM	N, NE, SE, S
Eucinostomus gula (Quoy & Gaimard, 1824)	FBM	N, NE, SE, S
Eucinostomus havana (Nichols, 1912)	BM	N, NE
Eucinostomus lefroyi (Goode, 1874)	M	N, NE, SE, S
Eucinostomus melanopterus (Bleeker, 1863)	FBM	N, NE, SE, S
Eugerres brasilianus (Cuvier, 1830)	M	N, NE, SE, S
Gerres cinereus (Walbaum, 1792)	FBM	N, NE, SE
Family Glymostomatidae		
Ginglymostoma cirratum (Bonnaterre, 1788)	M	N, NE, SE, S
Family Globiesocidae		
Gobiesox barbatulus Starks, 1913	M	SE, S
Gobiesox strumosus Cope, 1870	BM	N, NE, SE, S
Tomicodon fasciatus (Peters, 1859)	M	N, NE, SE
Family Gobiidae		
Awaous tajasica (Lichtenstein, 1822)	FBM	N, NE, SE, S
Bathygobius soporator (Valenciennes, 1837)	FBM	N, NE, SE, S
Ctenogobius boleosoma (Jordan & Gilbert, 1882)	FBM	N, NE, SE, S
Ctenogobius shufeldti (Jordan & Eigenmann, 1887)	FBM	N, NE, SE, S

(continued)

Table 11.1 (continued)

Taxon	Habitat	Distribution
Ctenogobius smaragdus (Valenciennes, 1837)	FBM	N, NE, SE, S
Ctenogobius stigmaticus (Poey, 1860)	M	N, NE, SE, S
Evorthodus lyricus (Girard, 1858)	FBM	N, NE, SE, S
Gobioides broussonnetii Lacepède, 1800	FBM	N, NE, SE, S
Gobionellus oceanicus (Pallas, 1770)	FBM	N, NE, SE, S
Gobionellus stomatus Starks, 1913	B	NE, SE, S
Gobiosoma hemigymnum (Eigenmann & Eigenmann, 1888)	M	N, NE, SE, S
Gnatholepis thompsoni Jordan, 1904	M	N, NE, SE
Microgobius meeki Evermann & Marsh, 1899	M	N, NE, SE, S
Microgobius carri Fowler, 1945	BM	N, NE, SE
Priolepis dawsoni Greenfield, 1989	M	N, NE, SE
Family Gymnotidae		
Gymnotus carapo Linnaeus, 1758	F	N, SE
Family Gymnuridae		
Gymnura altavela (Linnaeus, 1758)	BM	N, NE, SE, S
Gymnura micrura (Bloch & Schneider, 1801)	BM	N, NE, SE, S
Family Haemulidae		
Anisotremus surinamensis (Bloch, 1791)	M	N, NE, SE, S
Anisotremus virginicus (Linnaeus, 1758)	M	N, NE, SE, S
Brachygenys chrysargyrea Günther, 1859	M	N, NE
Boridia grossidens Cuvier, 1830	M	SE, S
Conodon nobilis (Linnaeus, 1758)	FBM	N, NE, SE, S
Genyatremus luteus (Bloch, 1790)	BM	N, NE, SE, S
Haemulon atlanticus Carvalho, Marceniuk, Oliveira & Wosiacki, 2020	M	N, NE, SE, S
Haemulon aurolineatum Cuvier, 1830	M	N, NE, SE, S
Haemulon flavolineatum (Desmarest, 1823)	M	N, NE
Haemulon parra (Desmarest, 1823)	M	N, NE, SE
Haemulon plumierii (Lacepède, 1801)	M	N, NE, SE
Haemulon squamipinna Rocha & Rosa, 1999	M	NE, SE
Haemulopsis corvinaeformis (Steindachner, 1868)	BM	N, NE, SE, S
Orthopristis rubra (Cuvier, 1830)	BM	SE, S
Orthopristis scapularis Fowler, 1915	BM	N, NE
Paranisotremus moricandi (Ranzani, 1842)	M	N, NE, SE
Pomadasys ramosus (Poey, 1860)	FBM	N, NE, SE, S
Rhonciscus crocro (Cuvier, 1830)	FBM	N, NE, SE, S
Family Hemiodontidae		
Hemiodus unimaculatus (Bloch, 1794)	F	N
Family Hemiramphidae		
Hemiramphus balao Lesueur, 1821	BM	N, NE, SE
Hemiramphus brasiliensis (Linnaeus, 1758)	BM	N, NE, SE, S
Hyporhamphus roberti (Valenciennes, 1847)	BM	N, NE, SE, S
Hyporhamphus unifasciatus (Ranzani, 1841)	BM	N, NE, SE, S

(continued)

Table 11.1 (continued)

Taxon	Habitat	Distribution
Family Heptapteridae		
Pimelodella altipinnis (Steindachner, 1864)	F	N
Pimelodella australis Eigenmann, 1917	F	S
Pimelodella cristata (Müller & Troschel, 1849)	F	N, NE
Pimelodella lateristriga (Lichtenstein, 1823)	F	SE
Rhamdia quelen (Quoy & Gaimard, 1824)	F	N, NE, SE, S
Family Holocentridae		
Holocentrus adscensionis (Osbeck, 1765)	M	N, NE, SE, S
Family Hypopomidae		
Brachyhypopomus janeiroensis (Costa & Campos-da-Paz, 1992)	F	SE
Family Kyphosidae		
Kyphosus sectatrix (Linnaeus, 1758)	FBM	N, NE, SE
Kyphosus vaigiensis (Quoy & Gaimard, 1825)	M	N, NE, SE
Family Labridae		
Halichoeres poeyi (Steindachner, 1867)	M	NE, SE, S
Family Labrisomidae		
Labrisomus nuchipinnis (Quoy & Gaimard, 1824)	M	N, NE, SE, S
Malacoctenus delalandii (Valenciennes, 1836)	M	N, NE, SE, S
Family Lebiasinidae		
Nannostomus beckfordi Günther, 1872	F	N, NE
Family Lobotidae		
Lobotes surinamensis (Bloch, 1790)	FBM	N, NE, SE, S
Family Loricariidae		
Acanthicus hystrix Spix & Agassiz, 1829	F	N
Hypostomus affinis (Steindachner, 1877)	F	SE
Hypostomus plecostomus (Linnaeus, 1758)	F	N, NE
Hypostomus punctatus Valenciennes, 1840	F	SE
Hypostomus watwata Hancock, 1828	F	N, NE
Loricaria cataphracta Linnaeus, 1758	F	N
Loricaria parnahybae Steindachner, 1907	F	N, NE
Loricariichthys anus (Valenciennes, 1835)	F	S
Parotocinclus cearensis Garavello, 1977	F	NE
Pterygoplichthys joselimaianus (Weber, 1991)	F	N, NE
Schizolecis guntheri (Miranda Ribeiro, 1918)	F	SE
Family Lutjanidae		
Lutjanus alexandrei Moura & Lindeman, 2007	M	N, NE
Lutjanus analis (Cuvier, 1828)	FBM	N, NE, SE, S
Lutjanus apodus (Walbaum, 1792)	BM	N, NE
Lutjanus buccanella (Cuvier, 1828)	M	N, NE
Lutjanus campechanus (Poey, 1860)	M	N, NE
Lutjanus cyanopterus (Cuvier, 1828)	BM	N, NE, SE
Lutjanus griseus (Linnaeus, 1758)	FBM	N, NE, SE, S

(continued)

Table 11.1 (continued)

Taxon	Habitat	Distribution
Lutjanus jocu (Bloch & Schneider, 1801)	FBM	N, NE, SE
Lutjanus synagris (Linnaeus, 1758)	M	N, NE, SE, S
Lutjanus vivanus (Cuvier, 1828)	M	N, NE, SE
Ocyurus chrysurus (Bloch, 1791)	M	N, NE, SE
Family Megalopidae		
Megalops atlanticus Valenciennes, 1847	FBM	N, NE, SE, S
Family Monacanthidae		
Aluterus heudelotii Hollard, 1855	M	N, NE, SE, S
Aluterus monoceros (Linnaeus, 1758)	M	N, NE, SE, S
Aluterus schoepfii (Walbaum, 1792)	M	N, NE, SE, S
Cantherhines macrocerus (Hollard, 1853)	M	N, NE, SE
Cantherhines pullus (Ranzani, 1842)	M	N, NE, SE
Monacanthus ciliatus (Mitchill, 1818)	M	N, NE, SE, S
Stephanolepis hispida (Linnaeus, 1766)	M	N, NE, SE, S
Stephanolepis setifer (Bennett, 1831)	M	N, NE, SE
Family Mugilidae		
Mugil brevirostris Miranda Ribeiro, 1915	FBM	SE, S
Mugil cephalus Linnaeus, 1758	FBM	N, NE, SE, S
Mugil curema Valenciennes, 1836	FBM	N, NE, SE, S
Mugil curvidens Valenciennes, 1836	M	N, NE, SE, S
Mugil gaimardianus Desmarest, 1831	BM	N, NE, SE, S
Mugil hospes Jordan & Culver, 1895	BM	N, NE, SE, S
Mugil incilis Hancock, 1830	FB	N, NE, SE, S
Mugil liza Valenciennes, 1836	FBM	N, NE, SE, S
Mugil rubrioculus Harrison et al., 2007	BM	N, NE, SE, S
Mugil trichodon Poey, 1875	FBM	N, NE, SE
Family Mullidae		
Mullus argentinae Hubbs & Marini, 1933	M	SE, S
Pseudupeneus maculatus (Bloch, 1793)	M	N, NE, SE, S
Upeneus parvus Poey, 1852	M	N, NE, SE, S
Family Muraenidae		
Gymnothorax funebris Ranzani, 1839	M	N, NE, SE
Gymnothorax moringa (Cuvier, 1829)	M	N, NE, SE
Gymnothorax ocellatus Agassiz, 1831	M	N, NE, SE, S
Gymnothorax vicinus (Castelnau, 1855)	M	N, NE, SE
Family Muraenosocidae		
Cynoponticus savanna (Bancroft, 1831)	BM	N, NE, SE, S
Family Myliobatidae		
Aetobatus narinari (Euphrasen, 1790)	BM	N, NE, SE, S
Rhinoptera bonasus (Mitchill, 1815)	BM	N, NE, SE, S
Family Narcinidae		
Narcine brasiliensis (Olfers, 1831)	M	NE, SE, S

(continued)

Table 11.1 (continued)

Taxon	Habitat	Distribution
Family Ogcocephalidae		
Ogcocephalus nasutus (Cuvier, 1829)	M	N, NE
Ogcocephalus vespertilio (Linnaeus, 1758)	M	N, NE, SE, S
Family Ophichthidae		
Ahlia egmontis (Jordan, 1884)	M	N, NE, SE
Echiophis intertinctus (Richardson, 1848)	M	N, NE, SE, S
Myrichthys ocellatus (Lesueur, 1825)	M	N, NE, SE, S
Myrophis punctatus Lütken, 1852	FBM	N, NE, SE, S
Ophichthus cylindroideus (Ranzani, 1839)	M	N, NE, SE, S
Ophichthus gomesii (Castelnau, 1855)	BM	N, NE, SE, S
Ophichthus ophis (Linnaeus, 1758)	M	N, NE, SE, S
Family Opistognathidae		
Opistognathus cuvierii Valenciennes, 1836	M	NE, SE
Family Ostracidae		
Acanthostracion polygonius Poey, 1876	M	N, NE, SE, S
Acanthostracion quadricornis (Linnaeus, 1758)	M	N, NE, SE, S
Lactophrys trigonus (Linnaeus, 1758)	M	N, NE, SE
Lactophrys triqueter (Linnaeus, 1758)	M	N, NE, SE
Family Paralichthydae		
Citharichthys arenaceus Evermann & Marsh, 1900	FBM	N, NE, SE, S
Citharichthys cornutus (Günther, 1880)	M	N, NE, SE, S
Citharichthys dinoceros Goode & Bean, 1886	M	N, NE, SE, S
Citharichthys macrops Dresel, 1885	M	N, NE, SE, S
Citharichthys spilopterus Günther, 1862	FBM	N, NE, SE, S
Cyclopsetta chittendeni Bean, 1895	M	N, NE, SE
Etropus crossotus Jordan & Gilbert, 1882	BM	N, NE, SE, S
Etropus longimanus Norman, 1933	M	SE, S
Paralichthys brasiliensis (Ranzani, 1842)	BM	N, NE, SE, S
Paralichthys orbignyanus (Valenciennes, 1839)	B	NE, SE, S
Paralichthys patagonicus Jordan, 1889	M	NE, SE, S
Paralichthys tropicus Ginsburg, 1933	M	NE
Syacium micrurum Ranzani, 1842	M	N, NE, SE
Syacium papillosum (Linnaeus, 1758)	M	N, NE, SE, S
Family Pempheridae		
Pempheris schomburgkii Müller & Troschel, 1848	M	N, NE, SE
Family Percophidae		
Percophis brasiliensis Quoy & Gaimard, 1825	M	SE, S
Family Pimelodidae		
Brachyplatystoma filamentosum (Lichtenstein, 1819)	F	N
Brachyplatystoma platynemum Boulenger, 1898	F	N
Brachyplatystoma rousseauxii (Castelnau, 1855)	F	N
Brachyplatystoma vaillantii (Valenciennes, 1840)	F	N, NE

(continued)

Table 11.1 (continued)

Taxon	Habitat	Distribution
Hypophthalmus edentatus Spix & Agassiz, 1829	FB	N
Hypophthalmus marginatus Valenciennes, 1840	F	N
Parapimelodus nigribarbis (Boulenger, 1889)	F	SE, S
Pimelodus blochii Valenciennes, 1840	FB	N, NE, SE, S
Pimelodus maculatus Lacepède, 1803	F	SE, S
Pinirampus pirinampu (Spix & Agassiz, 1829)	F	N
Platystomatichthys sturio (Kner, 1858)	F	N
Propimelodus eigenmanni (Van der Stigchel, 1946)	F	N
Pseudoplatystoma fasciatum (Linnaeus, 1766)	F	N, NE
Family Pinguipedidae		
Pinguipes brasilianus Cuvier, 1829	M	SE, S
Family Pleuronectidae		
Oncopterus darwinii Steindachner, 1874	M	N, NE, SE, S
Family Poeciliidae		
Cnesterodon decemmaculatus (Jenyns, 1842)	BM	SE, S
Phalloceros caudimaculatus (Hensel, 1868)	BM	SE, S
Phalloceros harpagos Lucinda, 2008	F	SE, S
Phalloptychus januarius (Hensel, 1868)	F	SE, S
Poecilia reticulata Peters, 1859[a]	FB	N, NE, SE, S
Poecilia sarrafae Bragança & Costa, 2011	F	NE
Poecilia vivipara Bloch & Schneider, 1801	FB	N, NE, SE, S
Tomeurus gracilis Eigenmann, 1909	F	N, NE
Xiphophorus hellerii Heckel, 1848[a]	F	SE, S
Family Polynemidae		
Polydactylus oligodon (Günther, 1860)	FB	N, NE, SE, S
Polydactylus virginicus (Linnaeus, 1758)	FB	N, NE, SE, S
Family Pomacanthidae		
Pomacanthus arcuatus (Linnaeus, 1758)	M	N, NE, SE
Pomacanthus paru (Bloch, 1787)	M	N, NE, SE, S
Family Pomacentridae		
Abudefduf saxatilis (Linnaeus, 1758)	M	N, NE, SE, S
Stegastes fuscus (Cuvier, 1830)	M	NE, SE, S
Stegastes variabilis (Castelnau, 1855)	M	NE, SE, S
Family Pomatomidae		
Pomatomus saltatrix (Linnaeus, 1766)	BM	N, NE, SE, S
Family Potamotrygonidae		
Potamotrygon motoro (Müller & Henle, 1841)	F	N, NE
Potamotrygon orbignyi (Castelnau, 1855)	F	N
Family Priacanthidae		
Priacanthus arenatus Cuvier, 1829	M	N, NE, SE, S
Family Pristidae		
Pristis pristis (Linnaeus, 1758)	FBM	N

(continued)

Table 11.1 (continued)

Taxon	Habitat	Distribution
Family Pristigasteridae		
Chirocentrodon bleekerianus (Poey, 1867)	BM	N, NE, SE
Odontognathus mucronatus Lacepède, 1800	FBM	NE, SE
Pellona castelnaeana Valenciennes, 1847	FB	N
Pellona flavipinnis (Valenciennes, 1837)	FB	N, NE
Pellona harroweri (Fowler, 1917)	BM	N, NE, SE, S
Family Prochilodontidae		
Prochilodus argenteus Spix & Agassiz, 1829	F	N, NE
Prochilodus nigricans Spix & Agassiz, 1829	F	N, NE, SE, S
Prochilodus brevis Steindachner, 1875	F	NE
Prochilodus vimboides Kner, 1859	F	SE
Family Rachycentridae		
Rachycentron canadum (Linnaeus, 1766)	M	N, NE, SE, S
Family Rajidae		
Atlantoraja platana (Günther, 1880)	M	SE, S
Rioraja agassizii (Müller & Henle, 1841)	M	SE, S
Family Rhamphichthyidae		
Rhamphichthys marmoratus Castelnau, 1855	F	N
Rhamphichthys rostratus (Linnaeus, 1766)	F	N
Family Rhinobatidae		
Pseudobatos horkelii (Müller & Henle, 1841)	M	N, NE, SE, S
Pseudobatos percellens (Walbaum, 1792)	M	N, NE, SE, S
Zapteryx brevirostris (Müller & Henle, 1841)	M	NE, SE, S
Family Rivulidae		
Anablepsoides cearensis (Costa & Vono, 2009)	F	NE
Atlantirivulus santensis (Köhler, 1906)	F	SE, S
Kryptolebias ocellatus (Hensel, 1868)	FB	SE
Family Scaridae		
Nicholsina usta (Valenciennes, 1840)	M	N, NE, SE
Scarus trispinosus Valenciennes, 1840	M	NE, SE
Sparisoma amplum (Ranzani, 1841)	M	NE, SE
Sparisoma axillare (Steindachner, 1878)	M	N, NE, SE, S
Sparisoma radians (Valenciennes, 1840)	M	N, NE, SE
Family Sciaenidae		
Bairdiella goeldi (Marceniuk *et al*., 2019)	BM	N, NE, SE, S
Ctenosciaena gracilicirrhus (Metzelaar, 1919)	M	N, NE, SE, S
Cynoscion acoupa (Lacepède, 1801)	FBM	N, NE, SE, S
Cynoscion guatucupa (Cuvier, 1830)	M	SE, S
Cynoscion jamaicensis (Vaillant & Bocourt, 1883)	BM	N, NE, SE, S
Cynoscion leiarchus (Cuvier, 1830)	BM	N, NE, SE, S
Cynoscion microlepidotus (Cuvier, 1830)	BM	N, NE, SE, S
Cynoscion similis Randall & Cervigón, 1968	BM	N

(continued)

Table 11.1 (continued)

Taxon	Habitat	Distribution
Cynoscion steindachneri (Jordan, 1889)	FBM	N, NE
Cynoscion virescens (Cuvier, 1830)	BM	N, NE, SE, S
Isopisthus parvipinnis (Cuvier, 1830)	BM	N, NE, SE, S
Larimus breviceps Cuvier, 1830	FBM	N, NE, SE, S
Lonchurus lanceolatus (Bloch, 1788)	BM	N
Macrodon ancylodon (Bloch & Schneider, 1801)	BM	N, NE
Macrodon atricauda (Günther, 1880)	BM	SE, S
Menticirrhus cuiaranensis (Marceniuk et al., 2020)	BM	N, NE, SE
Menticirrhus gracilis (Cuvier, 1830)	BM	SE, S
Menticirrhus martinicensis (Cuvier, 1830)	FBM	N, NE, SE, S
Micropogonias furnieri (Desmarest, 1823)	FBM	N, NE, SE, S
Nebris microps Cuvier, 1830	BM	N, NE, SE, S
Odontoscion dentex (Cuvier, 1830)	M	N, NE, SE
Pachypops fourcroi (Lacepède, 1802)	F	N, NE
Paralonchurus brasiliensis (Steindachner, 1875)	BM	N, NE, SE, S
Pareques lineatus (Cuvier, 1830)	M	N, NE, SE
Plagioscion auratus (Castelnau, 1855)	F	N, NE
Plagioscion squamosissimus (Heckel, 1840)	F	N, NE
Plagioscion surinamensis (Bleeker, 1873)	F	N
Pogonias cromis (Linnaeus, 1766)	BM	N, NE
Pogonias courbina (Lacepède, 1803)	BM	SE, S
Stellifer brasiliensis (Schultz, 1945)	BM	NE, SE, S
Stellifer cervigoni Chao et al., 2021	M	N
Stellifer collettei Chao et al., 2021	M	N, NE, SE
Stellifer gomezi Chao et al., 2021	BM	N, NE, SE, S
Stellifer menezesi Chao et al., 2021	BM	NE, SE
Stellifer microps (Steindachner, 1864)	BM	N, NE
Stellifer musicki Chao et al., 2021	BM	N, NE
Stellifer naso (Jordan, 1889)	BM	N, NE
Stellifer punctatissimus (Meek & Hildebrand, 1925)	BM	N, NE, SE
Stellifer rastrifer (Jordan, 1889)	BM	N, NE, SE, S
Stellifer stellifer (Bloch, 1790)	BM	N, NE, SE, S
Umbrina canosai Berg, 1895	BM	N, NE, SE, S
Umbrina coroides Cuvier, 1830	BM	N, NE, SE, S
Family Scombridae		
Acanthocybium solandri (Cuvier, 1832)	M	N, NE, SE, S
Scomberomorus brasiliensis Collette et al., 1978	M	N, NE, SE, S
Scomberomorus cavalla (Cuvier, 1829)	M	N, NE, SE, S
Scomberomorus maculatus (Mitchill, 1815)	M	N, NE
Scomberomorus regalis (Bloch, 1793)	M	N, NE, SE
Family Scorpaenidae		
Scorpaena brasiliensis Cuvier, 1829	M	N, NE, SE, S

(continued)

Table 11.1 (continued)

Taxon	Habitat	Distribution
Scorpaena isthmensis Meek & Hildebrand, 1928	M	N, NE, SE, S
Scorpaena plumieri Bloch, 1789	M	N, NE, SE, S
Family Serranidae		
Acanthistius brasilianus (Cuvier, 1828)	BM	SE, S
Alphestes afer (Bloch, 1793)	M	N, NE, SE, S
Diplectrum formosum (Linnaeus, 1766)	M	N, NE, SE, S
Diplectrum radiale (Quoy & Gaimard, 1824)	BM	N, NE, SE, S
Dules auriga Cuvier, 1829	M	NE, SE, S
Epinephelus adscensionis (Osbeck, 1765)	M	N, NE, SE
Epinephelus itajara (Lichtenstein, 1822)	BM	N, NE, SE, S
Epinephelus marginatus (Lowe, 1834)	M	SE, S
Epinephelus morio (Valenciennes, 1828)	M	N, NE, SE, S
Hyporthodus nigritus (Holbrook, 1855)	M	N, NE, SE, S
Hyporthodus niveatus (Valenciennes, 1828)	M	N, NE, SE, S
Mycteroperca acutirostris (Valenciennes, 1828)	M	NE, SE, S
Mycteroperca bonaci (Poey, 1860)	M	N, NE, SE, S
Mycteroperca interstitialis (Poey, 1860)	M	SE, S
Mycteroperca microlepis (Goode & Bean, 1879)	M	N, NE, SE, S
Mycteroperca tigris (Valenciennes, 1833)	M	N, NE, SE, S
Rypticus randalli Courtenay, 1967	M	N, NE, SE, S
Rypticus saponaceus (Bloch & Schneider, 1801)	M	N, NE, SE
Serranus atrobranchus (Cuvier, 1829)	M	N, NE, SE, S
Serranus flaviventris (Cuvier, 1829)	M	N, NE, SE, S
Family Serrasalmidae		
Pristobrycon calmoni (Steindachner, 1908)	F	N
Pristobrycon striolatus (Steindachner, 1908)	F	N
Pygocentrus nattereri Kner, 1858	F	N, NE
Serrasalmus rhombeus (Linnaeus, 1766)	F	N, NE
Serrasalmus spilopleura Kner, 1858	F	N, NE
Family Sparidae		
Archosargus probatocephalus (Walbaum, 1792)	BM	N, NE, SE, S
Archosargus rhomboidalis (Linnaeus, 1758)	BM	N, NE, SE, S
Calamus penna (Valenciennes, 1830)	M	N, NE, SE, S
Calamus pennatula Guichenot, 1868	M	N, NE, SE
Diplodus argenteus (Valenciennes, 1830)	M	N, NE, SE, S
Family Sphyraenidae		
Sphyraena barracuda (Edwards, 1771)	BM	N, NE, SE, S
Sphyraena guachancho Cuvier, 1829	M	N, NE, SE, S
Sphyraena picudilla Poey, 1860	M	N, NE, SE, S
Sphyraena tome Fowler, 1903	M	SE, S
Family Sphyrnidae		
Sphyrna tiburo (Linnaeus, 1758)	BM	N, NE, SE, S

(continued)

Table 11.1 (continued)

Taxon	Habitat	Distribution
Sphyrna lewini (Griffith & Smith, 1834)	BM	N, NE, SE, S
Sphyrna mokarran (Rüppell, 1837)	BM	N, NE, SE, S
Sphyrna zygaena (Linnaeus, 1758)	BM	N, NE, SE, S
Family Sternopygidae		
Distocyclus conirostris (Eigenmann & Allen, 1942)	F	N
Eigenmannia macrops (Boulenger, 1897)	F	N
Eigenmannia trilineata López & Castello, 1966	F	SE
Eigenmannia virescens (Valenciennes, 1836)	F	N, NE, SE, S
Rhabdolichops caviceps (Fernández-Yépez, 1968)	F	N
Sternopygus macrurus (Bloch & Schneider, 1801)	F	N, NE, SE, S
Family Stromateidae		
Peprilus crenulatus Cuvier, 1829	M	N, NE, SE, S
Peprilus xanthurus (Quoy & Gaimard)	BM	N, NE, SE, S
Family Synbranchidae		
Synbranchus marmoratus Bloch, 1795	FB	NE, SE, S
Family Syngnathidae		
Bryx dunckeri (Metzelaar, 1919)	M	N, NE, SE, S
Cosmocampus elucens (Poey, 1868)	M	N, NE, SE, S
Hippocampus erectus Perry, 1810	M	N, NE, SE, S
Hippocampus patagonicus Piacentino & Luzzatto, 2004	M	SE, S
Hippocampus reidi Ginsburg, 1933	BM	N, NE, SE, S
Micrognathus erugatus Herald & Dawson, 1974	M	NE
Microphis brachyurus (Bleeker, 1854)	FBM	N, NE, SE, S
Microphis lineatus (Kaup, 1856)	FBM	N, NE, SE, S
Pseudophallus brasiliensis Dawson, 1974	FB	N, NE, SE, S
Pseudophallus mindii (Meek & Hildebrand, 1923)	FB	N, NE, SE, S
Syngnathus folletti Herald, 1942	M	NE, SE, S
Syngnathus pelagicus Linnaeus, 1758	M	N, NE, SE, S
Family Synodontidae		
Synodus bondi Fowler, 1939	M	N, NE, SE, S
Synodus intermedius (Spix & Agassiz, 1829)	M	N, NE, SE, S
Synodus poeyi Jordan, 1887	M	N, NE
Trachinocephalus myops (Forster, 1801)	M	N, NE, SE, S
Family Tetraodontidae		
Canthigaster figueiredoi Moura & Castro, 2002	M	N, NE, SE, S
Colomesus asellus (Müller & Troschel, 1849)	FBM	N, NE
Colomesus psittacus (Bloch & Schneider, 1801)	FBM	N, NE
Lagocephalus laevigatus (Linnaeus, 1766)	BM	N, NE, SE, S
Lagocephalus lagocephalus (Linnaeus, 1758)	M	N, NE, SE, S
Sphoeroides greeleyi Gilbert, 1900	BM	N, NE, SE, S
Sphoeroides spengleri (Bloch, 1785)	BM	N, NE, SE, S
Sphoeroides testudineus (Linnaeus, 1758)	FBM	N, NE, SE, S

(continued)

Table 11.1 (continued)

Taxon	Habitat	Distribution
Sphoeroides tyleri Shipp, 1972	M	N, NE, SE, S
Family Triakidae		
Galeorhinus galeus (Linnaeus, 1758)	M	SE, S
Family Trichiuridae		
Trichiurus lepturus Linnaeus, 1758	BM	N, NE, SE, S
Family Triglidae		
Prionotus nudigula Ginsburg, 1950	M	SE, S
Prionotus punctatus (Bloch, 1793)	BM	N, NE, SE, S
Family Uranoscopidae		
Astroscopus sexspinosus (Steindachner, 1876)	M	SE, S
Astroscopus y-graecum (Cuvier, 1829)	M	N, NE, SE, S

Habitats: *B* brackish water, *F* fresh water, *M* marine. Brazilian coastal regions: *N* north (AP and PA states), *NE* northeast (MA, PI, CE, RN, PB, PE, AL, SE, and BA states), *SE* southeast (ES, RJ, and SP states), *S* south (PR, SC, and RS states)
Occurrence refers to the presence in the 37 localities shown in Fig. 11.1
[a]Nonnative species

Table 11.2 Summary of the number of fish species found in estuaries, mangroves, and salt marshes and percentages in relation to 604 species along the Brazilian coast

	n	%
N	54	8.94
N, NE	75	12.42
N, NE, SE	63	10.45
N, NE, SE, S	262	43.43
N, SE	1	0.17
N, SE, S	2	0.33
NE	15	2.48
NE, SE	15	2.48
NE, SE, S	21	3.48
S	17	2.81
SE	13	2.15
SE, S	66	10.93
F	139	23.01
B	2	0.33
M	224	37.09
FB	37	6.13
BM	126	20.86
FBM	75	12.42
Species	604	
Families	123	
Orders	39	
Genera	324	

Habitats: *B* brackish water, *F* fresh water, *M* marine. Brazilian coastal regions: *N* north, *NE* northeast, *SE* southeast, *S* south

(Sarmento-Soares et al. 2014; Teixeira et al. 2017; Ywamoto 2019). Other families that are absent in the north region are Opistognathidae (also absent in the south); Crenuchidae (also absent in the southeast), and Arhynchobatidae, Clariidae, Clinidae, Congridae, Gagidae, Percophidae, Pinguipedidae, Rajidae, and Triakidae (also absent in the northeast).

The south region did not shelter individuals of the families Apteronotidae, Auchenipteridae, Kyphosidae, Labridae, Pempheridae, and Synbranchidae. The families Aspredinidae, Lebiasinidae, Potamotrygonidae, and Serrasalmidae are absent in the south as well as in the southeast.

The southeast regions show only one exclusive family, Hypopomidae, represented by the bluntnose knifefish *Brachyhypopomus janeiroensis*. The family encompasses other species in diverse freshwater habitats, but this is the only recorded in an estuarine region despite its freshwater preference.

11.3 Final Remarks

The current global problem of human waste in aquatic environments significantly affects all ecosystems where fish occur. Most mangrove areas in the country are closely related to urban centers where anthropogenic waste is discarded directly into the river or estuary, which frequently leads to waste accumulation in the complexity in between the mangrove tangle and sediment. All organisms are under the effects of this disturbance that has been reinforced in recent decades. This is yet another reason to consider scientific surveys of fish species for future comparisons. The appearance or disappearance of certain species may not be related only to environmental changes but also due to the direct human influence in modifying coastal ecosystems, especially through landfills, chemical pollution, toxic metals, and domestic waste.

Despite the occasional efforts and species-listing studies, the complexity of estuaries, mangroves, and salt marshes of the Brazilian coast still requires studies for sustainable management and conservation, always taking into consideration human impacts.

Appendix

The following six plates show 90 of the most common fish species in Brazilian mangroves (Plates 11.1, 11.2, 11.3, 11.4, 11.5, and 11.6).

Plate 11.1 Fish species, from left to right. First row: *Achirus declivis* | *Achirus lineatus* | *Albula vulpes* | *Anchoa januaria* | *Anchoa spinifer*. Second row: *Anchoa tricolor* | *Anchovia clupeoides* | *Anchoviella lepidentostole* | *Archosargus rhomboidalis* | *Atherinella brasiliensis*. Third row: *Bairdiella goeldi* | *Bathygobius spporator* | *Caranx hippos* | *Caranx latus* | *Cathorops spixii*. The bars correspond to a scale of 10 cm. *Exception to scale indicated in the plate

Plate 11.2 Fish species, from left to right. First row: *Centropomus paralellus* | *Centropomus undecimalis* | *Cetengraulis edentulus* | *Chaetodipterus faber* | *Chilomycterus spinosus*. Second row: *Chloroscombrus crysurus* | *Citharichthys arenaceus* | *Citharichthys spilopterus* | *Colomesus psittacus* | *Conodon nobilis*. Third row: *Ctenogobius boleosoma* | *Ctenogobius smaragdus* | *Cynoscion acoupa* | *Cynoscion leiarchus* | *Cynoscion microlepidotus*. The bars correspond to a scale of 10 cm. *Exception to scale indicated in the plate

Plate 11.3 Fish species, from left to right. First row: *Dactylopterus volitans* | *Diapterus auratus* | *Diapterus rhombeus* | *Diplectrum radiale* | *Elops smithi*. Second row: *Etropus crossotus* | *Eucinostomus argenteus* | *Eucinostomus gula* | *Eucinostomus melanopterus* | *Eugerres brasilianus*. Third row: *Genidens barbus* | *Genidens genidens* | *Genyatremus cavifrons* | *Geophagus brasiliensis* | *Gobionellus oceanicus*. The bars correspond to a scale of 10 cm

Plate 11.4 Fish species, from left to right. First row: *Haemulopsis corvinaeformis* |*Harengula clupeola* | *Hiporamphus unifasciatus* | *Isopistus parvipinnis* | *Lagocephalus laevigatus*. Second row: *Larimus breviceps* | *Lutjanus jocu* | *Lutjanus synagris* | *Lycengraulis grossidens* | *Menticirrhus martinicensis*. Third row: *Menticirrhus gracilis* | *Micropogonias furnieri* | *Mugil curema* | *Mugil incilis* | *Mugil liza*. The bars correspond to a scale of 10 cm

Plate 11.5 Fish species, from left to right. First row: *Oligoplites palometa* | *Oligoplites saliens* | *Opisthonema oglinum* | *Oreochromis niloticus* | *Orthopristis scapularis*. Second row: *Paralichthys brasiliensis* | *Paralichthys orbignyanus* | *Pellona harroweri* | *Poecilia vivipara** | *Polydactylus virginicus*. Third row: *Prionotus punctatus* | *Rypticus randalli* | *Sardinella brasiliensis* | *Sciades herzbergii* | *Scomberomorus brasiliensis*. The bars correspond to a scale of 10 cm. *Exception to scale indicated in the plate

Plate 11.6 Fish species, from left to right. First row: *Selene setapinnis* | *Selene vomer* | *Sparisoma axillare* | *Sphoeroides greeleyi* | *Sphoeroides testudineus*. Second row: *Stellifer rastrifer* | *Stellifer stellifer* | *Strongylura marina* | *Strongylura timucu* | *Symphurus tesselatus*. Third row: *Synodus bondi* | *Trachinotus carolinus* | *Trachinotus falcatus* | *Trichiurus lepturus* | *Trinectes paulistanus*. The bars correspond to a scale of 10 cm. *Exceptions to scale indicated in the plate

References

Andrade-Tubino MF, Ribeiro ALR, Vianna M (2008) Space-temporal organization of demersal fish assemblages in Brazilian estuarine ecosystems: a synthesis. Oecol Aust 12(4):640–661

Andreata JV, Marca AG (2013) Variações espaciais e temporais e estrutura da comunidade de peixes da Lagoa Rodrigo de Freitas, Rio de Janeiro, Brasil. Rev Bras Zool 15(1,2,3):219–246

Andreata JV, Saad AM, Barbiéri LR (1989) Associação e distribuição das espécies de peixes na laguna de Marapendi, Rio de Janeiro, no período de março de 1985 a fevereiro de 1987. Mem Inst Oswaldo Cruz 84(IV):45–51

Andreata JV, Manzano FV, Baptista MGS, Teixeira DE, Oliveira LOV, Longo MM, Freret NV, Valois AS (2002a) Assembléia de peixes da lagoa Rodrigo de Freitas, Rio de Janeiro. Bioikos 16(1):19–28

Andreata JV, Meurer BC, Baptista MGS, Manzano FV, Teixeira DE, Longo MM, Freret NV (2002b) Composição da Assembléia de peixes da Baía da Ribeira, Angra dos Reis, Rio de Janeiro, Brasil. Rev Bras Zool 19(4):1139–1146

Araújo FG, Azevedo MC (2001) Assemblages of southeast-south Brazilian coastal systems based on the distribution of fishes. Estuar Coast Shelf S 52(6):729–738

Araújo FG, Cruz-Filho AG, Azevedo MCC, Santos ACA (1998) Estrutura da comunidade de peixes demersais da Baía de Sepetiba, RJ. Rev Bras Biol 58:417–430

Araújo ARR, Souza JM, Lima RDCD, Abreu EFS, Virgens FF, Barbosa JM (2017) Diversidade da fauna aquática do estuário do rio Japaratuba, estado de Sergipe, Brasil. Acta Fisher Aquat Resour 5(1):33–42

Azevedo MCC, Araújo FG, Cruz-Filho AG, Pessanha ALM, Silva MA, Guedes APP (2007) Demersal fishes in a tropical bay in southeastern Brazil: Partitioning the spatial, temporal and environmental components of ecological variation. Estuar Coast Shelf S 75(4):468–480

Azpelicueta MDLM, Delpiani SM, Cione AL, Oliveira C, Marceniuk AP, Díaz de Astarloa JM (2019) Morphology and molecular evidence support the validity of *Pogonias courbina* (Lacepède, 1803) (Teleostei: Sciaenidae), with a redescription and neotype designation. PLoS One 14(6):e0216280

Barbanti B, Caires R, Marceniuk AP (2013) A ictiofauna do Canal de Bertioga, São Paulo, Brasil/ The ichthyofauna of the Bertioga Channel, São Paulo, Brazil. Biota Neotrop 13(1):276–291

Barletta M, Barletta-Bergan A, Saint-Paul U, Hubold G (2003) Seasonal changes in density, biomass, and diversity of estuarine fishes in tidal mangrove creeks of the lower Caeté Estuary (northern Brazilian coast, east Amazon). Mar Ecol Prog Ser 256:217–228

Barletta M, Barletta-Bergan A, Saint-Paul U, Hubold G (2005) The role of salinity in structuring the fish assemblages in a tropical estuary. J Fish Biol 66(1):45–72

Barreiros JP, Branco JO, Júnior FF, Machado L, Hostim-Silva M, Verani JR (2009) Space–time distribution of the ichthyofauna from Saco da Fazenda Estuary, Itajaí, Santa Catarina, Brazil. J Coast Res 25(5):1114–1121

Barros DF, Torres MF, Frédou FL (2011) Ictiofauna do estuário de São Caetano de Odivelas e Vigia (Pará, Estuário Amazônico). Biota Neotrop 11(2):367–373

Basilio TH, Godinho WO, Araújo ME, Neto MADAF, Faria VV (2009) Ictiofauna do estuário do rio Curu, Ceará, Brasil. Arq Cienc Mar 42(2):81–88

Bastos RF, Lippi DL, Gaspar ALB, Yogui GT, Frédou T, Garcia AM, Ferreira BP (2022) Ontogeny drives allochthonous trophic support of snappers: Seascape connectivity along the mangrove-seagrass-coral reef continuum of a tropical marine protected area. Estuar Coast Shelf S 264:1–7591

Blaber SJ, Barletta M (2016) A review of estuarine fish research in South America: what has been achieved and what is the future for sustainability and conservation? J Fish Biol 89(1):537–568

Bolzan MS (2014) Peixes em Áreas Rasas do Estuário do rio São Mateus, Espírito Santo. Master thesis, Universidade Federal do Espírito Santo

Caires RA, Santos W, Machado L, Oliveira C, Cerqueira NN, Rotundo MM, Oliveira C, Marceniuk AP (2019) The Tonkin weakfish, *Cynoscion similis* (Sciaenidae, Perciformes), an endemic species of the Amazonas-Orinoco Plume. Acta Amazon 49:197–207

Carvalho CO, Marceniuk AP, Oliveira C, Wosiacki WB (2020) Integrative taxonomy of the species complex *Haemulon steindachneri* (Eupercaria; Haemulidae) with a description of a new species from the western Atlantic. Zoology 141:125–782

Carvalho-Filho A, De Oliveira J, Soares C, Araripe J (2019) A new species of snook, *Centropomus* (Teleostei: Centropomidae), from northern South America, with notes on the geographic distribution of other species of the genus. Zootaxa 4671(1):81–92

Carvalho-Neta RNF, Castro ACL (2008) Diversidade das assembléias de peixes estuarinos da Ilha dos Caranguejos, Maranhão. Arq Cienc Mar 41(1):48–57

Castro ACL, Castro KDD, Porto HLR (2010) Distribuição da assembléia de peixes na área de influência de uma indústria de alumínio na Ilha de São Luis–MA. Arq Cienc Mar 43(2):71–78

Catelani PA, Petry AC, Di Dario F, Dos Santos VLM, Mincarone MM (2014) Fish composition (Teleostei) of the estuarine region of the Macaé River, southeastern Brazil. Check List 10(4): 927–935

Cattani AP, Ribeiro GC, Marcon E (2016) Fish assemblage dynamics in the Ratones River Mangrove, State of Santa Catarina, Brazil. Panam J Aquat Sci 11(4):324–335

Chao NL, Carvalho-Filho A, Santos JDA (2021) Five new species of Western Atlantic stardrums, *Stellifer* (Perciformes: Sciaenidae) with a key to Atlantic *Stellifer* species. Zootaxa 4991(3): 434–466

Chaves P, Bouchereau JL (2000) Use of mangrove habitat for reproductive activity by the fish assemblage in the Guaratuba Bay, Brazil. Oceanol Acta 23(3):273–280

Claudino MC, Pessanha ALM, Araújo FG, Garcia AM (2015) Trophic connectivity and basal food sources sustaining tropical aquatic consumers along a mangrove to ocean gradient. Estuar Coast Shelf S 167:45–55

Contente RF (2013) Padrões Ecológicos Locais e Multidecadais da Ictiofauna no Estuário Cananéia-Iguape. Masther Thesis, Universidade de São Paulo. 186 pp

Contente RF, Stefanoni MF, Spach HL (2011) Fish assemblage structure in an estuary of the Atlantic Forest biodiversity hotspot (southern Brazil). Ichthyol Res 58(1):38–50

Corrêa MDO, Uieda VS (2007) Diet of the ichthyofauna associated with marginal vegetation of a mangrove forest in southeastern Brazil. Iheringia Ser Zool 97(4):486–497

Costa MF, Camara MR (2012) Diversity, abundance and seasonal variation of the ichthyofauna associated with shrimp farming in a tropical estuary in northeastern Brazil. Mar Biodivers 42(4): 471–479

Dallevo-Gomes CI, Mattox GM, Toledo-Piza M (2020) Taxonomic review of the pipefish genus *Pseudophallus* Herald, with the description of a new species (Syngnathiformes: Syngnathidae). Zootaxa 4859(1)

Dias JF, Gonçalves AM, Fernandez WS, Silbiger HLN, Fiadi CB, Schmidt TCDS (2011) Ichthyofauna in an estuary of the Mataripe area, Todos os Santos Bay, Bahia, Brazil. Braz J Oceangr 59(1):75–95

Dorenbosch M, van Riel MC, Nagelkerken I, van der Velde G (2004) The relationship of reef fish densities to the proximity of mangrove and seagrass nurseries. Estuar Coast Shelf S 60(1):37–48

Favero JMD, Dias JF (2015) Juvenile fish use of the shallow zone of beaches of the Cananéia-Iguape coastal system, southeastern Brazil. Braz J Oceanogr 63(2):103–114

Figueiredo JM, Marceniuk AP, Feijo A, Siccha-Ramirez R, Ribeiro GS, Oliveira C, Rosa RS (2021) Taxonomy of *Centropomus* Lacépède, 1802 (Perciformes: Centropomidae), with focus on the Atlantic species of the genus. Zootaxa 4942(3)

Garcia AM, Vieira JP (1997) Abundance and diversity of fish assemblages inside and outside a bed of *Ruppia maritima* L. in the Patos Lagoon estuary (RS, Brazil). Atlantica 19:161–181

Garcia AM, Vieira JP (2001) O aumento da diversidade de peixes no estuário da Lagoa dos Patos durante o episódio El Niño 1997–1998. Atlantica 23:85–96

Garcia AM, Vieira JP, Winemiller KO (2003) Effects of 1997–1998 El Niño on the dynamics of the shallow-water fish assemblage of the Patos Lagoon Estuary (Brazil). Estuar Coast Shelf S 57(3): 489–500

Gasalla MA, Rossi-Wongtschowski CLDB (2004) Contribution of ecosystem analysis to investigating the effects of changes in fishing strategies in the South Brazil Bight coastal ecosystem. Ecol Model 172(2-4):283–306

Giarrizzo T (2007) Importance of mangroves for fish. Bases for the conservation and sustainable management of mangrove ecosystems in North Brazil. PhD dissertation, Universität Bremen

Giarrizzo T, Silva AJS, Lameira EC, Almeida JBA, Isaac V, Saint-Paul U (2006) Weight-length relationships for intertidal fish fauna in a mangrove estuary in Northern Brazil. J Appl Ichthyol 22(4):325–327

Giglio VJ, Freitas MO (2013) Caracterização da pesca artesanal com rede de camboa na Reserva Extrativista do Cassurubá, Bahia. Biotemas 26(2):249–259

Guedes D, Vasconcelos-Filho AL, Macedo R (2005) Ictiofauna do infralitoral adjacente às margens do Canal de Santa Cruz–Itapissuma, Pernambuco. Bol Tecn Cient CEPENE Tamandaré – PE 13(2):65-75

Hostim-Silva M, Costa Lima A, Spach HL, Passos AC, Souza JD (2013) Estuarine demersal fish assemblage from a transition region between the tropics and the subtropics of the South Atlantic. Biotemas 26(3):153–161

Huxham M, Kimani E, Augley J (2004) Mangrove fish: a comparison of community structure between forested and cleared habitats. Science 60(4):637–647

Jimenez EA, Asano-Filho M, Frédou FL (2013) Fish bycatch of the laulao catfish *Brachyplatystoma vaillantii* (Valenciennes, 1840) trawl fishery in the Amazon Estuary. Braz J Oceanogr 61(2):129–140

Joyeux JC, Giarizzo T, Macieira RM, Spach HL, Vaske T Jr (2009) Length-weight relationships for Brazilian estuarine fishes along a latitudinal gradient. J Appl Ichthyol 25(3):350–355

Krishnamurthy K, Jeyaseelan P, Sultan Ali MA (1984) The distribution of fishes in the global mangroves. Curr Sci India 53(17):901–906

Laegdsgaard P, Johnson CR (1995) Mangrove habitats as nurseries: unique assemblages of juvenile fish in subtropical mangroves in eastern Australia. Mar Ecol Prog Ser 126:67–81

Laegdsgaard P, Johnson CR (2001) Why do juvenile fish utilise mangrove habitats? J Exp Mar Biol Ecol 257(2):229–253

Lamas RA, Rossi-Wongtschowski CLDB, Contente RF (2016) Checklist of the fish fauna of the Araçá Bay, São Sebastião Channel, northern coast of São Paulo, Brazil. Check List 12(6):1–11

Lima MAT (2010) Composição da ictiofauna demersal do estuário do Rio de Contas, Bahia, Brasil. Dissertação da Universidade Estadual de Santa Cruz. 75p

Littmann MW, Lundberg JG, Rocha MS (2021) Revision of the South American catfish genus *Hypophthalmus* (Siluriformes, Pimelodidae) with descriptions of two new species from the Amazon and Orinoco Basins. Proc Acad Natl Sci Phila 167(1):171–223

Loebmann D, Vieira JP (2005) Distribuição espacial e abundância das assembléias de peixes no Parque Nacional da Lagoa do Peixe, Rio Grande do Sul, Brasil. Rev Bras Zool 22(3):667–675

Lopes PRD, Oliveira-Silva JT, Ferreira-Melo ASA (1998) Contribuição ao conhecimento da ictiofauna do manguezal de Cacha Pregos, Ilha de Itaparica, Baía de Todos os Santos, Bahia. Rev Bras Zool 15(2):315–325

Lourenço RCG (2016) Assembleia de peixes de estuários negativos do Nordeste brasileiro: atualização taxonômica, padrões espaço-temporais e aspectos funcionais. Master thesis, Universidade Federal do Ceará

Marceniuk AP, Caires R, Siccha-Ramirez R, Oliveira C (2016) Review of the harvestfishes, genus *Peprilus* (Perciformes: Stromateidae), of the Atlantic coast of South America. Zootaxa 4098(2): 311–332

Marceniuk AP, Caires RA, Rotundo MM (2017) The icthyofauna (Teleostei) of the Rio Caeté estuary, northeast Pará, Brazil, with a species identification key from northern Brazilian coast. Panam J Aquat Sci 12(1):31–79

Marceniuk AP, Caires RA, Rotundo MM, Cerqueira NNCD, Siccha-Ramirez R, Wosiacki WB, Oliveira C (2020) Taxonomic revision of the *Menticirrhus americanus* (Linnaeus, 1758) and *M. littoralis* (Holbrook, 1847) (Percomorphacea: Sciaenidae) species complexes from the western Atlantic. Zootaxa 4822(3):301–333

Martins TO, Vendel AL (2014) Fishes collected with an artisanal fish trap in Barra de Camaratuba estuary, northeastern Brazil. Check List 10(6):1260–1263

Melo FAG, Vianna JQ, Araújo TM, Dutra EA (2016) Lista atualizada das espécies de peixes do estuário dos rios Timonha e Ubatuba, Piauí e Ceará, nordeste brasileiro. Bol Soc Bras Ictiol 118: 9–14

Mumby PJ, Edwards AJ, Arias-González JE, Lindeman KC, Blackwell PG, Gall A, Gorczynska MI, Harborne AR, Pescod CL, Renken H, Wabnitz CCC, Llewellyn G (2004) Mangroves enhance the biomass of coral reef fish communities in the Caribbean. Nature 427:533–536

Nagelkerken I, van der Velde G, Gorissen MW, Meijer GJ, Van't Hof T, den Hartog C (2000) Importance of mangroves, seagrass beds and the shallow coral reef as a nursery for important coral reef fishes, using a visual census technique. Estuar Coast Shelf S 51(1):31–44

Neves LM, Teixeira TP, Araújo FG (2011) Structure and dynamics of distinct fish assemblages in three reaches (upper, middle and lower) of an open tropical estuary in Brazil. Mar Ecol 32(1): 115–131

Paiva ACG, Araújo ME (2016) Environmental characterization and spatial distribution of fish fauna in estuaries in the State of Pernambuco, Brazil. Trop Oceanogr 38(1):1679–3013

Paiva AC, Lima MF, Souza JR, Araújo MED (2009) Spatial distribution of the estuarine ichthyofauna of the Rio Formoso (Pernambuco, Brazil), with emphasis on reef fish. Zoologia 26(2): 266–278

Paiva ACG, Tarso Chaves P, Araújo ME (2013) Distribution of estuarine fish fauna along the coast of Brazil. Trop Oceanogr 41(1-2):1–36

Parenti P (2019) An annotated checklist of the fishes of the family Haemulidae (Teleostei: Perciformes). Iranian J Ichthyol 6(3):150–196

Passos ACD, Contente RF, Araújo CCVD, Daros FALDM, Spach HL, Abilhôa V, Fávaro LF (2012) Fishes of Paranaguá estuarine complex, southwest Atlantic. Biota Neotrop 12(3): 226–238

Pereira AML, Rocha FMR (2015) A pesca no estuário do Timonha e Ubatuba (PI/CE)/Comissão Ilha Ativa. Sieart, Parnaíba

Petean FF, Naylor G, Lima SM (2020) Integrative taxonomy identifies a new stingray species of the genus *Hypanus* Rafinesque, 1818 (Dasyatidae, Myliobatiformes), from the Tropical Southwestern Atlantic. J Fish Biol 97(4):1120–1142

Pickett BD, Wallace EM, Ridge PG, Kauwe JS (2020) Lingering taxonomic challenges hinder conservation and management of global bonefishes. Fisheries 45(7):347–358

Pinheiro MDSS (2010) Ciclo de vida e estrutura de uma assembléia de peixes teleósteos em um manguezal da Raposa, Maranhão, Brasil. PhD dissertation, Universidade Estadual Paulista

Pinheiro-Jr JDR, Castro ACL, Gomes LN (2017) Estrutura da comunidade de peixes do estuário do rio Anil, Ilha de São Luís, Maranhão. Arq Cienc Mar 38(1-2):29–37

Ramos LA, Vieira JP (2001) Composição específica e abundância de peixes de zonas rasas dos cinco estuários do Rio Grande do Sul, Brasil. Bol Inst Pesca 27(1):109–121

Ramos JAA, Barletta M, Dantas DV, Lima ARA, Costa MF (2011) Influence of moon phase on fish assemblages in estuarine mangrove tidal creeks. J Fish Biol 78(1):344–354

Reis-Filho JA, Nunes JDACD, Ferreira A (2010) Estuarine ichthyofauna of the Paraguaçu river, Todos os Santos bay, Bahia, Brazil. Biota Neotrop 10(4):301–311

Ribeiro GC, Cattani AP, Hostim-Silva CL, Passos AC, Soeth M, Cardoso RO, Spach HL (2018) Marine ichthyofauna of Santa Catarina Island, Southern Brazil: checklist with comments on the species. Biota Neotrop 19(3):e20180684

Rocha MLFD, Dias JF (2015) Inventory of Chondrichthyes and Actinopterygii species collected in the central coast of São Paulo State, Brazil. Biota Neotrop 15(2):1–9

Rooker JR, Dance MA, Wells JD, Quigg A, Hill RL, Appeldoorn RS, Ferreira BP, Boswell KM, Sanchez PJ, Moulton DL, Kitchens LL, Rooker GJ, Aschenbrenner A (2018) Seascape connectivity and the influence of predation risk on the movement of fishes inhabiting a back-reef ecosystem. Ecosphere 9(4):e02200

Rotundo MM, Gaulia LA, Cardoso GS, Carminatto AA, Henrique HS, Reigada ALD, Ramires M, Barrella W, Petrere-Junior M (2020) Ichthyofauna from Santos-São Vicente upper estuary: a study before and during fire at Santos port terminal. Res Soc Develop 9(10):e6269108901

Santos JAP, Schmiegelow JMM, Rotundo MM, Barrella W (2015) Composição e variação temporal da assembleia de peixes do alto sistema estuarino de Santos, São Paulo, Brasil. Bol Inst Pesca 41(4):945–959

Sarmento-Soares LM, Ingenito LFS, Duboc LF, Martins-Pinheiro RF, Borçato R, Silva JP (2014) Primeiros registros de *Kryptolebias ocellatus* (Hensel) (Cyprinodontiformes, Rivulidae) para riachos de Mata Atlântica no Espírito Santo. Bol Soc Bras Ictiol 111:15–19

Siliprandi CC, Tuset VM, Lombarte A, Farré M, Rossi-Wongtschowski CLDB (2019) Optimal fishing samplers to reveal the morphological structure of a fish assemblage in a subtropical tidal flat. Neotrop ichthyol 17(1):e170168

Silva Jr MGD (2012) Ecologia da ictiofauna do estuário do rio Paciência, Ilha do Maranhão-Brasil. PhD Dissertation, 165 pp, Universidade Federal do Pará. 165p

Silva DR, Paranhos R, Vianna M (2016) Spatial patterns of distribution and the influence of seasonal and abiotic factors on demersal ichthyofauna in an estuarine tropical bay. J Fish Biol 89(1):821–846

Silva VEL, Teixeira EC, Batista VS, Fabré NN (2018) Spatial distribution of juvenile fish species in nursery grounds of a tropical coastal area of the South-Western Atlantic. Acta Ichthyol Piscat 48(1):9–18

Soares-Filho AA (1996) Ictiofauna da região média da boca do estuário do rio Jaguaribe (Ceará-Brasil): composição, distribuição e aspectos bioecológicos. PhD dissertation, Universidade Federal do Ceará

Souza-Conceição JM, Spach HL, Bordin D, Frisanco D, Costa MDP (2013) The role of estuarine beaches as habitats for fishes in a Brazilian subtropical environment. Neotrop Biol Conserv 8(3): 121–131

Tavera J, Acero A, Wainwright PC (2018) Multilocus phylogeny, divergence times, and a major role for the benthic-to-pelagic axis in the diversification of grunts (Haemulidae). Mol Phylogenet Evol 121:212–223

Teixeira FK, Ramos TPA, Paiva RECD, Távora MA, Lima SMQ, Rezende CF (2017) Ichthyofauna of Mundaú river basin, Ceará State, Northeastern Brazil. Biota Neotrop 17(1):e20160174

Vendel AL, Chaves PDT (2006) Use of an estuarine environment (Barra do Saí lagoon, Brazil) as nursery by fish. Rev Bras Zool 23(4):1117–1122

Viana AP, Frédou FL (2014) Ichthyofauna as bioindicator of environmental quality in an industrial district in the amazon estuary, Brazil. Braz J Biol 74(2):315–324

Vieira JP (2006) Ecological analogies between estuarine bottom trawl fish assemblages from Patos Lagoon, Rio Grande do Sul, Brazil and York River, Virginia, USA. Rev Bras Zool 23(1): 234–247

Vilar CC, Spach HL, Santos LO (2011) Fish fauna of Baía da Babitonga (southern Brazil), with remarks on species abundance, ontogenic stage and conservation status. Zootaxa 2734(1):40–52

Vilar CC, Joyeux JC, Giarrizzo T, Spach HL, Vieira JP, Vaske-Jr T (2013) Local and regional ecological drivers of fish assemblages in Brazilian estuaries. Mar Ecol Prog Ser 485:181–197

Xavier JHDA, Cordeiro CAMM, Tenório GD, Diniz ADF, Júnior P, Nunes EP, Rosa IL (2012) Fish assemblage of the Mamanguape Environmental Protection Area, NE Brazil: abundance, composition and microhabitat availability along the mangrove-reef gradient. Neotrop Ichthyol 10(1):109–122

Ywamoto EV (2019) Taxonomic revision and biogeography of *Atlantirivulus santensis* Köhler, 1906 (Rivulidae, Cyprinodontiformes). Master thesis, Universidade Estadual Paulista

Chapter 12
Mangrove and Salt Marsh Migratory and Resident Birds

Patricia Luciano Mancini, Thiago Vernaschi Vieira da Costa, Rafael Antunes Dias, Luís Fábio Silveira, and Fabio Schunck

12.1 Importance of Mangroves and Salt Marshes to Birds

Mangrove ecosystems are highly productive, representing important nutrient sources for both terrestrial and aquatic food webs and serving as breeding and resting ground for many animals, including birds (Luther and Greenberg 2009). Bird diversity in mangroves is mostly related to habitat heterogeneity, especially due to plant species richness, the density of the understory, and food resource distribution (Nagelkerken et al. 2008; Mohd-Azlan et al. 2015). The aerial roots of mangroves provide substrates on which many species live, including algae, tunicates, sponges, and bivalves. Many infaunal and epifaunal species, together with prawns, crabs, and fish, dwell on the soft substrate and may be predated by birds (Nagelkerken et al. 2008). In Brazil, mangrove-resident bird species such as the scarlet ibis (*Eudocimus ruber*) and the yellow-crowned night heron (*Nyctanassa violacea*) feed mainly on fiddler crabs in mangroves (Olmos and Silva e Silva 2001), while the tricolored egret (*Egretta tricolor*) feeds on small fish (*Poecilia* spp.) (Martinez 2010).

P. L. Mancini (✉)
Instituto de Biodiversidade e Sustentabilidade, Universidade Federal do Rio de Janeiro, Macaé, RJ, Brazil
e-mail: patmancinibr@yahoo.com.br

T. V. V. da Costa
Instituto de Recursos Naturais, Universidade Federal de Itajubá, Itajubá, MG, Brazil

R. A. Dias
Departamento de Ecologia, Zoologia e Genética, Universidade Federal de Pelotas, Pelotas, RS, Brazil

L. F. Silveira
Seção de Aves - Museu de Zoologia, Universidade de São Paulo, São Paulo, SP, Brazil

F. Schunck
Comitê Brasileiro de Registros Ornitológicos - CBRO, São Paulo, Brazil

Y. Schaeffer-Novelli et al. (eds.), *Brazilian Mangroves and Salt Marshes*, Brazilian Marine Biodiversity, https://doi.org/10.1007/978-3-031-13486-9_12

The mangrove trees and canopy provide important habitats for bird species to rest, roost, and nest. All three abovementioned species and others, such as herons and passerines, nest in mangrove trees (Olmos et al. 2001; Mancini et al. 2018), highlighting the importance of this ecosystem as breeding sites. Some mangrove sites such as the Maranhão State and Santos-Cubatão area, São Paulo State, shelter several thousands of bird nests every year (Martinez and Rodrigues 1999; Silva e Silva 2007) (see Chap. 3, Maps 3 and 14, respectively).

On the Brazilian northern coast, roughly one thousand semipalmated plovers (*Charadrius semipalmatus*) were recorded perched on the red mangrove prop roots during high tide (Rodrigues 2007). Therefore, mangroves also provide high-tide refuge for birds feeding in nearby areas (Rodrigues 2007; Valente et al. 2011). Every year hundreds of thousands of migratory birds such as plovers, sandpipers, terns, and other species fly to Brazilian mangroves after the breeding season. They come mainly from North America, such as the gray plover (*Pluvialis squatarola*) and whimbrel (*Numenius hudsonicus*), both migrating from northern Canada and the Arctic (Sick 1997).

Mangroves are also important as a stopover or wintering grounds because birds need to restore their energy by feeding and resting in these areas to continue their migration. In northern Brazil, the Amazon River and its mangroves harbor about 50% of the North American population of migratory gray plovers, 70% of the population of ruddy turnstones (*Arenaria interpres*), around 50% of the population of willets (*Tringa semipalmata*), and 43% of the population of whimbrels (Morrison and Ross 1989). Also, in the Brazilian state of Sergipe, in the city of Aracaju, about 100,000 individuals of 18 different shorebird species yearly aggregate in mangrove areas (Barbieri 2007) (see Chap. 3, Map 10). Furthermore, there are records of site fidelity of the semipalmated sandpiper (*Calidris pusilla*) in three mangrove wintering areas in the Brazilian Amazon (Rodrigues et al. 2007) and semipalmated plover on the northern coast of São Paulo State (Olmos and Silva e Silva 2001). Thus, mangrove areas have an important role in supporting the maintenance of these species' migratory routes. Strips of mangroves provide habitat for diverse faunal species to rest, find shelter, and feed, especially in regions deprived of inland vegetation (Linneweber and Lacerda 2002).

Wading and aquatic birds often nest and rear their young in large colonies in mangroves, taking advantage of the relative inaccessibility of the forest canopy to terrestrial predators. Mangrove areas are also used by many bird species as roosting sites, especially for heron species such as the snowy egret (*Egretta thula*), little blue heron (*E. caerulea*), black-crowned night heron, and great egret (*Ardea alba*) (Olmos and Silva e Silva 2001; Mestre et al. 2007; Mancini et al. 2018). This means that mangroves are key to aggregations of several bird species for feeding, roosting, as dormitory sites, and breeding purposes.

At the same time, birds are also a key component of this ecosystem, due to their ecological role in the dynamics of mangroves (Acevedo and Aide 2008; Mohd-Azlan et al. 2015). They oxygenate the soil while feeding and fertilize these sites by releasing nutrients into the water column through their feces and food waste, distributing nutrients within the mangrove food chain (Onuf et al. 1977; Navedo et al. 2015). Moreover, they control prey populations and weed seeds, regulate competition through grazing, and consume invertebrate pests (e.g., golden apple snails or zebra mussels) while facilitating the colonization by less competitive plants and invertebrates (Nagelkerken et al. 2008; Green and Elmberg 2013). They also host exclusive parasites and disperse seeds and invertebrates, linking plants and organisms in remote marshes (Nagelkerken et al. 2008; Green and Elmberg 2013).

Salt marsh ecosystems are particularly important for birds due to their high primary productivity, acting as prime feeding sites and offering roosting and nesting opportunities for a series of resident and migratory species and occasional visitors from adjacent habitats (Hughes 2004; Greenberg et al. 2014). The total number of bird species in southern Brazilian salt marshes is considered relatively high. At the Saco da Mangueira, in the Patos Lagoon estuary, Rio Grande do Sul State, 89 species were recorded in salt marshes, including mud- and sandflats and open water at the edges of marshes (Dias and Maurício 1998) (see Chap. 3, Map 17). At the mouth of the same estuary, three salt marshes harbored 66, 85, and 87 species (Dias et al. 2017).

Variations in species richness between individual salt marshes in that region are largely driven by their distance to the ocean, with the more species-poor sites being found near the mouth of the estuary, which reflects the progressive influence of abiotic stress upon the avifauna (Dias et al. 2017). Within salt marshes, zonation and environmental gradients with adjacent freshwater marshes and grasslands play an important role in shaping avian diversity (Greenberg et al. 2014). In South Brazil, the more densely vegetated and less-flooded high marsh (at or above the mean high-tide line) harbors mostly passerines and rails (and some shorebirds when overgrazed by livestock or recently burned). The low marsh (below the mean high-tide line) and associated mud- and sandflats and open-water habitats are used by ducks, swans, grebes, flamingos, cormorants, egrets, herons, coots, gulls, terns, and shorebirds (Resende and Leeuwenberg 1987; Dias and Maurício 1998; Bencke et al. 2003; Dias et al. 2011, 2017).

Despite the relatively high number of species found in salt marshes, few of them use these wetlands as breeding sites, namely, small passerines, ducks, raptors, shorebirds, and rails (Bencke et al. 2003; Maurício et al. 2013; Greenberg et al. 2014). Likewise, there are no salt marsh-specialist species in South America, despite the high richness in the continent (Greenberg et al. 2014). One of the possible explanations refers to the low contrast between South American salt marshes and the adjacent open-vegetation environments, which minimizes isolation and speciation (Isacch et al. 2014).

The main uses of southern Brazilian salt marshes and associated estuarine habitats by birds are feeding and roosting (Greenberg et al. 2014; Dias et al. 2017). Salt marshes and adjacent waters support large populations of fishes and invertebrates that constitute major food sources for carnivorous birds, such as grebes, cormorants, egrets, herons, shorebirds, gulls, and terns. The seagrass *Ruppia maritima* Linnaeus and other aquatic plants growing in shallow waters along the margins of marshes are consumed by herbivorous ducks, swans, and coots, and the seeds of some salt marsh plants are eaten by a few granivorous species (Isacch et al. 2014; Greenberg et al. 2014; Dias et al. 2017). Salt marshes also constitute a habitat for terrestrial invertebrates, which in turn are prey for carnivorous passerines. Larger carnivorous birds, such as falcons and harriers, hunt birds and other small animals in the marshes, and scavengers such as caracaras and vultures patrol the vegetation and the water margins searching for dead fish, crabs, and shellfish.

Large concentrations of birds have been recorded in Brazilian salt marshes. For example, at least 800 individuals of buff-breasted sandpipers (*Calidris subruficollis*), 688 white-rumped sandpipers (*Calidris fuscicollis*), and 545 American golden plovers (*Pluvialis dominica*), all Nearctic migrants, were recorded in high densities feeding in overgrazed salt marshes and adjacent grasslands at Ilha da Torotama, in the Patos Lagoon estuary (RS) (Lanctot et al. 2002; Bencke et al. 2006; Dias et al. 2011). In transition areas with freshwater wetlands at the same locality, flocks of approximately 400 individuals of lesser yellowlegs (*Tringa flavipes*), 100 white-rumped sandpipers, and 200 pectoral sandpipers (*Calidris melanotos*) have been registered (Dias et al. 2011). Thousands of black-necked swans (*Cygnus melancoryphus*) gather in shallow bays surrounded by salt marshes at Patos Lagoon estuary during dry summers and autumns, a substantial part of the population inhabiting the country (Bencke et al. 2006). Nearly 500 barn swallows (*Hirundo rustica*) were seen foraging over salt- and neighboring freshwater marshes of the Ilha da Torotama (Dias et al. 2011). Salt marshes at the mouth of the Patos Lagoon estuary also harbor important concentrations of birds, especially gulls and terns, e.g., >5000 common terns (*Sterna hirundo*) (Dias et al. 2011, 2017).

The salt marshes at the Peixe Lagoon (RS) are also important areas for birds. Large numbers of buff-breasted sandpipers and American golden plovers use overgrazed marshes and adjacent grasslands, and thousands of Hudsonian godwits (*Limosa haemastica*), red knots (*Calidris canutus*), sanderlings (*Calidris alba*), and common terns feed and especially roost in mud- and sandflats on the margin of the marshes. Hundreds of black-necked swans and Chilean flamingos (*Phoenicopterus chilensis*) use the lagoon waters to feed (Bencke et al. 2006).

Bird communities inhabiting Brazilian mangroves and salt marshes have never been comprehensively reviewed. Here we present a compilation of published data together with our unpublished field observations to characterize this particular avifauna, its diversity patterns, how they are adapted to live in these ecosystems, the main threats, conservation strategies, and knowledge gaps.

12.2 Bird Adaptations to Live in Mangroves and Salt Marshes

Mangroves and salt marshes are unique and dynamic ecosystems marked by high- and low-tide levels. All exclusive and typical mangrove animal species are subject to dynamic, often extreme, environmental conditions, and birds are the terrestrial group with the most adaptations related to their mobility and feeding habits (Hutchings and Saenger 1987).

Mangrove bird assemblages comprise both terrestrial and aquatic species, which explore resources in diverse ways. Hence, all mangrove microhabitats, i.e., arboreal strata, mudflats, sandflats, and salt flats, are used for foraging by a wide range of bird species with different morphologies and behaviors. For instance, gleaning, bark-foraging, and flycatching insectivores (e.g., Tyrannidae species) are adapted to feed on prey items that piscivorous, carnivorous, or species that probe directly in mudflats are not able to explore (e.g., Scolopacidae species). In countries like Malaysia and Australia, bird assemblages exhibit zonation of prey exploration in mangrove trees. In such cases, different branching patterns and structures of the foliage seem to differ enough so that birds could specialize to different mangrove trees (Noske 1995, 1996; Luther and Greenberg 2009). However, in Brazil, no studies have been conducted on feeding adaptations or vertical zonation patterns in mangroves.

In general, mangrove-exclusive bird species feed primarily on insects (~50%), followed by a smaller proportion that feeds on crabs, nectar, and fish (Lefebvre and Poulin 1997; Luther and Greenberg 2009). Roughly 20% of the bird species restricted to mangroves have larger bills than related subspecies or sister taxa inhabiting inland habitats (Grenier and Greenberg 2005; Luther and Greenberg 2009). There is a consistent pantropical pattern of longer and deeper bills in passerine birds restricted to mangroves and salt marshes (Grenier and Greenberg 2005; Greenberg and Olsen 2010; Luther and Greenberg 2011; Greenberg et al. 2012). Longer and slender bills are correlated with a wider foraging-niche breadth and are advantageous for probing in small cracks and crevices, where many prey items can be found. Long bills are also likely useful for probing in mud and among mangrove roots where other prey may be abundant (Luther and Greenberg 2011). Bill size is also related to temperature, as birds living in habitats with higher temperatures tend to have larger bills than birds living in cooler climates and inland terrestrial habitats (Greenberg et al. 2012; Luther and Greenberg 2014). Bills might play an important thermoregulatory role, as reported for tidal marsh sparrows from hot and exposed dune/salt marsh environments. The bill expels excess body heat in these unbuffered, freshwater-limited environments and potentially may reduce water loss (Greenberg et al. 2012).

In salt marshes, some morphological characteristics facilitate bird activities (e.g., foraging, evading predators, and intraspecific communication). The most conspicuous is the morphology of the bill and legs of shorebirds using the marsh vegetation or adjacent mud- and sandflats. Rails have long legs and toes and slender bodies to move through the dense herbaceous vegetation, using colorful frontal shields for visual communication in the dark environment (Sick 2001). Some passerines adapted to live in grasslands also occur in the tall, dense vegetation of salt marshes and have strong legs, long and curved claws, and long tails used for balance. Most are carnivorous and use long and thin bills to hunt in the vegetation, some also hop or walk on the mud, and a few are granivorous and use thick bills to crush seeds of salt marsh plants. In coastal salt marshes, North American sparrow populations have comparatively longer and thinner bills than their inland counterparts, which could be an adaptation that increases the consumption of marine invertebrates at the decrease in seed availability (Greenberg et al. 2012). In the bay-capped wren-spinetail

(*Spartonoica maluroides*), a species strongly associated with South American salt marshes, there are differences in bill shape and plumage coloration between coastal and inland marsh populations. These differences may be explained by the selective pressures related to the adaptation of the beak shape to explore prey in salt marshes, whereas the change in plumage pigmentation (melanism) may improve camouflage and assist in predator evasion (Cardoni et al. 2013). Overall, in tidal marshes of North America, birds often show a high degree of local morphological differentiation (Chan et al. 2006; Greenberg et al. 2006), but this topic has not yet been investigated in Brazilian salt marshes.

12.3 Bird Diversity Across Mangroves and Salt Marshes

The species richness of birds in Brazilian mangroves and salt marshes has never been properly investigated and the available information is scattered throughout the literature. The data presented herein were obtained through the compilation of 81 published studies on the matter, encompassing 56 peer-reviewed articles, 16 books, and nine book chapters. Data available in online platforms, e.g., Wikiaves 2021, Xeno-Canto, e-Bird, and Taxeus, was not included for not specifying habitat type in their records. For the same reason, we have not included records of museum specimens, as the habitat type in which birds were collected was frequently omitted, particularly in the older ones. We have also omitted a large part of avian studies conducted in Brazilian mangroves and salt marshes for, again, not mentioning explicitly the habitat type of the records. Hence, in this chapter, we included only studies that clearly stated the presence of a species, or the number of species, in specific areas of mangroves and salt marshes, either in the results section itself or in a species list. Although some of these studies may have considered species from nearby habitats as being of mangroves and salt marshes, a fact that is hard to evaluate, the information presented in this chapter is the most objective compilation of the core avifauna of mangroves and salt marshes in Brazil based on available, published studies (see appendix for the list of the considered references).

The species (Table 12.1) were classified according to their conservation status at global (IUCN 2022) and national levels (MMA 2016, 2022) Nomenclature follows the Brazilian Ornithological Records Committee (Pacheco et al. 2021). Birds were classified into three categories, as follows:

Exclusive species (EXC): Resident species that occur exclusively in mangrove or salt marsh areas

Regular species (Re): Species that use mangrove and salt marsh habitats regularly, either throughout the year or on a seasonal basis (in case they are migratory), but that are also recorded in other habitat types, such as terrestrial forests, grasslands, freshwater wetlands, and marine environments, including beaches

Occasional species (O): Species occasionally recorded in mangrove and salt marsh habitats

Table 12.1 List of resident and migratory bird species recorded in mangroves and salt marshes along the Brazilian coast

Taxon	Common name (EN)	Common name (PT)	M	S	Mi
Rheiformes					
Rheidae (1)					
Rhea americana	Greater rhea	Ema		O	
Tinamiformes					
Tinamidae (1)					
Nothura maculosa	Spotted nothura	Codorna-amarela		O	
Anseriformes					
Anhimidae (1)					
Chauna torquata	Southern screamer	Tachã		O	
Anatidae (16)					
Dendrocygna bicolor	Fulvous whistling-duck	Marreca-caneleira	Re	O	
Dendrocygna viduata	White-faced whistling duck	Irerê	Re	O	
Dendrocygna autumnalis	Black-bellied whistling duck	Marreca-cabocla	Re		
Cygnus melancoryphus	Black-necked swan	Cisne-de-pescoço-preto		Re	
Coscoroba coscoroba	Coscoroba swan	Capororoca		Re	
Cairina moschata	Muscovy duck	Pato-do-mato	Re		
Amazonetta brasiliensis	Brazilian teal	Marreca-ananaí	O	Re	
Spatula versicolor	Silver teal	Marreca-cricri	O	Re	
Spatula platalea	Red shoveler	Marreca-colhereira		Re	
Spatula discors	Blue-winged teal	Marreca-de-asa-azul	O		PNW
Anas bahamensis	White-cheeked pintail	Marreca-toicinho	Re		
Anas georgica	Yellow-billed pintail	Marreca-parda	O	Re	
Anas flavirostris	Yellow-billed teal	Marreca-pardinha		Re	
Netta erythrophthalma	Southern pochard	Paturi-preta	O		
Netta peposaca	Rosy-billed pochard	Marrecão	O		
Oxyura vittata	Lake duck	Marreca-rabo-de-espinho	O		SACT
Phoenicopteriformes					
Phoenicopteridae (3)					
Phoenicopterus chilensis	Chilean flamingo	Flamingo-chileno		Re	SACT
Phoenicopterus ruber	American flamingo	Flamingo	O		NNTT
Phoenicoparrus andinus (**VU**[b])	Andean flamingo	Flamingo-dos-andes		Re	SACT
Podicipediformes					
Podicipedidae (4)					

(continued)

Table 12.1 (continued)

Taxon	Common name (EN)	Common name (PT)	M	S	Mi
Rollandia rolland	White-tufted grebe	Mergulhão-de-orelha-branca		Re	
Tachybaptus dominicus	Least grebe	Mergulhão-pequeno	O		
Podilymbus podiceps	Pied-billed grebe	Mergulhão-caçador	O	Re	
Podicephorus major	Great grebe	Mergulhão-grande	O	Re	
Columbiformes					
Columbidae (10)					
Columba livia	Rock pigeon	Pombo-doméstico	O		
Patagioenas picazuro	Picazuro pigeon	Pomba-asa-branca	O		
Patagioenas cayennensis	Pale-vented pigeon	Pomba-galega	O		
Leptotila verreauxi	White-tipped dove	Juriti-pupu	O		
Leptotila rufaxilla	Gray-fronted dove	Juriti-de-testa-branca	O		
Columbina passerina	Common ground-dove	Rolinha-cinzenta	Re		
Columbina minuta	Plain-breasted ground-dove	Rolinha-de-asa-canela	O		
Columbina talpacoti	Ruddy ground-dove	Rolinha-roxa	Re		
Columbina squammata	Scaled dove	Rolinha-fogo-apagou	O		
Columbina picui	Picui ground-dove	Rolinha-picuí	O		
Cuculiformes					
Cuculidae (7)					
Guira guira	Guira cuckoo	Anu-branco	Re		
Crotophaga major	Greater ani	Anu-coroca	O		
Crotophaga ani	Smooth-billed ani	Anu-preto	Re	O	
Tapera naevia	Striped cuckoo	Saci	O		
Piaya cayana	Squirrel cuckoo	Alma-de-gato	O		
Coccyzus melacoryphus	Dark-billed cuckoo	Papa-lagarta-acanelado	O		
Coccyzus minor	Mangrove cuckoo	Papa-lagarta-do-mangue	EXC		
Nyctibiiformes					
Nyctibiidae (1)					
Nyctibius griseus	Common potoo	Urutau	O		
Caprimulgiformes					
Caprimulgidae (6)					
Antrostomus rufus	Rufous nightjar	João-corta-pau	O		
Lurocalis semitorquatus	Short-tailed nighthawk	Tuju	O		
Nyctidromus albicollis	Common pauraque	Bacurau	O		
Hydropsalis torquata	Scissor-tailed nightjar	Bacurau-tesoura	O		

(continued)

Table 12.1 (continued)

Taxon	Common name (EN)	Common name (PT)	M	S	Mi
Podager nacunda	Nacunda nighthawk	Corucão	O	Re	
Chordeiles acutipennis	Lesser nighthawk	Bacurau-de-asa-fina	O		
Apodiformes					
Apodidae (5)					
Streptoprocne zonaris	White-collared swift	Taperuçu-de-coleira-branca	O		
Chaetura cinereiventris	Gray-rumped swift	Andorinhão-de-sobre-cinzento	O		
Chaetura meridionalis	Sick's swift	Andorinhão-do-temporal	Re		
Chaetura brachyura	Short-tailed swift	Andorinhão-de-rabo-curto	O		
Tachornis squamata	Fork-tailed palm-swift	Andorinhão-do-buriti	O		
Trochilidae (11)					
Ramphodon naevius	Saw-billed hermit	Beija-flor-rajado	O		
Colibri serrirostris	White-vented violetear	Beija-flor-de-orelha-violeta	O		
Polytmus guainumbi	White-tailed goldenthroat	Beija-flor-de-bico-curvo	O		
Chrysolampis mosquitus	Ruby-topaz hummingbird	Beija-flor-vermelho	O		
Thalurania glaucopis	Violet-capped woodnymph	Beija-flor-de-fronte-violeta	O		
Eupetomena macroura	Swallow-tailed hummingbird	Beija-flor-tesoura	O		
Aphantochroa cirrochloris	Sombre hummingbird	Beija-flor-cinza	O		
Chrysuronia versicolor	Versicolored emerald	Beija-flor-de-banda-branca	O		
Chrysuronia leucogaster	Plain-bellied emerald	Beija-flor-de-barriga-branca	Re		
Leucochloris albicollis	White-throated hummingbird	Beija-flor-de-papo-branco	O		
Chionomesa fimbriata	Glittering-throated emerald	Beija-flor-de-garganta-verde	O		
Opisthocomiformes					
Opisthocomidae (1)					
Opisthocomus hoazin	Hoatzin	Cigana	O		
Gruiformes					
Aramidae (1)					
Aramus guarauna	Limpkin	Carão	Re	Re	
Rallidae (22)					
Rallus longirostris	Mangrove rail	Saracura-matraca	EXC		

(continued)

Table 12.1 (continued)

Taxon	Common name (EN)	Common name (PT)	M	S	Mi
Porphyrio martinica	Purple gallinule	Frango-d'água-azul	O		
Rufirallus viridis	Russet-crowned crake	Sanã-castanha	O		
Laterallus flaviventer	Yellow-breasted crake	Sanã-amarela	O		
Laterallus melanophaius	Rufous-sided crake	Sanã-parda	O	Re	
Laterallus exilis	Gray-breasted crake	Sanã-do-capim	O		
Laterallus spilopterus **(EN[a]/VU[b])**	Dot-winged crake	Sanã-cinza		EXC	
Laterallus leucopyrrhus	Red-and-white crake	Sanã-vermelha	O		
Mustelirallus albicollis	Ash-throated crake	Sanã-carijó	O		
Neocrex erythrops	Paint-billed crake	Turu-turu	O		
Pardirallus maculatus	Spotted rail	Saracura-carijó		Re	
Pardirallus sanguinolentus	Plumbeous rail	Saracura-do-banhado		Re	
Amaurolimnas concolor	Uniform crake	Saracura-lisa	O		
Aramides ypecaha	Giant wood rail	Saracuruçu		Re	
Aramides mangle	Little wood rail	Saracura-do-mangue	Re		
Aramides cajaneus	Gray-necked wood rail	Saracura-três-potes	Re	O	
Aramides saracura	Slaty-breasted wood rail	Saracura-do-mato	O		
Porphyriops melanops	Spot-flanked gallinule	Galinha-d'água-carijó	O	O	
Gallinula galeata	Common gallinule	Galinha-d'água	Re	Re	
Fulica rufifrons	Red-fronted coot	Carqueja-de-escudo-vermelho		O	
Fulica armillata	Red-gartered coot	Carqueja-de-bico-manchado	O	Re	
Fulica leucoptera	White-winged coot	Carqueja-de-bico-amarelo		Re	
Charadriiformes					
Charadriidae (8)					
Pluvialis dominica	American golden-plover	Batuiruçu	O	Re	PNW
Pluvialis squatarola	Black-bellied plover	Batuiruçu-de-axila-preta	Re	Re	PNW
Vanellus chilensis	Southern lapwing	Quero-quero	O	Re	
Charadrius modestus	Rufous-chested dotterel	Batuíra-de-peito-tijolo	Re	Re	SACT
Charadrius semipalmatus	Semipalmated plover	Batuíra-de-bando	Re	Re	PNW

(continued)

Table 12.1 (continued)

Taxon	Common name (EN)	Common name (PT)	M	S	Mi
Charadrius wilsonia (**VU**[a])	Wilson's plover	Batuíra-bicuda	Re		
Charadrius collaris	Collared plover	Batuíra-de-coleira	Re	Re	
Charadrius falklandicus	Two-banded plover	Batuíra-de-coleira-dupla		Re	
Haematopodidae (1)					
Haematopus palliatus	American oystercatcher	Piru-piru	O	Re	
Recurvirostridae (2)					
Himantopus mexicanus	Black-necked stilt	Pernilongo-de-costas-negras	O		
Himantopus melanurus	White-backed stilt	Pernilongo-de-costas-brancas	Re	Re	
Scolopacidae (20)					
Numenius hudsonicus (**VU**[a])	American whimbrel	Maçarico-de-bico-torto	Re	Re	PNW
Limosa haemastica	Hudsonian godwit	Maçarico-de-bico-virado	O	Re	PNW
Limosa fedoa	Marbled godwit	Maçarico-marmóreo	O		OW
Arenaria interpres	Ruddy turnstone	Vira-pedras	Re	Re	PNW
Calidris canutus (**VU**[a])	Red knot	Maçarico-de-papo-vermelho	Re	Re	PNW
Calidris himantopus	Stilt sandpiper	Maçarico-pernilongo	O	Re	PNW
Calidris alba	Sanderling	Maçarico-branco	Re	Re	PNW
Calidris minutilla	Least sandpiper	Maçariquinho	Re		PNW
Calidris fuscicollis	White-rumped sandpiper	Maçarico-de-sobre-branco	Re	Re	PNW
Calidris subruficollis (**VU**[a])	Buff-breasted sandpiper	Maçarico-acanelado	O	Re	PNW
Calidris melanotos	Pectoral sandpiper	Maçarico-de-colete	O	Re	PNW
Calidris pusilla (**EN**[a])	Semipalmated sandpiper	Maçarico-rasteirinho	Re	Re	PNW
Limnodromus griseus (**EN**[a])	Short-billed dowitcher	Maçarico-de-costas-brancas	Re	O	PNW
Gallinago paraguaiae	South American snipe	Narceja	O	Re	
Phalaropus tricolor	Wilson's phalarope	Pisa-n'água	O	Re	PNW
Actitis macularius	Spotted sandpiper	Maçarico-pintado	Re	O	PNW
Tringa solitaria	Solitary sandpiper	Maçarico-solitário	Re	O	PNW
Tringa melanoleuca	Greater yellowlegs	Maçarico-grande-de-perna-amarela	Re	Re	PNW
Tringa semipalmata	Willet	Maçarico-de-asa-branca	Re	Re	PNW
Tringa flavipes	Lesser yellowlegs	Maçarico-de-perna-amarela	Re	Re	PNW

(continued)

Table 12.1 (continued)

Taxon	Common name (EN)	Common name (PT)	M	S	Mi
Jacanidae (1)					
Jacana jacana	Wattled jacana	Jaçanã	Re	O	
Rostratulidae (1)					
Nycticryphes semicollaris	South American painted-snipe	Narceja-de-bico-torto		Re	
Stercorariidae (1)					
Stercorarius parasiticus	Parasitic jaeger	Mandrião-parasítico		Re	
Laridae (17)					
Chroicocephalus maculipennis	Brown-hooded gull	Gaivota-maria-velha	O	Re	
Chroicocephalus cirrocephalus	Gray-hooded gull	Gaivota-de-cabeça-cinza	O	Re	
Leucophaeus atricilla	Laughing gull	Gaivota-alegre	O		PNW
Larus atlanticus	Olrog's gull	Gaivota-de-rabo-preto		Re	SACT
Larus dominicanus	Kelp gull	Gaivotão	Re	Re	
Rynchops niger	Black skimmer	Talha-mar	Re	Re	
Sternula antillarum	Least tern	Trinta-réis-miúdo	O	Re	PNW
Sternula superciliaris	Yellow-billed tern	Trinta-réis-pequeno	O	Re	
Phaetusa simplex	Large-billed tern	Trinta-réis-grande	O	Re	
Gelochelidon nilotica	Gull-billed tern	Trinta-réis-de-bico-preto	O	Re	
Chlidonias niger	Black tern	Trinta-réis-negro		O	PNW
Sterna hirundo	Common tern	Trinta-réis-boreal	Re	Re	PNW
Sterna dougallii (**VU**[a])	Roseate tern	Trinta-réis-róseo	O		PNW
Sterna hirundinacea (**VU**[a])	South American tern	*Trinta-réis-de-bico-vermelho*	Re	Re	
Sterna trudeaui	Snowy-crowned tern	Trinta-réis-de-coroa-branca	O	Re	
Thalasseus acuflavidus (***VU***[a])	Cabot's tern	Trinta-réis-de-bando	Re	Re	
Thalasseus maximus (**VU**[a])	Royal tern	Trinta-réis-real	Re	Re	
Eurypygiformes					
Eurypygidae (1)					
Eurypyga helias	Sunbittern	Pavãozinho-do-pará	O		
Ciconiiformes					
Ciconiidae (3)					
Ciconia maguari	Maguari stork	Maguari	O	Re	
Jabiru mycteria	Jabiru	Tuiuiú	O		
Mycteria americana	Wood stork	Cabeça-seca	O	Re	
Suliformes					
Fregatidae (1)					

(continued)

Table 12.1 (continued)

Taxon	Common name (EN)	Common name (PT)	M	S	Mi
Fregata magnificens	Magnificent frigatebird	Fragata	O	O	
Anhingidae (1)					
Anhinga anhinga	Anhinga	Biguatinga	O		
Phalacrocoracidae (1)					
Nannopterum brasilianum	Neotropic cormorant	Biguá	Re	Re	
Pelecaniformes					
Pelecanidae (1)					
Pelecanus occidentalis	Brown pelican	Pelicano	O		NNTT
Ardeidae (17)					
Tigrisoma lineatum	Rufescent tiger-heron	Socó-boi	Re		
Cochlearius cochlearius	Boat-billed heron	Arapapá	Re		
Botaurus pinnatus	Pinnated bittern	Socó-boi-baio	O	Re	
Ixobrychus exilis	Least bittern	Socoí-vermelho	O		
Ixobrychus involucris	Stripe-backed bittern	Socó-amarelo		O	
Nycticorax nycticorax	Black-crowned night-heron	Socó-dorminhoco	Re	Re	
Nyctanassa violacea	Yellow-crowned night-heron	Savacu-de-coroa	Re	Re	
Butorides striata	Striated heron	Socozinho	Re	Re	
Ardeola ralloides	Squacco heron	Garça-caranguejeira	Re		OW
Bubulcus ibis	Cattle egret	Garça-vaqueira	Re	Re	
Ardea cocoi	Cocoi heron	Garça-moura	Re	Re	
Ardea alba	Great egret	Garça-branca-grande	Re	Re	
Syrigma sibilatrix	Whistling heron	Maria-faceira	O	Re	
Egretta tricolor	Tricolored heron	Garça-tricolor	EXC		
Egretta gularis	Western reef heron	Garça-negra	O		OW
Egretta thula	Snowy egret	Garça-branca-pequena	Re	Re	
Egretta caerulea	Little blue heron	Garça-azul	Re	Re	
Threskiornithidae (7)					
Eudocimus ruber	Scarlet ibis	Guará	EXC		
Plegadis chihi	White-faced ibis	Caraúna	O	Re	
Mesembrinibis cayennensis	Green ibis	Coró-coró	O		
Phimosus infuscatus	Bare-faced ibis	Tapicuru	O	Re	
Theristicus caerulescens	Plumbeous ibis	Curicaca-real		O	
Theristicus caudatus	Buff-necked ibis	Curicaca	O		
Platalea ajaja	Roseate spoonbill	Colhereiro	Re	Re	

(continued)

Table 12.1 (continued)

Taxon	Common name (EN)	Common name (PT)	M	S	Mi
Cathartiformes					
Cathartidae (3)					
Coragyps atratus	Black vulture	Urubu-preto	Re	O	
Cathartes aura	Turkey vulture	Urubu-de-cabeça-vermelha	Re	O	
Cathartes burrovianus	Lesser yellow-headed vulture	Urubu-de-cabeça-amarela	Re	O	
Accipitriformes					
Pandionidae (1)					
Pandion haliaetus	Osprey	Águia-pescadora	Re	Re	PNW
Accipitridae (21)					
Elanus leucurus	White-tailed kite	Gavião-peneira	O		
Chondrohierax uncinatus	Hook-billed kite	Caracoleiro	O		
Leptodon cayanensis	Gray-headed kite	Gavião-gato	O		
Busarellus nigricollis	Black-collared hawk	Gavião-belo	O		
Rostrhamus sociabilis	Snail kite	Gavião-caramujeiro	Re	Re	
Ictinia plumbea	Plumbeous kite	Sovi	O		
Circus cinereus (**VU**[a])	Cinereous harrier	Gavião-cinza		Re	
Circus buffoni	Long-winged harrier	Gavião-do-banhado	O	Re	
Accipiter bicolor	Bicolored hawk	Gavião-bombachinha-grande	O		
Geranospiza caerulescens	Crane hawk	Gavião-pernilongo	O		
Buteogallus schistaceus	Slate-colored hawk	Gavião-azul	O		
Buteogallus aequinoctialis	Rufous crab hawk	Gavião-caranguejeiro	EXC		
Heterospizias meridionalis	Savanna hawk	Gavião-caboclo	O	O	
Amadonastur lacernulatus (**VU**[a,b])	White-necked hawk	Gavião-pombo-pequeno	O		
Urubitinga urubitinga	Great black hawk	Gavião-preto	Re		
Rupornis magnirostris	Roadside hawk	Gavião-carijó	Re		
Parabuteo unicinctus	Harris' hawk	Gavião-asa-de-telha	Re		
Geranoaetus albicaudatus	White-tailed hawk	Gavião-de-rabo-branco	O		
Leucopternis melanops	Black-faced hawk	Gavião-de-cara-preta	O		
Buteo nitidus	Gray-lined hawk	Gavião-pedrês	O		
Buteo brachyurus	Short-tailed hawk	Gavião-de-cauda-curta	O		
Strigiformes					
Tytonidae (1)					
Tyto furcata	American barn owl	Suindara	O		

(continued)

Table 12.1 (continued)

Taxon	Common name (EN)	Common name (PT)	M	S	Mi
Strigidae (6)					
Megascops choliba	Tropical screech-owl	Corujinha-do-mato	O		
Glaucidium brasilianum	Ferruginous pygmy owl	Caburé	O		
Athene cunicularia	Burrowing owl	Coruja-buraqueira	O		
Asio clamator	Striped owl	Coruja-orelhuda	O		
Asio stygius	Stygian owl	Mocho-diabo	O		
Asio flammeus	Short-eared owl	Mocho-dos-banhados		O	
Coraciiformes					
Alcedinidae (5)					
Megaceryle torquata	Ringed kingfisher	Martim-pescador-grande	Re	Re	
Chloroceryle amazona	Amazon kingfisher	Martim-pescador-verde	Re	Re	
Chloroceryle aenea	American pygmy kingfisher	Martim-pescador-miúdo	Re		
Chloroceryle americana	Green kingfisher	Martim-pescador-pequeno	Re	Re	
Chloroceryle inda	Green-and-rufous kingfisher	Martim-pescador-da-mata	O		
Galbuliformes					
Galbulidae (1)					
Galbula galbula	Green-tailed jacamar	Ariramba-de-cauda-verde	O		
Bucconidae (5)					
Monasa nigrifrons	Black-fronted nunbird	Chora-chuva-preto	O		
Notharchus tectus	Pied puffbird	Macuru-pintado	O		
Notharchus hyperrhynchus	White-necked puffbird	Macuru-de-testa-branca	O		
Notharchus macrorhynchos	Guianan puffbird	Macuru-de-pescoço-branco	O		
Nystalus maculatus	Spot-backed puffbird	Rapazinho-dos-velhos	O		
Piciformes					
Ramphastidae (2)					
Ramphastos toco	Toco toucan	Tucanuçu	O		
Ramphastos tucanus (**VU**[b])	White-throated toucan	Tucano-de-papo-branco	O		
Picidae (16)					
Picumnus exilis	Bahia piculet	Picapauzinho-de-pintas-amarelas	O		
Picumnus spilogaster (**VU**[b])	White-bellied piculet	Picapauzinho-de-pescoço-branco	O		

(continued)

Table 12.1 (continued)

Taxon	Common name (EN)	Common name (PT)	M	S	Mi
Picumnus pygmaeus	Spotted piculet	Picapauzinho-pintado	O		
Picumnus cirratus	White-barred piculet	Picapauzinho-barrado	O		
Picumnus temminckii	Ochre-collared piculet	Picapauzinho-de-coleira	O		
Melanerpes candidus	White woodpecker	Pica-pau-branco	O		
Veniliornis passerinus	Little woodpecker	Pica-pau-pequeno	O		
Veniliornis spilogaster	White-spotted woodpecker	Pica-pau--verde-carijó	O		
Campephilus melanoleucos	Crimson-crested woodpecker	Pica-pau-de-topete-vermelho	O		
Dryocopus lineatus	Lineated woodpecker	Pica-pau-de-banda-branca	O		
Celeus elegans	Chestnut woodpecker	Pica-pau-chocolate	O		
Celeus flavescens	Blond-crested woodpecker	Pica-pau-de-cabeça-amarela	O		
Piculus flavigula	Yellow-throated woodpecker	Pica-pau-bufador	O		
Colaptes punctigula	Spot-breasted woodpecker	Pica-pau-de-peito-pontilhado	O		
Colaptes melanochloros	Green-barred woodpecker	Pica-pau-verde-barrado	O		
Colaptes campestris	Campo flicker	Pica-pau-do-campo	O		
Falconiformes					
Falconidae (9)					
Herpetotheres cachinnans	Laughing falcon	Acauã	O		
Micrastur ruficollis	Barred forest-falcon	Falcão-caburé	O		
Micrastur semitorquatus	Collared forest-falcon	Falcão-relógio	O		
Caracara plancus	Southern caracara	Carcará	Re	Re	
Milvago chimachima	Yellow-headed caracara	Carrapateiro	Re	O	
Milvago chimango	Chimango caracara	Chimango	Re	Re	
Falco sparverius	American kestrel	Quiriquiri	O	O	
Falco femoralis	Aplomado falcon	Falcão-de-coleira	O	O	
Falco peregrinus	Peregrine falcon	Falcão-peregrino	O	Re	PNW
Psittaciformes					
Psittacidae (10)					
Brotogeris tirica	Plain parakeet	Periquito-rico	O		
Pionus maximiliani	Scaly headed parrot	Maitaca-verde	O		
Amazona farinosa	Mealy parrot	Papagaio-moleiro	O		
Amazona brasiliensis	Red-tailed parrot	Papagaio-de-cara-roxa	Re		
Amazona amazonica	Orange-winged parrot	Curica	Re		

(continued)

Table 12.1 (continued)

Taxon	Common name (EN)	Common name (PT)	M	S	Mi
Forpus xanthopterygius	Blue-winged parrotlet	Tuim	O		
Pyrrhura frontalis	Maroon-bellied parakeet	Tiriba-de-testa-vermelha	O		
Eupsittula aurea	Peach-fronted parakeet	Periquito-rei	O		
Thectocercus acuticaudatus	Blue-crowned parakeet	Aratinga-de-testa-azul	O		
Psittacara leucophthalmus	White-eyed parakeet	Periquitão	O		
Passeriformes					
Thamnophilidae (4)					
Formicivora grisea	White-fringed antwren	Papa-formiga-pardo	O		
Thamnophilus nigrocinereus	Blackish-gray antshrike	Choca-preta-e-cinza	O		
Thamnophilus caerulescens	Variable antshrike	Choca-da-mata	O		
Sclateria naevia	Silvered antbird	Papa-formiga-do-igarapé	O		
Scleruridae (1)					
Geositta cunicularia	Common miner	Curriqueiro		O	
Dendrocolaptidae (2)					
Xiphorhynchus guttatus	Buff-throated woodcreeper	Arapaçu-de-garganta-amarela	O		
Dendroplex picus	Straight-billed woodcreeper	Arapaçu-de-bico-branco	Re		
Xenopidae (1)					
Xenops minutus	Plain xenops	Bico-virado-miúdo	O		
Furnariidae (13)					
Furnarius figulus	Wing-banded hornero	Casaca-de-couro-da-lama	O		
Furnarius rufus	Rufous hornero	João-de-barro	O	O	
Phleocryptes melanops	Wren-like rushbird	Bate-bico		Re	
Limnornis curvirostris	Curve-billed reedhaunter	João-da-palha		O	
Cinclodes fuscus	Buff-winged cinclodes	Pedreiro-dos-andes		Re	SACT
Phacellodomus striaticollis	Freckle-breasted thornbird	Tio-tio		O	
Phacellodomus ferrugineigula	Orange-breasted thornbird	João-botina-do-brejo	O		
Asthenes hudsoni (**VU**[a])	Hudson's canastero	João-platino		Re	

(continued)

Table 12.1 (continued)

Taxon	Common name (EN)	Common name (PT)	M	S	Mi
Limnoctites sulphuriferus	Sulfur-throated spinetail	Arredio-de-papo-manchado		Re	
Spartonoica maluroides	Bay-capped wren-spinetail	Boininha		Re	
Certhiaxis cinnamomeus	Yellow-chinned spinetail	Curutié	Re	O	
Synallaxis ruficapilla	Rufous-capped spinetail	Pichororé	O		
Synallaxis spixi	Spix's spinetail	João-teneném	O	O	
Pipridae (2)					
Ilicura militaris	Pin-tailed manakin	Tangarazinho	O		
Chiroxiphia pareola	Blue-backed manakin	Tangará-príncipe	O		
Tityridae (3)					
Tityra inquisitor	Black-crowned tityra	Anambé-branco-de-bochecha-parda	O		
Pachyramphus rufus	Cinereous becard	Caneleiro-cinzento	O		
Pachyramphus polychopterus	White-winged becard	Caneleiro-preto	O		
Tachurisidae (1)					
Tachuris rubrigastra	Many-colored rush tyrant	Papa-piri		Re	
Rhynchocyclidae (7)					
Tolmomyias flaviventris	Yellow-breasted flycatcher	Bico-chato-amarelo	O		
Todirostrum maculatum	Spotted tody-flycatcher	Ferreirinho-estriado	O		
Todirostrum poliocephalum	Gray-headed tody-flycatcher	Teque-teque	O		
Todirostrum cinereum	Common tody-flycatcher	Ferreirinho-relógio	O		
Todirostrum chrysocrotaphum	Yellow-browed tody-flycatcher	Ferreirinho-de-sobrancelha	O		
Hemitriccus striaticollis	Stripe-necked tody-tyrant	Sebinho-rajado-amarelo	O		
Hemitriccus nidipendulus	Hangnest tody-tyrant	Tachuri-campainha	O		
Tyrannidae (37)					
Camptostoma obsoletum	Southern beardless-tyrannulet	Risadinha	Re		
Elaenia flavogaster	Yellow-bellied elaenia	Guaracava-de-barriga-amarela	O		
Elaenia mesoleuca	Olivaceous elaenia	Tuque	O		
Elaenia cristata	Plain-crested elaenia	Guaracava-de-topete-uniforme	O		

(continued)

Table 12.1 (continued)

Taxon	Common name (EN)	Common name (PT)	M	S	Mi
Phaeomyias murina	Mouse-colored tyrannulet	Bagageiro	O		
Phyllomyias fasciatus	Planalto tyrannulet	Piolhinho	O		
Pseudocolopteryx sclateri	Crested doradito	Tricolino		Re	
Pseudocolopteryx flaviventris	Warbling doradito	Amarelinho-do-junco		Re	
Serpophaga nigricans	Sooty tyrannulet	João-pobre	O	Re	
Serpophaga subcristata	White-crested tyrannulet	Alegrinho	O		
Attila rufus	Gray-hooded attila	Capitão-de-saíra	O		
Legatus leucophaius	Piratic flycatcher	Bem-te-vi-pirata	O		
Myiarchus swainsoni	Swainson's flycatcher	Irré	O		
Myiarchus ferox	Short-crested flycatcher	Maria-cavaleira	Re		
Myiarchus tyrannulus	Brown-crested flycatcher	Maria-cavaleira-de-rabo-enferrujado	O		
Pitangus sulphuratus	Great kiskadee	Bem-te-vi	Re	Re	
Philohydor lictor	Lesser kiskadee	Bentevizinho-do-brejo	O		
Machetornis rixosa	Cattle tyrant	Suiriri-cavaleiro	O	Re	
Myiodynastes maculatus	Streaked flycatcher	Bem-te-vi-rajado	O		
Megarynchus pitangua	Boat-billed flycatcher	Neinei	O		
Myiozetetes cayanensis	Rusty-margined flycatcher	Bentevizinho-de-asa-ferrugínea	O		
Myiozetetes similis	Social flycatcher	Bentevizinho-de-penacho-vermelho	Re		
Tyrannus melancholicus	Tropical kingbird	Suiriri	O	O	
Tyrannus savana	Fork-tailed flycatcher	Tesourinha	O	Re	
Tyrannus dominicensis	Gray kingbird	Suiriri-cinza	O		NNTT
Sublegatus modestus	Southern scrub-flycatcher	Guaracava-modesta	O		
Arundinicola leucocephala	White-headed marsh tyrant	Freirinha	O	O	
Fluvicola pica	Pied water tyrant	Lavadeira-do-norte	O		
Fluvicola albiventer	Black-backed water tyrant	Lavadeira-de-cara-branca	O		
Fluvicola nengeta	Masked water tyrant	Lavadeira-mascarada	Re		
Pyrocephalus rubinus	Vermilion flycatcher	Príncipe	O		
Heteroxolmis dominicanus (**VU**[a, b])	Black-and-white monjita	Noivinha-de-rabo-preto		O	

(continued)

Table 12.1 (continued)

Taxon	Common name (EN)	Common name (PT)	M	S	Mi
Myiophobus fasciatus	Bran-colored flycatcher	Filipe	O		
Contopus cinereus	Tropical pewee	Papa-moscas-cinzento	O		
Satrapa icterophrys	Yellow-browed tyrant	Suiriri-pequeno	O	Re	
Lessonia rufa	Austral negrito	Colegial	O	Re	SACT
Hymenops perspicillatus	Spectacled tyrant	Viuvinha-de-óculos	O	Re	
Vireonidae (4)					
Cyclarhis gujanensis	Rufous-browed peppershrike	Pitiguari	O		
Hylophilus pectoralis	Ashy-headed greenlet	Vite-vite-de-cabeça-cinza	O		
Vireo chivi	Chivi vireo	Juruviara	O		
Vireo altiloquus	Black-whiskered vireo	Juruviara-barbuda	O		NNTT
Corvidae (3)					
Cyanocorax caeruleus	Azure jay	Gralha-azul	Re		
Cyanocorax cristatellus	Curl-crested jay	Gralha-do-campo	O		
Cyanocorax cyanopogon	White-naped jay	Gralha-cancã	O		
Hirundinidae (10)					
Pygochelidon cyanoleuca	Blue-and-white swallow	Andorinha-pequena-de-casa	Re	Re	
Alopochelidon fucata	Tawny-headed swallow	Andorinha-morena		Re	
Stelgidopteryx ruficollis	Southern rough-winged swallow	Andorinha-serradora	Re		
Progne tapera	Brown-chested martin	Andorinha-do-campo	O	Re	
Progne chalybea	Gray-breasted martin	Andorinha-grande	O	O	
Tachycineta albiventer	White-winged swallow	Andorinha-do-rio	O		
Tachycineta leucorrhoa	White-rumped swallow	Andorinha-de-sobre-branco	O	Re	
Tachycineta leucopyga	Chilean swallow	Andorinha-chilena		Re	SACT
Riparia riparia	Bank swallow	Andorinha-do-barranco		Re	PNW
Hirundo rustica	Barn swallow	Andorinha-de-bando	O	Re	PNW
Troglodytidae (4)					
Troglodytes musculus	Southern house wren	Corruíra	Re	Re	

(continued)

Table 12.1 (continued)

Taxon	Common name (EN)	Common name (PT)	M	S	Mi
Cistothorus platensis	Sedge wren	Corruíra-do-campo		Re	
Cantorchilus leucotis	Buff-breasted wren	Garrinchão-de-barriga-vermelha	O		
Cantorchilus longirostris	Long-billed wren	Garrinchão-de-bico-grande	Re		
Polioptilidae (1)					
Polioptila plumbea	Tropical gnatcatcher	Balança-rabo-de-chapéu-preto	O		
Donacobiidae (1)					
Donacobius atricapilla	Black-capped donacobius	Japacanim	O		
Turdidae (4)					
Turdus flavipes	Yellow-legged thrush	Sabiá-una	O		
Turdus leucomelas	Pale-breasted thrush	Sabiá-barranco	O		
Turdus rufiventris	Rufous-bellied thrush	Sabiá-laranjeira	O		
Turdus amaurochalinus	Creamy-bellied thrush	Sabiá-poca	O		
Mimidae (1)					
Mimus saturninus	Chalk-browed mockingbird	Sabiá-do-campo	O		
Estrildidae (1)					
Estrilda astrild	Common waxbill	Bico-de-lacre	O		
Passeridae (1)					
Passer domesticus	House sparrow	Pardal	O		
Motacillidae (4)					
Anthus lutescens	Yellowish pipit	Caminheiro-zumbidor	O	Re	
Anthus furcatus	Short-billed pipit	Caminheiro-de-unha-curta		O	
Anthus correndera	Correndera pipit	Saminheiro-de-espora		Re	
Anthus hellmayri	Hellmayr's pipit	Caminheiro-de-barriga-acanelada		O	
Fringillidae (4)					
Spinus magellanicus	Hooded siskin	Pintassilgo	O		
Euphonia chlorotica	Purple-throated euphonia	Fim-fim	O		
Euphonia violacea	Violaceous euphonia	Gaturamo-verdadeiro	O		
Euphonia pectoralis	Chestnut-bellied euphonia	Ferro-velho	O		
Passerellidae (2)					
Ammodramus humeralis	Grassland sparrow	Tico-tico-do-campo	O	Re	

(continued)

Table 12.1 (continued)

Taxon	Common name (EN)	Common name (PT)	M	S	Mi
Zonotrichia capensis	Rufous-collared sparrow	Tico-tico	O	Re	
Icteridae (12)					
Leistes superciliaris	White-browed meadowlark	Polícia-inglesa-do-sul	O	Re	
Cacicus solitarius	Solitary black cacique	Iraúna-de-bico-branco	O		
Cacicus chrysopterus	Golden-winged cacique	Tecelão	O		
Cacicus cela	Yellow-rumped cacique	Xexéu	O		
Cacicus haemorrhous	Red-rumped cacique	Guaxe	O		
Icterus pyrrhopterus	Variable oriole	Encontro	O		
Molothrus bonariensis	Shiny cowbird	Chupim	O	Re	
Agelasticus thilius	Yellow-winged blackbird	Sargento	O	Re	
Agelasticus cyanopus	Unicolored blackbird	Carretão-do-oeste	O		
Chrysomus ruficapillus	Chestnut-capped blackbird	Garibaldi	O	Re	
Pseudoleistes guirahuro	Yellow-rumped marshbird	Chopim-do-brejo	O		
Pseudoleistes virescens	Brown-and-yellow marshbird	Dragão		Re	
Parulidae (5)					
Geothlypis aequinoctialis	Masked yellowthroat	Pia-cobra	Re	Re	
Setophaga pitiayumi	Tropical parula	Mariquita	O		
Setophaga petechia	Yellow warbler	Mariquita-amarela	O		NNTT
Myiothlypis rivularis	Neotropical river warbler	Pula-pula-ribeirinho	O		
Basileuterus culicivorus	Golden-crowned warbler	Pula-pula	O		
Thraupidae (33)					
Embernagra platensis	Great pampa-finch	Sabiá-do-banhado		Re	
Hemithraupis ruficapilla	Rufous-headed tanager	Saíra-ferrugem	O		
Tersina viridis	Swallow tanager	Saí-andorinha	O		
Dacnis cayana	Blue dacnis	Saí-azul	O		
Saltator maximus	Buff-throated saltator	Tempera-viola	O		
Coereba flaveola	Bananaquit	Cambacica	Re		
Volatinia jacarina	Blue-black grassquit	Tiziu	O		
Loriotus cristatus	Flame-crested tanager	Tiê-galo	O		

(continued)

Table 12.1 (continued)

Taxon	Common name (EN)	Common name (PT)	M	S	Mi
Tachyphonus rufus	White-lined tanager	Pipira-preta	O		
Tachyphonus coronatus	Ruby-crowned tanager	Tiê-preto	O		
Ramphocelus bresilia	Brazilian tanager	Tiê-sangue	Re		
Ramphocelus carbo	Silver-beaked tanager	Pipira-vermelha	O		
Sporophila frontalis (**VU**[a, b])	Buffy-fronted seedeater	Pixoxó	O		
Sporophila collaris	Rusty-collared seedeater	Coleiro-do-brejo	O	O	
Sporophila caerulescens	Double-collared seedeater	Coleirinho	O		
Sporophila albogularis	White-throated seedeater	Golinho	O		
Sporophila leucoptera	White-bellied seedeater	Chorão	O		
Sporophila angolensis	Chestnut-bellied seed-finch	Curió	O		
Thlypopsis sordida	Orange-headed tanager	Saí-canário	O		
Donacospiza albifrons	Long-tailed reed finch	Tico-tico-do-banhado	O	O	
Conirostrum bicolor	Bicolored conebill	Figuinha-do-mangue	EXC		
Sicalis flaveola	Saffron finch	Canário-da-terra	O		
Sicalis luteola	Grassland yellow-finch	Tipio	O	Re	
Haplospiza unicolor	Uniform finch	Cigarra-bambu	O		
Paroaria dominicana	Red-cowled cardinal	Cardeal-do-nordeste	O		
Paroaria gularis	Red-capped cardinal	Cardeal-da-amazônia	O		
Thraupis episcopus	Blue-gray tanager	Sanhaço-da-amazônia	O		
Thraupis sayaca	Sayaca tanager	Sanhaço-cinzento	Re		
Thraupis cyanoptera	Azure-shouldered tanager	Sanhaço-de-encontro-azul	O		
Thraupis palmarum	Palm tanager	Sanhaço-do-coqueiro	O		
Stilpnia peruviana (**VU**[b])	Black-backed tanager	Saíra-sapucaia	O		
Stilpnia cayana	Burnished-buff tanager	Saíra-amarela	O		
Tangara cyanocephala	Red-necked tanager	Saíra-militar	O		

Number in brackets refer to the number of species in each taxonomic family

M mangrove, *SM* salt marsh, *Mi* migratory, *EXC* exclusive, *Re* regular, *O* occasional, *PNW* Pan New World migrants, *SACT* South American cool, temperate migrants, *NNTT* Nearctic-Neotropical temperate-tropical, *OW* Old World migrants, *VU* vulnerable, *EN* endangered, *CR* critically endangered

[a]According to MMA (2016, 2022)

According to IUCN (2022)

Since mangroves and salt marshes are important areas for migratory birds, we highlighted migratory species using these ecosystems in Brazil. We defined migratory species following Somveille et al. (2018): "(...) those whose breeding and non-breeding distributions do not completely overlap." Migratory species were identified using references listed in the appendix and Somenzari et al. (2018). Only fully migratory species along the entire Brazilian coast were included, i.e., species breeding in south Brazil and that winter in the northern part of the country were not included in the list. Migratory birds were then attributed into the migration systems defined by Joseph (1997):

Pan New World migrants (PNW): Species that winter and summer between the geographical extremes of the South and North American continents

Nearctic-Neotropical temperate-tropical migrants (NNTT): Species breeding in temperate North America and migrating to the warm humid tropics

South American cool, temperate migrants (SACT): Species breeding in southernmost South America and that migrate towards the midlatitudes of the continent

Species breeding in the Old World (Africa and Eurasia) were indicated separately.

12.3.1 Mangrove Birds

In Brazilian mangroves there are 368 bird species distributed in 24 orders and 62 families, representing 19% of the species occurring in the country. The most representative families are Tyrannidae (34 out of 144 species in Brazil), Thraupidae (32/156), Rallidae (16/35), Accipitridae (20/47), Scolopacidae (20/36), Ardeidae (16/24), Picidae (16/57), Anatidae (12/26), Icteridae (11/42), Trochilidae (11/89), Psittacidae (10/87), and Falconidae (9/20). From this list, there are 117 species in common with birds recorded in salt marshes. Also, 87 species (23%) are regularly recorded in mangroves, in addition to the five exclusive species, representing 5% of the birds in Brazil. The most representative families of the regularly recorded species were Scolopacidae (13), Ardeidae (11), Anatidae (5), Accipitridae (4), Alcedinidae (4), Rallidae (4), and Charadriidae (4) (see Table 12.1).

12.3.1.1 Exclusive Species

Five species are included in this category: tricolored heron, scarlet ibis, rufous crab hawk (*Buteogallus aequinoctialis*), mangrove rail (*Rallus longirostris*), and mangrove cuckoo (*Coccyzus minor*). Another species, bicolored conebill (*Conirostrum bicolor*), is exclusive to mangroves in most of its range but also inhabits the *várzea* forests of the Amazon basin (Cohn-Haft et al. 2007). In mangroves, this species is frequently recorded and more common than the other exclusive ones. The tricolored egret and mangrove cuckoo show restricted geographical ranges in relation to the

Fig. 12.1 Mangrove-exclusive or nearly exclusive birds: tricolored heron (*Egretta tricolor*) (**a**), scarlet ibis (*Eudocimus ruber*) (**b**), rufous crab hawk (*Buteogallus aequinoctialis*) (**c**), bicolored conebill (*Conirostrum bicolor*) (**d**), mangrove rail (*Rallus longirostris*) (**e**), and mangrove cuckoo (*Coccyzus minor*) (**f**). Salt marsh-exclusive bird: dot-winged crake (*Laterallus spilopterus*) (**g**). The black-crowned night heron (*Nyctanassa violacea*) (**h**) and little wood-rail (*Aramides mangle*) (**i**) occur in both ecosystems in Brazil (Photos **a**, Robson Czaban; **b**, Daniel Mello; **c**, Guto Balieiro; **d**, Daniel Mello; **e**, Daniel Mello; **f**, Alexander Lees; **g**, Rafael Antunes Dias; **h**, Patricia Luciano Mancini; **i**, Robson Czaban)

distribution of mangroves along the Brazilian littoral, occurring only in the north and northeast regions, whereas the remaining species present broader distributions. The scarlet ibis has large populations in the north and northeastern regions, with a population in the southeast region currently expanding both northwards and southwards from Cubatão, São Paulo State. Rufous crab hawk presents local occurrence, with reduced populations; mangrove rail is frequent in the south and southeastern regions of the country, rarer in the northeast, and absent in the north (Vieira 2015) (Fig. 12.1).

12.3.1.2 Regular Species

As many as 87 species are included in this category. The yellow-crowned night heron and the little wood-rail (*Aramides mangle*) are largely restricted to mangroves. The former species is not considered exclusive to mangroves because there is a population in a salt marsh area in south Brazil (Gianuca et al. 2011). The little wood-rail is a typical mangrove species, although the populations from Northeast Brazil perform partial landward migrations during the rainy season, a movement still poorly known (Marcondes et al. 2014). A large part of the regular species is associated with aquatic habitats (families Anatidae, Ardeidae, Alcedinidae, among others), and only five passerines are associated with those environments, namely, the yellow-chinned spinetail (*Certhiaxis cinnamomeus*), short-crested flycatcher (*Myiarchus ferox*), masked water-tyrant (*Fluvicola nengeta*), southern rough-winged swallow (*Stelgidopteryx ruficollis*), and masked yellowthroat (*Geothlypis aequinoctialis*). Other species such as the black vulture (*Coragyps atratus*), ruddy ground-dove (*Columbina talpacoti*), and great kiskadee (*Pitangus sulphuratus*) are associated with a broad range of habitats, including open environments and urban areas. The azure jay (*Cyanocorax caeruleus*), long-billed wren (*Cantorchilus longirostris*), and Brazilian tanager (*Ramphocelus bresilia*) are frequently recorded in mangroves due to the direct connection between the latter and the coastal lowland forests and *restinga* forests where they dwell. Among sandy beach species, the migratory common tern and the resident collared plover (*Charadrius collaris*) are frequently recorded in mangroves, using sedimentary banks (muddy or preferably sandy) for resting. Some seabird species such as the South American tern (*Sterna hirundinacea*), Cabot's tern (*Thalasseus acuflavidus*), and royal tern (*Thalasseus maximus*) display local occurrences in mangroves, mainly in regions where there are sandbanks between or within mangroves.

12.3.1.3 Occasional Species

There are 275 species included in this category, represented by a wide range of families. The high species richness in this category may be explained by the broad latitudinal extension of mangroves along the Brazilian coast, enabling the occurrence of species from a broad range of adjacent habitat types, even sporadically. In northern Brazil, the least tern (*Sternula antillarum*), yellow-billed tern (*Sternula superciliaris*), large-billed tern (*Phaetusa simplex*), gull-billed tern (*Gelochelidon nilotica*), and roseate tern (*Sterna dougallii*) show local occurrences in mangroves, mainly in regions where there are sandbanks between or within mangrove areas. The toco toucan (*Ramphastos toco*) and curl-crested jay (*Cyanocorax cristatellus*) have expanded their ranges eastwards in the last decades, especially in São Paulo State, and have been locally recorded in mangrove areas (Silva e Silva and Olmos 2007; Mancini et al. 2018). Brown pelican (*Pelecanus occidentalis*) and western reef-heron (*Egretta gularis*) are vagrant (Somenzari et al. 2018). The remaining species present local and sporadic occurrences.

12.3.1.4 Migrant Species

In mangroves, 40 migrant species were recorded (including 27 in common with salt marshes), mainly of the families Scolopacidae (19), Charadriidae (4), and Laridae (4). There are 29 Pan New World migrants, 17 of which are frequently recorded in mangrove areas and 12 that use this habitat occasionally. Among the most frequent species are the shorebirds *Tringa* spp. and *Calidris* spp. and the osprey *Pandion haliaetus*. Six species are Nearctic-Neotropical temperate-tropical migrants such as the American flamingo (*Phoenicopterus ruber*), brown pelican, and great blue heron. The lake duck (*Oxyura vittata*), rufous-chested dotterel (*Charadrius modestus*), and austral negrito (*Lessonia rufa*) are South American cool, temperate migrants. Lastly, the squacco heron (*Ardeola ralloides*), western reef-heron, and marbled godwit (*Limosa fedoa*) are migrants from the Old World (Table 12.1).

12.3.1.5 Knowledge Gaps in Species Composition

The knowledge about mangrove birds in Brazil presents some important gaps regarding species composition. There are at least three species that occur in mangroves in adjacent French Guiana, namely, rufous-necked wood rail (*Aramides axillaris*), arrowhead piculet (*Picumnus minutissimus*), and northern scrub-flycatcher (*Sublegatus arenarum*) that may potentially occur in Brazil (Restall et al. 2006; Sigrist 2006; Robbins 2018; Taylor 2018; Winkler et al. 2018). Furthermore, there are species known to use mangroves in neighboring countries but that never have been reported for these environments in Brazil, e.g., red-legged honeycreeper (*Cyanerpes cyaneus*), carib grackle (*Quiscalus lugubris*), and Amazonian tyrannulet (*Inezia subflava*) (Ridgely and Tudor 1994; Restall et al. 2006). New taxonomic proposals, which often elevate some populations to the species level, may contribute to increasing avian species richness in Brazilian mangroves as well. In addition, some species have more than one subspecies with known or potential occurrence on the Brazilian coast and that may attain specific status after future studies, e.g., Wilson's plover (*Charadrius wilsonia*) and mangrove rail, the former even presenting two subspecies in North and Northeast Brazil (Grantsau and Lima 2008; Vieira 2015).

12.3.1.6 Conservation

Among the 368 species recorded in mangroves, 15 (4.1%) are currently considered endangered to some extent. Seven are globally threatened (IUCN 2022), nine are included in the Brazilian list of threatened taxa (MMA 2016, 2022) and two, Dot-winged Crake (*Laterallus spilopterus*) and buffy-fronted seedeater (*Sporophila frontalis*), are in both lists. Regarding the 87 species classified as regular mangrove users, seven are included in the Brazilian Red List: Wilson's plover, red knot, South American tern and Cabot tern are vulnerable; royal tern and semipalmated sandpiper and short-billed dowitcher (*Limnodromus griseus*) are endangered (MMA 2016, 2022).

Habitat loss is the main cause for all population declines (Morrison and Ross 1989; Mohr et al. 2008; Campos 2010; Sipinski et al. 2014; Schunck and Rodrigues 2018). The red-tailed parrot (*Amazona brasiliensis*) has a stable population in its main area of occurrence in the state of Paraná, but its restricted range (from the southern coast of São Paulo State to the northern coast of Santa Catarina State) makes it especially vulnerable to habitat loss (Sipinski et al. 2014). The same occurs with royal tern, whose breeding areas in Brazil are all located in São Paulo State (Mohr et al. 2008; Campos 2010). For the Pan New World migrant red knot, there was a population decline of 55% in Northeast Brazil in the last three decades (Morrison and Ross 1989; Schunck and Rodrigues 2018). This decline happened because of the fall in food availability at Delaware Bay, in the United States of America, together with habitat loss along migration routes (Baker et al. 2004; Morrison et al. 2004). Similarly, short-billed dowitcher has presented a population decline of roughly 86% in the last decades in North Brazil (Morrison and Ross 1989; Schunck and Almeida *in prep.*), due to habitat loss both in the wintering grounds and along the migration routes (Rodrigues and Carvalho 2011; Valente et al. 2011). The population of semipalmated sandpiper declined 79% in northern Brazil since the 1980s (Valente et al. 2011; Morrison et al. 2012; Schunck et al. in prep.). Wilsons's plover presents a resident subspecies in Brazil; nevertheless, some other populations migrate to Brazil, such as *Charadrius w. cinnamominus* (Wiersma et al. 2018), and in northern and northeastern Brazilian coast, these populations have been declining in the last decades (Rodrigues 2007; Oliveira in prep.), probably due to disturbance in their breeding grounds.

12.3.2 Salt Marsh Birds

In Brazilian salt marshes 163 bird species (18 orders and 39 families) were reported in the literature, representing 8% of the country's species list. The most representative families are Scolopacidae (18 out of 28 species in Brazil), Laridae (15/29), Tyrannidae (12/144), Rallidae (11/35), Ardeidae (11/24), Furnariidae (10/106), Anatidae (9/26), and Charadriidae (7/11) (see Table 12.1). From this list, 117 species also occur in mangroves. A total of 123 species occurs regularly in salt marshes, representing 75% of Brazilian salt marsh birds and 6% of the Brazilian avifauna. The most representative families of the regularly recorded species were Scolopacidae (18), Ardeidae (10), Sternidae (9), Rallidae (8), Anatidae (7), and Charadriidae (7).

12.3.2.1 Exclusive Species

The dot-winged crake (*Laterallus spilopterus*) is the only species exclusively recorded in salt marshes in Brazil (Bencke et al. 2003). This small, globally threatened bird inhabits mainly salt marshes covered by dense stands of denseflower cordgrass (*Spartina densiflora* Brongn). The Peixe and Patos lagoons (RS) are the only localities where this bird has been recorded in the country (Bencke et al. 2003). The species is found almost exclusively in halophytic vegetation throughout its range in southern South America, both in coastal and inland wetlands, and is one of the few birds strongly associated with salt marshes in the continent (Greenberg et al. 2014) (Fig. 12.1).

12.3.2.2 Regular Species

A total of 123 bird species are included in this category, 43 in common with regular mangrove species such as yellow-crowned night heron, roseate spoonbill (*Platalea ajaja*), and lesser yellowlegs. Nearly 65% of the regular species in this category belong to typical waterbird families (e.g., Scolopacidae, Anatidae, Ardeidae, among others). Among the remaining regular species, some are exclusive of wetland habitats, despite being members of families of predominantly terrestrial species, such as the furnariids wren-like rushbird (*Phleocryptes melanops*) and sulfur-throated spinetail (*Limnoctites sulphuriferus*), the tyrannids of the genus *Pseudocolopteryx*, and the icterids yellow-winged blackbird (*Agelasticus thilius*) and chestnut-capped blackbird (*Chrysomus ruficapillus*). Several grassland birds, aerial-feeding species, and habitat generalists complete the list. Most waterbirds of regular occurrence in salt marshes also use freshwater wetlands or beaches and lagoons in southern Brazil (Belton 1994). However, some of them are more abundant or frequent in salt marshes and associated habitats and probably rely more on this kind of wetland than other wetland types, namely, the Chilean flamingo, Andean flamingo (*Phoenicoparrus andinus*), yellow-crowned night-heron, little blue heron, semipalmated plover, Hudsonian godwit, willet, semipalmated sandpiper, Olrog's gull (*Larus atlanticus*), and bay-capped wren-spinetail (Barbieri 2008). The latter is also strongly associated with salt marsh habitats in Argentina and Uruguay and may be considered a salt marsh specialist after additional investigations (Greenberg et al. 2014). Despite being recorded in a broad range of grassland habitats, sometimes in expressive numbers, the buff-breasted sandpiper and grass wren (*Cistothorus platensis*) are particularly abundant in Brazilian salt marshes, where both attain high densities (Bencke et al. 2003).

12.3.2.3 Occasional Species

This category includes 39 species represented by a range of families. Included in this list are birds common in adjacent freshwater wetlands and grasslands and that occasionally venture into estuarine habitats, such as the white-faced whistling-duck (*Dendrocygna viduata*), spot-flanked gallinule (*Porphyriops melanops*), spotted nothura (*Nothura maculosa*), and Hellmayr's pipit (*Anthus hellmayri*), as well as species reliant on shrubs and trees to fulfill part of or their entire life cycles, e.g., rufous hornero (*Furnarius rufus*), Spix's spinetail (*Synallaxis spixi*), and tropical kingbird (*Tyrannus melancholicus*). These species are only found in salt marshes with the presence of woody vegetation, which is naturally rare. Species that are uncommon in south Brazil are also included among the occasional users of salt marshes, such as the magnificent frigatebird (*Fregata magnificens*), short-billed dowitcher, and black tern (*Chlidonias niger*) (Belton 1994).

12.3.2.4 Migrant Species

A total of 34 migrant species are reported in Brazilian salt marshes, mainly Scolopacidae (16) and Laridae (4), all of which also occur in mangroves. There are 27 Pan New World migrant species, of which 85% are regularly recorded in salt marshes. Among the most frequent species are the shorebirds *Tringa* spp. and *Calidris* spp. Seven species are South American cool, temperate migrants, all making frequent use of salt marshes (Table 12.1).

12.3.2.5 Conservation

Among the 163 species recorded in salt marshes, 13 are of conservation concern. One is globally threatened (IUCN 2022), ten are included in the Brazilian list of threatened taxa (MMA 2016, 2022), and two species, the dot-winged crake and the black-and-white monjita (*Heteroxolmis dominicanus*), are in both lists (MMA 2016, 2022; IUCN 2022). Of the three globally threatened species, only the Andean flamingo and dot-winged crake are salt marsh regulars. The former is mostly an Andean breeder found in Brazil in shallow waters and mudbanks adjacent to and, more rarely, within salt marshes in a few localities of Santa Catarina State and at Peixe Lagoon (Bencke et al. 2006; Ghizoni-Jr and Piacentini 2010). The species occurs in small and varying numbers throughout the year in Brazil, being more numerous in the winter, and may largely be composed of immature and nonbreeding individuals (Bencke et al. 2006; Ghizoni-Jr and Piacentini 2010). Disturbance by humans, coastal development, and hunting are the main threats to this flamingo in Brazil (Ghizoni-Jr and Piacentini 2010). Individuals at Peixe Lagoon are protected within the Lagoa do Peixe National Park, but since this protected area is not fully implemented, they are often disturbed by shrimp harvesters and tourists.

In Brazil, the dot-winged crake occurs only in salt marshes of the Peixe Lagoon, which are located entirely within the national park, plus a handful of individual salt marshes at the estuary of Patos Lagoon (Bencke et al. 2003; Dias et al. 2017). This resident species inhabits dense vegetation and is largely threatened by overgrazing and burning of salt marshes and, to a lesser extent, by coastal development (Bencke et al. 2003). Despite occurring in two protected areas (Parque Nacional da Lagoa do Peixe and Área de Proteção Ambiental da Lagoa Verde) (see Chap. 3, Map 17), overgrazing and burning are still major threats due to the low regulation enforcement in both areas (Bencke et al. 2003, 2006; Dias et al. 2017). Two of the regionally threatened species considered regular salt marsh users, semipalmated sandpiper and red knot, are found in large numbers in Brazilian salt marshes only at the mouth of the Peixe Lagoon. In this locality, semipalmated sandpiper occurs throughout the austral warm season, whereas red knot peaks in the late austral summer and early austral autumn (Resende and Leeuwenberg 1987; Bencke et al. 2006). Both species may be impacted by disturbance by shrimp harvesters, fishermen, and tourists, which are frequent at the mouth of the lagoon (Bencke et al. 2006).

Royal tern also occurs in large numbers in salt marshes at the mouths of the Patos and Peixe lagoons, especially in winter, where they use sandbanks for roosting and may also suffer disturbance from fishermen and tourists. The two most important areas used by the buff-breasted sandpiper in Brazil, Lagoa do Peixe National Park and Torotama Island, are used throughout the austral spring and summer (Resende and Leeuwenberg 1987; Bencke et al. 2003, 2006). This sandpiper is restricted to areas of stunted vegetation where overgrazing by domestic livestock maintains the grass low. Suitable habitat on Torotama Island is relatively stable, but with the full implementation of Lagoa do Peixe National Park and the removal of livestock from the protected area, this species is expected to lose a substantial area of habitat unless specific habitat management practices are put in place (Bencke et al. 2003, 2006).

Cinereus harrier (*Circus cinereus*) is regionally threatened by the loss of nesting habitat in freshwater wetlands (Bencke et al. 2003). This bird hunts over a variety of natural and man-made open vegetation habitats, especially grasslands and freshwater wetlands, and is comparatively rarer in salt marshes (Bencke et al. 2003). Likewise, populations of Hudson's canastero (*Asthenes hudsoni*) inhabiting salt marshes are comparatively less threatened than those using adjacent sandy grasslands, which are prone to exotic pine tree invasion and urban and wind energy development (Serafini 2013).

12.4 Threats to Birds in Mangroves and Salt Marshes

Brazilian mangroves have steadily decreased in the area (Magris and Barreto 2010; Schaeffer Novelli et al. 2016), which represents a major threat to its avifauna. On the other hand, salt marsh destruction has been less severe, remaining almost unchanged since 1947 in south Brazil (Marangoni and Costa 2009a, b). Although the role of habitat loss and fragmentation upon avian diversity in Brazilian mangroves has not been properly explored, studies in other countries have demonstrated that area reduction can lead to population declines and local extinctions, predominantly of insectivorous species (Alongi 2009; Buelow and Sheaves 2015; Lawson et al. 2017).

Studies evaluating the effects of mangrove fragmentation on birds have shown contradictory results, either emphasizing (Alongi 2009) or deemphasizing (Chacin et al. 2015) the role of isolation in reducing diversity. In salt marshes, bird diversity is positively associated with patch size and proximity to other patches, with large wetlands possibly serving as population sources for some species, while small, isolated marshes act as population sinks (Shriver et al. 2004; Powell 2006). The landscape context in which mangrove and salt marsh fragments are inserted may also influence species richness and composition since adjacent patches of unsuitable habitat usually correlate with less diversity (Lefebvre and Poulin 1997; Shriver et al. 2004).

Mangrove and salt marsh remnants in Brazil are subject to habitat modification from a series of anthropogenic activities. Avian diversity in mangroves is positively associated with habitat heterogeneity, especially of vegetation features and foraging habitats (Mohd-Azlan et al. 2015; Mancini et al. 2018). Although there are no studies evaluating how habitat modification influences mangrove birds in the country, it is likely that human-induced habitat homogenization, especially through logging, garbage disposal, and canalization (Olmos and Silva e Silva 2003; Valente et al. 2011; Schaeffer Novelli et al. 2016), is negatively affecting bird diversity.

Fishing and harvesting of aquatic resources are widespread along the Brazilian coastline. Direct effects include the damage to mangrove roots and propagules, plastic and oil pollution, and erosion from the wake of boats, while indirect effects are linked to the removal of keystone species known to influence forest structure and nutrient cycling (Schories et al. 2003; Nascimento et al. 2017). Harvesting may impact fish-, shrimp- and crab-eating birds or even the entire bird community indirectly via habitat modification (Nagelkerken et al. 2008; Mohd-Azlan et al. 2015). Trophic cascade effects are also known to affect salt marsh ecosystems. In North America, overfishing of predator species increased herbivory by crabs and led to marsh die-offs (Altieri et al. 2012). Crabs are also important primary consumers in Brazilian salt marshes (Alberti et al. 2007) and may indirectly influence the diversity of bird communities by altering vegetation structure, as observed in nearby Argentina (Cardoni et al. 2007). Nevertheless, how the harvesting of aquatic resources influences avian diversity in Brazilian estuaries remains to be properly evaluated.

The increasing use of estuarine habitats for leisure may also impact birds due to direct disturbance promoted by people and their pets. People walking or running, accompanied by unleashed dogs, driving all-terrain vehicles and boats disrupt avian behavior, especially feeding, resting, and breeding (Smit and Visser 1993; Borgmann 2011; Scarton 2018). This may cause alterations on how birds use key estuarine habitats, affecting food intake, resting/non-resting budgets, and breeding performance – all of which are expected to have negative consequences at the population level due to energetic and reproductive costs (Smit and Visser 1993; Borgmann 2011; Scarton 2018). In Brazil, such forms of disturbance are more likely to affect plovers, sandpipers, and terns (most of which are migratory) roosting in sandbars and mudflats, especially near cities, ports, and touristic destinations (Olmos and Silva e Silva 2003; Valente et al. 2011; Dias et al. 2013).

Invasive alien species, such as water buffalos in mangroves (Valente et al. 2011) and feral pigs and wild boars in salt marshes (Quintela et al. 2010) may additionally impact bird communities. Browsing by livestock alters the morphological structure of mangroves (Hoppe-Speer and Adams 2015) and may lead to habitat simplification and associated loss of avian diversity. Pigs uproot salt marsh vegetation, eliminating habitat for rails and salt marsh passerines reliant on dense, tall vegetation. Pigs may also prey on bird eggs and younglings, directly affecting the population of breeding birds such as rails and ducks.

Southern Brazilian salt marshes are also affected by poorly managed extensive livestock ranching, which impacts wetlands by decreasing aboveground plant material, altering biodiversity, and modifying ecosystem functioning (FNMA et al. 2009; Marangoni and Costa 2009a). Similar effects have been recorded in salt marshes subject to fire, which are frequently used to manage pastures in livestock ranching areas (Marangoni and Costa 2009a). In Argentinean salt marshes, overgrazing and burning impact birds by reducing the abundance of tall-vegetation specialists, including species of conservation concern as dot-winged crake and bay-capped wren-spinetail, while favoring a few short-vegetation species but without major influence on abundance and species richness (Isacch et al. 2004; Isacch and Cardoni 2011). However, tallgrass birds can maintain populations under low-intensity grazing and burning, which demonstrates that livestock ranching and avian conservation are compatible under proper management (Cardoni et al. 2012).

Pollution from various sources (e.g., sewage and urban wastewater, industrial and agricultural effluents, garbage and solid waste, airborne pollutants) is also a common threat to Brazilian estuarine ecosystems, particularly to mangroves and salt marshes inserted within or near urban and industrial areas (Olmos and Silva e Silva 2003; Marangoni and Costa 2009a; Magris and Barreto 2010; Valente et al. 2011; Schaeffer Novelli et al. 2016). Birds are vulnerable to a range of pollutants that can result in mortality or sublethal behavioral, reproductive, and physiological effects depending on the intrinsic toxicity of the pollutant and exposure (Burger and Gochfeld 2001). The effects of pollutants depend on whether the exposure is acute or chronic, and in aquatic birds, this usually occurs through the ingestion of food and water (Burger and Gochfeld 2001, 2004). Oil spills recorded in Brazilian mangroves (Olmos and Silva e Silva 2003; Valente et al. 2011) may have impacted birds by disrupting their thermal balance and via toxicological effects following ingestion (Jenssen 1994; Burger and Gochfeld 2001). Plastic pollution is widespread (Olmos and Silva e Silva 2003; Valente et al. 2011) and tends to increase in coastal environments, threatening birds through entanglement and ingestion of plastic litter (Derraik 2002). Chemicals from sewage and urban wastewater, as well as industrial and agricultural effluents, usually affect individuals and populations through chronic effects linked to neurotoxicity and endocrine disruption (Burger and Gochfeld 2001, 2004; Köhler and Triebskorn 2013). Bioaccumulation of pollutants is a major concern, especially because many birds in mangroves and salt marshes occupy high trophic levels (Burger and Gochfeld 2001, 2004). On top of that, chemical pollution from sewage discharges may also modify the physiognomy of salt marsh vegetation, indirectly altering the composition and abundance of bird assemblages (Cardoni et al. 2011).

The greatest potential threat to mangroves and salt marshes is climate change, which will impact estuarine habitats via alterations in temperature and rainfall regimes, increasing extreme weather events, storms, and high tides, higher oceanic carbon dioxide concentration, and sea-level rise (Schaeffer Novelli et al. 2016). The survival of bird populations under climate change will depend on how they adapt to climate change and track their preferred climate via dispersal (Sekercioglu et al. 2012). Migratory species are expected to be particularly susceptible because higher temperatures may influence the timing of migration, as well as the availability of food resources in breeding and nonbreeding areas (Wrona et al. 2006; Sekercioglu et al. 2012). Since many mangroves and salt marsh areas in Brazil are key staging or wintering sites for migratory sea- and shorebirds (Valente et al. 2011; Dias et al. 2013), the fate of these areas under climate change will have important implications for avian conservation on a global scale.

If higher temperatures indeed promote an increase in mangrove cover along the Brazilian coast, such habitat expansion could benefit mangrove birds. However, the predicted increase is expected to be accompanied by changes in vegetation physiognomy and diversity (Schaeffer Novelli et al. 2016), potentially altering the diversity of avian communities. Loss of mangrove area among other effects of climate change, on the other hand, would reduce avian diversity overall, implying important population declines (or even extinction) of typical mangrove species (e.g., rufous crab hawk, little wood-rail), as well as forest birds that use these habitats as a refuge because of the loss of other upland adjacent forest types (Nagelkerken et al. 2008; Luther and Greenberg 2009).

The extensive salt marshes in south Brazil are expected to shrink in area due to erosion by rising sea level and invasion by southward expanding mangroves (Schaeffer Novelli et al. 2016). This may lead to the regional extinction of exclusive species (e.g., dot-winged crake), as well as important population losses of birds that are particularly abundant in these marshes (e.g., bay-capped wren-spinetail, grass wren) (Bencke et al. 2003). Changes in bird community structure may already be taking place in southern Brazilian salt marshes, even though mangrove trees have not yet colonized these formations. For example, the recent expansion and breeding of mangrove birds (e.g., yellow-crowned night heron and little blue heron) in salt marshes of south Brazil have been linked to the local increase in air temperature (Gianuca et al. 2011, 2012). Such expansion suggests that mangrove bird species may move ahead of the vegetation shift by using other forest types to breed and roost.

12.5 Conservation Challenges and Initiatives

Mangroves represent a rare forest type found in intertidal coastal zones largely restricted to tropical and subtropical regions (Sandilyan and Katherisan 2012). This complex ecosystem has been facing increasing threats due to human activities in the last decades, and it is estimated that roughly 35% of the global cover of mangroves was lost between 1980 and 2000 (Valiela et al. 2001; ICMBio 2015). Such reduction may be responsible for increasing the risk of extinction of at least 40% of the animal species that are restricted to mangroves (Polidoro et al. 2010), and ca. 14 bird species that are threatened to some extent. Nevertheless, detailed studies regarding the effects of mangrove destruction on the avifauna are still insufficient, especially in the Neotropics.

Brazil is the second country in mangrove extension, encompassing nearly 10% of this ecosystem in the world (Hamilton and Casey 2016). Roughly 80% of the Brazilian mangroves are in legally protected areas in the three levels of governance (national, state, and municipal) (Ferreira and Lacerda 2016). Taking national protected areas into consideration, ca. 79% are designed to promote sustainable use, most under the categories of environmental protected areas (APA) and extractive reserves (RESEX) (Magris and Barreto 2010). However, continuous threats are known to impact Brazilian mangroves, such as the deterioration of water quality in freshwater effluents and coastal habitats, deforestation, use of salt flats for salt extraction, and shrimp farming, among others. Of all threats, aquaculture represents the highest one (Alongi 2002; Magris and Barreto 2010). Recent estimates indicate that roughly 500 km^2 of Brazilian mangroves were destroyed in the last 30 years (FAO 2007; Romanach et al. 2018), even though the Brazilian Forest Code defines these ecosystems as Areas of Permanent Preservation (APP) and imposes restrictions to their use and occupation. Thus, it is extremely important not only to create protected areas and elaborate management plans but also to strengthen law enforcement (for more information, see Chap. 16).

Globally, salt marsh areas have declined between 25% and 50% in cover (Duarte et al. 2008; Crooks et al. 2011). In Brazil, a large portion of salt marshes is considered either Areas of Permanent Preservation (APP) or of restricted use according to the Brazilian Forest Code (Dias et al. 2017). Although a few individual marshes in southern Brazil have been partially or entirely lost to urban and industrial development and shrimp farms up to the 1990s, law enforcement has maintained salt marsh area stable up to the present (Marangoni and Costa 2009a, b). Two protected areas conserve salt marshes in Brazil: the Lagoa Verde Environmental Protection Area (510 ha under municipal responsibility) and the Lagoa do Peixe National Park (34,000 ha under national responsibility) (FNMA et al. 1999; Bencke et al. 2006; Dias et al. 2017). The former has been poorly implemented and allows "sustainable development," meaning that it is managed not only for biodiversity conservation but also for human sustainable activities (Dias et al. 2017). Although included in a more restrictive category, the Lagoa do Peixe National Park is also not fully implemented and still has many private areas used for livestock ranching within its boundaries (Bencke et al. 2006). Particularly troublesome is the dilemma of the removal of livestock from the park since it provides habitat for some bird species reliant on shortgrass habitats while eliminating tallgrass salt marshes on which other birds depend (Bencke et al. 2006). Shrimp harvesters and fishermen still carry on with their activities in the park, and unregulated tourism is common (Bencke et al. 2006). Fully implementing these protected areas and including formal measures to ensure adequate habitat management and the protection of threatened species in management plans would benefit salt marsh birds (Bencke et al. 2006; Dias et al. 2017).

Large, virtually pristine salt marshes exist at Pequena Lagoon and Torotama Island and the establishment of protected areas could ensure their conservation (Dias et al. 2017). Law enforcement and environmental education campaigns would also benefit avian conservation, especially if the restrictions imposed by the Forest Code were reinforced (Dias et al. 2017). Moreover, awareness towards conservation should be stimulated through environment-friendly activities (e.g., birdwatching, organic farming, properly managed extensive livestock ranching), all of which have the potential to generate income for local communities if properly implemented (Dias et al. 2017).

Recently, the Chico Mendes Institute for Biodiversity Conservation (ICMBio), a branch of the Brazilian Ministry of the Environment, has promoted working groups to elaborate conservation plans for specific bird groups, such as mangrove birds and

Table 12.2 Summary of knowledge gaps in ecological and anthropogenic impact studies in Brazilian mangrove and salt marsh birds

Ecological studies
1. Lack of bird inventories and studies on avian habitat use
2. Lack of reviews on the information available in museum specimens and citizen science online platforms for a better understanding of the occurrence status and species conservation
3. Lack of studies on the ecology and natural history of exclusive and regular species for which the reduction in habitat quality or extent may be an important pressure
4. Limited knowledge on habitat and niche differentiation within these ecosystems
5. Lack of studies on avian morphological differentiation in Brazilian mangroves and salt marshes
6. Lack of studies on avian feeding adaptations or zonation patterns
7. Publication of relevant unpublished data carried out in mangroves and salt marshes (reducing the use of grey literature)
Anthropogenic impact studies
1. Long-term quantitative bird population studies, especially of sensitive taxa (e.g., threatened, exclusive, or migrant species)
2. Long-term monitoring and assessment of the effects of landscape changes on avian diversity
3. Evaluation of the impacts of water pollutants on birds
4. Evaluation of the impacts of aquatic resource harvesting on birds
5. Evaluation of the impacts of habitat modification (logging, garbage disposal, canalization, etc.) on birds
6. Evaluation of the impacts of mangrove and salt marsh destruction on birds
7. Evaluation of the impacts of climate change on birds and their habitats
8. Establishment of local conservation education programs and citizen science initiatives

migrant shorebirds. The National Plan for the Conservation of Migrant Shorebirds (ICMBio 2013) and the National Plan for Threatened Mangrove Species of Socio-economic Importance (ICMBio 2015) have as their main goals the proposal of effective conservation policies and the establishment of mechanisms that ensure the protection of the species under their scope and minimize the loss of their habitat.

12.6 Knowledge Gaps

Knowledge gaps are summarized in Table 12.2. Among the 69 consulted pieces of literature on Brazilian mangrove birds, 74% were scientific papers, 17% books, and 9% book chapters. A substantial part of these studies does not present detailed information on species specifically recorded in mangroves. Field studies represent the largest amount of information in the literature (93%), while the other 7% are large compilations (books), which are also based on field studies, museum specimens, and authors' observations (Ridgely and Tudor 1989, 1994; Stotz et al. 1996; Sick 1997; Willis and Oniki 2003; Sigrist 2006; Grantsau 2010). There is also important information on mangrove birds in dissertations and congress abstracts (gray literature). Virtually nothing is known about birds using salt marshes in Brazil other than what features in studies carried out in the Peixe and Patos lagoons (RS). The avifauna of salt marsh fringes in mangrove areas, for example, is unknown.

12.6.1 Field Studies

The published information on mangrove birds was collected in 16 Brazilian states that contain these habitats, with the number of studies varying according to the extension of habitat in each state. One of the most well-studied mangrove areas in Brazil is Mangue Seco in Bahia State, northeast Brazil, but the published studies do not mention the precise habitat where the bird species were recorded.

The largest part of the available studies presents qualitative data over quantitative or natural history data. Among the quantitative data, most studies focus on Nearctic migrants of the families Charadriidae, Laridae, Scolopacidae, and Sternidae, as well as Accipitridae, Anatidae, Ardeidae, Falconidae, Psittacidae, and Threskiornithidae (e.g., Olmos and Silva e Silva 2003; Galetti et al. 2006; Almeida and Rodrigues 2015; Souza and Rodrigues 2015). Natural history accounts are available only for a few species, such as the scarlet ibis, yellow-crowned night heron, Wilson's plover, orange-winged parrot (*Amazona amazonica*), and red-tailed parrot (e.g., Cunha et al. 2000; Gianuca 2007; Grose et al. 2013). The more incipient studies are those aimed at investigating the effects of human activities, e.g., water pollution, illegal mangrove occupation, and oil spills on coastal birds, including mangrove species (e.g., Olmos and Silva e Silva 2003; Rodrigues 2007; Valente et al. 2011), albeit not quantitatively evaluating those impacts (e.g., Barbieri 2001; Hvenegaard and Barbieri 2010).

Long-term bird population studies are still lacking in Brazilian mangroves and salt marshes. Such studies are crucial for a better understanding of population trends on both local and national scales. Quantitative studies that detect annual rates of changes in the average number of individuals in different areas, especially of sensitive taxa (e.g., threatened, exclusive, or migrant species), may reveal population declines that would go unnoticed with purely qualitative, presence/absence-based studies (see Lloyd and Doyle 2011). Likewise, natural history studies focused on elucidating avian ecology may also help to understand the proximal causes of eventual declines. This may be especially important for habitat-specialist species, for which the diminishing quality or extent of habitat surely is an important pressure.

Virtually nothing is known about the potential impacts of pollutants in urban and industrial wastewater on Brazilian mangrove and salt marsh bird communities. Long-term, wide-range monitoring initiatives as well as short-term, local studies targeting the effects of water quality on mangrove and salt marsh bird species are highly recommended. Likewise, environmental impact assessments (EIA), a national requirement ensuring that decision-makers understand the environmental impacts of their projects and plans, must be conducted more rigorously and taking into account their effects on mangrove and salt marsh birds. Unfortunately, human interference and its consequences to coastal ecosystems are still underestimated, and thorough population studies in mangroves and salt marshes, correlated with physical and other biological parameters, are critical to determine avian population trends and whether eventual declines are in course.

12.7 Final Remarks

Besides the significant advances in the knowledge of mangrove and salt marsh birds and the gaps mentioned above, additional procedures may improve our understanding of this ecological system. Ornithological studies conducted in mangrove and salt marsh areas (e.g., inventories, monitoring programs, specimen collection) must explicitly include a reference of the habitat where each species was recorded. This is valid for photographic, sound, or video records deposited in databases as well. In the case of published works, this information must be included either in the results section, in species accounts, or as details in species lists (see, e.g., Olmos and Silva e Silva 2003; Lees et al. 2014). Relevant information present in the grey literature should be formally published to better understand the occurrence and status of poorly known species in mangroves, such as red-and-white spinetail (*Certhiaxis mustelinus*), pileated finch (*Coryphospingus pileatus*), crested oropendola (*Psarocolius decumanus*), comb duck (*Sarkidiornis sylvicola*), and yellow-bellied seedeater (*Sporophila nigricollis*), among others.

Field studies in Brazilian mangroves are urgent along the whole coast, including quantitative data, which are crucial for short- and long-term populational monitoring initiatives. Detailed studies of the human impacts on mangrove birds are also fundamental from a conservation perspective. In addition to that, reviews of the information available in museum specimens and citizen science online platforms (e.g., Wikiaves, e-Bird, Xeno-Canto, Táxeus) are also important for a better understanding of the occurrence and conservation status of mangrove and salt marsh birds.

In some mangrove regions, bird assemblages exhibit zonation, with species even specializing on different mangrove tree species (Noske 1995, 1996; Luther and Greenberg 2009). Since habitat loss may have a differential effect upon distinct mangrove zones, some bird species may be more threatened than others by human activities within or nearby these ecosystems (Magris and Barreto 2010; Polidoro et al. 2010). Therefore, future studies should focus on detecting and documenting fine-scale habitat requirements of exclusive and regular species to better understand niche partitioning and the possible existence of avian zonation in Brazilian mangroves. This could shed light on how different species respond to different human activities and their potential negative effects. Another goal is the implementation of local conservation education programs and citizen science initiatives, with public engagement and the involvement of local communities, which are fundamental to set best practices for the conservation of mangrove and salt marsh birds.

Finally, the creation of new protected areas encompassing mangroves (ICMBio 2018) and salt marshes would be important to conserve birds and other organisms in Brazil. However, even the categorization of mangrove habitats as Areas of Permanent Protection (APP) has been constantly threatened by the current government. In 2020, the National Environment Council (CONAMA) approved the removal of these regulations, which would allow developers to clear large areas of natural habitats for tourism or real state enterprises. Fortunately, the Federal Supreme Court revoked the removal of these regulations following protests from researchers and environmentalists. Mangrove restoration projects are limited to less than

25 isolated attempts, rarely exceeding half a hectare and with high plant mortality rates (Menghini et al. 2018). Furthermore, data on restoration and monitoring remain unpublished or reduced to planting techniques (Rovai et al. 2012). Salt marsh restoration projects are also incipient and limited to small areas (Costa 2011).

Appendix

Mangrove Bird References Ridgely and Tudor (1989), Azevedo-Júnior (1992), Pacheco (1994), Ridgely and Tudor (1994), Martuscelli (1995), Rodrigues (1995), Stotz et al. (1996), Alves et al. (1997), Sick (1997), Silva et al. (1997), Azevedo-Júnior (1998), Hass et al. (1999), Cunha et al. (2000), Olmos (2000), Olmos and Silva-e-Silva (2001), Silva e Silva and Olmos (2002), Naka et al. (2002), Azevedo et al. (2003), Olmos and Silva e Silva (2003), Willis and Oniki (2003), Develey (2004), Araújo et al. (2006), Galetti et al. (2006), Lima (2006 apud Somenzari et al. 2018), Roda and Pereira (2006), Sigrist (2006), Silva e Silva (2007), Amorim and Piacentini (2007), Araújo and Nishida (2007), Dantas et al. (2007), Efe et al. (2007), Mestre et al. (2007), Rodrigues (2007), Rodrigues et al. (2007), Ruschi and Simon (2007), Silva e Silva and Olmos (2007), Almeida and Barbieri (2008), Pereira et al. (2008), Periquito et al. (2008), Rupp et al. (2008), Silva e Silva (2008), Souza et al. (2008), Fedrizzi and Carlos (2009), Maciel (2009), Piacentini et al. (2009), Aguiar et al. (2010), Grantsau (2010), Lima et al. (2010), Nunes et al. (2010), Almeida and Ferrari (2011), Girão and Albano (2011), Carvalho and Rodrigues (2011), Santos (2011), Xavier and Boss (2011), Lees et al. (2012), Pereira et al. (2012), Ghizoni-Jr et al. (2013), Gomes et al. (2013), Grose et al. (2013), Olmos et al. (2013), Bisinela et al. (2014), Lees et al. (2014), Melo et al. (2014), Pereira et al. (2014), Grose and Cremer (2015), Vieira (2015), Favoretti and Batalla (2017), and Mancini et al. (2018)

Salt Marsh Bird References Resende and Leeuwenberg (1987), Maurício and Dias (1996), Dias and Maurício (1998), FNMA et al. (1999), Maurício and Dias (2000), Bencke et al. (2003), Bencke et al. (2006), Dias et al. (2011), Nascimento (2011), Grimm (2013), Martinez-Curci et al. (2014), Greenberg et al. (2014), and Dias et al. (2017)

References

Acevedo MA, Aide TM (2008) Bird community dynamics and habitat associations in karsts, mangrove and *Pterocarpous forest* fragments in an urban zone in Puerto Rico. Caribb J Sci 44:402–416

Aguiar KMO, Naiff RH, Xavier B (2010) Aves da Reserva Biológica do Lago Piratuba, Amapá, Brasil. Ornithologia 4:1–14

Alberti J, Escapa M, Daleo P, Iribarne O, Silliman B, Bertness M (2007) Local and geographic variation in grazing intensity by herbivorous crabs in SW Atlantic salt marshes. Mar Ecol Prog Ser 349:235–243

Almeida BJM, Barbieri E (2008) Biodiversidade das aves do manguezal da 13 de julho em Aracaju, Sergipe. O Mundo da Saúde 32:317–328

Almeida BJM, Ferrari SF (2011) Occurrence and breeding record of the American Oystercatcher (*Haematopus palliatus palliatus*, Temminick 1820) in Sergipe, Northeastern Brazil. Ararajuba 19:405–408

Almeida BJM, Rodrigues AAF (2015) Abundância sazonal de aves limícolas em área costeira amazônica, praia de Panaquatira, golfão maranhense, Brasil. Ornithologia 8:38–42

Alongi DM (2002) Present status and future of the world's mangrove forest. Environ Conserv 29(3):331–349

Alongi DM (2009) Paradigm shifts in mangrove biology. In: Perillo GME, Wolanski EJ, Cahoon DR, Brinson MM (eds) Coastal wetlands: an integrated ecosystem approach. Elsevier, Duivendrecht, pp 615–640

Altieri AH, Bertness MD, Coverdale TC, Herrmann NC, Angelini C (2012) A trophic cascade triggers collapse of a salt marsh ecosystem with intensive recreational fishing. Ecology 93: 1402–1410

Alves VS, Soares ABA, Ribeiro AB (1997) Birds of Jequia mangrove system, Ilha do Governador, Baía da Guanabara, Rio de Janeiro, Brazil. In: Björn K, Lacerda LD, Diop HS (eds) Mangrove ecosystem studies in Latin America and Africa. UNESCO, Paris, pp 163–170

Amorim JF, Piacentini VQ (2007) Novas áreas de ocorrência de três Passeriformes no sul do Brasil. Lundiana 8(1):69–73

Araújo HFP, Nishida AK (2007) Considerações sobre colônias de garças (Ciconiiformes, Ardeidae) no estado da Paraíba-Brasil. Ornithologia 2:34–40

Araújo HFP, Rodrigues RC, Nishida AK (2006) Composição da avifauna em complexos estuarinos no estado da Paraíba, Brasil. Ararajuba 14:249–259

Azevedo MAG, Machado DA, Albuquerque JLB (2003) Aves de rapina da Ilha de Santa Catarina, SC: Composição, Frequência de ocorrência, Uso de habitat e Conservação. Ararajuba 11:75–81

Azevedo-Júnior SM (1992) Anilhamento de aves migratórias na Coroa do Avião, Igarassu, Pernambuco, Brasil. In: Cadernos Ômega da Universidade Federal Rural de Pernambuco, Série Biologia, vol 3. UFRPE, Recife, pp 31–47

Azevedo-Júnior SM (1998) As aves do Canal de Santa Cruz, Pernambuco, Brasil. In: Cadernos Ômega da Universidade Federal Rural de Pernambuco, Série Biologia, vol 5. UFRPE, Recife, pp 35–50

Baker AJ, Gonzáles PM, Piersma T, Niles LJ, Nascimento ILS, Atkinson PW, Clark NA, Minton CDT, Peck MK, Aarts G (2004) Rapid population decline in red knots: fitness consequences of decreased refuelling rates and late arrival in Delaware Bay. P Roy Soc B-Biol Sci 271:875–882

Barbieri E (2001) A expansão urbana e seu impacto sobre as aves da Lagoa Verde, Rio Grande (RS), Cadernos Faculdades Integradas São Camilo, vol 7. Faculdades Integradas São Camilo, São Paulo, pp 85–95

Barbieri E (2007) Seasonal abundance of shorebirds at Aracaju, Sergipe, Brazil. Wader Study Group Bull 113:34–41

Barbieri E (2008) Variação sazonal do gaivotão (*Larus dominicanus*) durante o ano de 2005 no estuário de Cananéia-Iguape-Ilha Comprida, São Paulo, Brasil. Biota Neotrop 8:21–26

Belton W (1994) Aves do Rio Grande do Sul: distribuição e biologia. UNISINOS, São Leopoldo

Bencke GA, Fontana CS, Dias RA, Maurício GN, Mähler JKF Jr (2003) Aves. In: Fontana CS, Bencke GA, Reis RE (eds) Livro vermelho da fauna ameaçada de extinção no Rio Grande do Sul. EDIPUCRS, Porto Alegre, pp 189–479

Bencke GA, Maurício GN, Develey PF, Goerck JM (2006) Áreas importantes para a conservação das aves no Brasil. Parte I – Estados do Domínio da Mata Atlântica. SAVE Brasil, São Paulo

Bisinela GC, Araújo TRA, Cremer MJ (2014) Ecologia reprodutiva de *Nyctanassa violacea*, *Egretta thula* e *Egretta caerulea* no ninhal do rio Pedreira, Santa Catarina. Biotemas 27:109–121

Borgmann KL (2011) A review of human disturbance impacts on waterbirds. Audubon California. http://www.sfbayjv.org/news-general.php

Buelow C, Sheaves M (2015) A birds–eye view of biological connectivity in mangrove systems. Estuar Coast Shelf S 152:33–43

Burger J, Gochfeld M (2001) Effects of chemicals and pollution on seabirds. In: Schreiber EA, Burger J (eds) Biology of marine birds. CRC Press, Boca Raton, pp 485–525

Burger J, Gochfeld M (2004) Marine birds as sentinels of environmental pollution. EcoHealth 1: 263–274

Campos FP (2010) *Thalasseus maximus* (Boddaert, 1783). In: Bressan PM, Kierulff MCM, Sugieda AM (eds) Fauna Ameaçada de extinção no Estado de São Paulo – Vertebrados, 1st edn. Fundação Parque Zoológico de São Paulo and Secretaria do Meio Ambiente, São Paulo, p 156

Cardoni DA, Isacch JP, Iribarne OO (2007) Indirect effects of the intertidal burrowing crab *Chasmagnathus granulatus* in the habitat use of Argentina's South West Atlantic salt marsh birds. Estuar Coast 30:382–389

Cardoni DA, Isacch JP, Fanjul ME, Escapa M, Iribarne OO (2011) Relationship between anthropogenic sewage discharge, marsh structure and bird assemblages in an SW Atlantic saltmarsh. Mar Environ Res 71:122

Cardoni DA, Isacch JP, Iribarne O (2012) Effects of cattle grazing and fire on the abundance, habitat selection, and nesting success of the Bay–Capped Wren–Spinetail (*Spartonoica maluroides*) in coastal saltmarshes of the Pampas Region. Condor 114(4):803–811

Cardoni DA, Greenberg R, Maldonado JE, Isacch JP (2013) Morphological adaptation to coastal marshes in spite of limited genetic structure in the Neotropical passerine *Spartonoica maluroides* (Aves:Furnariidae). Biol J Linn Soc 109:78–91

Carvalho DL, Rodrigues AAF (2011) Spatial and temporal distribution of migrant shorebirds (Charadriiformes) on Caranguejos Island in the Gulf of Maranhão, Brazil. Ararajuba 19:486–492

Chacin DH, Giery ST, Yeager LA, Layman CA, Langerhans RB (2015) Does hydrological fragmentation affect coastal bird communities? A study from Abaco Island, The Bahamas. Wetl Ecol Manag 23:551–557

Chan YL, Hill CE, Maldonado JE, Fleischer RC (2006) Evolution and conservation of tidal marsh vertebrates: molecular approaches. Stud Avian Biol-Ser 32:54–76

Cohn-Haft M, Naka LN, Fernandes AM (2007) Padrões de distribuição da avifauna da várzea dos rios Solimões e Amazonas. In: Albernaz AL (ed) Conservação da várzea: identificação e caracterização de regiões biogeográficas. IBAMA, Manaus, pp 287–323

Costa CSB (2011) Restoration of coastal salt marshes in Brazil using native salt marsh plants. In: Greipsson S (ed) Restoration ecology. Jones and Bartlett Publishers, Sudbury, pp 333–338

Crooks S, Herr D, Tamelander J, Laffoley D, Vandever J (2011) Mitigating climate change through restoration and management of coastal wetlands and near-shore marine ecosystems: challenges and opportunities, Environment department papers, vol 121. World Bank, Washington, DC

Cunha AHF, Rodrigues AAF, Martínez C (2000) Desenvolvimento de filhotes de Taquiri, *Nyctanassa violacea* (Ciconiiformes: Ardeidae) na Ilha do Cajual, Alcântara, Maranhão, Brasil. Bol Mus Paraense Emilio Goeldi Ser Zool 16(1):7–21

Dantas SM, Pereira GA, Farias GB, Brito MT, Periquito MC, Vasconcelos EST (2007) Registros relevantes de aves para o estado de Pernambuco, Brasil. Ararajuba 15:113–115

Derraik JGB (2002) The pollution of the marine environment by plastic debris: a review. Mar Pollut Bull 44(9):842–852

Develey PF (2004) As aves da Estação Ecológica Juréia-Itatins. In: OAV M, Duleba W (eds) Estação Ecológica Juréia-Itatins: Ambiente físico, flora e fauna. Ribeirão Preto (SP). Holos, Ribeirão Preto, pp 278–295

Dias RA, Maurício GN (1998) Lista preliminar da avifauna da extremidade sudoeste do Saco da Mangueira e arredores, Rio Grande, Rio Grande do Sul. Atualidades Ornitológicas 86:10–11

Dias RA, Gianuca D, Gianuca A, Gomes-Júnior A, Chiaffitelli R, Ferreira WLS (2011) Estuário da Lagoa dos Patos. In: Valente RM, Silva JMC, Straube FC, Nascimento JLX (eds) Conservação de Aves Migratórias Neárticas do Brasil. Conservation International, Belém, pp 335–341

Dias RA, Bastazini VAG, Gonçalves MSS, Bonow FC, Müller SC (2013) Shifts in composition of avian communities related to temperate-grassland afforestation in southeastern South America. Iheringia Ser Zool 103(1):12–19

Dias RA, Maurício GN, Bugoni L (2017) Birds of the Patos Lagoon Estuary and adjacent coastal waters, southern Brazil: species assemblages and conservation implications. Mar Biol Res 13: 108–120

Duarte C, Dennison W, Orth RW, Carruthers TB (2008) The Charisma of coastal ecosystems: addressing the imbalance. Estuar Coast 3(2):233–238

Efe MA, Azevedo MAG, Filippini A (2007) Avifauna da Estação Ecológica de Carijós, Florianópolis – SC. Ornithologia 2:1–13

Favoretti MR, Batalla JF (2017) Levantamento da avifauna no mangue do rio Juqueriquerê, Caraguatatuba, SP. Unisanta Bioscience 6:272–285

Fedrizzi EC, Carlos CJ (2009) Aves, Charadriiformes, Scolopacidae, *Calidris fuscicolis* (Vieillot, 1819) (White-rumped Sandpiper): documented records for the states of Piauí and Ceará in northeastern Brazil. Checklist 5:471–474

Ferreira AC, Lacerda LD (2016) Degradation and conservation of Brazilian mangroves, status and perspectives. Ocean Coast Manage 125:38–46

Food and Agriculture Organization of the United Nations – FAO (2007) The world's mangroves: 1980–2005. FAO, Rome

Fundo Nacional do Meio Ambiente, Universidade Federal do Rio Grande, Instituto Brasileiro do Meio Ambiente, Núcleo de Ecologia e Monitoramento Ambiental, Universidade Federal de Pelotas – FNMA, FURG, IBAMA, NEMA, UFPel (1999) Plano de Manejo do Parque Nacional da Lagoa do Peixe. Fundação Universidade do Rio Grande, Rio Grande

Fundo Nacional do Meio Ambiente, Universidade Federal do Rio Grande, Instituto Brasileiro do Meio Ambiente, Núcleo de Ecologia e Monitoramento Ambiental, Universidade Federal de Pelotas – FNMA, FURG, IBAMA, NEMA, UFPel (2009) Plano de Manejo do Parque Nacional da Lagoa do Peixe. Fundação Universidade do Rio Grande, Rio Grande

Galetti M, Schunck F, Ribeiro M, Paiva A, Toledo R, Fonseca L (2006) Distribuição e tamanho populacional do papagaio-de-cara-roxa (*Amazona brasiliensis*) no Estado de São Paulo. Ararajuba 14:239–247

Ghizoni-Jr IR, Piacentini VQ (2010) The Andean Flamingo *Phoenicoparrus andinus* (Philippi, 1854) in southern Brazil: is it a vagrant? Rev Bras Ornitol 18(3):263–266

Ghizoni-Jr IR, Faria FB, Vieira BP, Willrich G, Silva ES, Mendonça EM, Albuquerque JLB, Gass D, Ternes MH, Nascimento CE, Roos AL, Couto CCM, Serrão M, Serafini PP, Dias D, Fantacini F, Santi S, Souza MCR, Silva MS, Barcellos A, Albuquerque C, Espinola CRR (2013) Checklist da avifauna da Ilha de Santa Catarina. Atualidades Ornitológicas 171:50–75

Gianuca D (2007) Ocorrência sazonal e reprodução do socó-caranguejeiro *Nyctanassa violacea* no estuário da Lagoa dos Patos, novo limite sul da sua distribuição geográfica. Ararajuba 15:464–467

Gianuca D, Branco JO, Vooren CM (2011) Notes on breeding by Yellow–crowned Night Heron *Nyctanassa violacea* in southern Brazil. Cotinga 33(1):63–72

Gianuca D, Gianuca AT, Vooren CM (2012) Abundance, breeding and food of the Little Blue Heron *Egretta caerulea* (Aves, Ardeidae) in the Patos lagoon estuary, a recently colonized area in southern Brazil. Iheringia Ser Zool 102(1):19–25

Girão W, Albano C (2011) Ilha Grande. In: Valente RM, Silva JMC, Straube FC, Nascimento JLX (eds) Conservação de Aves Migratórias Neárticas no Brasil. Conservation International, Belém, pp 129–132

Gomes DN, Tavares AA, Cardoso CO, Santos AGS, Silva PC, Guzzi A (2013) Ocorrência de *Aratinga acuticaudata haemorrhous* Spix, 1824 na Área de Proteção Ambiental Delta do Parnaíba, Piauí, Brasil. Ornithologia 5:118–121

Grantsau R (2010) Guia completo para identificação das Aves do Brasil. Editora Vento Verde, São Carlos, 1249 pp

Grantsau R, Lima PC (2008) Uma nova subespécie de *Charadrius wilsonia* (Aves, Charadriiformes) para o Brasil. Atualidades Ornitológicas 142:4–5

Green AJ, Elmberg J (2013) Ecosystem services provided by waterbirds. Biol Rev 89:105–122

Greenberg R, Olsen B (2010) Bill size and dimorphism in tidal–marsh sparrows: island–like processes in a continental habitat. Ecology 91:2428–2436

Greenberg R, Elphick C, Nordby JC, Gjerdrum C, Spautz H, Shriver G, Schmeling B, Olsen B, Marra P, Nur N, Winter M (2006) Flooding and predation: trade-offs in the nesting ecology of tidal-marsh sparrows. Stud Avian Biol 32:96–109

Greenberg R, Danner RM, Olsen BJ, Luther D (2012) High temperatures explain bill size in salt marsh sparrows. Ecography 35:146–152

Greenberg R, Cardoni A, Ens B, Gan X, Isacch J, Koffijberg K, Loyn R (2014) The distribution and conservation of birds of coastal salt marshes. In: Maslo B, Lockwood J (eds) Coastal conservation. Cambridge University Press, Cambridge, pp 180–242

Grenier JL, Greenberg R (2005) A biogeographic pattern in sparrow bill morphology: parallel adaptation to tidal marshes. Evolution 59:1588–1595

Grimm R (2013) Santuário das aves: Parque Nacional da Lagoa do Peixe. Autor, Porto Alegre

Grose AV, Cremer MJ (2015) Aves migratórias no litoral norte de Santa Catarina, Brasil. Ornithologia 8:22–32

Grose AV, Cremer MJ, Moreira N (2013) Reprodução do guará *Eudocimus ruber* no estuário da baía da Babitonga, litoral norte de Santa Catarina. In: Annals of the XX Congresso Brasileiro de Ornitologia, Universidade de Passo Fundo, Passo Fundo, 4–7 November 2013

Hamilton SE, Casey D (2016) Creation of a high spatio-temporal resolution global database of continuous mangrove forest cover for the 21st century (CGMFC-21). Global Ecol Biogeogr 25: 729–738

Hass A, Matos RHR, Marcondes-Machado LO (1999) Ecologia reprodutiva e distribuição espacial da colônia de *Eudocimus ruber* (Ciconiiformes, Threskiornithidae) na Ilha do Cajual, Maranhão. Ararajuba 7:41–44

Hoppe-Speer SCL, Adams JB (2015) Cattle browsing impacts on stunted *Avicennia marina* mangrove trees. Aquat Bot 121:9–15

Hughes RG (2004) Climate change and loss of saltmarshes: consequences for birds. Ibis 146:21–28

Hutchings P, Saenger P (1987) Ecology of mangroves. University of Queensland Press, St Lucia

Hvenegaard GT, Barbieri E (2010) Shorebirds in the State of Sergipe, northeast Brazil: potential tourism impacts. Ararajuba 18:169–175

Instituto Chico Mendes de Conservação da Biodiversidade – ICMBio (2013). Sumário Executivo do Plano de Ação Nacional para Conservação das Aves Limícolas Migratórias. Brasília.

Instituto Chico Mendes de Conservação da Biodiversidade – ICMBio (2015) Plano de Ação Nacional para Conservação das Espécies Ameaçadas e de Importância Socioeconômica do Ecossistema Manguezal – PAN Manguezal. Diário Oficial da União 21, Brasília. http://www.icmbio.gov.br/portal/faunabrasileira/plano-de-acao-nacional-lista/2840-plano-de-acao-nacional-para-a-conservacao-dos-manguezais. Accessed 26 Mar 2018

Instituto Chico Mendes de Conservação da Biodiversidade – ICMBio (2018) Atlas dos Manguezais do Brasil. ICMBio, Brasília

International Union for Conservation of Nature – IUCN (2022) The IUCN Red List of Threatened Species. Version 2022-1.

Isacch JP, Cardoni DA (2011) Different grazing strategies are necessary to conserve endangered grassland birds in short and tall salty grasslands of the flooding Pampas. Condor 113:724–734

Isacch JP, Holz S, Ricci L, Martínez MM (2004) Postfire vegetation change and bird use of a salt marsh in coastal Argentina. Wetlands 24:235–243

Isacch JP, Cardoni DA, Iribarne OO (2014) Diversity and habitat distribution of birds in coastal marshes and comparisons with surrounding upland habitats in southeastern South America. Estuar Coast 37:229–239

Jenssen BM (1994) Review article: effects of oil pollution, chemically treated oil, and cleaning on thermal balance of birds. Environ Pollut 86(2):207–215

Joseph L (1997) Towards a broader view of Neotropical migrants: consequences of a re-examination of austral migration. Ornitol Neotrop 8:31–36

Köhler HR, Triebskorn R (2013) Wildlife ecotoxicology of pesticides: can we track effects to the population level and beyond? Science 341:759

Lanctot RB, Blanco DE, Dias RA, Isacch JP, Gill VA, Almeida JB, Delhey K, Petracci PF, Bencke GA, Balbueno R (2002) Conservation status of the buff-breasted sandpiper: historic and contemporary distribution and abundance in South America. Wilson Bull 114:44–47

Lawson LP, Fessl B, Vargas FH, Farrington HL, Cunninghame HF, Mueller JC, Nemeth E, Sevilla PC, Petren K (2017) Slow motion extinction: inbreeding, introgression, and loss in the critically endangered mangrove finch (*Camarhynchus heliobates*). Conserv Genet 18(1):159–170

Lees AC, Moura NG, Santana A, Aleixo A, Barlow J, Berenguer E, Ferreira J, Gardner TA (2012) Paragominas: a quantitative baseline inventory of an eastern Amazonian avifauna. Rev Bras Ornitol 20:93–118

Lees AC, Thompsom I, Moura NG (2014) Salgado Paraense: an inventory of a forgotten coastal Amazonian avifauna. Bol Mus Paraense Emílio Goeldi Cienc Natur 9:135–168

Lefebvre G, Poulin B (1997) Bird communities in Panamanian black mangroves: potential effects of physical and biotic factors. J Trop Ecol 13:97–113

Lima PC (2006) Aves do litoral norte da Bahia. Atualidades Ornitológicas 134:29

Lima LM, Schunck F, Siciliano S, Carlos CJ, Rennó B, Fonseca-Neto FP, Fedrizzi CE, Albano C, Moura JF (2010) Distribuição, abundância e sazonalidade de *Leucophaeus atricilla* (Charadriiformes: Laridae) no Brasil. Ararajuba 18:199–206

Linneweber V, Lacerda LD (2002) Mangrove ecosystems: function and management. Springer, Berlin

Lloyd JD, Doyle T (2011) Abundance and population trends of mangrove landbirds in southwest Florida. J Field Ornithol 82(2):132–139

Luther DA, Greenberg R (2009) Mangroves: a global perspective on the evolution and conservation of their terrestrial vertebrates. BioScience 59:602–612

Luther DA, Greenberg R (2011) The island syndrome in coastal wetland ecosystems: convergent evolution of large bills in mangrove passerines. Auk 128:201–204

Luther DA, Greenberg R (2014) Habitat type and ambient temperature contribute to bill morphology. Ecol Evol 4(5):699–705

Maciel E (2009) Aves do município do Rio de Janeiro. Technical Books Editora, Rio de Janeiro, 407 pp

Magris RA, Barreto R (2010) Mapping and assessment of protection of mangrove habitats in Brazil. Pan Am J Aquat Sci 5(4):546–556

Mancini PL, Reis-Neto A, Fischer LG, Silveira LF, Schaeffer Novelli Y (2018) Differences in diversity and habitat use of avifauna in distinct mangrove areas in São Sebastião. Ocean Coast Manage 164:79–91

Marangoni JC, Costa CSB (2009a) Diagnóstico ambiental das marismas no estuário da Lagoa dos Patos – RS. Atlântica 31(1):85–98

Marangoni JC, Costa CSB (2009b) Natural and anthropogenic effects on salt marsh over five decades in the Patos Lagoon (Southern Brazil). Braz J Oceanogr 57(4):345–350

Marcondes RS, Del-Rio G, Rego MA, Silveira LF (2014) Geographic and seasonal distribution of a little-known Brazilian endemic rail (*Aramides mangle*) inferred from occurrence records and ecological niche modeling. Wilson J Ornithol 126:663–672

Martinez C (2010) Trophic niche breadth and overlap of three egret species in a Neotropical mangrove swamp. Waterbirds 33:285–292

Martinez C, Rodrigues AAF (1999) Breeding biology of the Scarlet Ibis on Cajual Island, Northern Brazil. J Field Ornithol 70:558–566

Martinez–Curci NS, Azpiroz A, Gianuca A, Gianuca D, Simpson RE, Dias RA (2014) Willet (*Tringa semipalmata*) status update in southeastern South America. Ornitol Neotrop 25:135–144

Martuscelli P (1995) Ecology and conservation of the Red-tailed Amazon *Amazona brasiliensis* in south-eastern Brazil. Bird Conserv Int 5:225–240

Maurício GN, Dias RA (1996) Novos registros e extensões de distribuição de aves palustres e costeiras no litoral sul do Rio Grande do Sul. Ararajuba 4(1):45–71

Maurício GN, Dias RA (2000) New distributional information for birds in southern Rio Grande do Sul, Brazil, and the first record of the Rufous Gnateater *Conopophaga lineata* for Uruguay. Bull Br Ornithol Club 118:14–25

Maurício GN, Bencke GA, Repenning M, Machado DB, Dias RA, Bugoni L (2013) Review of the breeding status of birds in Rio Grande do Sul, Brazil. Iheringia Ser Zool 103(2):163–184

Melo DC, Ferreira AP, Leal LGPS, Souza EA, Loures-Ribeiro A, Sousa AEBA (2014) Distribution of the Orange-winged Parrot *Amazona amazonica* (Linnaeus, 1766) (Aves, Psittacidae) in the coastal region of the State of Paraíba, Brazil. Acta Sci Biol 36(3):307–312

Menghini RP, Rovai AS, Almeida R, Coelho-Junior C, Schaeffer-Novelli Y (2018) Restauração ecológica de manguezais. In: Instituto Chico Mendes de Conservação da Biodiversidade (ICMBio) Atlas dos Manguezais do Brasil. ICMBio, Brasília, pp 95–104

Mestre LAM, Krul R, Moraes V (2007) Mangrove bird community of Paranaguá Bay – Paraná, Brazil. Braz Arch Biol Techn 50:75–83

Ministério do Meio Ambiente – MMA 2022. Altera os Anexos da Portaria nº 443, de 17 de dezembro de 2014, da Portaria nº 444, de 17 de dezembro de 2014, e da Portaria nº 445, de 17 de dezembro de 2014, referentes à atualização da Lista Nacional de Espécies Ameaçadas de Extinção. Diário Oficial da União. 148. Seção 1. Publicado em 07/06/2022. Ministério do Meio Ambiente, Brasília. https://www.icmbio.gov.br/cepsul/destaques-e-eventos/704-atualizacao-da-lista-oficial-das-especies-ameacadas-de-extincao.html. Acesso em 14 de novembro 2022.

Ministério do Meio Ambiente – MMA (2016) Portaria N° 77, de 27 de julho de 2016. Atualiza e aprova o Plano de Ação Nacional para a Conservação das Aves Limícolas Migratórias – PAN Aves Limícolas Migratórias, contemplando cinco táxons ameaçados de extinção, estabelecendo seu objetivo geral, objetivos específicos, prazo de execução, abrangência e formas de implementação e supervisão (Processo SEI n° 02061.000023/2012-77). Diário Oficial da União 145, Brasília. http://www.icmbio.gov.br/portal/faunabrasileira/plano-de-acao-nacional-lista/3567-plano-de-acao-nacional-para-conservacao-das-aves-limicolas. Accessed 28 Mar 2018

Mohd–Azlan J, Noske RA, Lawes MJ (2015) The role of habitat heterogeneity in structuring mangrove bird assemblages. Diversity 7:118–136

Mohr LV, Efe MA, Bugoni L (2008) *Thalasseus maximus* (Boddaert, 1783). In: Machado ABM, Drummond GM, Paglia AP (eds) Livro vermelho da fauna brasileira ameaçada de extinção. Ministério do Meio Ambiente, Brasília and Belo Horizonte, pp 448–450

Morrison RIG, Ross RK (1989) Atlas of Nearctic shorebirds on the coast of South America, vol 2. Canadian Wildlife Service Special Publication, Ottawa

Morrison RIG, Ross RK, Niles LJ (2004) Declines in wintering populations of red knots in southern South America. Condor 106(1):60–70

Morrison RIG, Mizrahi DS, Ross RK, Ottema OH, de Pracontal N, Narine A (2012) Dramatic declines of Semipalmated Sandpipers on the Major wintering areas in the Guianas, Northern South America. Waterbirds 35(1):120–134

Nagelkerken I, Blaber SJM, Bouillon S, Green P, Haywood M, Kirton LG, Meynecke JO, Pawlik J, Penrose HM, Sasekumar A, Somerfield PJ (2008) The habitat function of mangroves for terrestrial and marine fauna: a review. Aquat Bot 89:155–185

Naka LN, Rodrigues M, Roos AL, Azevedo MAG (2002) Bird conservation on Santa Catarina Island. Bird Conserv Int 12:123–150

Nascimento JLX (2011) Parque Nacional da Lagoa do Peixe. In: Valente RM, Silva JMC, Straube FC, Nascimento JLX (eds) Conservação de Aves Migratórias Neárticas do Brasil, 1st edn. Conservation International, Belém, pp 321–324

Nascimento DM, Alves AGC, Alves RRN, Barboza RRD, Diele K, Mourão JS (2017) An examination of the techniques used to capture mangrove crabs, *Ucides cordatus*, in the Mamanguape River estuary, northeastern Brazil, with implications for management. Ocean Coast Manage 130:50–57

Navedo JG, Hahn S, Parejo M, Abad–Gómez JM, Gutiérrez JS, Villegas A, Sánchez–Guzmán JM, Masero JA (2015) Unravelling trophic subsidies of agroecosystems for biodiversity conservation: food consumption and nutrient recycling by waterbirds in Mediterranean rice fields. Sci Total Environ 511:288–297

Noske RA (1995) Ecology of mangrove forest birds in peninsular Malaysia. Ibis 137:250–263

Noske RA (1996) Abundance, zonation, and foraging ecology of birds in mangroves of Darwin Harbour, Northern Territory. Wildlife Res 1996:443–474

Nunes FC, Barbosa-Filho RC, Roos A, Mestre LAM (2010) The Cattle Egret (*Bubulcus ibis*) on Fernando de Noronha Archipelago: history and population trends. Ararajuba 18:1–13

Olmos F (2000) Dieta e biologia reprodutiva de *Eudocimus ruber* e *Egretta caerulea* (Aves, Ciconiiformes) nos manguezais de Santos-Cubatão, São Paulo. PhD dissertation, Universidade Estadual Paulista

Olmos F, Silva e Silva R (2001) The avifauna of a southeastern Brazilian mangrove swamp. Int J Ornithol 4(3/4):135–205

Olmos F, Silva e Silva R (2003) Guará: ambiente, flora e fauna dos manguezais de Santos-Cubatão. Empresa das Artes, São Paulo

Olmos F, Silva e Silva R, Prado A (2001) Breeding season diet of Scarlet Ibises *Eudocimus ruber* and Little Blue Herons *Egretta caerulea* in a Brazilian mangrove. Waterbirds 24:50–57

Olmos F, Rumsey S, Brickle N (2013) First documented record of Grey Kingbird, *Tyrannus dominicensis* (Passeriformes: Tyrannidae) in Brazil. Rev Bras Ornitol 21:133–135

Onuf CP, Teal JM, Valiela I (1977) Interactions of nutrients, plant growth, and herbivory in a mangrove ecosystem. Ecology 58:514–526

Pacheco JF (1994) O interessante gavião-asa-de-telha (*Parabuteo unicintus*) no Brasil: um gavião raro? Atualidades Ornitológicas 61:13

Pacheco JF, Silveira LF, Aleixo A, Agne CE, Bencke GA, Bravo G, Brito GRR, Cohn-Haft M, Maurício G, Naka LN, Olmos F, Posso S, Lees AC, Figueiredo LF, Carrano E, Guedes RC, Cesari E, Franz I, Schunck F, Piacentini VQ (2021) Annotated checklist of the birds of Brazil by the Brazilian Ornithological Records Committee, 2nd ed. Ornithol Res 29:94–105

Pereira GA, Whittaker A, Whitney BM, Zimmer KJ, Dantas SM, Roda AS, Bevier LR, Coelho G, Hoyer RC, Albano C (2008) Novos registros de aves para Pernambuco, Brasil, com notas sobre algumas espécies pouco conhecidas no Estado. Ararajuba 16:47–53

Pereira GA, Lobo-Araújo LW, Leal S, Medcratf J, Marantz CA, Toledo MT, Araújo HF, Albano C, Pinto T, Santos CHA, Serapião LCH, Silva GBM, Pioli D (2012) Important bird records from Alagoas, Pernambuco and Paraíba, north-east Brazil. Cotinga 34:17–22

Pereira GA, Rodrigues PP, Leal S, Periquito MC, Silva GBM, Menezes M, Correa GS, Sonntag FA, Almeida AEMN, Nunes PB (2014) Important bird records from Alagoas, Pernambuco and Paraíba, north-east Brazil, vol 2. Cotinga 36:45–50

Periquito MC, Pereira GA, Brito MT (2008) Aves no manguezal do Espaço Ciência, Olinda, Pernambuco. Atualidades Ornitológicas 145:36–38

Piacentini VQ, Ghizoni-Jr IR, Azevedo MAG, Carrano E, Borchardt-Jr CA, Amorim JF, Grose AV (2009) Ocorrência, expansão e distribuição do maçarico-de-cara-pelada *Phimosus infuscatus* (Lichtenstein, 1823) (Ciconiiformes: Threskiornithidae) no Estado de Santa Catarina, sul do Brasil. Rev Bras Ornitol 17:107–112

Polidoro BA, Carpenter KE, Collins L, Duke NC, Ellison AM, Ellison JC, Farnsworth EJ, Fernando ES, Kathiresan K, Joedam NE, Livingstone SR, Miyagi T, Moore GE, Nam VN, Ong JE, Primavera JH, Salmo SG III, Sanciangco JC, Sukardjo S, Wang Y, Yong WHY (2010) The loss of species: mangrove extinction risk and geographic areas of global concern. PLoS One 5(4): 1–10

Powell AN (2006) Are southern California's fragmented saltmarshes capable of sustaining endemic bird populations? Stud Avian Biol-Ser 32:198–204

Quintela FM, Santos MB, Oliveira SV, Costa RC, Christoff AU (2010) Javalis e porcos ferais (Suidae, *Sus scrofa*) na Restinga de Rio Grande, RS, Brasil: ecossistemas de ocorrência e dados preliminares sobre impactos ambientais. Neotropical Biology and Conservation 5(3):172–178

Resende SL, Leeuwenberg F (1987) Ecological studies of Lagoa do Peixe, Final report to World Wildlife Fund (WWF). WWF, Washington, DC

Restall R, Rodner C, Lentino M (2006) Birds of Northern South America: an identification guide. Volume 1: species accounts. Yale University Press, New Haven/London

Ridgely RS, Tudor G (1989) The birds of South America, vol I. The oscine passerines. University of Texas, Austin

Ridgely RS, Tudor G (1994) The birds of South America, vol II. The suboscine passerines. University of Texas, Austin

Robbins M (2018) Northern Scrub-flycatcher (*Sublegatus arenarum*). In: del Hoyo J, Elliott A, Sargatal J, Christie DA, de Juana E (eds) Handbook of the birds of the World Alive. Lynx Edicions, Barcelona

Roda AS, Pereira GA (2006) Distribuição recente e conservação das aves de rapina florestais do Centro Pernambuco. Ararajuba 14:331–344

Rodrigues AAF (1995) Ocorrência da reprodução de *Eudocimus ruber* na Ilha do Cajual, Maranhão, Brasil (Ciconiiformes: Threskiornithidae). Ararajuba 3:67–68

Rodrigues AAF (2007) Priority areas for conservation of migratory and resident waterbirds on the coast of Brazilian Amazonia. Rev Bras Ornitol 15:209–218

Rodrigues AAF, Carvalho DL (2011) Reentrâncias Paraenses. In: Valente RM, Silva JMC, Straube FC, Nascimento JLX (eds) Conservação de aves migratórias neárticas no Brasil. Conservação Internacional Brasil, Belém, pp 85–87

Rodrigues RC, Araújo HFP, Lyra-Neves RM, Telino-Jr WR, Botelho MCN (2007) Caracterização da Avifauna na Área de Proteção Ambiental de Guadalupe, Pernambuco. Ornithologia 2:47–61

Romanach SS, DeAngelis DL, Koh HL, Li Y, Teh SY, Barizan RSR, Zhai L (2018) Conservation and restoration of mangroves: global status, perspectives, and prognosis. Ocean Coast Manage 154:72–82

Rovai AS, Soriano-Sierra EJ, Pagliosa PR, Cintrón G, Schaeffer-Novelli Y, Menghini RP, Coelho-Jr C, Antunes-Horta P, Lewis RR, Simonassi JC, Andrade JÁ, Boscatto F, Dutra SJ (2012) Secondary succession impairment in restored mangroves. Wetl Ecol Manag 20(5):447–459

Rupp AE, Fink D, Silva GT, Zermiani M, Laps RR, Zimmermann CE (2008) Novas espécies de aves para o Estado de Santa Catarina, sul do Brasil. Biotemas 21:165–169

Ruschi PA, Simon JE (2007) Primeiro registro de *Agyrtria leucogaster* (Gmelin, 1788) (Aves: Trochilidae) para o Estado do Espírito Santo, Brasil. Ararajuba 15:451–452

Sandilyan S, Katherisan K (2012) Mangrove conservation: a global perspective. Biodivers Conserv 21(14):3523–3542

Santos MPD (2011) Cajueiro da Praia. In: Valente RM, Silva JMC, Straube FC, Nascimento JLX (eds) Conservação de Aves Migratórias Neárticas no Brasil, 1st edn. Conservation International, Belém, pp 125–128

Scarton F (2018) Disturbance of non–breeding waders by pedestrians and boats in a Mediterranean lagoon. Ardeola 65(2):209–220

Schaeffer Novelli Y, Soriano–Sierra EJ, Vale CC, Bernini E, Rovai AS, Pinheiro MAA, Schmidt AJ, Almeida R, Coelho-Jr C, Menghini RP, Martinez DI, Abuchahla GMO, Cunha–Lignon M, Charlier–Sarubo S, Shirazawa–Freitas J, Cintrón–Molero G (2016) Climate changes in mangrove forests and salt marshes. Braz J Oceanogr 64:37–52

Schories D, Barletta–Bergan A, Barletta M, Krumme U, Mehlig U, Rademaker V (2003) The keystone role of leaf–removing crabs in mangrove forests of North Brazil. Wetl Ecol Manag 11: 243–225

Schunck F, Rodrigues AAF (2018) *Calidris canutus* (Linnaeus, 1758). In: Ministério do Meio Ambiente (org) Livro Vermelho da Fauna Brasileira Ameaçada de Extinção, 1st edn. MMA, Brasília, pp 145–148

Sekercioglu CH, Primack RB, Wormworth J (2012) The effects of climate change on tropical birds. Biol Conserv 148:1–18

Serafini PP (2013) Plano de Ação Nacional para Conservação dos Passeriformes Ameaçados dos Campos Sulinos e Espinilho. ICMBio, Brasília

Shriver WG, Hodgman TP, Gibbs JP, Vickery PD (2004) Landscape context influences salt marsh bird diversity and area requirements in New England. Biol Conserv 119:545–553

Sick H (1997) Ornitologia Brasileira, 2nd edn. Nova Fronteira, Rio de Janeiro

Sick H (2001) Ornitologia Brasileira, 3rd edn. Nova Fronteira, Rio de Janeiro

Sigrist T (2006) Aves do Brasil: uma visão artística. Fosfertil, Uberaba

Silva e Silva R (2007) Guarás Vermelhos do Brasil. Avis Brasilis, Vinhedo

Silva e Silva R (2008) Aves de Fernando de Noronha. Avis Brasilis, Vinhedo

Silva e Silva R, Olmos F (2002) Osprey ecology in the mangroves of southeastern Brazil. J Raptor Res 36:328–331

Silva e Silva R, Olmos F (2007) Adendas e registros significativos para a avifauna dos manguezais de Santos-Cubatão, SP. Ararajuba 15:551–560

Silva JMC, Oren DC, Roma JC, Henriques LMP (1997) Composition and distribution patterns of the avifauna of an Amazonian upland savanna, Amapá, Brazil. Ornithol Monogr 48:743–762

Sipinski EAB, Abbud MC, Sezerban RM, Serafini PP, Boçon R, Manica LT, Guaraldo AC (2014) Tendência populacional do papagaio-de-cara-roxa (*Amazona brasiliensis*) no litoral do estado do Paraná. Ornithologia 6:136–143

Smit CJ, Visser GJM (1993) Effects of disturbance on shorebirds: a summary of existing knowledge from the Dutch Wadden Sea and Delta area. Wader Study Group Bull 68:6–19

Somenzari M, Amaral PP, Cueto VR, Guaraldo AC, Jahn AE, Lima DM, Lima PC, Lugarini C, Machado CG, Martinez J, Nascimento JLX, Pacheco JF, Paludo D, Prestes PN, Serafini PF, Silveira LF, Souza AEBAS, Souza NA, Souza MA, Telino-Jr WR, Whitney BM (2018) An overview of migratory birds in Brazil. Pap Avulsos Zool 58:e20185803

Somveille M, Manica A, Rodrigues AS (2018) Where the wild birds go: explaining the differences in migratory destinations across terrestrial bird species. Ecography 41:1–12

Souza AP, Rodrigues AAF (2015) Censo populacional do maçarico-de-peito-vermelho *Calidris canutus rufa* na praia de Panaquatira, Maranhão, Brasil, em um ciclo anual. Ornithologia 8:33–37

Souza EA, Nunes MFC, Roos AL, Araújo HFP (2008) Aves do Parque Nacional do Cabo Orange, Amapá: Guia de Campo. ICMBio/Cemave, Brasília

Stotz DF, Fitzpatrick JW, Parker TA III, Moskovits DK (1996) Neotropical birds: ecology and conservation. University of Chicago Press, Chicago, 478 pp

Taylor B (2018) Rufous-necked wood-rail (*Aramides axillaris*). In: del Hoyo J, Elliott A, Sargatal J, Christie DA, de Juana E (eds) Handbook of the birds of the World Alive. Lynx Edicions, Barcelona

Valente RM, Silva JMC, Straube FC, Nascimento JLX (2011) Conservação de aves migratórias neárticas no Brasil. Conservation International, Belém

Valiela I, Bowen JL, York JK (2001) Mangrove forests: one of the world's threatened major tropical environments: at least 35% of the area of mangrove forests has been lost in the past two decades, losses that exceed those for tropical rain forests and coral reefs, two other well-known threatened environments. BioScience 51(10):807–815

Vieira BP (2015) Population trends and conservation of the Mangrove Rail. Rev Bras Ornitol 23: 327–335

Wiersma P, Kirwan GM, Boesman P (2018) Wilson's Plover. In: del Hoyo J, Elliott A, Sargatal J, Christie DA, de Juana E (eds) Handbook of the birds of the World Alive. Lynx Edicions, Barcelona

WIKIAVES - A Enciclopédia das Aves do Brasil. Avalilable: https://www.wikiaves.com.br/index.php. Acess em: 11 nov. 2021. eBird - https://ebird.org/home Cornell Lab of Ornithology, Ithaca, New York. Xeno-Canto - https://xeno-canto.org/ Taxeus - https://www.taxeus.com.br/

Willis EO, Oniki Y (2003) Aves do Estado de São Paulo. Editora Divisa, Rio Claro, 398 pp

Winkler H, Christie DA, Sharpe CJ, Motis A (2018) Arrowhead Piculet (*Picumnus minutissimus*). In: del Hoyo J, Elliott A, Sargatal J, Christie DA, de Juana E (eds) Handbook of the birds of the World Alive. Lynx Edicions, Barcelona

Wrona FJ, Prowse TD, Reist JD, Hobbie JE, Lévesque LMJ, Vincent WF (2006) Climate change effects on aquatic biota, ecosystem structure and function. AMBIO 35(7):359–369

Xavier BF, Boss RL (2011) Estação Ecológica Maracá-Jipioca. In: Valente RM, Silva JMC, Straube FC, Nascimento JLX (eds) Conservação de aves migratórias neárticas no Brasil. Conservation International, Belém, pp 28–32

Chapter 13
Flagship Species: Manatees as Tools for Mangrove Conservation in Northeast Brazil

Iran C. Normande, Alexandra F. Costa, Clemente Coelho-Jr, José Ulisses dos Santos, and Richard J. Ladle

13.1 Introduction

Conservation actions, such as creating a new protected area or reintroducing an endangered species, are the expression of people's desire to preserve the elements of the natural world that they value (Ladle et al. 2011). The continued success of conservation, therefore, depends to a greater or lesser degree on public support (Kareiva and Marvier 2012). Such support is motivated by a range of factors, including the desire to preserve/exploit the economic value of nature, moral concerns about the treatment of sentient animals, or appreciation of nature's aesthetics (reviewed in Newman et al. 2017). Clearly, not all species or landscapes can

I. C. Normande (✉)
Instituto Chico Mendes de Conservação da Biodiversidade, Porto de Pedras, AL, Brazil

Instituto de Ciências Biológicas e da Saúde, Universidade Federal de Alagoas, Maceió, AL, Brazil
e-mail: iran.normande@icmbio.gov.br

A. F. Costa
Instituto Chico Mendes de Conservação da Biodiversidade, Porto de Pedras, AL, Brazil

Instituto Bicho D'água: Conservação Socioambiental, Belém, PA, Brazil

C. Coelho-Jr
Instituto de Ciências Biológicas, Universidade de Pernambuco, Recife, PE, Brazil

Instituto BiomaBrasil, Recife, PE, Brazil

J. U. dos Santos
Instituto Chico Mendes de Conservação da Biodiversidade, Porto de Pedras, AL, Brazil

R. J. Ladle
Instituto de Ciências Biológicas e da Saúde, Universidade Federal de Alagoas, Maceió, AL, Brazil

Y. Schaeffer-Novelli et al. (eds.), *Brazilian Mangroves and Salt Marshes*, Brazilian Marine Biodiversity, https://doi.org/10.1007/978-3-031-13486-9_13

mobilize similar levels of public support or affection, leading conservationists to foreground certain characteristics of biodiversity depending on the conservation outcomes they want to achieve.

Species that are particularly well suited to mobilize public support, raise awareness, and stimulate conservation actions are often referred to as "flagship species" (Heywood 1995; Verissimo et al. 2011). These species often share traits (e.g., large size, charisma, distinct physical appearance), though their choice as focal points for conservation initiatives ultimately depends on the specific conservation objectives (Verissimo et al. 2011). This principle was formalized by Barua et al. (2011) who identified a suite of ecological and cultural traits of flagship species associated with seven different types of conservation action (Table 13.1).

It becomes clear from Table 13.1 that a single flagship species may possess traits that predispose it to be used in several types of conservation strategies. Here, we argue that this is the case of the Antillean manatee (*Trichechus manatus manatus*) in Brazil, a large and charismatic aquatic mammal whose continued existence critically depends on the conservation of the highly threatened mangrove habitats that it uses. In the following sections, we will present a case for the Antillean manatee as a flagship species for mangrove conservation in Brazil, highlighting its symbolic value and the multiple conservation actions that it can support.

13.2 The West Indian Manatee (*Trichechus manatus*): Biogeography, Ecology, and Cultural Value

The West Indian manatee (*Trichechus manatus* Linnaeus 1785) is an aquatic mammal that inhabits tropical and subtropical areas of the Western Atlantic. It occurs in bays, lagoons, and estuaries (Folkens and Reeves 2002), ranging from Rhode Island, in the USA, to Alagoas in Northeastern Brazil (Albuquerque and Marcovaldi 1982) (see Chap. 3, Maps 1–9). Two subspecies of the West Indian manatee are currently recognized: the Florida manatee (*Trichechus manatus latirostris*) and the Antillean manatee (*Trichechus manatus manatus*) (Committee on Taxonomy 2016). The distribution of the latter stretches from the east coast of Mexico and Central America to the northern and northeastern coasts of South America and the Caribbean Sea (Lefebvre et al. 1989).

The distribution of Antillean manatees on the Brazilian coast is highly fragmented, with a particularly large gap between the subpopulation in the north of Alagoas State and south of Pernambuco State and the one in the west of Ceará and Maranhão states (Albuquerque and Marcovaldi 1982; Borobia and Lodi 1992; Luna et al. 2008) (Fig. 13.1). At one time manatees were relatively abundant along the Brazilian coast (Fig. 13.1), ranging as far south as Espírito Santo State (Whitehead 1977). However, a small population size and long periods of isolation have caused low genetic diversity among extant populations (García-Rodríguez et al. 1998; Vianna et al. 2006; Luna et al. 2012).

West Indian manatees are habitat generalists, occurring in lakes, rivers, estuaries, and shallow coastal waters where they feed on a wide variety of submerged, floating

Table 13.1 Criteria for selecting flagship species according to their intended uses

Criteria	Conservation awareness	Fund raising	Promoting ecotourism	Community-based conservation	Promotion of research	Species/habitat protection	Influencing policy
Geographic location	•		•	•		•	
Conservation status	•	•			•	•	•
Umbrella effect					•	•	
Represents other species	•					•	•
Recognizable	•	•					
Easily observed	•	•					
Physical appearance/ special traits	•	•	•				
Positive cultural significance		•		•			•
Traditional knowledge				•			
Economic value				•			
Scientific value					•		
Charisma	•	•	•				
Existing usage	•						•

Adapted from Barua et al. (2011)

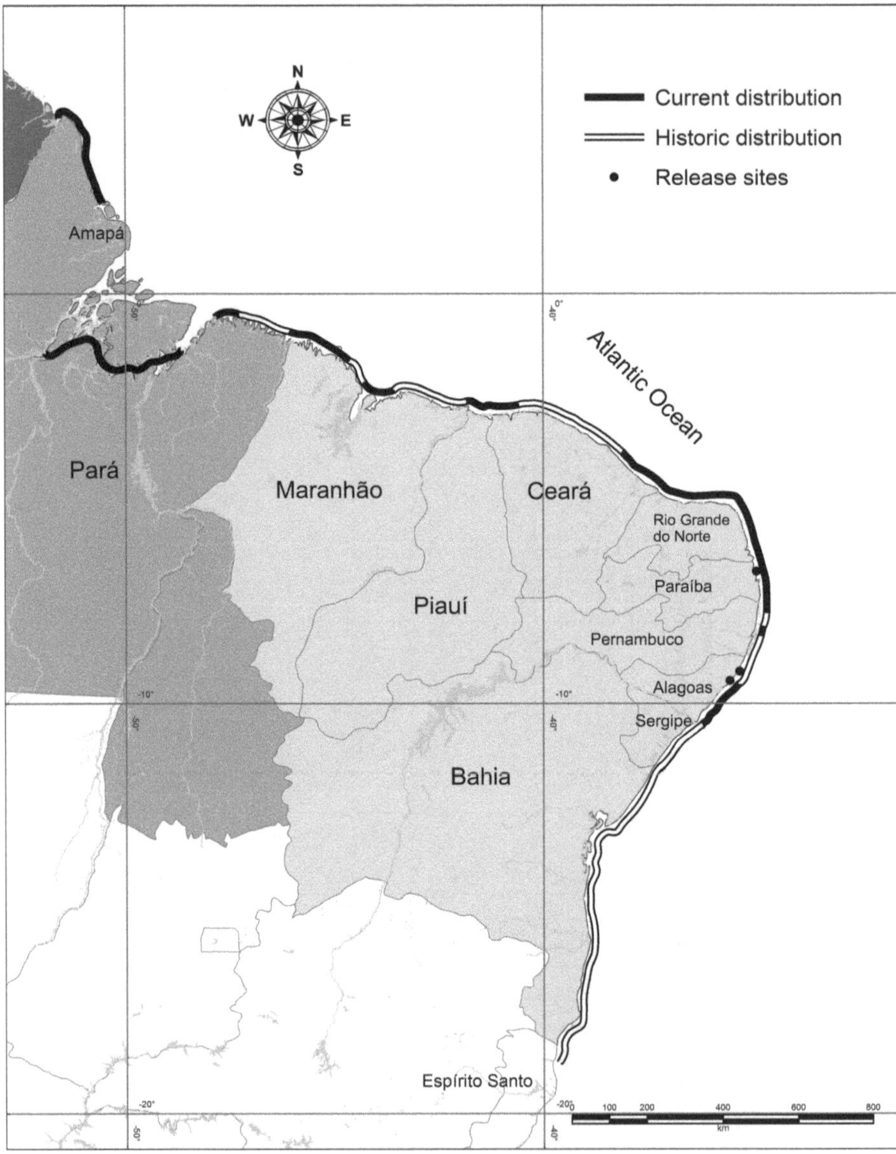

Fig. 13.1 Historical and current distribution of *T. manatus manatus* in Brazil. (Adapted from Normande et al. 2015). See Chap. 3, Maps: 1 Amapá, 2 Pará, 3 Maranhão, 4 Piauí, 5 Ceará, 6 Rio Grande do Norte, 7 Paraíba, 8 Pernambuco, 9 Alagoas, 10 Sergipe, 11 Bahia, and 12 Espírito Santo

and emergent vegetation (Fig. 13.2), including mangrove leaves (Spiegelberger and Ganslosser 2005) and roots (Normande, pers. obs.). Being mainly herbivorous, they need to spend between 6 and 8 h per day foraging (Marsh et al. 2011; Allen et al. 2017). A study conducted in Rio Grande do Norte, Paraíba, and Alagoas states (Borges et al. 2008) identified 17 species of macroalgae consumed by manatees,

Fig. 13.2 Manatees in a natural environment on the Tatuamunha River feeding on mangrove leaves, Porto de Pedras, Alagoas State, Northeast Brazil. (Photo: Clemente Coelho Jr) (see Chap. 3, Map 9)

including red algae, two species of marine phanerogams (i.e., *Halodule wrightii* and *Halophila* sp.), as well as cnidarians. Similarly, on the northern and northeast coasts of Brazil, manatees have been observed to consume the plants *Montrichardia arborescens*, *Spartina brasiliensis*, *Eichornia crassipes*, *Eleocharis* spp., *Crenea maritima*, *Cyperus* spp., and *Blutaparon portulacoides*, and the leaves of mangrove species *Avicennia* spp., *Laguncularia racemosa*, and *Rhizophora mangle* (Borges et al. 2008; Lins et al. 2014). There are even reports of manatees consuming fish, illustrating the opportunistic nature of the diet of this species (Sousa et al. 2013; Meirelles and Carvalho 2016).

As might be expected with such a generalist feeder, feeding preference appears to be very variable in time and space and is related to availability, nutritional value, and palatability of different types of vegetation (Meirelles et al. 2018). Even when mangrove vegetation does not make up a substantial proportion of their diet, manatees still frequently enter mangrove-lined estuaries and bays in search for food and freshwater (Normande et al. 2015). This is confirmed by stable isotope analysis of manatees from the north and northeast of Brazil which indicated that individuals predominantly graze in estuarine and freshwater environments (Ciotti et al. 2014). More generally, there is a large overlap in the historical distribution of mangroves and manatees in Brazil, with the former dominating coastal habitats as far south as Santa Catarina State and the latter being limited to the southern coast of Alagoas State nowadays (Fig. 13.1).

Environmental factors such as water temperature, depth, hydrological cycle, and proximity of freshwater sources can influence sirenian distributions within and between habitats (Irvine 1983; Reid et al. 1991; Oliveira-Gómez and Mellink 2005; Sheppard et al. 2006; Castelblanco-Martínez et al. 2009). In Florida, seasonal fluctuations in water temperature play an important role in determining habitat use of manatees, since they use warm water sites during winter (Whitehead 1977; Irvine 1983; Reid et al. 1991). However, in tropical and subtropical areas such as northeast Brazil, water temperature shows far less variability and is therefore unlikely to strongly influence manatee habitat use (Deutsch et al. 2003).

Despite more stable water temperatures in the tropics, seasonal migrations have been observed in Antillean manatees in Mexico (Colmenero-Rolón and Hoz-Zavala 1986), Honduras (Rathbun et al. 1983), and Trinidad (Reynolds III and Odell 1991). Reeves et al. (1988) also observed seasonal migrations in African manatees (*Trichechus senegalensis*), and Best (1983) and Arraut et al. (2017) recorded the same behavior in Amazonian manatees (*Trichechus inunguis*). In the latter cases, the populations inhabit freshwater systems far from the coast, and seasonal migration was associated with fluctuations in food availability and habitat accessibility caused by seasonal fluctuations in water level (Deutsch et al. 2003).

Coastal populations of Antillean manatees may show smaller-scale variation in habitat use. For example, Normande et al. (2015) observed more intensive use of estuaries than marine environments using radiotelemetry data from 21 reintroduced manatees in northeast Brazil. This pattern may be related to the increased concentration of freshwater sources in estuaries, though it may also be related to the presence of soft-release (acclimatization) enclosures that the manatees may associate with being fed as released manatees tend to spend some time using the area around the acclimatization facilities (Fig. 13.2). In Alagoas, Pernambuco, and Paraíba states, freshwater sources available to the manatees are mainly concentrated in rivers. This is somewhat different from Ceará and Rio Grande do Norte states, where freshwater springs in the sea floor are more abundant. Consequently, the manatees in the southernmost subpopulation are more dependent on estuaries and mangroves and more frequently observed in these ecosystems.

Mangrove forests provide an abundance of water sources for manatees, from natural springs to runoff from leaves and roots. They are also very effective at protecting the coastline from erosion and trapping sediment (Almeida et al. 2008). This, in turn, prevents the estuaries from the worst effects of sedimentation and ensures a sufficient water depth for manatees to access a large proportion of the habitat. It is also important to note that despite the high abundance of manatee food sources on the shallow inshore reefs (which act as excellent substrates for algae fixation), many of these areas are only accessible during high tides. This limited access may partly explain the relatively low frequency of observation of utilization of reefs in northeast Brazil (Normande et al. 2015). Again, this contrasts with Rio Grande do Norte State, where Paludo and Langguth (2002) noted that manatees predominantly use reefs that are densely colonized by algae.

Mangrove ecosystems may also act as important nursery habitats for manatees. In northeast Brazil, manatee calves have been observed in the estuary of the Maracaípe River in Pernambuco (Lima et al. 2005) and the Timonha-Ubatuba complex on the border between the states of Ceará and Piauí (Magnus Machado Severo, pers. comm.) (see Chap. 3, Map 4). The coasts of the neighboring states of Rio Grande do Norte and Ceará are characterized by high rates of neonate stranding (Balensiefer et al. 2017). This is generally attributed to the degraded state of the local estuaries, with heavily silted rivers restricting the access of pregnant females into the estuaries and forcing them to give birth in the open sea (Meirelles 2008).

13.3 Manatee Conservation: Threats and Actions

The West Indian manatee is classified as vulnerable by IUCN (2019). In Brazil, the situation is more alarming, and the species is considered endangered by the federal government (Luna et al. 2018). The entire Brazilian population has been estimated at only 500 individuals based on questionnaires with fishermen and coastal residents (Lima 1999; Luna 2001). IUCN (2012) suggests that this number may be as low as 200, although they did not provide details on the methods used to estimate population size. It should be noted, however, that despite their large size, manatees are very difficult to survey (as are most marine mammals). Indirect population estimates based on extrapolation of genetic data (Luna et al. 2012) suggest that the Brazilian population of *T. manatus* could have as high as 1,000 individuals.

The most recent direct population estimate of Brazilian manatees was made by aerial surveys and covered more than 1500 km of coast, from the border between the states of Alagoas and Sergipe to the border between the states of Piauí and Maranhão (Alves et al. 2015). This study estimated an average of 1104 individuals along the surveyed coast, although the data is likely to cause an underestimation due to the low detectability of manatees in locations with turbid waters such as estuaries and within mangroves. The highest density of individuals was observed in the estuary complex of Timonha-Ubatuba and Cardoso-Camurupim Rivers, formed by a group of islands with well-preserved estuaries and bays, with five animals found within the estuarine complex (Alves et al. 2015). This result confirms the enormous importance of protected areas for the Brazilian manatee population, especially those that are large enough to protect one or more estuarine complexes in their totality such as the Delta do Parnaíba Environmental Protected Area (Maranhão, Piauí, and Ceará States) and the Costa dos Corais Environmental Protected Area (Pernambuco and Alagoas States) (see Chap. 3, Maps 4 and 8, respectively).

Although there is a lack of baseline data, there are good reasons to believe that the Brazilian manatee population was orders of magnitude larger in the precolonial period. Such large, docile animals were easy targets for predatory hunting and there was a ready market for manatee meat, skin, and oil during the early colonization of the country (ICMBio 2011). Manatee hunting is now almost nonexistent on the northeast coast, although it is still practiced in the north of Brazil where it may account for as much as 86% of recorded mortalities (Luna 2001). Other threats to manatees include accidental death by getting entangled in fishing gear (Parente et al. 2004; Meirelles 2008) or by collisions with motorized boats (Borges et al. 2007).

Habitat loss is also a major threat to extant populations. Loss of mangroves and degradation of estuaries may be especially important since these provide manatees with clean water to drink and calm conditions to feed and reproduce. Mangroves are particularly vulnerable to the deleterious effects of the growth of socioeconomic activities and the disorderly expansion of urban centers, which impose severe changes on the quality of estuarine waters, bays, lagoons, and coastal lagoons (see Chap. 16). The recent approval of the New Forest Code, Federal Law no. 12,651/2012, may put further pressure on Brazil's mangroves, since the new law allows the use of 35% of mangroves in the northeast for shrimp farming (Rovai et al. 2012; Schaeffer-Novelli et al. 2012). The loss of mangroves is also associated with increased silting and pollution, further reducing the quality of the habitat for manatees (Lima et al. 2011).

Habitat degradation due to the loss of mangroves is currently considered the main threat to the conservation of Antillean manatees in Brazil (Campos et al. 2003). The impact of this degradation can be seen in an increased frequency of stranding of dependent pups. Such strandings are probably caused by females being excluded from traditional nursing grounds within estuaries and giving birth in suboptimal habitats in the open sea. This increases the probability of separation between mother and calf and ultimately leads to the stranding of neonates and juveniles (Lima 1999). As many as 83% of manatee deaths in Ceará State were classified as dependent offspring, with death through entanglement in fishing gear representing only 12.5% of deaths (Meirelles 2008).

In a global context, sirenian conservation initiatives are primarily focused on the creation and implementation of protected areas and on introducing measures to reduce illegal hunting, such as environmental education and inspection activities. The expansion of scientific knowledge on distribution, habitat use, and population parameters is also a priority in national conservation action plans (ICMBio 2011, 2018). A recent expansion of small purpose-built rehabilitation centers for dependent pups (e.g., in Puerto Rico and Belize) is being recorded, with the purpose of rehabilitation and release of rescued individuals. Specialized rehabilitation centers have also been constructed in Brazil, supported by on-call rescue teams that bring in stranded orphaned cubs for treatment and rehabilitation. The federal conservation agency (Instituto Chico Mendes de Conservação da Biodiversidade [ICMBio]) works in partnership with these organizations to release rehabilitated animals into ecologically appropriate locations that are closest to the stranding sites.

In 1980, the Brazilian federal government created the *Peixe-boi* ("Manatee") Project to carry out research and actions that would reduce the threat of extinction to Antillean manatees in the country. Among the many actions carried out over 42 years of the project, two are particularly worthy of note: (i) rescue and rehabilitation of stranded dependent pups, followed by release (and monitoring) in a natural environment, and (ii) extensive environmental education initiatives aimed at reducing intentional hunting (Luna and Passavante 2010).

13.3.1 The Brazilian Antillean Manatee Reintroduction Program

The manatee reintroduction program was initiated in 1994 to connect isolated populations, minimizing inbreeding depression and loss of diversity through genetic

Table 13.2 Criteria used to choose release sites for the Brazilian Antillean Manatee Reintroduction Program

Criterion	Description
Food availability	Seagrass; macroalgae; mangroves
Availability of freshwater	Preserved and accessible rivers and estuaries; sources of fresh water
Logistic support	Unit for the Conservation of aquatic mammals (CMA, in Portuguese); partners
Human occupation	Low motorboat traffic; distance from urban areas
Presence of manatees	Native animals; introduced animals
Protected area	Protected area created; preserved habitat for species
Origin	Puppies stranded region; genetics

Adapted from Lima et al. (2007)

drift and recolonizing parts of the historical distribution of the subspecies (Lima et al. 2007). Release sites were selected at the beginning of the program using criteria based on the availability of food and fresh water, the existence of protected areas, release logistics, and level of human occupation (Table 13.2). From 1994 to 2019, 46 rescued and rehabilitated individuals have been released at three different sites (Fig. 13.1): Paripueira and Porto de Pedras (Fig. 13.3) in Alagoas State and Rio Tinto in Paraíba State, with a success rate of approximately 76% (Normande et al. 2015) (see Chap. 3, Maps 7 and 9).

A total of six reintroduced females gave birth to 13 pups, with one individual "Lua" giving birth to five pups (three alive and two dead) (Attademo et al. 2022). These pups were born and raised in a natural environment assisting in restocking populations. The offspring of reintroduced females were all born in Alagoas, in the estuaries of the Manguaba, Tatuamunha, São Miguel, and Santo Antônio rivers. These sites were used by females for both parturition and parental care (ICMBio, unpublished data) (see Chap. 3, Map 9).

Among the three release sites, only Porto de Pedras did not contain an extant free-living population of manatees; this area is considered a historical occurrence site and is located between two isolated populations (Lima 1999). In this way, the releases at Porto de Pedras (Figs. 13.1 and 13.3) are considered reintroductions, while the releases at Paripueira and Tinto rivers (Fig. 13.1) are examples of reinforcement (sensu IUCN 1998) of native manatee populations.

Before release, rehabilitated manatees received radio transmitters with VHF and satellite technology to monitor their post-release movements. This monitoring aims at assessing the adaptation of individuals to the environment and enabling veterinary interventions in case of debilitated individuals and collecting information on movement and habitat use by released animals. Such information increases scientific knowledge about this subspecies and its ecological relationships, as well as contributes to the evaluation of the effectiveness of the management program for conservation (Lima et al. 2007).

Fig. 13.3 Aerial view of the semi-captivity assembled in the mangrove of Tatuamunha River, Alagoas State. The manatees kept in natural environment are being prepared for reintroduction after a period of adaptation. A small boat called *jangada* used for manatee watching. (Photo: Caio Salles/ Projeto Verde Mar)

13.3.2 Community Conservation and Ecotourism

Ecotourism based on manatee watching on the Tatuamunha River near the release site at Porto de Pedras began to develop in the late 1990s. It has been a very successful strategy, promoting conservation and providing opportunities for social inclusion and income generation for local communities (Normande et al. 2015). The first trips were organized informally with local fishermen, who would act as guides and take tourists on their rafts ("jangadas") to view manatees in sheltered areas of the inshore reef and the mangrove-lined estuary. These guides were native to the municipalities of Porto de Pedras and São Miguel dos Milagres and typically had very basic formal education. With the initiation of the release program (see above), manatees became even easier to locate and there was a growing realization that organized manatee watching represented a substantial source of supplementary income for local communities.

The first trips visited a range of habitats where the manatees had a higher chance to be found and were completely unregulated, with tourists frequently swimming around and feeding and touching the manatees. This led to increasing levels of habituation to human presence, a potentially negative process that can lead to harmful behaviors such as approaching motorized boats and making individuals more vulnerable to hunting. With the development of the release program and the associated increase in opportunities for manatee viewing, it became apparent that there was an urgent need to develop an educational, participative process to encourage best practices among the ecotourism providers. This was achieved through extensive dialogue with the local community at every step in the process, with the goal of ensuring that the manatee watching was environmentally and economically sustainable, causing the least possible disturbance to the released and wild individuals. It is important to note that not all members of the local community were in favor of the release program, with some fishermen complaining that the increased population of manatees was damaging their nets.

The first steps towards the formalization of manatee watching in Porto de Pedras began in 2006, when technicians of the "Peixe-Boi" Project, under the responsibility of the Brazilian government, began to train the local guides and to formulate a set of normative practices. From 2007 to 2009 there were three training courses for manatee watching guides. Initially, there were 16 participants in the training program, though this number expanded to a fixed number of 20 accredited guides, and other supporting staff, currently operating in the area. Between 2009 and 2010 a formal set of rules and procedures, logistics, training, accreditation, and division of responsibility was established after extensive negotiations between guides, municipal government, and ICMBio, the latter responsible for project management. In 2013, the standards were revised and adapted for publication in the Coral Coast Environmental Protection Area Management Plan, where the release site is inserted.

An important aspect of the development of manatee ecotourism in Porto de Pedras was the creation of the Association of Manatee Watching Tourism Guides in 2009. This organization is responsible for representing guides, marketing, and conducting daily tours (see Fig. 13.4). The association currently provides up to 10 departures and a maximum of 70 people a day all year round. Significantly, it provides substantial livelihood benefits, being the main source of income for more than 50 local families, in addition to developing educational projects, joint community biodiversity monitoring, and assisting ICMBio and other partners in the management and conservation of the manatees, mangroves, and reefs.

13.4 Final Remarks

Based on the criteria suggested by Barua et al. (2011) (Table 13.1), manatees are an ideal flagship species for mangrove conservation in northeast Brazil. First, they are strongly associated with mangrove ecosystems, spending a high proportion of their time in or around estuaries (Normande et al. 2015). Second, manatees can act as an umbrella species, whose protection serves to protect many co-occurring species (Roberge and Angelstam 2004). In northeast Brazil, conserving the mangroves is essential for providing nursing and feeding areas for manatees. Moreover, manatee

Fig. 13.4 *Jangada* during a manatee watching boat trip in Tatuamunha river, Porto de Pedras, Alagoas, Brazil. (Photo: Rafael Munhoz)

conservation also provides a strong justification for conserving the local reefs. Third, the manatee is widely perceived as an endangered species, proving a strong justification for prioritizing the protection of its ecosystem. Fourth, given this species is recognizable, is easily observed, and has unique morphological and behavioral traits, the interactions with manatees produce memorable experiences. Fifth, manatees in northeast Brazil typically have positive cultural associations and, due to their frequent interactions with local fishermen, there is considerable local ecological knowledge within coastal communities. Finally, they have a relevant scientific value, especially in the fields of ecology and animal behavior.

In conclusion, the presence of a large, charismatic marine mammal within the mangrove habitats of northeast Brazil provides a unique opportunity for raising conservation awareness, promoting ecotourism and scientific research, mobilizing conservation funding, and engaging local populations in the conservation of mangroves.

Acknowledgments We would like to acknowledge the Instituto Chico Mendes de Conservação da Biodiversidade Conservation, Fundação Toyota do Brasil, Fundação SOS Mata Atlântica, and Fundo Brasileiro para a Biodiversidade for current logistics and financial support and all professionals that dedicated their lives to manatee conservation in Brazil.

References

Albuquerque C, Marcovaldi GM (1982) Ocorrência e distribuição do peixe-boi marinho no litoral brasileiro (Sirenia, Trichechidade, *Trichechus manatus*, Linnaeus 1758). In: Abstracts of Simpósio Internacional sobre a Utilização de Ecossistemas Costeiros: Planejamento, Poluição e Produtividade, Rio Grande

Allen AC, Beck CA, Bonde RK, Powell JA, Gomez NA (2017) Diet of the Antillean manatee (*Trichechus manatus manatus*) in Belize, Central America. J Mar Biol Assoc UK 98(7): 1831–1840

Almeida R, Coelho C Jr, Corets E (2008) Impactos Humanos Sobre os Manguezais in: Guia Didático: Os Maravilhosos Manguezais do Brasil. Papagaya Editora, Vitória, pp 1–3

Alves MD, Kinas PG, Marmontel M, Borges JCG, Costa AF, Schiel N, Araújo ME (2015) First abundance estimate of the Antillean manatee (*Trichechus manatus manatus*) in Brazil by aerial survey. J Mar Biol Assoc UK 96(SI4):955–966

Arraut EM, Arraut JL, Marmontel M, Mantovani JE, Novo EMLM (2017) Bottlenecks in the migration routes of Amazonian manatees and the threat of hydroelectric dams. Acta Amaz 47:7–18

Attademo FLN, Normande IC, Sousa GP, Costa AF, Borges JCG, de Alencar AEB, Foppel EFdaC, Luna FdeO (2022) Reproductive success of Antillean manatees released in Brazil: implications for conservation. J Mar Biol Assoc UK 1–8. https://doi.org/10.1017/S0025315422000443

Balensiefer DC, Loffler F, Attademo N, Sousa P, Carlos A, Carneiro G, Emı A, Silva DL (2017) Three decades of Antillean manatee (*Trichechus manatus manatus*) stranding along the Brazilian coast. Trop Conserv Sci 10:1–9

Barua M, Root-Bernstein M, Ladle RJ, Jepson P (2011) Defining flagship uses is critical for flagship selection: a critique of the IUCN climate change flagship fleet. Ambio 40:431–435

Best RC (1983) Apparent dry season fasting in Amazonian manatees (Mammalia: Sirenia). Biotropica 15:61–64

Borges JCG, Emmanuel G, Miranda C (2008) Identificação de itens alimentares constituintes da dieta dos peixes-boi marinhos (*Trichechus manatus*) na região Nordeste do Brasil. Biotemas 21: 77–81

Borges JCG, Vergara-Parente JE, de Carvalho Alvite CM, Marcondes MCC, de Lima RP (2007) Embarcações motorizadas: uma ameaça aos peixes-bois marinhos (*Trichechus manatus*) no Brasil. Biota Neotrop 7(3):199–204

Borobia M, Lodi L (1992) Recent observations and records of the west Indian manatee *Trichechus manatus* in northeastern Brazil. Biol Conserv 59:37–43

Campos AA, Monteiro AQ, Monteiro-Neto C, Pollete M (2003) A Zona Costeira do Ceará: diagnóstico para a Gestão Integrada, 248 pp. AQUASIS, Fortaleza

Castelblanco-Martínez DN, Bermúdez-Romero AL, Gómez-Camelo IV, Rosas FCW, Trujillo F, Zerda-Ordoñez E (2009) Seazonality of habitat use, mortality and reproduction of the vulnerable Antillean manatee *Trichechus manatus manatus* in the Orinoco River, Colombia: implications for conservation. Oryx 43(2):235–242

Ciotti LL, Luna FO, Secchi ER (2014) Intra-and interindividual variation in d13C and d15N composition in the Antillean manatee *Trichechus manatus manatus* from northeastern Brazil. Mar Mamm Sci 30:1238–1247

Colmenero-Rolón L, Hoz-Zavala M (1986) Distribución de los manatíes, situación y su conservación en México. An del Inst Biol 56:955–1020

Committee on Taxonomy (2016) List of marine mammal species and subspecies. https://www.marinemammalscience.org/species-information/list-marine-mammal-species-subspecies/

Deutsch CJ, Reid JP, Bonde RK, Easton DE, Kochman HI, O'Shea TJ (2003) Seasonal movements, migratory behavior, and site fidelity of West Indian manatees along the Atlantic coast of the United States. Wildlife Monogr 151:1–77

Folkens PA, Reeves RR (2002) Guide to marine mammals of the world. National Audubon Society, New York City

García-Rodríguez AI, Bowen BW, Domning D, Mignucci-Giannoni AA, Marmontel M, Montoya-Ospina RA, McGuire PM (1998) Phylogeography of the West Indian manatee (*Trichechus manatus*): how many populations and how many taxa? Mol Ecol 7:1137–1149

Heywood VH (1995) Global biodiversity assessment. Cambridge University Press, Cambridge

Instituto Chico Mendes de Conservação da Biodiversidade – ICMBio (2011) Plano de ação nacional para a conservação dos sirênios: peixe-boi-da-amazônia *Trichechus inunguis* e peixe-boi-marinho *Trichechus manatus*. ICMBio, Brasília

Instituto Chico Mendes de Conservação da Biodiversidade – ICMBio (2018) Plano Nacional para a Conservação do Peixe-boi marinho. Portaria n° 249 de 04 de abril de 2018. ICMBio, Brasília

Irvine AB (1983) Manatee metabolism and its influence on distribution in Florida. Biol Conserv 25: 315–334

International Union for Conservation of Nature – IUCN (1998) Guidelines for re-introduction. Prepared by the IUCN/SSC Re-introduction Specialist Group. IUCN, Gland

International Union for Conservation of Nature – IUCN (2012) IUCN Red List of Threatened Species, version 2011

International Union for Conservation of Nature – IUCN (2019) The IUCN Red List of Threatened Species, version 2019–1. http://www.iucnredlist.org. Accessed 21 Mar 2019

Kareiva P, Marvier M (2012) What is conservation science? Bioscience 62:962–969

Ladle RJRJ, Jepson P, Gillson L (2011) Social values and conservation biogeography. In: Ladle RJ, Whittaker RJ (eds) Conservation biogeography. Wiley-Blackwell, Oxford, pp 13–30

Lefebvre LW, O'Shea TJ, Rathbun GB, Best RC (1989) Distribution, status, and biogeography of the West Indian manatee. In: Woods CA (ed) Biogeography of the West Indies: past, present and future. Sandhill Crane Press, Gainesville, pp 567–620

Lima RP (1999) Peixe-boi marinho (*Trichechus manatus*): distribuição, status de conservação e aspectos tradicionais ao longo do litoral Nordeste do brasil. IBAMA, Brasília

Lima RP, Alvite CMC, Vergara-Parente JE (2007) Protocolo de reintrodução de peixes-bois marinhos no Brasil. IBAMA/MMA & Instituto Chico Mendes, São Luís

Lima RP, Alvite CMC, Vergara-Parente JE, Castro DF, Paszkiewicz E, Gonzalez M (2005) Reproductive behavior in a captive-released manatee (*Trichechus manatus manatus*) along the Northeastern Coast of Brazil and the life history of her first calf born in the wild. Aquat Mamm 31:420–426

Lima RP, Paludo D, Soavinski RJ, Silva KG, Oliveira EMA (2011) Levantamento da distribuição, ocorrência e status de conservação do Peixe-Boi Marinho (*Trichechus manatus*, Linnaeus, 1758) no litoral nordeste do Brasil. Nat Resour 1:41–57

Lins ALFA, Gurgel ESC, Bastos MNC, Sousa MEM, Emin-Lima R (2014) Which aquatic plants of the intertidal zone do manatees of the Amazon estuary eat? Sirenews 62:11–12

Luna FO (2001) Distribuição, status de conservação e aspectos tradicionais do peixe-boi marinho (*Trichechus manatus*) no litoral Norte do Brasil. Master Thesis, 122 pp, Universidade Federal de Pernambuco

Luna FO, Araújo JP, Passavante JZ, Mendes PP, Pessanha M, Soavinski RJ, Oliveira EM (2008) Ocorrência do peixe-boi marinho (*Trichechus manatus manatus*) no litoral norte do Brasil. Bol Mus Biol Mello Leitão 23(3):37–49

Luna FO, Balensiefer DC, Fragoso AB, Stephano A, Attademo FLN (2018) Trichechus manatus. In: Livro Vermelho da Fauna Brasileira Ameaçada de Extinção. ICMBio/MMA, Brasília, pp 103–109

Luna FO, Bonde RK, Attademo FLN, Saunders JW, Meigs-Friend G, Passavante JZO, Hunter ME (2012) Phylogeographic implications for release of critically endangered manatee calves rescued in Northeast Brazil. Aquat Conserv 22:665–672

Luna FO, Passavante JZO (2010) Projeto peixe-boi/ICMBio: 30 anos de conservação de uma espécie ameaçada. ICMBio, Brasília

Marsh H, O'Shea TJ, Reynolds JE III (2011) Ecology and conservation of Sirenia: dugongs and manatees. Cambridge University Press, New York

Meirelles ACO (2008) Mortality of the Antillean manatee, *Trichechus manatus manatus*, in Ceará State, North-Eastern Brazil. J Mar Biol Assoc UK 88:1133–1137

Meirelles ACO, Carvalho VL (2016) Peixe-boi-marinho - Biologia e Conservação no Brasil, 1st edn. Bambu Editora e Artes Gráficas, São Paulo

Meirelles ACO, Carvalho VL, Marmontel M (2018) West Indian manatee *Trichechus manatus* in South America: distribution, ecology and health assessment. In: Rossi-Santos M, Finkl C (eds) Advances in marine vertebrate research in Latin America. Springer, Cham, pp 263–291

Newman JA, Varner G, Linquist S (2017) Defending biodiversity: environmental science and ethics. Cambridge University Press, Cambridge

Normande I, Luna F, Malhado A, Borges J, Viana Junior P, Attademo F, Ladle R (2015) Eighteen years of Antillean manatee Trichechus manatus manatus releases in Brazil: Lessons learnt. Oryx, 49(2):338–344. https://doi.org/10.1017/S0030605313000896

Oliveira-Gómez LD, Mellink E (2005) Distribution of the Antillean manatee (*Trichechus manatus manatus*) as a function of habitat characteristics, in Bahía de Chetumal, México. Biol Conserv 121:127–133

Paludo D, Langguth A (2002) Use of space and temporal distribution of *Trichechus manatus manatus* Linnaeus in the region of Sagi, Rio Grande do Norte State, Brazil (Sirenia, Trichechidae). Rev Bras Zool 19:205–215

Parente CL, Vergara-Parente JE, Lima RP (2004) Strandings of Antillean manatees, *Trichechus manatus manatus*, in northeastern Brazil. Lat Am J Aquat Mamm 3(1):69–75

Rathbun GB, Powell JA, Cruz G (1983) Status of the West Indian manatee in Honduras. Biol Conserv 26:301–308

Reeves RR, Tuboku-Metzger D, Kapindi RA (1988) Distribution and exploitation of manatees in Sierra Leone. Oryx 22:75–84

Reid JP, Rathbun GB, Wilcox JR (1991) Distribution patterns of individually identifiable West Indian manatees (*Trichechus manatus*) in Florida. Mar Mamm Sci 7(2):180–190

Reynolds JE III, Odell DK (1991) Manatees and dugongs, 192 pp. Facts on File Inc, New York

Roberge J, Angelstam P (2004) Usefulness of the umbrella species concept as a conservation tool. Conserv Biol 1:76–85

Rovai AS, Menghini RP, Schaeffer-Novelli Y, Molero GC, Coelho-Jr C (2012) Protecting Brazil's coastal wetlands. Science 335(6076):1571–1572

Schaeffer-Novelli Y, Rovai AS, Coelho-Jr C, Menghini RP, Almeida R (2012) Alguns impactos do PL30/2011 sobre os manguezais brasileiros. In: Código Florestal e a Ciência: o que nossos legisladores ainda precisam saber. Comitê Brasil, Brasília, pp 18–27

Sheppard JK, Preen AR, Marsh H, Lawler IR, Whiting SD, Jones RE (2006) Movement heterogeneity of dugongs, *Dugong dugon* (Müller), over large spatial scales. J Exp Mar Biol Ecol 334: 64–83

Sousa MEM, Martins BML, Fernandes MEB (2013) Meeting the giants: the need for local ecological knowledge (LEK) as a tool for the participative management of manatees on Marajó Island, Brazilian Amazonian coast. Ocean Coast Manag 86:53–60

Spiegelberger T, Ganslosser U (2005) Habitat analysis and exclusive bank feeding of the Antillean manatee (*Trichechus manatus manatus* L. 1758) in the Coswine Swamps of French Guiana, South America. Trop Zool 18:1–12

Verissimo D, MacMillan DC, Smith RJ (2011) Toward a systematic approach for identifying conservation flagships. Conserv Lett 4:1–8

Vianna JA, Bonde RK, Caballero S, Giraldo JP, Lima RP, Clark A, Rodríguez-Lopez MA (2006) Phylogeography, phylogeny and hybridization in trichechid sirenians: implications for manatee conservation. Mol Ecol 15:433–447

Whitehead PJP (1977) The former southern distribution of New World manatees (*Trichechus* spp.). Biol J Linn Soc 9:165–189

Part IV
Conservation Strategies

Chapter 14
The Eco-history of Brazilian Mangroves

Arthur Soffiati Netto and Armando S. Reis-Neto

14.1 Precolonial Era

About 5000 years ago, in the second half of the Holocene, Brazil's coastline was in a process of stabilization. Through time, rivers and estuaries have reshaped themselves and the same happened to mangroves. Currently, the sea level and, indeed the coastline, is well above the point where it was in the Pleistocene (11,700 years ago) and below the point where it was in the Holocene Climate Optimum (9000–5000 years ago). Nowadays, there is clear evidence that the sea level is rising due to glacier melting and ocean volume expansion, processes that are accelerated by human activities (IPCC 2014). Mangroves, as resilient ecosystems, follow the coastal reconfiguration in tropical and subtropical zones.

At the beginning of the Holocene (11,500 years ago), many Paleolithic societies, archaic societies that lived from collecting, fishing, and hunting, responded to the challenges of natural climate change with the domestication of plants (agriculture) and livestock (pasture), which required profound changes in social structure. Some of those societies evolved to the Neolithic and became sedentary. States and city boundaries as well as their specific administrative conditions did not exist at the time.

In the third millennium Before the Common Era (BCE),[1] some Neolithic societies became more complex, originating structured societies, or civilizations, based on the social division of labor and distinction between city and countryside. In these

[1] All years and centuries in this chapter will be expressed according to the Common Era (CE) or Before the Common Era (BCE) standard.

A. S. Netto (✉)
Instituto de Ciências da Sociedade e Desenvolvimento Regional, Universidade Federal Fluminense, Campos dos Goytacazes, RJ, Brazil

A. S. Reis-Neto
Instituto Bioma Brasil, São Paulo, São Paulo, Brazil

Y. Schaeffer-Novelli et al. (eds.), *Brazilian Mangroves and Salt Marshes*, Brazilian Marine Biodiversity, https://doi.org/10.1007/978-3-031-13486-9_14

societies, the production system generated surpluses, leading to the development of a merchant's class. Trade routes reached an extensive range over the globe but did not change the primitive production system. Therefore, in the fifteenth century, there were innumerable Paleolithic and Neolithic societies on almost every continent. From East to West, the main civilizations around the globe were the Sino-Japanese, Hinduist, Islamic, Orthodox Christianist, Roman Christianist (Western), Andean, and Mexican (Toynbee 1986). From all of them, only Western civilizations faced a unique phenomenon throughout the history of humanity: Commercial activity began to influence the current feudal system and created a new market society based on the production of goods. This means that productive structures progressively began to produce goods for profit.

Until then, the Paleolithic, Neolithic, and civilized societies had a balanced relationship with nature. In fact, there were environmental crises produced by human activities, nevertheless always reversible and on a local scale (Toynbee 1978). In order to obtain growing profits, Western civilizations needed to expand their territories to look for raw materials (i.e., commodities) and to search new consumer markets. The first manifestation of these necessities was the Christian Crusades between the eleventh and thirteenth centuries CE, which had an undercover aim to control maritime commerce in the west, then under the Muslim domain. This attempt failed, and Europeans returned to their geographical limits. The second attempt took place in the fifteenth century, with the Western expansion through the Atlantic Ocean. At that time, mangroves were already known by Antiquity authors, such as Eratosthenes, Megasthenes, Ptolemy, Nearchus, Aristobulis, and Pliny (Schneider 2011). As Portuguese navigators began to explore the West coast of Africa, they have also been in touch with this ecosystem. Ahmad Ibn-Mājid, an Arab navigator who helped Vasco da Gama reach India in 1598, described mangroves as a very dangerous area for navigation (Cliff 2011).

In the ancient Eastern World, there were narratives about amazing marine forests. Greeks already knew mangrove areas from three regions: The Red Sea (known by the ancient Greeks as "Arabic Golf"), the Arabian Sea (commonly known to Graeco-Romans as "Erythraean Sea"), and the Persian Gulf (Schneider 2011). Alexander the Great's party described mangroves in the Red Sea in the travel report to India (328–325 BCE).

14.2 Colonial Era

The Portuguese expedition of Pedro Álvares Cabral to the West Indies, in 1500 CE, reached the Brazilian coastline somewhere in Bahia State (see Chap. 3, Map 11). The caravels anchored, probably, in the mouth of rivers Coroa Vermelha and Mutari. Today, in this same region, a few remnants of the larger mangroves described at that time still exist. Back then, three documents were produced by members of this expedition, introducing Brazil to the Western world. The longest and most detailed one is the letter written by Pero Vaz de Caminha to the king of Portugal. The experienced pilots already knew mangroves from the African coastline. However, Caminha, who was not a navigator, probably did not know mangroves and briefly mentioned the existence of this ecosystem in Brazil. He wrote about seamen looking for shellfish and only finding "thick and short shrimps." There is a strong indication that these shrimps were in fact crabs, very common in mangroves (Pereira 1999).

Human groups from lands other than Europe preceded the Europeans in America's occupation. Whatever the hypothesis to explain the arrival of the first human group to the American continent, the occupation process was slow, adjusting to ecosystem limits, at a different pace when compared to the European's dominance during colonization. On one side, natives have adapted their lives to the ecosystem's capacity, while Europeans adapted ecosystems to their interests. Humans have always used mangroves for food collecting and shelter, a long time before Europeans arrived in America (Schaeffer Novelli and Cintrón-Molero 1999). A specialist concluded that intertropical estuaries, where mangroves are found, are rich in mollusks, crustaceans, and fishes, allowing a self-sufficient way of life concerning terrestrial environments (Yesner 1980). That way, a typical mangrove culture was born, configuring this ecosystem as a reliable source for human livelihood (Figute 1993). Evidence from precolonial civilizations is scarce, mainly represented as shell middens ("sambaquis" in Brazilian Portuguese).

Comparisons with the remaining archaic people today and some documentation led us to conclude that the *sambaqui*-building societies had a sacred and ritualized view of nature. Despite this, studies on this subject are still lacking, either because the documentation does not allow conclusions beyond a certain period or because the scientific community is more interested in the tangible aspects of culture (Wiener 1876; Lhering 1903; Hurt and Blasi 1960; Salles Cunha 1965; Duarte 1968; Kneip 1974; Beltrão 1995; Gaspar 2000).

The Western spirit had desacralized the world as globalization expanded. All ecosystems gradually became sources and stocks of commodities in a global economy and, sometimes, areas for disposing waste. In the sixteenth century, this "westernization" of the New World was still crawling, still free from the gears of the global economy. Brazilian mangroves' first description is authored by the Jesuit Priest José de Anchieta in a letter written in 1560 (Anchieta 1989). Still, in the same century, Pero de Magalhães, another Portuguese, wrote about mangroves, probably referring to the Captaincy of Ilhéus, Bahia State, and the Captaincy of São Vicente, São Paulo State (Gandavo 2008). Magalhães' book was written between 1558 and 1572. Yet another Portuguese, Gabriel Soares de Sousa, described the mangrove vegetation and how native people called them "sereíba," pointing it out as good wood for fuel and house and mill construction, besides feeding crabs with their leaves. He mentioned another tree, the "canapaúba," whose branches twine to the tide level, allowing oyster cultivation. Certainly, he was referring to "siriba," "siribeira," or "siriúba" (i.e., the black mangrove genus *Avicennia*) and the red mangrove genus *Rhizophora* (Sousa 1938). In his words, mangroves started to show economic value to the Portuguese.

In the same sixteenth century, there were testimonies from the French André Thévet, who described trees standing between the river's mouth and the sea, laden with oysters that fed the Amerindians (Thévet 1944). There was also the Priest Fernão Cardim, who described trees that give excellent firewood and tannin, commonly used in tanning leather. In mangroves, wrote Cardim, live mosquitos, crabs, oysters, black rats, and parrots. The most common crab, according to Cardim, was the swamp ghost crab *Ucides cordatus*, which served as food for natives and slaves. He also described the blue land crab *Cardisoma guanhumi* as a big crab capable of cutting off a human leg. During thunderstorms, these crabs would leave their burrows and become so threatening that it was necessary to shoot them with fire guns. He made a pertinent remark that the blue land crab lived in the mangrove terrestrial margins, while "aratu" (*Aratus pisonii*) lived on trees. The almost 12 species of crabs counted by Cardim were used as food by natives (Cardim 1925). In the last decade of the sixteenth century, the English corsair Anthony Knivet, who was a prisoner of Salvador Correia de Sá in Rio de Janeiro, described that he caught crabs from deep holes in the mud by the seashore. At that time, Guanabara Bay was covered in lush mangroves (Knivet 2007).

The historian Sérgio Buarque de Holanda underpinned that Europeans made use of indigenous technology to handle the unknown ecosystems (Holanda 2014). This remark is very likely valid for mangroves as well. Poor white men and African slaves learned with natives how to collect food in the mangroves. Overall, Europeans described mangroves as strange environments, with no comparison to anything existing in Europe, equaling them to swamps. Just at the end of that century, some reports showed the economic value of mangroves as a source of goods and benefits.

In the seventeenth century, European knowledge about Brazilian nature and native people advanced greatly. In 1614, the French missionary Claude D'Abeville, who served Equinoctial France, a colony established in Maranhão State (see Chap. 3, Map 3), found the mangroves already known by French navigators who went to the Antarctic French in the previous century. In his reports, he cited the unique plants of this ecosystem and called them *apparituriers*, from the Tupi language term *aparahiwa* ("curved tree"). This chronicler presented a detailed description showing the tight relationship between this ecosystem and crustaceans, highlighting *uçá*, *aratu*, and other crabs (Abeville 1975).

Franciscan friar Vicente do Salvador, born in Bahia State, wrote about the history of Brazil in 1627 and described some mollusks and crustaceans living in mangroves (Salvador 1918). But it is only in the Dutch domain in Northeast Brazil, between 1630 and 1654, under the government of John Maurice of Nassau (1637–1644), that we find a more detailed representation of mangroves. Nassau came to Brazil followed by artists and naturalists. Until then, mangroves were exclusively described through words. There is a strong indication that the first pictured representation of the Brazilian mangrove is a red mangrove *(Rhizophora mangle)* painted by Frans Post, in 1638, reproduced by Caspar Barlaeus in his book called *The Dutch Domain* (Fig. 14.1) (Lago 2006).

George Marcgrave, one of the most reputable scientists in Nassau's court, made a brief description of mangroves in his book *Brazil's Natural History* (Marcgrave 1942). Europeans surely had the opportunity to see natives consuming mangrove goods. Another naturalist in the court, Guilherme Piso, described plants that were

Fig. 14.1 First depiction of Brazilian mangroves is a red mangrove (*Rhizophora mangle*) painted by Frans Post, in 1638

useful to men in his book from 1648. For example, the salt expelled by the leaves of "siribeira" (*Avicennia* spp.) could be used to salt a soup with two or three leaves. The bark of "cereibuna" (probably *Laguncularia racemosa*) was used by fishermen to weave ropes. The red mangrove reportedly produced propagules that were eaten by crabs, which is illustrated in his book, and its roots were used as medicine for treating poisoning by the spines of the toadfish *Thalassophryne nattereri* (Piso 1957).

Yet in the seventeenth century, a conflict about local mangroves was triggered in Rio de Janeiro between city residents and the Jesuits (Viera Fazenda 1921; Vivaldo Coaracy 1961). The Jesuits seized all mangrove areas, banishing their use by the citizens. The municipal council appealed to the King of Portugal in 1677, under the claim that mangrove wood was widely used for construction, fuelwood, and coal to supply sugar mills at Guanabara Bay (see Chap. 3, Map 13). It was also used to supply wood for ships on their journeys. They adduced that it has never had any restriction, from the state or the church to the use of mangrove trees, and thus, the city has grown with energy and raw materials taken in this ecosystem. No reference has been made about fauna extraction practiced by the common people. The Portuguese crown gave cause to the city residents, arguing that lands washed by tides belonged to the royal government. Then, we can conclude that the seventeenth century consolidated a trend of social dispute for the use of mangrove resources that started in the late sixteenth century.

Mangroves started to be explored not only by indigenous people, poor people, and slaves in an extractive economy but were also used in more government-structured activities that generated social conflicts. This trend accentuated even more in the following centuries. In 1711, Father André João Antonil published *Brazil's Culture and Opulence*, a valuable document about the Brazilian economy and colonial society. Living in Bahia State, he demonstrated the importance of mangroves to the rich and the poor, attesting that the appropriation of this ecosystem started to generate social conflicts. For the rich, mangroves were an essential source of firewood to sugar mills and potteries. For the poor, mangroves mainly provided food that depended on the maintenance of healthy natural habitats. This set up a conflict of interest between rich and poor. Antonil was probably the first author in Brazil to comment that mangrove suppression caused the death or vanishing of animals (Antonil 1976).

Throughout the eighteenth century, the state of Bahia witnessed conflicts of different interests over the mangroves. The government occasionally took the side of mill owners and tanneries and at times the side of fishermen and collectors, as reported in vast official correspondence. At that time, the Portuguese Colonial Empire ranged from Brazil to East Timor, including many sites in Africa, India (Goa, Diu, and Damão), and China (Macao). Malacca, present-day Jakarta (Indonesia), had already been lost to the Dutch.

The most prominent figure of the eighteenth century was, undoubtedly, Sebastião José de Carvalho e Melo, the Marquis of Pombal. He was the State Secretary of the Portuguese Crown during the reign of King Joseph I of Portugal between 1750 and 1777. His project was to unify and strengthen the Portuguese Colonial Empire through several reforms. Among them, he had slavery abolished in the metropolis, although keeping it in the colonies; he granted to all colonies' inhabitants the condition of subjects, including indigenous people, and handled with a strong grip the rebellions in the colonies (Almeida 2011). Mangroves were present in all colonies of the Portuguese Colonial Empire, and the marquis' project could also be understood as indirectly establishing a political unit upon that ecosystem.

In Brazil, an important document about this matter is the Royal License of 1760, which stated that only red mangroves with no bark could be cut down (see Chap. 16). This determination benefited tanneries (Soffiati 2006), as a good example of a public policy that confirmed the market trend on mangrove exploitation. There is a lack of information about whether this policy was also applied to other Portuguese colonies.

The nineteenth century politically began with the opening of ports to friendly nations in 1808, ordered by the Regent Prince John VI of Portugal. This act attended a British demand during the British isolation in the Continental Blockade imposed by Napoleon I of France. This was also the first step towards Brazilian independence, as the Portuguese Colonial Empire's seat was transferred to Rio de Janeiro. The end of the port's monopoly facilitated the entry of naturalists in Brazil. One of them was Maximilian zu Wied-Neuwied, a German noble and naturalist, who promoted a scientific expedition through the Brazilian coastline from Rio de Janeiro to Salvador between 1815 and 1817. In his travel journal, mangroves are barely described, suggesting that this ecosystem was not favorably seen by Europeans, always excited to know the majestic Amazonian and Atlantic forests. There are strong indications that the repulse to mangrove forests has a European point of view in its roots, as they were compared to Europe's swamps. Nevertheless, Maximilian described Bahia State's coastline as filled with mangroves, from the genera *Conocarpus* (not a true mangrove per se) and *Avicennia*, highlighting that the barks were coveted by the tanning industry that required a large contingent of slaves. This activity suggested a considerable capital investment and high profitability. Like every naturalist of his time, Maximilian classified plants and animals by Carl Linnaeus' binary nomenclature. In that way, species were not seen as single elements but grouped in parentage associations. The author also reported that mangroves from south Bahia State sustained owners of sugar mills and that the fishermen were opposed to mangrove tree extraction for firewood, as they depended on what the standing forest provided them with. This fact was also known by the Portuguese Regent Prince, then in Rio de Janeiro (Wied-Neuwied 1989).

In 1817, Manuel Aires de Casal published the book *Corografia Brasílica* ("Brazilian Chorography" or "Historical-Geographical Relation of the Kingdom of Brazil"), reporting mangroves in a colonial chronicler style but lacking in detail (Aires de Casal 1817). A year before, the English trader John Luccock took note of mangroves when touring Guanabara Bay, Rio de Janeiro State. Luccock reinforced the European trend to consider mangroves as a putrid, malodorous, and infected environment (Luccock 1975). A couple of years later, the French naturalist Augustin de Saint-Hilaire described the Rio de Janeiro coastline and commented on an increase of fishing in mangrove areas; however, the relationship between animal productivity and the presence of mangroves was not clearly established (Saint-Hilaire 1974).

14.3 Postcolonial Era

After the independence from Portugal in 1822, the Brazilian Empire experienced two reigns until the proclamation of the Brazilian Republic in 1889. In the first reign (1822–1831), two German naturalists made a great impact on the European audience about findings in the New World; they were Johann Baptist Ritter von Spix and Carl Friedrich Philipp von Martius. The latter wrote *Flora Brasiliensis*, where he dedicated a whole page to the mangroves of Ubatuba (Fig. 14.2), on the northern coast of São Paulo State (see Chap. 3, Map 14). He has described them as an admirable maritime forest of viviparous trees. Martius also highlighted the unpleasant odors that exude from the pasty and putrid mud. The author distinguished mangrove functioning as we understand nowadays but neglected its importance to surrounding communities; only the economic value for construction, tanning, and medicine was expressed in his work (von Martius 1833).

Another important aspect of this period was the promulgation of the law of November 15, 1831, appointing the marine land (coastal territory) as public

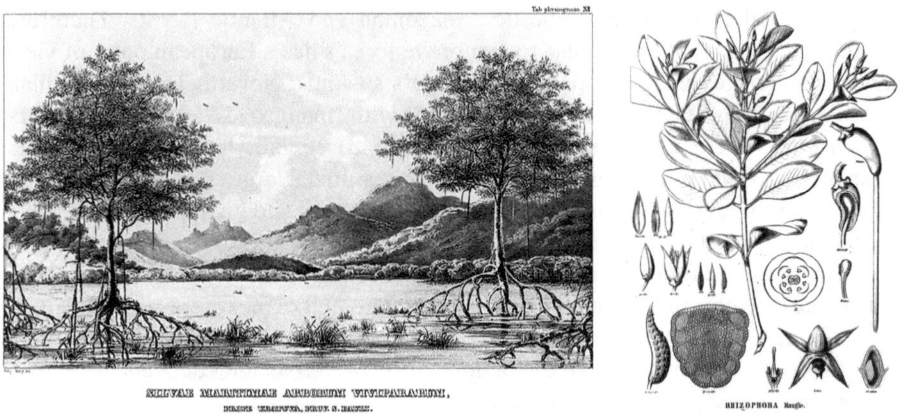

Fig. 14.2 Left: Depiction of the landscape in Ubatuba, São Paulo State, and the mangroves in the foreground. Right: Vegetative structures of *Rhizophora mangle* (from the same author). Both images were extracted from Flora Brasiliensis (von Martius 1833)

propriety, encompassing 33 m landward from the line of the average high tide in that year, including mangroves. This law is still in force today.

In the second reign, yet another two Germans who visited Brazil deserve mention. The first is the physician Robert Christian Avé-Lallemant, who traveled from south to north of Brazil. In his book *Journey through Northern Brazil in 1859*, mangroves were mentioned only a few times; in Camamu Bay, Bahia State, the explorer sighted some stretches of the coast completely flat and covered by thriving mangroves. He contributed to the dissemination in Europe of the idea that the tropical environment induces an indolent life. To him, the coastal inhabitants and residents of the mangrove's surroundings were stimulated to laziness because of the easy way to obtain food (Avé-Lalement 1961).

The second German to be mentioned is the naturalist Karl Hermann Konrad Burmeister, who, passing through the Guanabara Bay in 1850, in direction to the mountainous region of Rio de Janeiro and Minas Gerais States, mentioned the existence of mangroves and their crustacean decapod inhabitants (Burmeister 1980).

The Brazilians Francisco Freire Alemão (1961), Pedro Soares Caldeira (1884), and Emílio Joaquim da Silva Maia (1835) had an important contribution to the relevance of mangroves. Maia linked the integrity of mangroves and public health. Caldeira, a journalist, was the most incisive of them all when pointing out the relevance of mangrove ecosystems. He was a self-taught writer who exposed with vehemence the mangroves uprooting in Guanabara Bay, attributing to mangroves the pandemics and the endemic focus of yellow fever. While the Imperial Government looked after the nearby Tijuca Forest, Caldeira advocated for the protection of mangroves because of their importance to the fisherfolk.

Overall, the nineteenth-century naturalists failed in grasping the importance of mangroves. Meanwhile, extensive areas had served the economic pressure, providing timber, tannins, ground for crab and shellfish collection, and fishing. The trend to the destruction of this ecosystem continues throughout the centuries.

14.4 Republic of Brazil to Actual Times

In Republican times (1889 onwards), the voices in defense of mangroves and their sustainable use became more frequent, although not yet listened to by lawmakers and governmental organs. Hermann Lüderwaldt brought to light a robust study on mangroves in Santos, São Paulo State, in 1919. This is likely the first book that considered Brazilian mangroves in a systemic view, going beyond the observations made by Martius a century earlier. Lüderwaldt also noticed the animal exploitation by natives (Lüderwaldt 1919).

In a phytosociology course held by the Brazilian National Museum in Rio de Janeiro, in 1932, the botanist Alberto José de Sampaio did not consider mangroves as a Brazilian ecosystem because of their pantropical distribution (Sampaio 1945). Nevertheless, the scientist Frederico W. Freise addressed the issue in the I (Brazilian) National Congress on Fisheries when the author alerted the scientific community about the importance of mangroves to the fisheries. Freire's understanding of mangrove ecosystems was very similar to the present day's (Freise 1934). However, two prominent scientists from that time, João Moreira da Rocha and Melo Leitão, were very skeptical towards his thesis. The first considered the relation between mangroves and fisheries as merely an abstractedness not worth deeper investigations, while the latter suggested that the cutting of red mangroves for paper production should be subsidized. On the other hand, Freise stressed that mangroves are valuable reservoirs of sea life and stock of tannin for the industry. He proposed that the government could allow logging for industrial usage, maintaining a buffer zone of 40 meters from the riverbanks. So, the conflict between business owners and subsistence used by communities became present also in Academia.

Gerlach (1958) wrote *Die Mangroveregion tropischer Küsten als Lebensraum* with the main goal of describing nematodes. Nevertheless, the author ends up describing a systemic view on the mangrove ecosystems of Cananéia, on the southern coast of São Paulo State. This publication can be considered a milestone in the perception of mangroves as ecosystems.

In the 1970s, mangroves began to be recognized as an ecosystem worldwide. The pioneering work carried out by scientists like Watson, Dansereau, Egler, and Chapman inspired a group of international and Brazilian scientists to dedicate their efforts to investigating this specific tropical environment. In the late 1960s, the notable Lugo, Snaedaker, McNae, Tomlinson, and Walsh, among others, have exceptionally enriched the knowledge about mangrove ecosystems. Gilberto Cintrón has associated himself with Brazilian and foreign scientists, being a link between them and promoting diverse scientific production. Among Brazilians, Lamberti was the first to publish a consistent work on mangroves. In the 1970s, many scientists took an interest in mangroves, highlighting Yara Schaeffer-Novelli. Since then, a vast contribution has been made by several scientists all over the country. Marta Vanucci has also given important contributions in her descriptions of these ecosystems (Vanucci 1999).

Regarding the researcher's political view, most of the studies are dedicated to the description of structure, function, and disturbances (natural and human-induced) in mangroves. A systematic reading of Brazilian papers shows that only one researcher proposed the complete suppression of mangroves, if necessary, but the majority defends their conservation, leaving room for artisanal fisheries, sustainable fauna exploitation, ecotourism, and even apiculture (in the north). There is also strong consensus regarding silviculture, aquaculture, and agriculture, rarely considering ecotourism, apiculture, and traditional exploitation.

Only recently the voice of popular societal classes, living or not near mangroves, started being taken into consideration. Mangroves have been known as an unsanitary environment since colonial times, but this vision has started to change since the 1970s when traditional communities who had their livelihoods intimately related to mangroves started to gain voice and notoriety. These people have an ill-registered history. There are shreds of evidence that poor communities have lived near mangroves since the sixteenth century, practicing extractive economy and incorporating indigenous knowledge on sustainable fishing and catching. Also, the 1970s witnessed crescent attention to ecosystem biodiversity, and this included mangrove ecosystems.

In the 1980s, mangroves and communities living on them were finally "discovered" by the scientific community. Since then, many works have been published, NGOs focusing on those coastal ecosystems have been created, and meetings were held in many cities all over the country. Table 14.1 summarizes the views on mangroves from all periods mentioned above, from precolonial to actuality.

However, traditional communities, collectors, and fishermen decedents from indigenous populations, "caboclos,"[2] "quilombolas," and women from all over the Brazilian coastline are threatened by the global trade economy. Pushed by the global economy, such mangrove-dependent communities have swapped traditional methods of collecting and fishing for other predatory methods. Many of them found alternative jobs outside this ecosystem-community relationship because of market pressures. There are even some situations in which the subsistence job is related to mangrove suppression, such as in shrimp farms. In urban perimeters, these poor workers, attracted by the promise of good jobs and easy money, migrate to suburbs and sometimes do not even get the promised job. It is common for these migrants to end up in the marginalized spectrum of society, pushed to being homeless or involved in diverse illicit activities (Soffiati 2016).

The more appropriate solution to mangrove protection and the continuity of traditional communities who live on them may be the establishment of new Extractive Reserves, a mechanism figuring in Brazilian Environmental Law, considering that this kind of legally protected reserve is exclusively created and managed by the own community who has evolved intimately connected to the ecosystem's functioning (see Chap. 16).

[2] *Caboclo* is one of the names given to the mixture of native Brazilians with white Europeans. *Quilombola* is the name given to people who still preserve strong traces of African cultures and live in communities called "quilombo."

Table 14.1 Synthesis of the descriptions about mangrove ecosystems that have influenced the current view on the ecosystem in Brazil, highlighting the period of each description, the nationality of the authors, the location where the description has been developed, and the main view on the mangrove ecosystem regarding each period

Authors	Location	View on mangroves
IV century BCE – Alexander the great travels to India		
Graeco-Roman	Red Sea Arabian Sea Persian Gulf	Mystic forest
XV century CE – First European navigators in Africa		
Portuguese	West Africa	Obstacle to navigation Spread of diseases
XVI century CE – Europeans arrive in the New World (the Americas)		
Portuguese Dutch French	Brazilian coast (BA, RJ, MA)	Obstacle to navigation Food source (fish and others) Source of timber Firewood (mills)
XVII century – Exploitation of New World's natural resources		
Portuguese French Dutch	Brazilian coast (BA, SP, MA)	Culinary Source of tannins Source of timber Social conflicts (church vs commoners) Spread of diseases First depiction
XVIII century CE – Naturalists' descriptions and first regulations		
Portuguese Dutch	Brazilian coast (BA, RJ, SP)	Economic relevance Social conflicts (millers vs commoners) Law: Mangrove bark
XIX century CE – Naturalists' descriptions and new trends of exploitation		
Portuguese Dutch French German	Brazilian coast (BA, SP, MA, RJ)	Published: *Flora Brasiliensis* (Ubatuba) Law: Marine terrains are public Mangrove suppression in RJ: Social conflicts and public health
XX century CE – Science		
Scientific community	Worldwide	Mangrove relevance as ecosystem Published: *Manguesaes de Santos* Dignity of "mud people" despite impoverishment Globalization and growing threats

BA Bahia, *RJ* Rio de Janeiro, *MA* Maranhão, *SP* São Paulo.

14.5 Final Remarks

Throughout Brazilian history, since European colonization, the relationship between human societies and mangroves has been marked by conflicts. On the one hand, we have seafood collectors and fishermen who depend directly on the ecosystem goods and services in their natural condition. On the other hand, many economic interests are related to tanneries, agriculture, aquaculture, and urbanization that use mangrove territories to reproduce market interests through exploration and predatory methods. To summarize the Brazilian mangrove protection actions, it was only in 1884 when Pedro Soares Caldeira assumed an explicit defense of this ecosystem. And more recently, in the 1970s, scientists became more interested in understanding the complexity of the mangrove ecological process and its relationship with human activities. The production of scientific knowledge leads to a change in the popular sense to value mangroves as a biologically and culturally rich ecosystem. However, it still is a threatened ecosystem in Brazil and is surrounded by different social and economic interests. Mangrove protected areas should be a priority to governments, and societies should consider better the consequences of economic growth and its externalities (impacts) on the ecosystem to determine how it is going to be written in the future chapters of mangrove ecological history and the biological and social legacy to the next generations.

References

Abeville C (1975) História da missão dos padres capuchinhos na ilha do Maranhão e suas circunvizinhanças. Edusp/Itatiaia, São Paulo/Belo Horizonte

Aires de Casal M (1817) Corografia Brasílica. Impressão Regia, Rio de Janeiro

Alemão FF (1961) Papéis da expedição ao Ceará: viagem a Vila Velha, e Barra do Ceará. In: Damasceno D, Cunha W (eds) Anais da Biblioteca Nacional 81. Biblioteca Nacional, Rio de Janeiro

Almeida ACL (2011) Inconfidência no império: Goa de 1787 e Rio de Janeiro de 1794. Sete Letras, Rio de Janeiro

Anchieta J (1989) Carta fazendo a descrição das inúmeras coisas naturais, que se encontram na província de S. Vicente hoje S. Paulo: Cartas inéditas. Fundação Getúlio Vargas/Instituto de Documentação, Rio de Janeiro

Antonil AJ (1976) Cultura e opulência do Brasil por suas drogas e minas, 239 pp. Melhoramentos, Brasília

Avé-Lalement R (1961) Viagem pelo norte do Brasil no ano de 1859, vol 2. Instituto Nacional do Livro, Rio de Janeiro

Beltrão M (1995) Arqueologia do Estado do Rio de Janeiro. Arquivo Público do Estado do Rio de Janeiro, Niterói

Burmeister H (1980) Viagem ao Brasil. Edusp, São Paulo

Caldeira PS (1884) O corte do mangue: breves considerações sobre o antigo e atual estado da Baía do Rio de Janeiro, consequências da destruição da árvore denominada mangue, método bárbaro da pesca e decadência desta indústria. Tipografia Imp. e Const. de J. Villeneuve & C., Rio de Janeiro

Cardim F (1925) Tratados da terra e gente do Brasil. J. Leite & Cia, Rio de Janeiro

Cliff N (2011) The last crusade. The Epic Voyages of Vasco da Gama. Atlantic Books, London

Coaracy V (1961) O Rio de Janeiro no século 17. Instituto Nacional do Livro, Rio de Janeiro

Duarte P (1968) O sambaqui visto através de alguns sambaquis. Instituto de Pré-História, São Paulo

Fazenda JV (1921) Antiqualhas e memórias do Rio de Janeiro. Imprensa Nacional, Rio de Janeiro

Figute I (1993) Ecossistemas costeiros e homens pré-históricos. ACIESP, São Paulo

Freise FW (1934) A importância da conservação dos mangues como viveiros de peixes. Ministério da Agricultura, Rio de Janeiro

Gandavo PM (2008) Tratado da Terra do Brasil: história da província Santa Cruz, a que vulgarmente chamamos Brasil. Edições do Senado Federal, Brasília

Gaspar M (2000) Sambaqui: arqueologia do litoral brasileiro, 89 pp. Jorge Zahar Editora, Rio de Janeiro
Gerlach SA (1958) Die Mangroveregion tropischer Küsten als Lebensraum. Z Morph u Ökol Tiere 46:636–730
Holanda SB (2014) Monções e capítulos de expansão paulista. Companhia das Letras, São Paulo
Hurt WR, Blasi O (1960) O sambaqui do Macedo, A.52.B – Paraná, Brasil. Universidade do Paraná, Departamento de Arqueologia 2, Curitiba
Intergovernmental Panel for Climate Change – IPCC (2014) Synthesis report. Contribution of working groups I, II and III to the Fifth Assessment Report of the Intergovernmental Panel on Climate Change. IPCC, Geneva
Kneip LM (1974) Sambaqui do Forte: identificação espacial das atividades humanas e suas implicações (Cabo Frio, RJ, Brasil). Master Thesis, Universidade de São Paulo
Knivet A (2007) As incríveis aventuras e estranhos infortúnios de Anthony Knivet. Jorge Zahar Editora, Rio de Janeiro
Lago PBC (2006) Frans Post {1612–1680} Obra completa. Capivara, Rio de Janeiro
Lhering HV (1903) As origens dos sambaquis. IGSP, São Paulo
Luccock J (1975) Notas sobre o Rio de Janeiro e partes meridionais do Brasil. Edusp/Itatiaia, São Paulo/Belo Horizonte
Lüderwaldt H (1919) Os manguesaes de Santos. Rev Mus Paulista, São Paulo
Maia EJS (1835) Discurso sobre os males que tem produzido no Brasil, o corte das matas, e sobre os meios de os remediar. Tipografia Fluminense de Brito & Companhia, Rio de Janeiro
Marcgrave G (1942) História Natural do Brasil. Museu Paulista, São Paulo
Pereira PR (1999) Os três únicos testemunhos do descobrimento do Brasil. Lacerda, Rio de Janeiro
Piso G (1957) História natural e médica da Índia Ocidental. Instituto Nacional do Livro, Rio de Janeiro
Saint-Hilaire A (1974) Viagem pelo Distrito dos Diamantes e Litoral do Brasil. Edusp/Itatiaia, São Paulo/Belo Horizonte
Salles Cunha EM (1965) Sambaquis do litoral carioca. Instituto Brasileiro de Geografia e Estatística/Conselho Nacional de Geografia, Rio de Janeiro
Salvador V (1918) História do Brasil: 1500–1627. Fundação Biblioteca Nacional, Rio de Janeiro
Sampaio AJ (1945) Fitogeografia do Brasil. Companhia Editora Nacional, São Paulo
Schaeffer Novelli Y, Cintrón-Molero G (1999) Brazilian mangroves: a historical ecology. Ciência e Cultura (São Paulo)
Schneider P (2011) The discovery of tropical mangroves in Graeco-Roman antiquity: science and wonder. J Hakluyt Soc:1–16
Soffiati A (2006) O manguezal na história e na cultura do Brasil. Faculdade de Direito de Campos, Campos dos Goytacazes
Soffiati A (2016) Da mão que captura o caranguejo à globalização que captura o manguezal. In: Soffiati A (ed) Tempo e espaço nos manguezais. Autografia, Rio de Janeiro
Sousa GS (1938) Tratado descritivo do Brasil em 1587, 3rd edn. Companhia Editora Nacional, São Paulo
Thévet A (1944) Singularidades da França Antártica. Companhia Editora Nacional, São Paulo
Toynbee A (1978) A humanidade e a mãe Terra – Uma história narrativa do mundo. Ed. Zahar, Rio de Janeiro
Toynbee A (1986) Um estudo da História, 2nd edn. Editora Universidade de Brasília, São Paulo
Vanucci M (1999) Manguezais e nós – uma síntese de percepções. EDUSP, São Paulo
von Martius CFP (1833) Flora Brasiliensis. Sumptibus J. G. Cottae, Stuttgart
Wied-Neuwied M (1989) Viagem ao Brasil. Companhia Editora Nacional, São Paulo
Wiener C (1876) Estudos sobre os sambaquis do sul do Brasil. Arquivo do Museu Nacional Museu Nacional, Rio de Janeiro
Yesner DR (1980) Maritime hunters-gatherers: ecology and prehistory. Curr Anthropol 21:727–750

Chapter 15
Sociocultural Valuation of Mangroves: Subsidies for Public Policies Towards the Conservation of Brazilian Coastal Wetlands

Luciana S. Queiroz, Sergio Rossi, and Antônio Jeovah A. Meireles

15.1 Introduction

Mangroves are rich, diverse, and complex ecosystems at the interface between terrestrial, estuarine, and marine systems in coastal zones present in the tropical and subtropical regions of 123 countries (Barbier et al. 1997; Spalding et al. 2010). These ecosystems provide at least US$ 1.6 billion each year in ecosystem services, supporting coastal livelihoods of communities with raw materials and food, coastal protection, soil erosion control, water purification, maintenance of fisheries, carbon sequestration, and recreation, education, and research possibilities (Costanza et al. 1997; Barbier et al. 2011). Some worldwide assessments have considered mangroves as a subset of other coastal ecosystems in the economic evaluations of ecosystem services (ES). However, the contribution of mangrove ecosystems to the aggregate economic value is often hard to disentangle. The possible pitfall in

L. S. Queiroz (✉)
Institut de Ciència I Tecnologia Ambientals, Universitat Autònoma de Barcelona, Edifici ICTA-ICP, Cerdanyola del Vallés, Catalunya, Spain

Department of Geography, Federal University of Ceará, Pici University Campus Building 911, Fortaleza, CE, Brazil

S. Rossi
Institut de Ciència I Tecnologia Ambientals, Universitat Autònoma de Barcelona, Edifici ICTA-ICP, Cerdanyola del Vallés, Catalunya, Spain

Dipartimento di Scienze e Tecnologie Biologiche e Ambientali, Università del Salento, Lecce, Italy

A. J. A. Meireles
Department of Geography, Federal University of Ceará, Pici University Campus Building 911, Fortaleza, CE, Brazil

Y. Schaeffer-Novelli et al. (eds.), *Brazilian Mangroves and Salt Marshes*, Brazilian Marine Biodiversity, https://doi.org/10.1007/978-3-031-13486-9_15

such large-scale studies is that there is considerable overlap with several other ecosystem types, possibly leading to double counting. For instance, mangroves are either combined with tidal marshes (wetlands) in Costanza et al. (2014) or divided into "tropical forests," "coastal systems," and "coastal wetlands" in de Groot et al. (2012).

In South America, mangroves have been exploited by society for thousands of years, but it has been in the last 400 years when a systematic transformation of these wetlands has taken place (López-Angarita et al. 2016). The mangroves that we now see are far different from the original ones, being in general younger and less structured and biodiverse (Valiela et al. 2001). In fact, this is a long story of interaction, marked by profound transformation over the last few centuries due to the shift from pre-Columbian to colonial management. Since 1990, despite the increasingly positive attitude towards mangroves and their inclusion in protected areas and conservation policies, mangrove cover has continued to decline due to expanding human activities (e.g., agriculture, aquaculture, and coastal development), even in the presence of laws prohibiting their removal (López-Angarita et al. 2016). Remnant mangroves are severely threatened, with up to 40% of the mangrove plant species being susceptible to extinction in some regions (Polidoro et al. 2010). This loss and degradation may seriously undermine the ability of mangroves to provide valuable ES for present and future generations (Feller et al. 2010). Alongi (2002) predicted that in the 25 years following his study, shrimp aquaculture, together with overfishing and other intensive practices, would be the greatest threats to mangrove conservation. Duke et al. (2007) reinforced this point of view, setting out a very bleak prospect for one of the world's greatest providers of biodiversity and ecosystem services at local and global levels. Stemming this loss is urgent and requires better management of intact and damaged mangrove ecosystems, including restoration efforts. It also calls for systematic assessments of current "stocks" and "flows" of ES to ensure the sustainable use of these resources (Bateman et al. 2013). Since mangroves have not received their due share of conservation, these ecosystems have been greatly reduced and fragmented over the last decades due to excessive exploitation and "human development" (Giri et al. 2011).

Despite the cultural, ecological, and economic importance of mangroves and legislation designed to protect this frontier, land-to-sea transitional ecosystems worldwide, these forests are in serious decline. Over the last 20 years, mangroves have suffered degradation and an annual loss of between 0.16% and 0.39% due to rapid coastal development (Hamilton and Casey 2016). Extensive loss has left degraded and highly fragmented mangroves in many parts of the world (Giri et al. 2011; Hamilton and Casey 2016). These fragments may have limited potential to deliver services in the future (Barbier et al. 2011; Lee et al. 2014).

Brazil is one of the countries that has been severely affected by shrimp farming, which represents the greatest threat to the country's mangrove conservation (Queiroz et al. 2013a). During the last 40 years, industrial shrimp farming in Brazil has experienced intense growth. The first shrimp culture experiments were carried out at the beginning of the 1970s, but they have failed due to technical problems and a lack of appropriate knowledge. At the end of the 1990s, the activity expanded rapidly, and shrimp farming became a relevant export industry, due to government assistance, public-bank financing, university technical collaboration, and legislative permissiveness. The vertiginous growth of the industry has been accompanied by a profound transformation in natural resources, causing the degradation of mangroves. Behind the numbers and high shrimp-production rates for export lies a context of conversion of extensive coastal areas (fragile and fundamental environmental systems like wetlands and mangroves) into production zones (shrimp farms), generating social, economic, and environmental impacts (Barbier and Strand 1998; Rönnbäck 1999; Polidoro et al. 2010). In the present chapter, we will present a brief description of how the ecosystem services have been evaluated under the economy's lenses, what is the relevance of cultural and non-tangible ecosystem services, and how they have been managed in Brazil. The chapter also discusses the noncompatibility of the industrial exploitation of mangroves (especially by the shrimp farm industry) with the maintenance of the local economy and biodiversity.

15.2 Ecosystem Services: The Sociocultural Approach

Over the past few decades, increasing efforts addressed the topic of the link between ecosystems and human well-being. Gradually, humans began to perceive (and forcefully face) that changes imposed on nature by their activities provoke effects, direct or indirect, on all components of well-being. From these perceptions, the concept of ecosystem services (ES) arose, aiming at bringing to light the relationships between people and nature embedded in daily life and to mobilize environmental conservation and management. ES have been defined as the direct and indirect contributions of ecosystems to human well-being: in synthesis, ES are benefits people obtained from nature, directly enjoyed, consumed, or used (Paoli et al. 2017). From this original definition, the concept has been later applied and interpreted in multiple and often contested ways and raised several significant questions of scientific and ethical nature (Jax et al. 2013).

Ecosystem services are essential for human well-being, but the links between ecosystem services and human well-being are complex, diverse, context-dependent, and complicated by the need to consider different spatial and temporal scales to assess them properly (Paoli et al. 2017). Human society has and will always be faced with the decision of how to manage ecosystems for sustainability. This is also true for the mangrove ecosystem that has often been converted to alternate use, based solely on economic consideration by policymakers (James et al. 2013). One main reason for mangrove deforestation is that wetlands throughout the world are still considered to have little or no value, or even sometimes to have a negative value (Mitsch and Gosselink 1993; López-Angarita et al. 2016). Probably the main problem in this sense is a lack of appreciation of the multiple functions of the ecosystem and associated services (James et al. 2013; Arias-González et al. 2017). Several services have been estimated for mangroves (see Table 15.1), being those related with direct economic benefits or with biogeochemical cycles – these counting

Table 15.1 Ecosystem services provided by mangroves identified in the literature review and by Cumbe community informants

Services	Characterization
Regulation/production of gases	Regulation of atmospheric chemical composition: SO_2 levels, CO_2/O_2 balance.
Climate regulation	Global temperature, precipitation, and biological processes that mediate local and global climatic phenomena (greenhouse effect).
Water supply	Water storage and retention (aquifer and reservoir dynamics).
Coastal protection against extremes	Buffering of ecosystem responses associated with environmental fluctuations (protection against storms, control of fine sediment production, and controlled environmental variability by vegetation structure).
Hydrological regulation	Regulation of hydrological flows integrated with watersheds (water for agricultural and industrial activities; transportation of people, food, etc.).
Erosion control and sediment retention	Soil conservation within the ecosystem (prevention of slides and other processes of material removal).
Soil formation	Soil formation process (weathering of rocks and accumulation of organic material).
Nutrient cycling	Storage, internal recycling, processing, and acquisition of nutrients (fixation of N, P, and other elements of the nutrient cycle).
Material and energy dissipation	Recuperation, removal, and control of excess nutrients and organic compounds (control of contaminants).
Pollination	Movement of gametes for population reproduction.
Biological control	Regulation of trophic dynamic of populations.
Biodiversity regulation	Biological interactions between organisms and with abiotic components of ecosystems.
Refuge	Habitat for resident and migratory populations (stopover, nursery, and feeding areas for migratory birds).
Food production	Part of gross primary production transformed into food (fish, mollusks, crustaceans, and subsistence of activities).
Primary production	Part of gross primary production transformed into raw materials (lumber, fuel, and forage).
Genetic resources	Production of materials and biological products for medicine, scientific materials, acquisition of genes resistant to pests, and ornamental species.
Recreation/tourism	Carrying out leisure activities (fishing, boat cruises meals with family and friends, games, etc.) and opportunities for various tourist activities.
Aesthetics	Mangroves as part of the coastal scenery.
Inspiration for culture and art	Mangroves are the motive and inspiration for artistic creations.
Spiritual	Many fisherfolk and indigenous communities recognize mangroves as sacred.
Maintenance of traditional ecological knowledge	In mangroves, traditional activities are carried out, which are important for the maintenance of autochthonic and ancestral knowledge.

(continued)

Table 15.1 (continued)

Services	Characterization
Science and environmental education	Important spaces for the development of scientific research and environmental education actions.
Creation and maintenance of social relationships	In mangroves, interpersonal relations are built and/or strengthened with people from the same community, neighboring communities, and visitors.
Personal satisfaction	The relation with mangroves generates sentiments of personal satisfaction for the communities, such as strength to live, richness (not from a monetary point of view), pride, and liberty.
Mental and physical relaxation	Using mangroves for resting, reflection, and/or physical activities for mental well-being and relaxation, functioning as therapy.

The results were obtained in the Cumbe community, with community participation in the definition of ecological services and compared with the scientific literature (Queiroz et al. 2017)
Modified from Queiroz et al. (2017)
Sources: Schaeffer-Novelli (1989), Barbier et al. (1997), Costanza et al. (1997), De Groot et al. (2002), MEA (2005a, b), McLeod and Salm (2006), Rivera and Cortés (2007), Kumar (2010), De Groot et al. (2010), Meireles and Campos (2010), and Fransan-Sanchez (2019)

with much more accurate calculations in literature (Saenger 1999; Walters et al. 2008; Hussain and Badola 2010; Warren-Rhodes et al. 2011).

Mangroves, however, also provide cultural ecosystem services that are defined by MEA (2005a) as "non-material benefits that people obtain from ecosystems through spiritual enrichment, cognitive development, reflection, recreation, and aesthetic experiences." This approach defines cultural ES as the interactions between environmental spaces (i.e., physical settings such as coasts, woodlands, allotments) and the cultural or recreational practices that take place within them. This places cultural ES in a geographic or site-specific context. In this framework, cultural benefits (in terms of experiences), identities, and capabilities are seen to arise from the mutually reinforcing relationships between the environment and the cultural practices (Fish et al. 2016). Thus, most of these services operate outside the market system and are integrally linked to the way of life, traditions, and other community-specific values (NRC 2004). Even though the cultural dimensions of well-being are multifaceted and complex (Russell et al. 2013), many studies highlight the importance of considering the cultural benefits of the environment to human well-being in environmental decision-making (e.g., Satz et al. 2013; Fish and Church 2014). This approach presents some of the most compelling reasons for ecosystem conservation; these benefits are considered a fundamental component of all current ES frameworks (Chan et al. 2011). Neglecting cultural services provided by ecosystems excludes considerations that often matter to vulnerable and otherwise underrepresented communities (Satz et al. 2013; Queiroz et al. 2017). It is thus of fundamental importance to understand how people perceive mangroves and to use this vision as another essential element in making such social-ecological systems sustainable in the long-term perspective (Kittinger et al. 2012; Gould et al. 2014; Queiroz et al. 2017).

Despite the abovementioned, cultural ES remains poorly understood as they are commonly subjective and have multifaceted and complex dimensions (Russell et al. 2013). Much of the coastal wetland valuation literature is focused on economic value, the social and cultural values not being directly ascribable to the ecological or the economic domain (Chiesura and De Groot 2003). The complexity of the perception of landscape and well-being by the community should be considered in the ecosystem service quantification, even if the quantitative tools used are new (Queiroz et al. 2017). A rigorous application of methods to quantify noneconomic values of mangroves is still lacking (James et al. 2013; Thiagarajah et al. 2015; Hsieh et al. 2015). Besides, decision-making processes should not neglect the experience of local communities (Raheem et al. 2012; Peres et al. 2016). In this sense, the concept of cultural ES offers a powerful way of conveying that natural systems underpin a range of benefits for the people (Fish and Church 2014). This approach presents some of the most compelling reasons for ecosystem conservation being considered a fundamental component of all current ES frameworks (Chan et al. 2011). However, there is no doubt that this social value of coastal wetlands is seldom captured by policy- and decision-making actors (Turner et al. 2000).

15.3 Calculating the Importance of Local Economy: The Fisheries Example

Another problem to be faced towards mangrove sustainability is the understanding of the real impact of local economies in mangrove ecosystems and the surrounding areas. How important are, for example, artisanal fisheries in the local context? Can we reliably calculate it? Small-scale fishing (SSF), significant in mangrove areas, is a highly productive sector accounting for more than 50% of the world's annual fish catch (FAO 2017). Concerning local communities, SSF plays a dynamic and diversified economic role, is typically respectful of local natural resources, and seeks sustainable habitat exploitation. They incorporate the values and traditions of the areas where they happen, favoring cohesive social processes that contribute to global cultural enrichment (FAO 2017). Although their importance has been demonstrated, artisanal fisheries are disappearing in many places (Tesfamichael et al. 2014). This is especially evident in coastal areas near large urban areas, where ca. 50% of the human population lives (Small and Nicholls 2003). Strategies for managing and recognizing the importance of artisanal fisheries are still very weak, once information is scarce or even nonexistent (Salas et al. 2007).

The local economies of traditional communities obey their logic but are nevertheless relevant to the capitalist societies in which they operate. Many authors stress that traditional communities are important agents in nature conservation (e.g., Saenger 1999; Rönnbäck et al. 2007), and so it is with mangroves and traditional communities. This reflects another important aspect of traditional cultures, which is an approach to natural resource management that is marked by respect for the system's natural cycles and exploitation that considers the capacity of animal and plant species for recuperation (Hussain and Badola 2010). Monocultures such as shrimp aquaculture have led to relegation of the traditional economy to a second and third level compared to the alleged progress associated with intensive shrimp farming (López-Angarita et al. 2016). In Brazil, thousands of families survive on artisanal fishing, but little is known about how and how much they contribute to the economic sustainability of local populations (Diele et al. 2005; Aburto-Oropeza et al. 2008; Walters et al. 2008; Hussain and Badola 2010). Data on artisanal fisheries are sparse, incomplete, and biased, and their position within a country's economic and social framework being very difficult to ascertain (Hussain and Badola 2010; Hellebrandt et al. 2014). That is the main reason for the invisibility of this economic sector.

Brazil is now facing a complicated fisheries scenario, in which biodiversity and renewable resources are threatened by the lack of appropriate management policies (Amaral and Jablonski 2005; Pinheiro et al. 2015). Some efforts to recognize the economic and social role of artisanal fishing have been made, including the creation of the Fisheries and Aquaculture Especial Secretariat (SEAP) in 2003. SEAP's main objective was establishing sustainable measures for exploring marine and fluvial renewable resources. Despite the measures adopted since 2000, efforts have not been regular, and the work has not yielded an ordered and effective body of data, especially for artisanal fisheries (Dias-Neto and Dias 2015). Data on Brazil's artisanal fisheries show that the sector is responsible for ca. 65% of the country's seafood production, employing 957,000 people (99.2% of the officially registered fishermen in the country) (IBAMA 2007; MPA 2012). Precise and reliable (comparable) data are, however, almost nonexistent, and understanding the current importance of this sector requires semiquantitative tools.

In fact, there are very few studies based on direct monitoring using economic and social tools, but there is a consensus that they are urgently needed to understand the potential loss of tangible and nontangible ES (Saenger 1999; Queiroz et al. 2017). Coastal communities are economically dependent on artisanal fishing (Kuhl and Sheridan 2009; Hussain and Badola 2010), but it is important to highlight that fishermen's know-how and practices are essential in any attempt to preserve the ecosystem because they are the people who best understand the seasonal cycles of renewable resources and the system's carrying capacity (Yates and Schoeman 2014). However, most of the time their voice is disregarded in top-down management and strategies (Saenger 1999). Mangroves are fundamental to the way of life of traditional communities (Diegues and Arruda 2001; Kuhl and Sheridan 2009; Hussain and Badola 2010; Queiroz et al. 2017), but the economy generated by local people working on the place (e.g., artisanal fisheries) is almost invisible in the official statistics (Queiroz et al. 2020). Artisanal fishing in Latin America is mostly maintained by the efforts of fishermen rather than through the support of official bodies (Acosta 1996). This is a low-investment economic sector that generates a variety of activities while producing food for local and regional markets. One of the few studies that include an in-depth economic evaluation of artisanal fisheries (and other services) is the one by Hussain and Badola (2010). The authors calculated that in areas surrounded by mangroves, income may be as high as US$ 44 per work hour. In areas where mangroves are not present, the rate drops to US$ 3 per hour. Nonetheless, care needs to be taken when considering these numbers, as each community is different, and the renewable resources may differ widely.

15.4 Exploitation of Mangroves for Aquaculture and Other Monocultures

Human society has and will always be faced with the decision of how to manage ecosystems for sustainability. This is also true for the mangrove ecosystem that has often been converted to human use, based solely on economic consideration by policymakers (James et al. 2013). People tend to forget that mangrove ecosystem conservation deserves special attention because of the number of people living within 10 km of significant mangrove areas, estimated to be 120 million by 2015 globally (UNEP 2014). The bulk of this population resides in developing countries in Latin America, Asia, and West and East Africa and is significantly dependent on mangrove resources for daily sustenance and livelihood.

Approximately 26% of mangrove forests worldwide are degraded due to overexploitation for fuelwood and timber production (Valiela et al. 2001). On the other hand, 38% of degraded mangrove areas are attributed to the conversion to industrial shrimp aquaculture (Ellison 2008), which makes this industry one of the most important causes for mangrove degradation and suppression (FAO 2010). In Brazil, the shrimp industry is considered the greatest threat to mangrove conservation (Queiroz et al. 2013a).

The shrimp industry argued that its expansion in tropical and subtropical areas would increase the supply of food, decrease the pressure on fish stocks, increase foreign exchange earnings, and provide food for countries in need, developing the policy idea of the "Blue Revolution" (Costa-Pierce 2002). Shrimp aquaculture has emerged as a major cause of the destruction of mangroves, coastal landscapes, and the transformation of livelihoods in areas where there has been intensive development (Barbier and Strand 1998; Rönnbäck 1999; Alongi 2002; Shanahan et al. 2003; Polidoro et al. 2010; Queiroz et al. 2013b; Queiroz 2014). The results on mangrove and adjacent ecosystems seem to be linked to poverty, food insecurity, displacement of communities, and pollution of drinking water, as well as poor conditions and impacts on the health of workers in the shrimp industry itself (Bailey 1988; Beveridge et al. 1994; Flaherty and Karnjanakesorn 1995; Dewalt et al. 1996; Stonich et al. 1997; Stanley 1998; Kautsky et al. 2000). This industry favors the destruction of habitats formerly used for artisanal fishery by reducing the possibility to perform extractive subsistence activities, endangering food security, and transforming the dynamics of life in traditional communities (EMBRAPA 2004; IBAMA 2005; Meireles et al. 2007; Meireles and Queiroz 2010; Warren-Rhodes et al. 2011; Montserrat 2011; Montserrat et al. 2011; Queiroz 2014).

Traditional cultures develop a small-scale commodity mode of production, opposite from the capitalist mode of shrimp production. For example, in the case of industrial aquaculture, the workforce but also nature itself becomes a commodity, transforming the broad perception of the mangrove habitat. These two societies (industrial and artisanal) have different rationales, presenting a set of social goals, which are consciously and differently developed to achieve a very distant set of objectives. One mode of production is oriented for monetary profit (aquaculture), where traditional collective solidarity disappears and therefore natural resources are degraded. The other (artisanal fisheries/collection) still belongs to a society whose goal is the maintenance of that collective solidarity and not the accumulation of assets and income. In this way of life, the natural resources on which they depend upon are preserved (Godelier 1984). Therefore, between these two types of society, there is a fundamental difference in the conception and representation of nature and its resources.

In Brazil, the average productivity of farmed shrimp reached 6084 kg/ha/year in 2003 (ABCC 2004; Rocha et al. 2004). Dote Sá (2010) established the average productivity of 12,194 kg/ha/year for shrimp farming developed in the environment of Jaguaribe River (Ceará State) which is higher than that of any Brazilian state, including Ceará itself (7676 kg/ha/year) (ABCC 2004; Rocha et al. 2004) (see Chap. 3, Map 5). If shrimp are sold at 2.34 EUR/kg, the economic profit generated from the commercialization of shrimp would be about 28,533 EUR/ha/year. When comparing this value with the economic gains generated by mangroves per hectare at 7120 EUR/ha/year, it is possible to conclude that in the very short term, the shrimp industry seems very appealing. However, part of the mangrove is deeply affected by this type of activity (see below), and the distribution of benefits is much lower among the community. Profits generated per hectare of shrimp farm are difficult to match by the economic values that a hectare of mangrove can provide, being apparently superior. It is widely recognized that shrimp aquaculture generates medium-term environmental damage of high importance because they must physically occupy and displace part of the natural resources of an area. This process of occupation and installation produces a range of biochemical changes in the ground, causing soil waterproofing and making them unusable (Alongi 2002; Shanahan et al. 2003; IBAMA 2005; Rivera-Ferre 2009; Polidoro et al. 2010; Queiroz et al. 2013a).

Many reasons make shrimp overexploitation and mangrove conservation noncompatible. For example, it is well known that mangroves are breeding grounds for many commercially important fish species (Robertson and Duke 1990). Wild shrimp spend a considerable amount of their life cycle within estuaries. The estuarine habitat provides nutrient-rich waters, and the mangrove rhizosphere provides shelter from predators. Any disturbance to this ecosystem by mangrove conversion results in a smaller fish population and lower incomes for fisherfolks and the health of the ecosystem (Spaninks and Van Beukering 1997). Several studies estimated that nearly 80% of fish catches in tropical coastal areas are directly or indirectly related to mangrove health (Costanza et al. 1997; Field et al. 1998; Sathirathai 2003; Ellison 2008; Polidoro et al. 2010). If we consider mangrove ES that gives economic benefit, it can be argued that their economic value would be estimated at ca. 10,000 EUR/ha/

year (Costanza et al. 1997). The lack of ownership and clear land-use policies, as well as the underestimation of other services nonvisible with conventional tools, has made mangroves vulnerable to an economy based on short-term economic growth. This is a common fact not only in this case but in many others where the opinion of traditional communities is not taken into consideration (Yates and Schoeman 2014).

Predatory practices, especially those related to achieving high productivity per hectare, have been widely adopted by aquaculture ventures and have generated social and environmental impacts with disastrous results, widely studied in Brazil (IBAMA 2005; EMBRAPA 2004; Queiroz et al. 2013a; Queiroz 2014; Lacerda et al. 2021). The search for high productivity in this context is revealing the true essence of the shrimp industry. While shrimp farms achieve temporary high productivity, they generate impacts such as the inevitable release of exotic species – in Brazil shrimp farming is based on exotic species (Lacerda et al. 2021) – competition with native species, and release of untreated effluents into water bodies, thus causing the decline of local species' stocks and making artisanal fishing unfeasible, thus causing the impoverishment of communities and jeopardizing their livelihoods.

In this context, the shrimp industry, behind the high productivity figures reported from its marketing, hides an unsustainable practice revealed when examined from the socio-ecological point of view. The development of this activity involves outsourcing high costs to society and the environment – disease, pollution, and poverty – while a minority appropriates the profits obtained, thus establishing a serious ecological and environmental conflict and reproducing environmental injustice (Martínez-Alier 2007; Acselrad et al. 2009; Meireles and Queiroz 2010). Furthermore, they lie about their own productivity when numbers demonstrate a decrease in terms of productivity and benefits. Queiroz et al. (2013a) showed that the official numbers presented by the shrimp aquaculture in the state of Ceará were virtually impossible, because the productivity repeated in different years the same number, rounded to the higher during several years. The opacity of the data given by this industry makes its practice not only difficult but also questionable.

Another point to consider is that this type of activity is favoring exclusively small societal groups at the expense of the impoverishment of traditional communities, reflecting a model characterized by the concentration of power and the appropriation of spaces and natural resources – that is the basis of environmental injustice. Traditional communities are led to territorial exclusion and insecurity caused by the impossibility of continuing traditional practices (Shanahan et al. 2003; C-CONDEM 2007; Montserrat 2011; Montserrat et al. 2011).

15.5 The Coast of Ceará State as Case Study

The shrimp aquaculture industry in NE Brazilian mangroves (e.g., Ceará State) developed upon *apicuns* (i.e., salt flats), drastically reducing fluvial and coastal mangrove forests (Table 15.2 and Figs. 15.1 and 15.2). Data shows that the shrimp industry expansion is the main driver for mangrove regression in this region

Table 15.2 Observed area of mangrove forests, salt flats, and shrimp pond in the state of Ceará, Northeast Brazil

	Mangrove forests (ha)			Salt flats (ha)			Shrimp ponds (ha)		
Estuary[a]	–1988	1988–1998	1998–2008	–1988	1988–1998	1998–2008	–1988	1988–1998	1998–2008
Timonha/Ubatuba	4,652.05	4,844.26	5,220.12	3.954.00	3,883.09	3,332.07	947.06	1,022.89	1,295.80
Remédios	446.88	587.02	536.14	1,345.94	1,239.91	921.06	0	0	484.05
Coreaú	3,281.78	3,432.34	3,626.31	3,416.13	2,991.88	1,848.66	416.70	799.41	1,486.39
Acaraú/Zumbi	3,266.75	3,673.36	3,448.15	3,711.93	3,205.11	2,553.28	154.11	309.10	1,768.33
Aracatimirim	68.54	134.75	137.65	379.18	352.14	329.42	0	0	33.41
Aracatiacu	931.03	743.20	844.20	714.48	900.03	461.29	0	0	309.59
Mundaú	1,254.95	1,028.58	1,205.46	660.98	595.68	389.70	0	148.21	311.07
Curu	91.09	91.44	132.00	134.35	128.44	65.26	0	0	465.79
Ceará	790.83	825.44	1,066.47	420.32	510.73	269.11	189.70	203.66	169.02
Cocó	857.16	728.84	905.25	210.43	102.86	64.89	129.18	72.15	53.61
Pacoti	570.49	608.00	850.09	452.18	385.79	275.02	0	44.41	14.91
Mal Cozinhado	63.92	52.86	69.54	154.78	143.96	166.44	72.65	73.83	72.90
Choró	109.96	90.85	113.17	571.72	654.72	442.87	109.89	91.79	349.45
Pirangi	209.23	174.19	160.44	3,079.84	3,127.91	2,010.95	971.10	1,017.32	1,485.02
Jaguaribe	995.37	885.10	994.41	528.97	366.67	365.64	688.23	957.80	1,811.29
Barra Grande	109.14	80.64	87.51	180.94	99.96	70.26	466.36	592.56	618.14
Total	17,699.17	17,980.87	19,396.91	19,916.10	18,688.88	13,565.92	4,144.98	5,333.13	10,728.77

Sources: AQUASIS (internal document); Thiers et al. (2017)
[a]See Fig. 15.1 and Chap. 3, Map 5 for the locations

Fig. 15.1 Estuaries of mangrove occurrence in the state of Ceará listed in Table 15.2. (Source: research database)

(Fig. 15.3). Thiers et al. (2017) found a discrepancy between official data and the data acquired through diverse methods and algorithms used to analyze the system's transformation. In fact, such changes deeply affected many traditional communities socially and economically, such as the Quilombo do Cumbe (Aracati, CE), increasing land ownership conflicts (Leroy and Meireles 2013). The quality of life of shrimp aquaculture workers is another worrisome aspect of this industry, including poor work conditions. In some cases, human rights violations have taken place (Meireles and Queiroz 2010; Queiroz 2014) (Fig. 15.3).

The Quilombo do Cumbe (Fig. 15.2) community has been practicing their own traditional management of mangroves on which their livelihoods depend; this makes this community an interesting case study. The research by Queiroz (2014) and Queiroz et al. (2017) identified and characterized the value of mangrove ES based on both the existent literature and the community perception and analyzed how these ES are embedded into the community's livelihood.

The abovementioned community has 621 inhabitants, whose livelihoods directly depend on mangroves. Their main activities are fishing, gathering shellfish (gleaning) and collecting crabs, developing a natural resources management system through a close relationship to natural cycles, and bordering a somewhat complicity

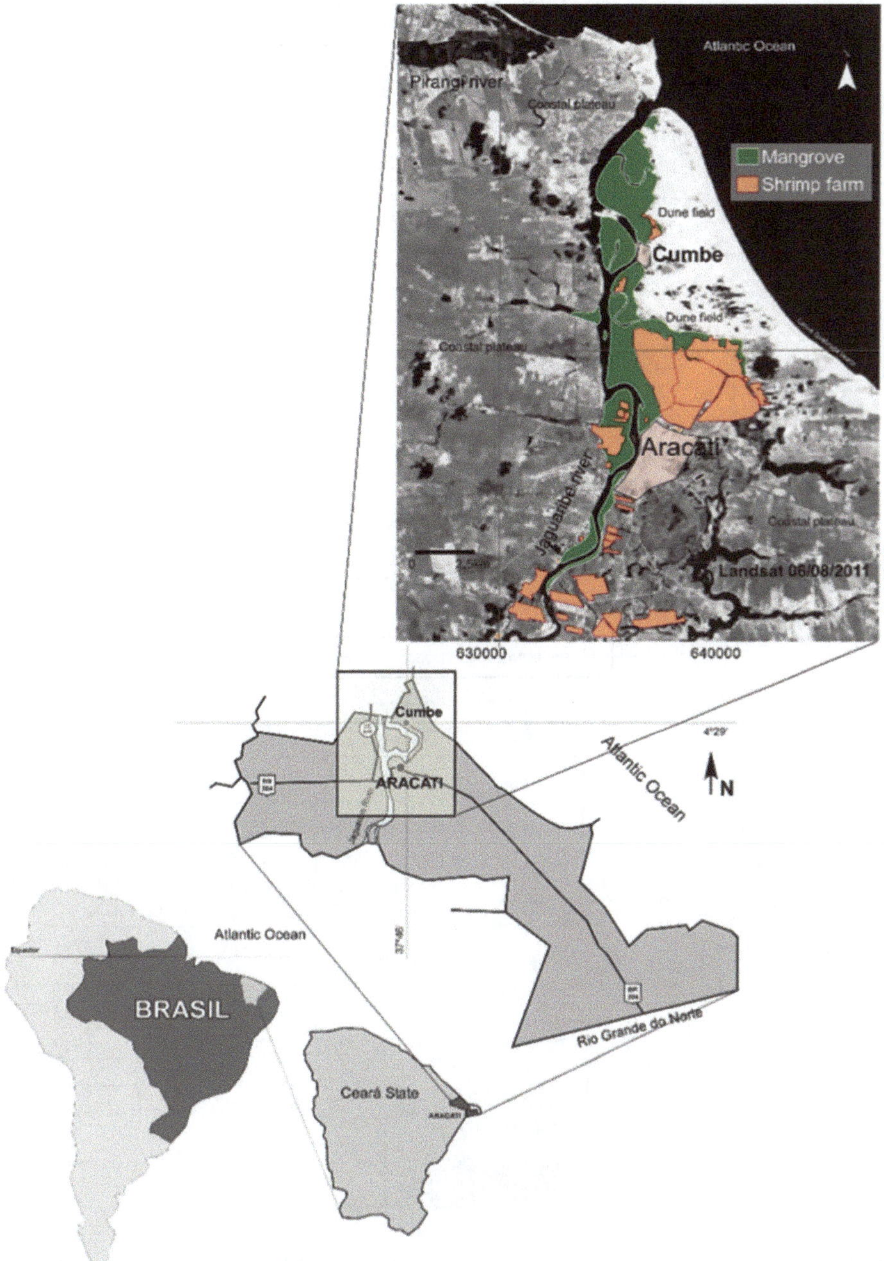

Fig. 15.2 Location of the Quilombo do Cumbe, Ceará State, Brazil. Green, mangrove forest; orange, shrimp aquaculture ponds

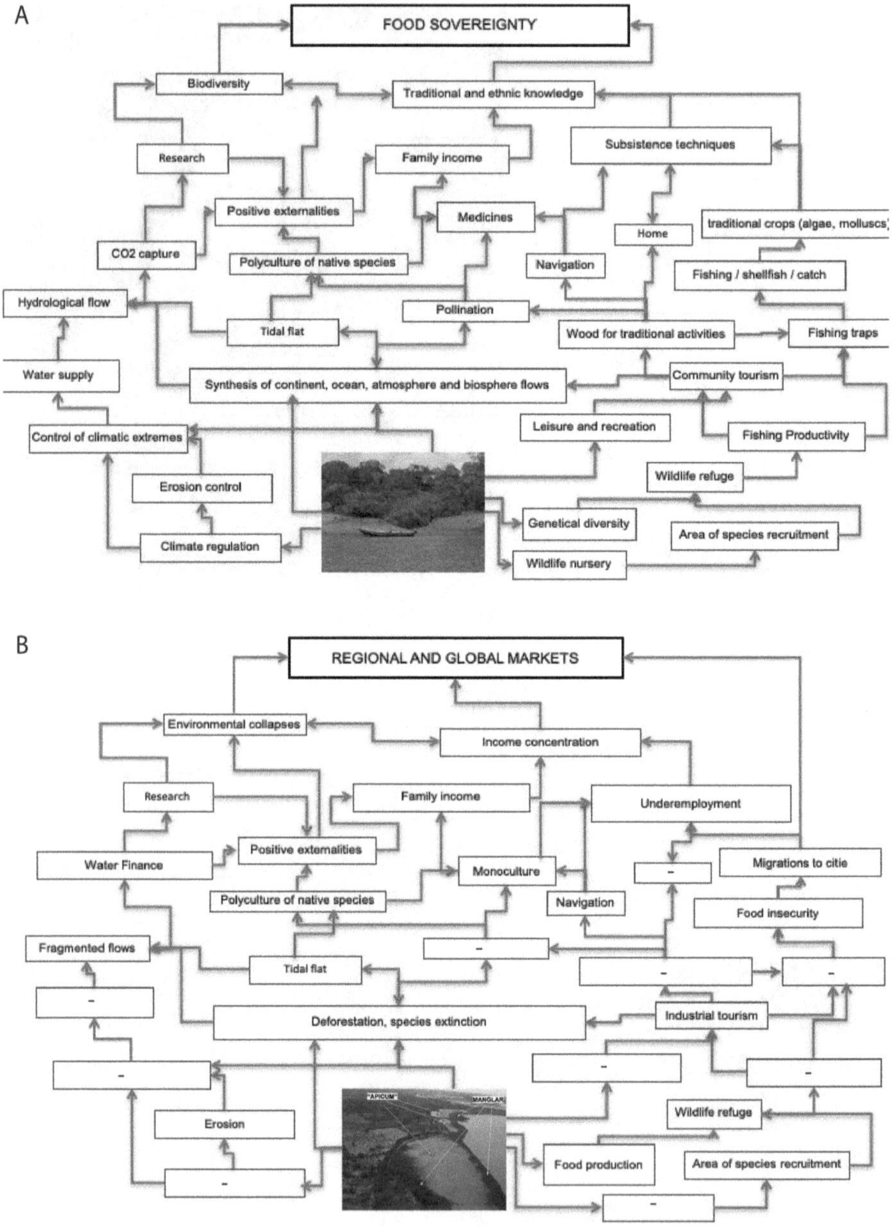

Fig. 15.3 Main mangrove ecosystem services and functions related with food sovereignty (**a**); threats to mangroves converted for shrimp farming (with extinct or fragmented components) (**b**). (Source: Queiroz et al. 2013a)

with nature. These "quilombolas"[1] maintain strong economic and symbolic ties with land and sea through continuous observation and accumulated knowledge. The Jaguaribe River basin, in which Quilombo do Cumbe is located, is the largest river basin in Ceará with an area of 72,645 km^2. A large portion of all local shrimp farms (44.2%) directly affected the mangroves; 63.6% of the farms caused serious damage to the riparian forest of endemic carnaúba palms (*Copernicia prunifera*) (Queiroz et al. 2013b).

It was thus possible to describe a set of ecosystem services altered by mangrove degradation (Fig. 15.2). Queiroz et al. (2017) demonstrated that societal relationships were altered and that the community dynamics were broken after the ingress of the shrimp farming industry. Interestingly, this study identified the joy of locals being closely linked to the proximity to the forest. For the *quilombolas*, mangroves were considered spaces for meditation and reflection: "The mangrove is the best place to hear the noise of the wind. It is a place for my thoughts. The truth is that at times I get stressed in the city, but I never get that way when I am in the mangrove." The testimony is in line with studies on coastal environments evidencing that people living closer to the coast self-report higher levels of health and personal fulfillment (e.g., Wheeler et al. 2012). For the studied community, the mangrove constituted a critical aspect of their worldview and their sense of belonging. A fisherman stated: "Mangroves mean everything to me, they are life. I feel privileged to be part of it, to live close to it, to open my window and see this landscape makes me feel well and happy because it is from where I draw sustenance for myself and my family." This perception of mangroves held by the fishermen of the Quilombo do Cumbe links ecosystem functions, services, and well-being, highlighting the prominent role of cultural services (James et al. 2013; Bell et al. 2015; Hsieh et al. 2015; Thiagarajah et al. 2015).

Another important point to highlight is the loss of food sovereignty caused by shrimp farming. It has been demonstrated in this area that the earnings of people living from fisheries or clam harvesting are potentially higher than those of people working on shrimp ponds (Queiroz 2014). The search for excessive productivity in a short-elapsed time causes systemic collapse and inhibits other future uses of this system (Alongi 2002). The shrimp aquaculture industry has often been presented as "one of the most lucrative economical activities" in Brazil, but the truth is that no more than 5% of the benefits return to the local people (Queiroz 2014). The decline and the consequences of the business failure, with the inevitable abandonment of the shrimp ponds, came without reaching the expected earnings for the investors. The main problem was that the coastal and fluvial communities lost the mangrove ecosystem welfare, their economic inputs, and part of their ways of living. The

[1] Up to a hundred years later from the signing of the Áurea Law (Lei Áurea) that freed the enslaved in Brazil, *quilombos* were considered places with large concentrations of enslaved African or Afro-Brazilians who rebelled against and escape from the colonial regime. With the Federal Constitution of 1988, the term *quilombo* had its concept expanded so that today it is considered any area occupied by communities that remain from the former *quilombos*. Source: http://www.palmares.gov.br/?p=19099 (accessed on June 2021).

food sovereign has been deeply affected by this boom-and-bust industry (Acselrad et al. 2009). Overall, the NE of Brazil has been considered an emblematic case study of an exponential monoculture that is bound to fail in its structure, but not without spoiling a rich, complex coastal landscape.

15.6 Final Remarks

This chapter intended to show the relevance of social and cultural valuation of mangrove ES in management and decision-making and the importance of considering local users' perceptions in conservation policies. Primarily the case study in Quilombo do Cumbe contributes to the advancement in the theoretical framework and methodological approach of sociocultural valuation of the ecosystem services. The study captured the importance of locally identified cultural services that are context-specific, in such a community that sees beyond monetary value. In order to accomplish that, further research should employ valuation surveys and participatory methods such as focus groups and participant observation to gather information and actively involve target communities – whose design should be informed by both international and local studies. Furthermore, such studies have implications for mangrove conservation. The fishers of the Quilombo do Cumbe community maintain strong symbolic ties with land and sea through continuous observation and interpretation of natural cycles for the sake of the sustainable management of mangroves. Such understanding and close relationship with mangroves lead, intentionally or unintentionally, to their environmental protection, as it has been shown in other areas (Walters 2004). Nevertheless, fishers' comprehension and perceptions of mangroves have not been considered in past and current government management policies in the coastal area. Instead, shrimp aquaculture has been prioritized over artisanal mangrove exploitation, which led to rapid degradation of mangrove habitat and resources (Queiroz et al. 2013b). It is necessary, as highlighted in other coastal wetlands, to include social value in policy- and decision-making (James et al. 2013). Such an approach responds to the United Nations Sustainable Development Goals of improving human well-being and promoting the conservation of marine ecosystems (United Nations 2015). We, therefore, suggest that mangrove conservation and management should embrace such complexity by considering community perceptions of ecosystems and well-being as an indispensable criterion for confronting the key challenges in conservation.

The chapter aimed at demonstrating that economic goods derived from direct extraction of natural resources can be as important as other types of exploitation such as shrimp aquaculture (Rönnbäck 1999). A rigorous application of methods to capture noneconomic values is still lacking and decision-making processes should not neglect the experience of autochthonous populations (Raheem et al. 2012).

References

ABCC – Associação Brasileira dos Criadores de Camarão (2004) Projeto executivo para apoio político ao desenvolvimento do camarão marinho cultivado), Recife, Brasil. www.mcraquacultura.com.br/arquivos/projeto/Marco04.pdf. Accessed 20 Dec 2005

Aburto-Oropeza O, Ezcurra E, Danemann G, Valdez V, Murray J, Sala E (2008) Mangroves in the Gulf of California increase fishery yields. P Natl A Sci USA 105:10456–10459

Acosta VG (1996) Historia y Desastres en América Latina, vol 1. La Red/Ciesas, Mexico

Acselrad H, Campello CA, Mello GNB (2009) O que é justiça ambiental? Editora Garamond, Rio de Janeiro

Alongi DM (2002) Present state and future of the world's mangrove forests. Environ Conserv 29(3): 331–349

Amaral ACZ, Jablonski S (2005) Conservation of marine and coastal biodiversity in Brazil. Conserv Biol 19(3):625–631

Arias-González JE, Rivera-Sosa A, Zaldívar-Rae J, Alva-Bausrto C, Cortés-Useche C (2017) The animal forest and its socio-ecological connections to land and coastal ecosystems. In: Rossi S, Bramanti L, Gori A, Orejas C (eds) Marine animal forests. Springer, Cham, pp 1209–1240

Bailey C (1988) The social consequences of tropical shrimp mariculture development. Ocean Shoreline Manag 11(1):31–44

Barbier EB, Strand I (1998) Valuing mangrove fishery linkages: a case study of Campeche, Mexico. Environ Res Econ 12:151–166

Barbier EB, Acreman M, Knowler D (1997) Economic valuation of wetlands: a guide for policy makers and planners. Ramsar Convention Bureau, Gland

Barbier EB, Hacker SD, Kennedy C, Koch EW, Stier AC, Silliman BR (2011) The value of estuarine and coastal ecosystem services. Ecol Monogr 81(2):169–193

Bateman IJ, Harwood AR, Mace GM, Watson RT, Abson DJ, Andrews B, Binner A, Crowe A, Day BH, Dugdale S, Fezzi C, Foden J, Hadley D, Haines-Young R, Hulme M, Kontoleon A, Lovett AA, Munday P, Pascual U, Paterson J, Perino G, Sen A, Siriwardena G, van Soest D, Termansen M (2013) Bringing ecosystem services into economic decision-making: land use in the United Kingdom. Science 341:45–50

Bell SL, Phoenix C, Lovell R, Wheeler BW (2015) Seeking everyday wellbeing: the coast as a therapeutic landscape. Soc Sci Med 142:56–67

Beveridge MCM, Ross LG, Kelly LA (1994) Aquaculture and biodiversity. Ambio 23:497–498

Chan KMA, Goldstein J, Satterfield T, Hannahs N, Kikiloi K, Naidoo R, Vadeboncoeur N, Woodside U (2011) Cultural services and non-usevalues. In: Kareiva P, Tallis H, Ricketts TH, Daily GC, Polasky S (eds) Natural capital: theory and practice of mapping ecosystem services. Oxford University Press, Oxford, pp 206–228

Chiesura A, De Groot R (2003) Critical natural capital: a socio-cultural perspective. Ecol Econ 44: 219–231

Corporación Coordinadora Nacional para la Defensa del Ecosistema Manglar – C-CONDEM (2007) Certificando la Destrucción: análisis integral de la certificación orgánica a la acuacultura industrial de camarón en Ecuador. C-CONDEM, Quito

Costanza R, d'Arge R, de Groot RS, Farber S, Grasso M, Hannon B, Limburg K, Naeem S, O'Neill RV, Paruelo J, Raskin RG, Sutton P, Van den Belt M (1997) The value of the world's ecosystem services and natural capital. Nature 387:253–260

Costanza R, de Groot R, Sutton P, van der Ploeg S, Anderson SJ, Kubiszewski I, Farber S, Turner K (2014) Changes in the global value of ecosystem services. Global Environ Chang 26:152–158

Costa-Pierce BA (2002) The 'Blue Revolution' – Aquaculture must go green. World Aquac 33(4): 4–5

de Groot RS, Wilson MA, Boumans RMJ (2002) A typology for the classification, description and valuation of ecosystem functions, goods and services. Ecol Econ 41(3):393–408 de Groot RS, Alkemade R, Braat L, Hein L, Willemen L (2010) Challenges in integrating the concept of ecosystem services and values in landscape planning, management and decision making. Ecol Complex 7(3):260–272

de Groot RS, Brander L, van der Ploeg S, Costanza R, Bernard F, Braat L, Christie M, Crossman N, Ghermandi A, Hein L, Hussain S, Kumar P, McVittie A, Portela R, Rodriguez LC, ten Brink P, van Beukering P (2012) Global estimates of the value of ecosystems and their services in monetary units. Ecosyst Serv 1:50–56

Dewalt BR, Vergne P, Hardin M (1996) Shrimp aquaculture development and the environment: people, mangroves and fisheries on the Gulf of Fonseca, Honduras. World Dev 24(7): 1193–1208

Dias-Neto J, Dias JFO (2015) O uso da biodiversidade aquatica no Brasil: uma avaliacao com foco na pesca. Ministério do Meio Ambiente and Instituto Brasileiro do Meio Ambiente e dos Recursos Naturais Renováveis, Brasília

Diegues AC, Arruda RSV (2001) Saberes tradicionais e biodiversidade no Brasil. Ministério do Meio Ambiente, Brasília, 176 pp

Diele K, Koch V, Saint-Paul U (2005) Population structure, catch composition and CPUE of the artisanally harvested mangrove crab Ucides cordatus (Ocypodidae) in the Caeté estuary, North Brazil: indications for overfishing? Aquat Living Resour 18(2):169–178

Dote-Sá T (2010) Carcinicultura ecológica y sostenible: Un nuevo enfoque para la producción lucrativa, la preservación del ambiente y el desarrollo social en el área de entorno del estuario de río Jaguaribe, Ceará – Brasil. PhD dissertation, Universidad Metropolitana de Asunción, 227 pp

Duke NC, Meynecke JO, Dittmann S, Ellison AM, Anger K, Berger U, Cannicci S, Diele K, Ewel KC, Field CD, Koedam N, Lee SY, Marchand C, Nordhaus I, Dahdouh-Guebas F (2007) A world without mangroves? Science 317(5834):41–42

Ellison AM (2008) Managing mangroves with benthic biodiversity in mind: moving beyond roving banditry. J Sea Res 59:2–15

Empresa Brasileira de Pesquisa Agropecuária – EMBRAPA (2004) Questões ambientais da carcinicultura de águas interiores: o caso da bacia do baixo Jaguaribe/Ceará. EMBRAPA, Fortaleza, 51 pp

Feller IC, Lovelock CE, Berger U, McKee KL, Joye SB, Ball MC (2010) Biocomplexity in Mangrove ecosystems. Annu Rev Mar Sci 2:395–417

Field CB, Osborn JG, Hoffman LL, Polsenberg JF, Ackerly DD, Berry JA, Björkman O, Held A, Matson PA, Mooney HA (1998) Mangrove biodiversity and ecosystem function. Glob Ecol Biogeogr Lett 7(1):3–14

Fish R, Church A (2014) Cultural ecosystem services: stretching out the concept. J Inst Environ Sci:31–44

Fish R, Church A, Winter M (2016) Conceptualising cultural ecosystem services: a novel framework for research and critical engagement. Ecosyst Serv 21(B):208–217

Flaherty M, Karnjanakesorn G (1995) Marine shrimp aquaculture and natural resource degradation in Thailand. Environ Manage 19(1):27–37

Food & Agriculture Organization of the United Nations – FAO (2010) The state of world fisheries and aquaculture 2009. FAO, Rome

Food & Agriculture Organization of the United Nations – FAO (2017) The state of world fisheries and aquaculture 2016. FAO, Rome

Fransan-Sanchez P (2019) Evaluation of mangrove connectivity using biological models with different larval dispersal. PhD thesis dissertation, Universidade de Sao Paolo, 87 pp

Giri C, Ochieng E, Tieszen LL, Zhu Z, Singh A, Loveland T, Masek J, Duke N (2011) Status and distribution of mangrove forests of the world using earth observation satellite data. Glob Ecol Biogeogr 20:154–159

Godelier M (1984) L'idéel et le matériel. Fayard, Paris

Gould RK, Ardoin NM, Woodside U, Satterfield T, Hannahs N, Daily GC (2014) The forest has a story: cultural ecosystem services in Kona. Hawai'i. Ecol Soc 19(3):55

Hamilton SE, Casey D (2016) Creation of a high spatio-temporal resolution global database of continuous mangrove forest cover for the 21st century (CGMFC-21). Global Ecol Biogeogr 25: 729–738

Hellebrandt D, Allison EH, Delaporte A (2014) Segurança alimentar e pesca artesanal: análise crítica de iniciativas na América Latina. Desenvolv Meio Ambiente 32:7–27

Hsieh HL, Lin HJ, Shih SS, Chen CP (2015) Ecosystem functions connecting contributions from ecosystem services to human wellbeing in a mangrove system in Northern Taiwan. Int J Env Res Pub He 12:6542–6560

Hussain SA, Badola R (2010) Valuing mangrove benefits: contribution of mangrove forests to local livelihoods in Bhitarkanita Conservation Area, East Coast of India. Wetl Ecol Manag 18:321–331

Instituto Brasileiro do Meio Ambiente e dos Recursos Naturais Renováveis – IBAMA (2007) Estatística da pesca Brasil. Grandes regiões e unidades de federações. Ministério do Meio Ambiente, Brasília DF

Instituto Brasileiro do Meio Ambiente e Recursos Renováveis – IBAMA (2005) Diagnóstico da Carcinicultura no Estado do Ceará – Relatório final. Ministério do Meio Ambiente, Brasília DF, 177 pp

James GK, Adegoke JO, Osagie S, Ekechukwu S, Nwilo P, Akinyede J (2013) Social valuation of mangroves in the Niger Delta region of Nigeria. Int J Biodivers Sci Ecosyst Serv Manage 9:311–323

Jax K, Barton DN, Chan KMA, de Groot R, Doyle U, Eser U, Görg C, Gómez-Baggethun E, Griewald Y, Haber W, Haines-Young R, Heink U, Jahn T, Joosten H, Kerschbaumer L, Korn H, Luck GW, Matzdorf B, Muraca B, Neßhöver C, Norton B, Ott K, Potschin M, Rauschmayer F, von Haaren C, Wichmann S (2013) Ecoststem services and ethics. Ecol Econ 93:260–268

Kautsky N, Rönnbäck P, Tadengren M, Troell M (2000) Ecosystem perspectives on management of disease in shrimp pond farming. Aquaculture 191:145–161

Kittinger JN, Finkbeiner EM, Glazier EW, Crowder LB (2012) Human dimensions of coral reef social-ecological systems. Ecol Soc 17(4):17

Kuhl L, Sheridan MJ (2009) Stigmatized property, clams, and community in coastal Ecuador. Ecol Environ Anthropol 5:1–22

Kumar P (ed) (2010) The economics of ecosystems and biodiversity: ecological and economic foundations: an output of TEEB: the economics of ecosystems and biodiversity. Earthscan, London

Lacerda LD, Ward RD, Godoy MDP, de Andrade Meireles AJ, Borges R, Ferreira AC (2021) 20-years cumulative impact from Shrimp farming on Mangroves of Northeast Brazil. Front For Glob Change 4:653096. https://doi.org/10.3389/ffgc.2021.653096

Lee SY, Primavera JH, Dahdouh-Guebas F, McKee K, Bosire JO, Cannicci S, Diele K, Fromard F, Koedam N, Marchand C, Mukherjee N, Record S, Mendelssohn I (2014) Ecological role and services of tropical mangrove ecosystems: a reassessment. Glob Ecol Biogeogr 23(7):726–743

Leroy JP, Meireles AJA (2013) Povos Indígenas e Comunidades Tradicionais: os Visados Territórios dos Invisíveis. In: Porto MF, Pacheco T, Leroy JP (orgs) Injustiça Ambiental e Saúde no Brasil – O Mapa de Conflitos. Fiocruz, Rio de Janeiro, pp 115–132

López-Angarita J, Roberts CM, Tilley A, Hawkins JP, Cooke RG (2016) Mangroves and people: lessons from a history of use and abuse in four Latin American countries. Forest Ecol Manag 368:151–162

Martínez-Alier J (2007) La defensa de los manglares contra la industria camaronera. Ecol Polit 32: 41–48

McLeod E, Salm RV (2006) Managing mangroves for resilience to climate change. IUCN, Gland

Meireles AJA, Campos AA (2010) Componentes Geomorfológicos, Funções e Serviços Ambientais de Complexos Estuarinos no Nordeste do Brasil. Rev ANPEGE 4:89–107

Meireles AJA, Queiroz L (2010) A monocultura do camarão: danos socioambientais à base da vida comunitária tradicional no litoral do Nordeste brasileiro. In: Zhouri A, Laschefski K (eds) Desenvolvimento e Conflitos Ambientais. Universidade Federal de Minas Gerais (UFMG), Belo Horizonte, pp 224–249

Meireles AJA, Cassola RS, Vanini STE, Queiroz LS (2007) Impactos ambientais decorrentes das atividades da carcinicultura ao longo do litoral cearense, nordeste do Brasil. Mercator 12:83–106

Millennium Ecosystem Assessment – MEA (2005a) Ecosystems and human well-being: bio-diversity synthesis, Millennium Ecosystem Assessment Series. Island Press, Washington, DC

Millennium Ecosystem Assessment – MEA (2005b) Ecosystems and human well-being: wetlands and water synthesis, Millennium Ecosystem Assessment Series. World Resources Institute, Washington, DC

Ministérie da Pesca e Aquicultura – MPA (2012) Boletim Estatístico da Pesca e Aquicultura – 2011. MPA, DEMOC, CGMIP, Brasília

Mitsch W, Gosselink J (1993) Wetlands. Van Nostrand Reinhold, New York

Montserrat MM (2011) Conflictos socioambientales de la acuicultura del camarón en Centroamérica: un análisis desde la justicia ambiental. Ecol Polit 41:90–96

Montserrat MM, Ortega-Cerdà M, Villegas EM, Ramos-Martín J (2011) Conflictos socioambientales de la acuicultura del camarón en Centroamérica. Un análisis desde la justicia ambiental. Fundació Ent, Vilanova i la Geltrú

National Research Council – NRC (2004) Valuing ecosystem services: toward better environmental decision-making. National Academy Press, Washington, DC

Paoli C, Montefalcone M, Morri C, Vassallo P, Bianchi CK (2017) Ecosystem functions and services of the marine animal forests. In: Rossi S, Gori A, Bramanti L (eds) Marine animal forests: the ecology of Benthic biodiversity hotspots. Springer, Cham, pp 1271–1312

Peres CM, Xavier LY, Santos CR, Turra A (2016) Stakeholders perceptions of local environmental changes as a tool for impact assessment in coastal zones. Ocean Coastal Manage 119:135–145. https://doi.org/10.1016/j.ocecoaman.2015.10.005

Pinheiro HT, Di Dario F, Gerhardinger LC, Melo MRS, Moura RL, Reis RE, Vieira F, Zuanon J, Rocha LA (2015) Brazilian aquatic biodiversity in peril. Science 350(6264):1043–1044

Polidoro BA, Carpenter KE, Collins L, Duke NC, Ellison AM, Ellison JC, Farnsworth EJ, Fernando ES, Kathiresan K, Koedam NE, Livingstone SR, Miyagi T, Moore GE, Nam VN, Ong JE, Primavera JH, Salmo SG III, Sanciangco JC, Sukardjo S, Wang Y, Yong JWH (2010) The loss of species: mangrove extinction risk and geographic areas of global concern. PLoS ONE 5: e10095

Queiroz L (2014) Industrial shrimp aquaculture and mangrove ecosystems: a multidimensional analysis of a socioenvironmental conflict in Brazil. PhD Dissertation, 174 pp, Universitat Autònoma de Barcelona

Queiroz L, Meireles J, Rossi S (2013a) Serviços ecossistêmicos costeiros e comunidades tradicionais. Rev ANPEGE 8(10):153–167

Queiroz L, Rossi S, Meireles J, Coelho C Jr (2013b) Shrimp aquaculture in the state of Cearà during the period 1970–2012: trends of the privatization of mangrove forest in Brazil. Ocean Coast Manag 73:54–62

Queiroz L, Rossi S, Calvet-Mir L, Ruíz-Mallén I, Betroz S, Prat J, Meireles AJA (2017) Neglected ecosystem services: highlighting the socio-cultural perception of mangroves in decision-making processes. Ecosyst Serv 26:137–145

Queiroz L, Rossi S, Mercader AT, Serra-Pompei C, Vide-Pifarré D, Carrasco-Domínguez J, Monrabà J, Carol MJ, Burriel MC, Briansó-Martínez M, Meireles AJA (2020) The economic and social framework of artisanal fishing in the state of Ceará, Brazil. GeoSaberes 11:180–198

Raheem N, Colt S, Fleishman E, Talberth J, Swedeen P, Boyle KJ, Rudd M, Lopez RD, Crocker D, Bohan D, O'Higgins T, Willer C, Boumans RM (2012) Application of non-market valuation to California's coastal policy decisions. Mar Policy 36:1161–1166

Rivera ES, Cortés IS (2007) Las experiencias del Instituto Nacional de Ecología en la valoración económica de los ecosistemas para la toma de decisiones. Instituto Nacional de Ecología, México. Gac Ecol 84:93–105

Rivera-Ferre MG (2009) Can export-oriented aquaculture in developing countries be sustainable and promote sustainable development? The shrimp case. J Agr Environ Ethic 22:301–321

Robertson AJ, Duke NC (1990) Mangrove fish-communities in tropical Queensland, Australia: spatial and temporal patterns in densities, biomass, and community structure. Mar Ecol 104: 369–379

Rocha IP, Rodrigues J, Amorim LAA (2004) Carcinocultura brasileira em 2003. Rev ABCC. http://www.abccam.com.br

Rönnbäck P (1999) The ecological basis for economic value of seafood production supported by mangrove ecosystems. Ecol Econ 29:235–252

Rönnbäck P, Crona I, Ingwall L (2007) The return of ecosystem goods and services in replanted mangrove forests: perspectives from local communities. Environ Conserv 4:313–324

Russell R, Guerry AD, Balvanera P, Gould RK, Basurto X, Chan KMA, Klain S, Levine J, Tam J (2013) Humans and nature: how knowing and experiencing nature affect well-being. Annu Rev Environ Resour 38:473–502

Saenger P (1999) Sustainable management of mangroves. In: Rais J, Dutton IM, Pantimena L, Dahuri R, Plouffe J (eds) Integrated coastal and marine resource management: proceedings of international symposium, National Institute of Technology (ITN) Malang in association with Bakosurtanal and Proyek Pesisir, Batu, 25–27 November 1999

Salas S, Chuenpagdee R, Seijo JC, Charles A (2007) Challenges in the assessment and management of small-scale fisheries in Latin America and the Caribbean. Fish Res 87:5–16

Sathirathai S (2003) Environmental governance: the case of industrial waste and pollution management in Thailand under the existing context of globalization. In: Abstracts of politics of the commons: articulating development and strengthening local practices, Chiang Mai, 11–14 July 2003

Satz D, Gould RK, Chan KMA, Guerry A, Norton B, Satterfield T, Halpern BS, Levine J, Woodside U, Hannahs N, Basurto X, Klain S (2013) The challenges of incorporating cultural ecosystem services into environmental assessment. Ambio 42:675–684

Schaeffer-Novelli Y (1989) Perfil dos ecossistemas litorâneos brasileiros, com especial ênfase sobre o sistema manguezal. Publicação Especial do Instituto Oceanográfico 7:1–16

Shanahan M, Thornton C, Trent S, Williams J (2003) Risky business: Vietnamese shrimp aquaculture impacts and investments. Environmental Justice Foundation, London

Small C, Nicholls RJ (2003) A global analysis of human settlement in coastal zones. J Coast Res 19: 584–599

Spalding MD, Kainuma M, Collins L (2010) World Atlas of mangroves. Earthscan, London

Spaninks F, van Beukering O (1997) Economic valuation of mangrove ecosystems: potential and limitations, CREED working paper 14. International Institute for Environmental Development, Amsterdam

Stanley DL (1998) Explaining persistent conflict among resource users: the case of Honduran mariculture. Soc Natur Resour 11:267–278

Stonich S, Bort J, Ovares L (1997) Globalization of shrimp mariculture: the impact on social justice and environmental quality in Central America. Soc Nat Resour 10:161–179

Tesfamichael D, Pitcher TJ, Pauly D (2014) Assessing changes in fisheries using fishers' knowledge to generate long time series of catch rates: a case study from the Red Sea. Ecol Soc 19(1): 18

Thiagarajah J, Wong SK, Richards DR, Friess DA (2015) Historical and contemporary cultural ecosystem service values in the rapidly urbanizing city state of Singapore. Ambio 44:666–677

Thiers PRL, Meireles AJA, Santos JO (2017) Ecossistema manguezal na costa oeste cearense: preservação permeada de meias verdades. Imprensa Universitária da Universidade Federal do Ceará, Fortaleza, 126 pp

Turner RK, Van den Bergh JCJM, Soderqvist T, Barendregt A, Van der Straaten J, Maltby E, Van Ierland EC (2000) Ecological-economic analysis of wetlands: scientific integration for management and policy. Ecol Econ 35:7–23

United Nations Environment Programme – UNEP (2014) The importance of mangroves to people: a call to action. In: van Bochove J, Sullivan E, Nakamura T (eds) United Nations Environment programme. World Conservation Monitoring Centre, Cambridge

Valiela I, Bowen JL, York JK (2001) Mangrove forests: one of the world's threatened major tropical environments. Bioscience 51:807–815

Walters BB (2004) Local management of mangrove forests in the Philippines: successful conservation or efficient resource exploitation? Hum Ecol 32:177–195

Walters BB, Rönnbäck P, Kovacs JM (2008) Ethnobiology, socio-economics and management of mangrove forests: a review. Aquat Bot 89(2):220–236

Warren-Rhodes K, Schwarz A, Boyle LN (2011) Mangrove ecosystem services and the potential for carbon revenue programmes in Solomon Islands. Environ Conserv 38(04):485–496

Wheeler BW, White M, Stahl-Timmins W, Depledge MH (2012) Does living by the coast improve health and wellbeing? Health Place 18:1198–1201

Yates KL, Schoeman DS (2014) Incorporating the spatial access priorities of fishers into strategic conservation planning and marine protected area design: reducing cost and increasing transparency. ICES J Mar Sci 72(2):587–594

Chapter 16
Mangrove and Salt Marsh Protected Areas in Brazil

Anders J. Schmidt, Jesus Manuel Delgado-Mendez, and Marília Cunha-Lignon

16.1 Introduction

After centuries of disrespect for mangrove and salt marsh ecosystems, there has been an increased recognition in recent decades of their importance and the need for their conservation. Current initiatives concerning environmental education, rehabilitation of degraded areas, and creation of protected areas (PAs), among others, can be readily accessed online using keywords such as "mangrove swamp" or "salt marsh" – leading to many sites containing information about the ecological, social, economic, and cultural roles of those ecosystems. That readily available knowledge of their value has not, however, been accompanied by a decrease in threats to their functioning, preservation, and sustainable use – at least not in Brazil.

Threats to mangroves and salt marshes in Brazil are considered by many authors as not widespread, but locally intensive (Spalding et al. 2010). To a certain extent, the containment of the further spread of environmental threats in Brazil has been attributed to the existence of relatively strong environmental laws that regard mangroves and salt marshes as areas to be preserved. Nonetheless, this chapter will show that the existing environmental legislation is currently being modified to suit goals of economic growth – this only reinforces the importance of creating new, representative, well-connected, and properly managed PAs (see SCBD 2010) so mangroves and salt marshes can be adequately protected.

A. J. Schmidt (✉)
Centro de Formação em Desenvolvimento Territorial, Universidade Federal do Sul da Bahia, Campus Paulo Freire, Teixeira de Freitas, BA, Brazil

J. M. Delgado-Mendez
Centro de Ciências Agrárias, Ambientais e Biológicas, Universidade Federal do Recôncavo da Bahia, Cruz das Almas, BA, Brazil

M. Cunha-Lignon
Campus de Registro, Universidade Estadual Paulista, Registro, SP, Brazil

Y. Schaeffer-Novelli et al. (eds.), *Brazilian Mangroves and Salt Marshes*, Brazilian Marine Biodiversity, https://doi.org/10.1007/978-3-031-13486-9_16

16.2 Threats to Mangroves and Salt Marshes

The need for PAs is closely related to actual and potential threats to local ecosystems. Brazilian mangroves and salt marshes suffer from anthropic pressure, and the administrators of those wetland PAs consistently report ongoing threats to their integrity (Delgado-Mendez 1989; Schmidt 2012a). The most recurring and generalized threat to wetlands is pollution originating in residential areas. Pollution by solid waste, mostly plastics, is not limited to urban wetlands. Even rural zones and isolated traditional populations face this type of pollution that is aggravated by the overall deficiency in garbage collection and treatment. Pollution by garbage remains difficult to control within PAs, as plastics may come from elsewhere floating on tidal currents or rivers that form the estuaries. The arriving plastics are easily trapped by mangrove forests and salt marshes (Marangoni and Costa 2009; Viehman et al. 2011; van Bijsterveldt et al. 2021). Marsh plants may be disturbed by shading effects of debris (Viehman et al. 2011), mangrove seedlings may be compromised by dislodgement and entanglement (Gorman and Turra 2016), and mangrove trees may be stressed by the covering of aerial roots or even die by suffocation by plastic (van Bijsterveldt et al. 2021). The deterioration of plastic in fragments makes this problem even more complex and a high abundance of microplastics has been recorded in Brazilian wetlands (Silva and de Sousa 2021; Zamprogno et al. 2021). In Brazilian estuaries, many cases of ingestion of microplastics by zooplankton, crabs, fish, and sea turtles have been reported and injuries caused to organisms and trophic transfer are a matter of concern (e.g., Santana et al. 2017; Dantas et al. 2020; Macieira et al. 2021).

In addition to the arrival of floating garbage, unprotected mangroves and salt marshes areas are often subject to unregulated garbage dumps at their upper zones. At these zones, plants and animals are directly affected not only by a thick layer of solid waste but also by the resulting leachate, including acidic water with high concentrations of chlorides and nutrients that contribute to the eutrophication together with sewage (Araújo and Maciel 1979; Seeliger and Costa 1988; Marangoni and Costa 2009).

Runoff and untreated sewage from urban or rural areas reach the mangroves often daily, and the high concentrations of organic material in those effluents reduce dissolved oxygen contents to lethal levels for many animals. Araújo et al. (2021) assessed the recent contamination levels by domestic effluents in mangrove sediments along the Brazilian coast. Mangroves from Natal, Maceió, and Rio de Janeiro metropolitan areas showed higher fecal contamination, due to the intense contribution of untreated domestic effluents. Mangrove trees may die when their lenticels are obstructed by particulate material carried with sewage waters (Maciel 1991). Both sewage and leachate can also contain heavy metals, causing contamination and bioaccumulation in mangroves and salt marshes (Seeliger and Costa 1988; Machado et al. 2002; Pinheiro et al. 2012; Vasconcelos et al. 2021; Costa et al. 2022; Tognella et al. 2022). This is particularly dangerous for humans as many animals such as crabs and fish are exploited for food. Likewise, untreated sewage waters may carry hepatitis and cholera, which can contaminate humans directly or through the ingestion of filtering shellfish among other seafood (Maciel 1991; Costa 2009; Denadai et al. 2015). Litter can also cause serious problems in terms of human public health. In Brazil, cases of yellow fever, dengue, and chikungunya may increase and spread due to the proliferation of mosquito larvae in the freshwater that accumulates on solid waste (Löwy 2017).

The diseases associated with degraded wetlands have traditionally been cited by public administrators as a means to justify the "urbanization" of irregular human occupations on mangroves and salt marshes. That occupation generally begins with the removal of vegetation and the establishment of precarious stilt houses on the bare mud and ends with the full conversion of the wetlands into urban areas. Human occupation adjacent to mangroves inevitably initiates a series of collateral effects, including the development of dense road networks and illegal wood extraction. Although mangrove deforestation is currently low, it remains a recurrent threat even within many PAs in Brazil (Schmidt 2012a). Many traditional populations do not have access to electricity or natural gas, so mangrove timber is widely used for cooking, as well as for building houses, boats, and material for artisanal fishing (Paludo and Klonowski 1999; Spalding et al. 2010).

In addition to the widespread problems mentioned above, each region of Brazil deals with specific environmental impacts of economic activities in coastal areas. According to PA managers (Schmidt 2012a), mangroves in the highly industrialized Southeast Brazil face intense threats linked to oil pollution and industrial effluents. The north of the country is particularly more affected by vegetation clearing by fire for agriculture. Impacts from tourism are more common in the south and northeast regions (Schmidt 2012a). Northeastern mangroves are the most affected by salt extraction and shrimp aquaculture. Both activities have directly affected mangroves by converting their upper zones (e.g., hypersaline tidal flats) into salt pans or shrimp ponds (Schaeffer-Novelli et al. 2016). The latter also indirectly impact mangroves through pollution inputs from effluents and associated loss of ecosystem services, including reductions in primary productivity, carbon storage, resilience to other environmental stressors, the efficiency of the estuarine filter, and biodiversity (Lacerda et al. 2021).

All those kinds of impacts threaten the natural resources and biodiversity of mangroves and salt marshes in Brazil. Since Brazilian environmental legislation is increasingly permissive for such impacts, the alternative strategy is to invest in a well-connected and effective system of protected areas. This is in line with the Post-2020 Global Biodiversity Framework, which proposes in the "Aichi Target 11" an expansion of protected areas to 30% of the earth's surface by 2030 (see SCBD 2018).

16.3 A Historical Review of Protected Areas in Brazil

Efforts for the conservation of mangroves in Brazil dates to colonial times, aiming to maintain the harvesting of firewood, house and boat construction, and extraction of tannic acids for tanning leather and fishing nets (Maciel 2001, see Chap. 14). The uncontrolled cutting of mangrove trees, together with the occupation of the borders of the ecosystem for housing, led to the imposition of the first regulations governing their use to be established by the Portuguese crown. In 1743, the Portuguese crown prohibited the cutting of red mangroves (*Rhizophora* spp.) for burning and, in 1760, prohibited cutting any trees of that species whose bark had not been previously used to produce tannin – a product exported from Brazil to Portugal and then sold to other countries. The economic motivation for protecting natural areas in Brazil at that time was purely motivated by extractivism.

An ecological motivation for the legal preservation of mangrove swamps (but not salt marshes) only emerged in the first Brazilian Forest Code (Brasil 1934). That set of laws did not specifically cite mangrove ecosystems but called out for permanent preservation of *protective forests* (i.e., avoiding erosion) – a category in which mangrove forests were included. The 1934 code also specified the category of *remnant forests*, which were those with biological and aesthetic attributes that justified their protection. That policy resulted in the creation of the first federal PA in the country, in 1937 – the Itatiaia National Park (Rio de Janeiro State) – and the first PA including a fragment of mangrove forest – the Monte Pascoal National and Historical Park, in 1961 (Bahia State) (Figs. 16.1 and 16.2, see Chap. 3, Maps 13 and 11, respectively).

The second forest code was created in 1965 (Brasil 1965) and designated certain vegetation types as having permanent preservation status, including mangrove ecosystems. However, it is worth mentioning that the language used left loopholes for alternative interpretations. Salt marshes were likewise protected under the category of vegetation bordering bodies of water, all subject to permanent protection.

The creation of the first PAs in Brazil followed the experiences of the United States of America and the recommendations of the World Commission of PAs of the International Union for Conservation of Nature (IUCN), involving specific legal instruments for each newly established management category. The first Brazilian initiatives to systematize PAs only began in the 1970s (Pádua 2011; Pureza et al. 2015) and the first peak of no-take PAs, including some mangrove areas, occurred during the early 1980s (Figs. 16.1 and 16.2), coinciding with the weakening of the military dictatorship. The new democratic republic started in 1985 when the number of PAs linked to sustainable use of natural resources started to increase. However, most of them did not comprise mangrove and salt marshes. The first PA with a focus on salt marshes – the Lagoa do Peixe National Park (Rio Grande do Sul State) – was created in 1986 (Figs. 16.1 and 16.2, see Chap. 3, Map 17).

In 1989 changes in the governmental agencies culminated in the creation of the Brazilian Institute of Environment and Renewable Natural Resources (IBAMA). The IBAMA then undertook a revision of existing conservation policies and

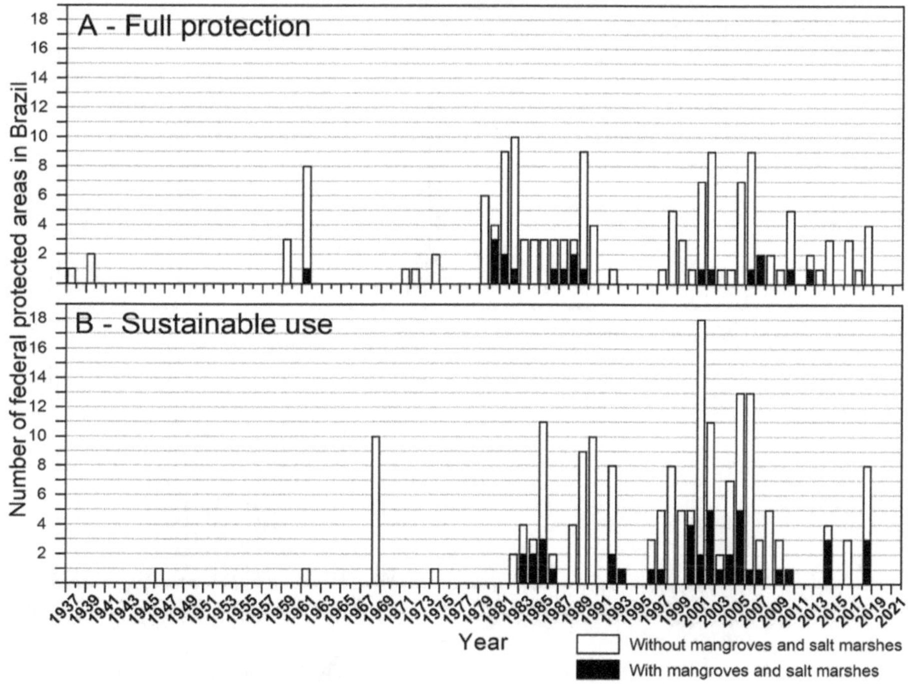

Fig. 16.1 Noncumulative amount of federal protected areas of the no-take (or full protection) (**a**) and sustainable use (**b**) categories, with and without mangroves and salt marshes, created over the years in Brazil. (Source: Ministry of Environment of Brazil, Department of Protected Areas)

consolidated them into a proposed law that was presented to congress in 1992 (Pádua 2011; Pureza et al. 2015). The middle of the 1990s is marked by a sharp reduction in the creation of PAs and by heated discussions concerning the different categories of PAs. On one side, environmentalists defended the idea of PAs without any human presence, while, on the other hand, social scientists and social activists defended the presence of humans, especially traditional populations, in those areas (Pádua 2011). After numerous debates and public hearings, the national system of PAs came into law in 2000 (Brasil 2000).

16.4 *Sistema Nacional de Unidades de Conservação da Natureza* (National System of Protected Areas, SNUC)

A relatively large number of categories of PAs were created in order to satisfy many different interests. These categories are divided into two large groups: no-take and sustainable use (Brasil 2000, 2002a). The no-take areas (aka full protection) group

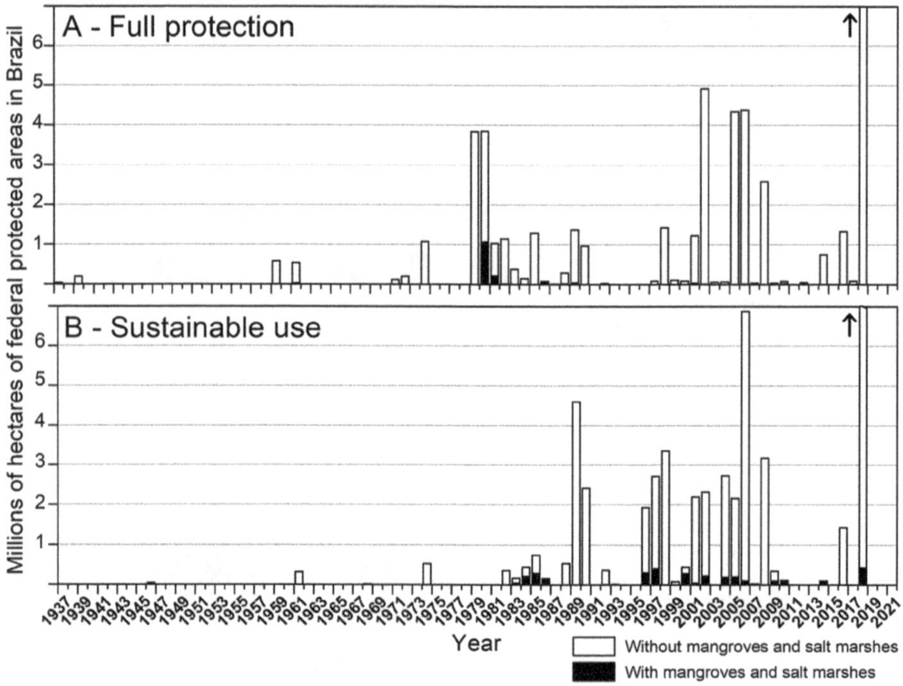

Fig. 16.2 Noncumulative coverage by federal-protected areas of the no-take (or full protection) (**a**) and sustainable use (**b**) categories, with and without mangroves and salt marshes, created over the years in Brazil. Source: Ministry of Environment of Brazil, Department of Protected Areas. The arrows indicate the large protected areas created in 2018. The areas covered by mangroves and salt marshes are smaller than those indicated by the bars, since many protected areas have only fragments of these ecosystems

includes five PA categories (Table 16.1) greatly influenced by the IUCN. The same did not hold for the sustainable use areas' categories (Table 16.2).

No-take or full protection areas aim at the integral conservation or preservation of biodiversity. They constitute areas in which no human settlements are allowed, in which only research or ecotourism activities can take place.

The categories of *Biological Reserve* and *Ecological Station* are analogous to the IUCN's category of Strict Nature Reserve (Ia) (see Dudley 2008). The two categories differ slightly concerning research permits, with Ecological Station accepting a certain degree of intervention or modification to that end, while a Biological Reserve would only allow research projects that require very limited environmental alterations. Public visitation to both is only permitted (exceptionally) for educational purposes. *National Parks*, on the other hand, are open to tourism (visitation and recreation). This type of protected area is similar to IUCN's Category II, with the condition that no private properties are allowed within its limits. Human settlements are allowed in two No-take protected area categories, preferably under temporary conditions. The first is *Wildlife Refuge*, designed to maintain conditions that

Table 16.1 No-take protection areas according to the Sistema Nacional de Unidades de Conservação da Natureza (National System of Protected Areas, SNUC) (Brasil 2000)

No-take areas (EN)	Áreas de Preservação Permanente (PT)	Goals and characteristics
Ecological Station	Estação Ecológica (EE)	Aims at full nature conservation and research development. Must be representative of Brazilian ecosystems. At least 90% of its area must be of integral protection of biota.
Biological Reserve	Reserva Biológica (REBIO)	Aims at integral biota conservation and adjacent attributes. Does not allow for human intervention except for occasional necessary ecosystem management.
National Park	Parque Nacional (PN)	Aims at the conservation of ecosystems of extreme ecological relevance or scenic beauty. Research, recreational, ecotourism, and education activities are allowed. There are subcategories: State Park ("Parque Estadual," PE, managed by the states) and Municipal Park ("Parque Municipal," PM, managed by the municipal authorities).
Natural Monument	Monumento Natural (MN)	Aims at protecting unique natural elements that may be singular, rare, or representative in aesthetics, cultural, or ecological aspects. They may occur within private property and allow for visitation under specific controlled settings.
Wildlife Refuge	Refúgio de Vida Silvestre (REVIS)	Aims at protecting natural spaces essential for the existence or reproduction of local or migratory species.

guarantee the existence or reproduction of certain resident or migratory animal species. These refuges can be situated in either public or private areas and may vary greatly in terms of size and aim. Consequently, some larger Wildlife Refuges are equivalent to IUCN's Wilderness Areas (Ib), while the smaller areas correspond to IUCN's Habitat/Species Management Areas (IV). The other Brazilian no-take protected area that allows human settlements is the *Natural Monument*, which was designed to preserve unique natural sites or landscapes of great natural beauty, equivalent to IUCN's Category III (see Dudley 2008).

The sustainable use group comprises seven distinct categories, four of which can be considered subdivisions of IUCN's Category VI (Protected Area with Sustainable Use of Natural Resources). *Sustainable Development Reserves* and *Extractive Reserves* are categories that consider traditional extractivist populations, although the areas used by them are of state's ownership. The differences between the two categories are quite subtle, with the former having a more conservationist nature and the latter assuming a more societal-oriented posture. *National Forests* and *Fauna Reserves* focus on research towards innovative management techniques for flora and fauna, respectively. *Areas of Relevant Ecological Interest* are more focused on conservation than sustainable use, although still classified within the sustainable use type of protected area. This category covers smaller areas with little or no human occupation and showing exceptional natural characteristics or the presence of rare species. As such, that type of area would better fit in the IUCN's Category IV. PAs of this category can be created on private or public lands. The same is true for

Table 16.2 Sustainable use areas according to the Sistema Nacional de Unidades de Conservação da Natureza (National System of Protected Areas, SNUC) (Brasil 2000)

Sustainable use areas (EN)	Áreas de Uso Sustentável (PT)	Goals and characteristics
Environmental Protection Area	Área de Proteção Ambiental (APA)	Aims at biological conservation while managing and ensuring human occupation and/or utilization of natural resources in a sustainable way. They correspond to extensive areas that present biotic, abiotic, aesthetic, or cultural attributes essential to the human populations that inhabited within them.
Area of Relevant Ecological Interest	Área de Relevante Interesse Ecológico (ARIE)	Aims at the protection of small areas up to 5000 ha of extension. They must present outstanding natural characteristics or show the occurrence of rare biota.
National Forest	Floresta Nacional (Flona)	Aims at promoting multiple sustainable uses of forest resources, besides research also on methods of better forest resources. There are subcategories: State Forest ("Floresta Estadual," managed by the states) and Municipal Forest ("Floresta Municipal," managed by the municipal authorities).
Extractive Reserve	Reserva Extrativista (RESEX)	Aims at protecting means of life and culture of local populations within a sustainable framework. It is managed by local communities among other civilian stakeholders and by determined state representatives.
Fauna Reserve	Reserva de Fauna	Aims at protecting native and endemic species, resident or migratory, enabling research and technical studies aiming at their sustainable use.
Sustainable Development Reserve	Reserva de Desenvolvimento Sustentável (RDS)	Aims at the conservation of territories inhabited by traditional communities whose subsistence depends on sustainable methods for the use of natural resources. Those areas are of public domain.
Private Natural Heritage Reserve	Reserva Particular do Patrimônio Natural (RPPN)	Aims at general conservation of biodiversity in a certain private land, including activities of tourism, research, education, and recreation. These private lands are irreversibly destined to this goal. It is the most abundant category of protected area in Brazil.

Environmental Protection Areas, which comprise extensive areas of important ecological, aesthetic, or cultural relevance – nevertheless under considerable human occupation or use. They were designed with the goal of regulating the processes of human occupation and guarantee the sustainable use of its natural resources, equivalently IUCN's Category V (Protected Landscape/Seascape). Finally, the last category is *Private Natural Heritage Reserve*, in which private lands are designated to conserve biological diversity; the only activities permitted within their limits are research and tourism (recreational and educational). The same setting can be found at IUCN's Category VI (PA with Sustainable Use of Natural Resources), although they differ by not necessarily demonstrating associated traditional or cultural values and are generally relatively small (see Dudley 2008).

Except for Private Natural Heritage Reserves (which is not the focus of this chapter), the federal protected areas are currently managed by the Instituto Chico Mendes de Conservação da Biodiversidade (ICMBio), a governmental board dismembered from IBAMA in 2007. With that dismemberment, IBAMA became dedicated exclusively to environmental licensing and monitoring, while ICMBio assumed full responsibility for the conservation of threatened species and the management of protected areas. Brazilian states and municipalities follow the abovementioned categories and are allowed to create additional ones. Besides, the system gets more and more complex when governmental and nongovernmental institutions at all levels get involved in the management.

There are currently 334 federal protected areas that are managed by ICMBio, covering a total of 171,294,897 ha (ICMBio 2021). From those, 62 (4,683,773 ha) include mangrove or salt marsh ecosystems (Schmidt and Gomes 2016; Ministério do Meio Ambiente 2021). Moreover, 31% of them are under no-take areas, while 69% are in sustainable use. National Parks predominate (42%) in the no-take category, followed by Ecological Stations and Biological Reserves (both with 21%) and Wildlife Refuges (16%). In the sustainable use group, there are prominently more Extractive Reserves (60%), followed by Environmental Protection Areas (30%) (Fig. 16.3) (see ICMBio 2018).

Brazilian PAs harbor more mangroves than salt marshes. Although marsh plants may occur associated with mangroves in lower mudflats and the landward boundaries (Schmidt et al. 2013), representative salt marshes that are independent of mangroves occur in only four PAs, all in South Brazil.

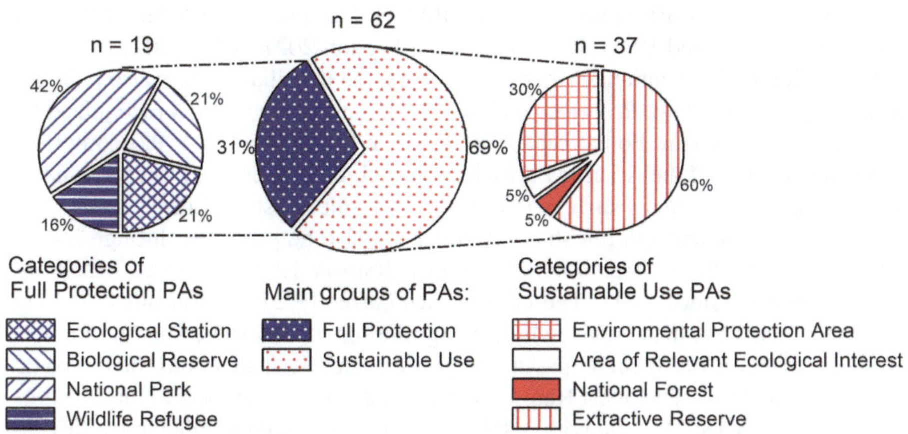

Fig. 16.3 Distribution of Brazilian federal protected areas (PAs) in the groups of no-use (or full protection) and sustainable use (center) and in their categories (left and right, respectively). (Modified from Schmidt and Gomes (2016) and updated until 2021)

Although the number of PAs is an indicator of conservation effort, it is important to consider that the specific coverage of salt marshes and mangroves varies greatly between them. For example, the oceanic island of Fernando de Noronha has only a 0.89-ha mangrove (Barcellos et al. 2011), equivalent to 0.05% of the protected area of its National Park (see Chap. 3, Map 18). At the other extreme, the Piratuba Lake Biological Reserve, at Amapá State, covers 107,454 hectares of mangroves, which is equivalent to 27.38% of its protected area (Leão et al. 2018a) (see Chap. 3, Map 1). Based on ICMBio data, the average area of mangroves protected by a single federal protected area is 9213 ± 18,417 ha. Projecting this average on the whole number of federal PAs results in an approximate value of 561,988 ha of protected mangroves, which is equivalent to 46% of the total coverage of this ecosystem in Brazil (1,398,966 ha according to CENIMA 2014 apud. Leão et al. 2018b).

Compared to mangroves, salt marshes are even less protected. A study carried out in 2009, encompassing federal and state reserves, estimated that at the time only 0.6% of the 12,149 hectares of Brazilian salt marshes were protected (MMA 2010). After this study, no federal PA with a focus on salt marshes has been created, so the scenario is unlikely to have changed much.

Salt marshes' effective protection is restricted to the southern region: one PA with 36,716 ha in Rio Grande do Sul State and three PAs covering 160,055 ha in Santa Catarina State. The south region also includes nine mangrove PAs (541,711 ha), similar to the southeast region, which also presents nine PAs totaling 535,360 ha. Both the area and the number of protected mangroves increase towards lower latitudes. The northeast region (08°03′S), with the longest coastline in Brazil, shelters around 50% of the country's mangrove area and presents 28 PAs covering 2,094,404 ha. Fifty percent of this northeastern protected area is in Maranhão State (02°31′S), whose mangroves achieve great extent because of the local macrotidal regime. Large tidal ranges also contribute to the vast mangrove forests of the northern region, comprising 30% of the coverage of this ecosystem in Brazil. Northern mangroves are protected by 15 PAs (1,475,582 ha) distributed through two states (Amapá and Pará) (Dale and Cunha-Lignon 2021; MMA 2021).

In addition to the mentioned categories of PAs, the Brazilian system also includes the concept of *Biosphere Reserve* – an integrated management model recognized by UNESCO. The Atlantic Forest Biosphere Reserve (established in 1992) comprises important portions of the Atlantic Forest biome, including threatened forest remnants and associated ecosystems such as mangroves and salt marshes (Corrêa 1995).

Both mangrove and salt marsh ecosystems can also be protected through instruments created by the *Convention on Wetlands* (Ramsar 1975), to which Brazil has been a signatory since 1996. The country has currently 27 Ramsar sites, totaling 26,794,455 ha; 12 of these sites include mangroves or salt marshes. The first Ramsar site (recognized in 1993) encompassing important salt marsh areas was the Lagoa do Peixe National Park. The latest Ramsar site created was the Amazon Estuary and its mangroves, in 2018. This site with 3,850,253 ha includes the world's largest fluvial-maritime archipelago (Marajó archipelago) (Ramsar 2021) (see Chap. 3, Map 2). It is composed of 23 PAs, constituting the world's largest continuous portion of

Table 16.3 List of the 13 Ramsar sites along the Brazilian coast

Ramsar sites	State	Year	Link
Cape Orange National Park	AP	2013	https://rsis.ramsar.org/ris/2190
Amazon Estuary and its Mangroves	AP|CE	2018	https://rsis.ramsar.org/ris/2337
Reentrâncias Maranhenses Environmental Protection Area	MA	1993	https://rsis.ramsar.org/ris/640
Baixada Maranhense Environmental Protection Area	MA	2000	https://rsis.ramsar.org/ris/1020
Parcel de Manuel Luiz State Marine Park	MA	2000	https://rsis.ramsar.org/ris/1021
Atol das Rocas Biological Reserve[a]	RN	2015	https://rsis.ramsar.org/ris/2259
Fernando de Noronha Marine National Park	PE	2018	https://rsis.ramsar.org/ris/2333
Abrolhos Marine National Park	BA	2010	https://rsis.ramsar.org/ris/1902
Cananéia-Iguape-Peruíbe Environmental Protection Area	SP	2017	https://rsis.ramsar.org/ris/2310
Guaraqueçaba Ecological Station	PR	2017	https://rsis.ramsar.org/ris/2305
Guaratuba State Environmental Protection Area	PR	2017	https://rsis.ramsar.org/ris/2317
Peixe Lagoon National Park	RS	1993	https://rsis.ramsar.org/ris/603
Taim Ecological Station	RS	2017	https://rsis.ramsar.org/ris/2298

Brazilian states: Amapá (AP), Amapá-Pará-Maranhão-Piauí-Ceará (AP–CE), Ceará (CE), Rio Grande do Norte (RN), Pernambuco (PE), Bahia (BA), São Paulo (SP), Paraná (PR), and Rio Grande do Sul (RS). See Chap. 3, Maps 1–18

[a]No salt marshes or mangroves present

mangroves under legal protection, extending over 700 km of the coastline, and protecting nearly 70% of the country's ecosystem (Dale and Cunha-Lignon 2021). Table 16.3 shows a list of the 13 Ramsar sites located in the Brazilian coast.

16.5 Controversies

Pressure on the Brazilian government to consolidate and amplify its system of PAs has been supported by its adhesion to the Convention on Biological Diversity, which advocates a "particular attention to No-take areas" (Brasil 2002b). Nonetheless, after the ratification of the National System of PAs in 2000, there was a notorious allocation of public resources towards the creation of sustainable use PAs (often containing mangroves and salt marshes) (Figs. 16.1 and 16.2). These PAs are less costly to create as they require fewer expropriation costs, but their effectiveness in nature conservation will strongly depend on the efficiency of management (discussed in the next section) and on the number of resource users. Populations located within sustainable use PAs are usually larger than those within no-take PAs which makes environmental conservation in the former more challenging due to greater exploitation of resources (Dourojeanni and Pádua 2013). However, a high potential for sustainability has been reported for artisanal fisheries in Extractive Reserves with a large supply of natural resources due to the vast coverage of mangroves, disproportionately larger than the area occupied by human populations (e.g., Diele et al. 2005). On the other hand, highly urbanized areas are relatively often found in other sustainable use PAs, mainly from the category Environmental Protection Areas (Fig. 16.3). Although these PAs might be useful tools for integrated coastal management (e.g., Ferreira et al. 2006), they have largely proven to be superfluous in terms of biodiversity conservation (Delgado-Mendez 2003; Dourojeanni and Pádua 2013). The low effectiveness of many sustainable use areas does not necessarily mean that they are dispensable, as it can be said that any PA is better than none. Nonetheless, the expansion of sustainable use PAs from 2000 to 2009 had an undesirable consequence – masking the lack of full protection PAs (Delgado-Mendez 2003; Dourojeanni and Pádua 2013).

From 2010 to 2017, the no-take group once again surpassed the sustainable use group in the number of reserves created (15 × 8) and in the coverage area (2,267,846 ha × 1,653,318 ha). But this trend was not followed by the establishment of reserves with mangroves and salt marshes which, in this period, resulted in four new sustainable use PAs (173,485 ha) against two new full protection PAs (51,928 ha). The strategy of creating large sustainable use PAs with massive mangrove cover was adopted by the government again in 2018, but even so the overall protection of this ecosystem in this period was much lower than in the last decade.

The Aichi Target of 10% of marine protected areas before 2020 was only achieved through the creation of two Environmental Protection Areas and two Natural Monuments, all without mangroves or salt marshes (Figs. 16.1 and 16.2). These large PAs (90,278,190 ha in total) were created around oceanic islands with few or no inhabitants, making Brazil jump from 1.5% to 25% of protected coastal and marine areas, but leaving unprotected many other more priority sites for conservation (Giglio et al. 2018).

In terrestrial and inland water areas, Brazil also reached the Aichi Target (in this case, 17% of protected areas until 2020), but the strategy was different. Since 2013, Brazilian authorities managed to mask the stagnation of the SNUC over *Indigenous Lands, Legal Reserves* (i.e., a certain percentage of mandatory preservation within private rural properties), and *Permanent Preservation Areas* in the area calculation within the scope of the Strategic Plan for Biodiversity (SCBD 2010). Although Indigenous Lands are important to safeguarding the culture and continuity of native peoples, they largely lack adequate management plans for biodiversity conservation (Dourojeanni and Pádua 2013). Additionally, the sustainable long-term use of their resources is questionable due to the demographic growth of indigenous populations and changes in their societal dynamics (Terborgh and Peres 2002). Legal Reserves and Permanent Preservation Areas have a long history of degradation and questionable use, in addition to often being small and scattered, hindering their management and offering low connectivity to fauna and flora (Pacheco et al. 2018).

The achievement of the Aichi Target in 2018 together with the entry of a new government even less committed to environmental issues in 2019 resulted in the complete halt in the creation of federal PAs, with or without mangroves and salt marshes, at least until 2021. Regarding mangroves, the lack of PAs could be supplied by their status as "Permanent Preservation Area" (Brasil 1965), but the 2012 Brazilian Forest Code allowed the conversion of salt flats, i.e., upper-tidal hypersaline sections of the ecosystem, to be converted into shrimp farms and salt extraction ponds despite vehement opposition by environmentalists (Metzger et al. 2010; Schaeffer-Novelli et al. 2012).

With a weak forest code, the permanent preservation status of mangroves was still reinforced by Resolution 303/2002 of the National Council for the Environment (CONAMA), the main advisory body of the Ministry of the Environment. However, in 2019, a government decree reduced from 23 to 4 the number of chairs for NGOs, universities, and traditional peoples in this council. The CONAMA, with government supremacy, revoked Resolution 303/2002 in 2020, leaving mangroves even more unprotected and vulnerable to be exploited by the minority interested in shrimp farming and salt exploitation. Finally, the resolution was subject of litigation before the Federal Supreme Court and had its effect reversed.

16.6 On the Support to Protected Areas in Brazil

Promises of the Brazilian government to expand and improve PAs have not been followed by a proportional increase in investments in infrastructure, personal and necessary resources for conservation. As an example, the total investment in environmental conservation and preservation represented only 25% of the overall funds disbursed by the Ministry of the Environment, reaching a low of 9% in 2017 (MTCGU 2018). That lack of investment is not only due to the priority given to other governmental agendas, but also a reflection of the poor distribution of resources within the structure of the Ministry of the Environment itself. Approximately 80% of the investment on the ICMBio is linked to its headquarters in Brasília, which comprised four directories, 41 coordinators, and 230 other civil servants (MTCGU 2018; ICMBio 2021). The PAs spread throughout the country had an average of three environmental analyst staff each (ICMBio 2021), which corresponded to an overwhelming average of approximately 81,000 ha of PAs per agent. The lack of agents makes it difficult to fulfill the basic requirements of the PAs: Currently, 24% of them do not have a formalized council and 60% still do not have a management plan.

The administration of extensive PAs is made even more difficult by the lack of adequate infrastructure and equipment (Onaga and Drumond 2007). A good example is the lack of motorized boats in extensive protected wetlands. Boats are available in about half of the federal mangrove PAs (Schmidt 2012b). The lack of appropriately trained human resources adds to the equation, making nearly impossible the essential activities of surveillance, control, and environmental monitoring. Concerning the latter activity, ICMBio has been investing in the elaboration of a mangrove biodiversity monitoring program in federal PAs since 2007 (Schmidt

2012b; Schmidt and Gomes 2016), but it has failed to implement it yet. Monitoring is crucial to collect data on the current situation of mangroves within different categories of PAs, reinforcing their importance and understanding how they are managed (Dale and Cunha-Lignon 2021).

Due to the lack of investments by the government, much of the administration of Brazilian PAs depends on NGOs, private companies, and international agencies. Despite the increasing regulatory mechanisms that make such an alternative investment difficult, these funds remain useful to maintaining the vital functions of many PAs, such as the continuity of their regular board meetings.

Other functions, such as scientific research and monitoring, are largely dependent on NGOs and, principally, universities. Those partnerships need to be improved and adjusted, as the research undertaken is usually linked to dissertations, theses, or funded projects with an average duration of up to 2–4 years. It is necessary to ensure the long-term continuity of those research projects to effectively aid environmental monitoring and adequately contribute to the management and operationality of the PAs (Cunha-Lignon et al. 2015; Schmidt and Gomes 2016).

PAs can also be benefited from closer ties to universities through graduate courses that promote the development of human resources for conservation and resource management. Environmental analysts working in the PAs are currently recruited through predominantly objective exams for public employment that demand only a college degree – but without specific requirements for training in environmental administration. As such, people from many different educational backgrounds end up hired before any training on environmental management. In the meanwhile, these managers face many additional difficulties (Onaga and Drumond 2007), mostly because new positions normally become available in PAs in isolated regions of the country amid social conflicts. This results in high withdrawal rates and transfer requests. Worse than that are the cases in which coordinators, directors, and other authorities within ICMBio are indicated as a political strategy to back up governmental agendas for the environment. In recent years, authorities from the Ministry of the Environment have pressured, persecuted, and even dismissed good managers of PAs engaged in environmental conservation and defense of traditional populations.

All in all, the great challenge for the PAs is to achieve autonomy through (i) appropriate recruitment strategies and (ii) government investments compatible with their needs. This will enable sustainable management so that partnerships and other investments will no longer be indispensable, but rather a plus for improving the performance of the PAs.

16.7 Final Remarks

The process of abandonment and distortion of the national system of PAs has both ideological and political roots. Ideologically, there is a conflict between two fallacies; on one hand, there are those who defend the presence of people in PAs and who postulate that no-take is an illusion perpetrated by the "modern myth of untouched nature" and that areas containing traditional populations can be more efficiently conserved than those supposedly untouched areas (e.g., Diegues 2001; Stevens 2014). On the other hand, many environmentalists consider the belief that traditional populations living in PAs contribute to their conservation is an illusion perpetrated by the myth of the "ecologically noble savage" and that deliberate conservation by native peoples either does not occur or is a side effect of factors such as low population density and simple exploitation technologies (e.g., Redford 1991; Hames 2007; Dourojeanni and Pádua 2013). That polarization molded the Brazilian multicategory system of PAs and a bias towards the insertion of humans into protected ecosystems appears to have dominated in the last decades (Delgado-Mendez 2007, 2008; Almeida and Jardim 2018; Costa and Seabra 2020; Pereira et al. 2021). The trend is corroborated by the increasing cases of changes in categories of existing PAs, from more restrictive to more permissive, which is alarming in determined ecosystems (Bernard et al. 2014).

While that ideological debate continues, the political explanation for the stagnation of the national system of PAs takes on an obscure note: many politicians seem to consider some PA categories as impairments to economic growth. There has been an apparent political motivation for weakening the protection status of mangrove ecosystems in the new 2012 Forest Code, as many Brazilian politicians take much interest in shrimp farming and salt exploitation. By no longer having total protection under the forest code and an effective National Council for the Environment, hopes for the continued conservation of mangroves are now deposited in PAs. However, new areas are not being created at a satisfactory rate, and those that already exist lack fundamental investments and infrastructure. Worse than that is the fact that all those problems are masked by a presumed adherence to the goals of international agreements.

All abovementioned problems have been counterbalanced, however, by social activism and the growing recognition of the importance of wetlands to human well-being. There is a hope that our society can put aside ideological differences and work together to generate real changes in the Brazilian political system, thus improving measures for effective environmental protection and conservation in the country.

References

Almeida AF, Jardim MAG (2018) Mudanças socioeconômicas e ambientais resultants das políticas públicas de desenvolvimento socioeconômico no litoral do Nordeste do Pará, Brasil. Desenvolv Meio Ambiente 49:108–129

Araújo DSD, Maciel NC (1979) Os Manguezais do Recôncavo da Baía de Guanabara. Série Técnica 10/79, 1/3. FEEMA, Rio de Janeiro

Araújo MP, Hamacher C, Farias CO, Soares LG (2021) Fecal sterols as sewage contamination indicators in Brazilian mangroves. Mar Pollut Bull 165:112149

Barcellos RL, Coelho-Júnior C, Lins SRRM, Silva MS, Camargo PB, Travassos PEPF (2011) Island beaches morphological and sedimentary short term variations – the case of SE Fernando de Noronha Island, South Atlantic, Brazil. J Integr Coast Zone Manag 11(4):471–478

Bernard E, Penna LA, Araújo E (2014) Downgrading, downsizing, degazettement, and reclassification of protected areas in Brazil. Conserv Biol 28(4):939–950

Brasil (1934) Decree no 23,793, January 23, 1934. Aprova o Código Florestal

Brasil (1965) Law nº 4,771, September 15, 1965. Institui o Novo Código Florestal

Brasil (2000) Law nº 9,985, July 18, 2000. Regulamenta o art. 225, § 1o, incisos I, II, III e VII da Constituição Federal, institui o Sistema Nacional de Unidades de Conservação da Natureza e dá outras providências

Brasil (2002a) Decree nº 4,340, August 22, 2002. Regulamenta artigos da Lei n. 9.985 que dispõe sobre o Sistema Nacional de Unidades de Conservação

Brasil (2002b) Decree nº 4,339, August 22, 2002. Institui princípios e diretrizes para a implementação da Política Nacional da Biodiversidade

Corrêa FA (1995) A Reserva da Biosfera da Mata Atlântica. Roteiro para o Entendimento de seus Objetivos e seu Sistema de Gestão. In: Série Cadernos da Reserva da Biosfera da Mata Atlântica, vol 2. Conselho Nacional da Reserva da Biosfera da Mata Atlântica, São Paulo

Costa CSB (2009) Impactos Antrópicos e Ameaças à Conservação das Marismas Sul-americanas. In: Abstracts of the 2nd Congresso Brasileiro de Biologia Marinha, ABBN, Armação de Búzios, 24–28 May 2009

Costa ECP, Seabra VS (2020) Dinâmicas naturais e atividades socioeconômicas na planícies costeira da Lagoa de Araruama – RJ. R Formação 27(52):353–382

Costa PG, Marube LC, Artifon V, Escarrone AL, Hernandes JC, Zebral YD, Bianchini A (2022) Temporal and spatial variations in metals and arsenic contamination in water, sediment and biota of freshwater, marine and coastal environments after the Fundão dam failure. Sci Total Environ 806(3):151340

Cunha-Lignon M, Almeida R, Lima NGB, Galvani E, Menghini RP, Coelho-Jr C, Schaeffer-Novelli Y (2015) Monitoramento de Manguezais: abordagem integrada frente às alterações ambientais. Anais do VII CBUC – Trabalhos Técnicos 2015, Curitiba, pp 1–17

Dale MV, Cunha-Lignon M (2021) A reflection on mangroves within Brazilian protected areas. Book of abstracts – VLIZ marine science day. Oostende 85:42–43

Dantas NCFM, Duarte OS, Ferreira WC, Ayala AP, Rezende CF, Feitosa CV (2020) Plastic intake does not depend on fish eating habits: identification of microplastics in the stomach contents of fish on an urban beach in Brazil. Mar Pollut Bull 153:110959

Delgado-Mendez JM (1989) Situação Atual das Unidades de Conservação do Sistema Brasileiro. Technical report: Programa Nacional do Meio Ambiente. SEMA/IBDF, Brasilia

Delgado-Mendez JM (2003) Principais Problemas Existentes na Implementação de Áreas de Proteção Ambiental: Cenários e Recomendações. In: Guaryassu SMS (ed) Gerenciamento de Áreas de Proteção Ambiental do Brasil. Fundação O Boticario, Curitiba, pp 42–53

Delgado-Mendez JM (2007) Áreas Protegidas de Brasil y la Presencia de Comunidades Humanas: estamos preparados para eso? Annals of II Congreso Latinoamericano de Parques Nacionales y Otras Áreas Protegidas. UICN, Bariloche, 30 September–6 October 2007

Delgado-Mendez JM (2008) Proteção de Áreas Naturais e Desenvolvimento Social: percepções de um conflito na gestão de Unidades de Conservação de Proteção Integral. PhD Dissertation, Universidade de São Paulo

Denadai MR, Jacobucci DFC, Fontana I, Taniguchi S, Turra A (2015) Assessment of contamination of the beach clam *Tivela mactroides*: implications for food safety of a recreational and subsistence marine resource in Caraguatatuba Bay, Brazil. Int J Food Contam 2:6

Diegues AC (2001) O Mito Moderno da Natureza Intocada. HUCITEC, São Paulo

Diele K, Koch V, Saint-Paul U (2005) Population structure, catch composition and CPUE of the artisanally harvested mangrove crab *Ucides cordatus* (Ocypodidae) in the Caeté estuary, North Brazil: indications for overfishing? Aquat Living Resour 18:169–178

Dourojeanni MJ, Pádua MTJ (2013) Arcas à Deriva: unidades de conservação do Brasil. Technical Books, Rio de Janeiro

Dudley N (2008) Guidelines for applying protected area management categories. IUCN, Gland

Ferreira BP, Messias LT, Maida M (2006) The environmental municipal councils as an instrument in coastal integrated management: the *Área de Proteção Ambiental Costa dos Corais* (AL/PE) experience. J Coast Res 39:1003–1007

Giglio VJ, Pinheiro HT, Bender MG, Bonaldo RM, Costa-Lotufo LV, Ferreira CEL, Floeter SR, Freire A, Gasparini JL, Joyeux JC, Krajewski JP, Lindner A, Longo GO, Lotufo TMC, Loyola R, Luiz OJ, Macieira RM, Magris RA, Mello TJ, Quimbayo JP, Rocha LA, Segal B, Teixeira JB, Vila-Nova DA, Vilar CC, Zilberberg C, Francini-Filho RB (2018) Large and remote marine protected areas in the South Atlantic Ocean are flawed and raise concerns: comments on Soares and Lucas. Mar Policy 96:13–17

Gorman D, Turra A (2016) The role of mangrove revegetation as a means of restoring macrofaunal communities along degraded coasts. Sci Total Environ 566–567:223–229

Hames R (2007) The ecologically noble savage debate. Annu Rev Anthropol 36:177–190

Instituto Chico Mendes de Conservação da Biodiversidade – ICMBio (2018) Atlas dos Manguezais do Brasil. ICMBio, Brasília, 176 p

Instituto Chico Mendes de Conservação da Biodiversidade – ICMBio (2021) Painel Dinâmico. http://qv.icmbio.gov.br/QvAJAXZfc/opendoc2.htm?document=painel_corporativo_6476. Accessed 18 June 2021

Lacerda LD, Ward R, Godoy MP, Meireles J, Borges R, Ferreira AR (2021) 20-years cumulative impact from shrimp farming on mangroves of Northeast Brazil. Front For Glob Change 4: 653096

Leão AR, Cruz AT, Oliveira DAS, Fumi M (2018a) Incentivo à conservação e ao uso sustentável do manguezal: O Projeto Manguezais do Brasil. In: ICMBio – Instituto Chico Mendes de Conservação da Biodiversidade. Atlas dos Manguezais do Brasil. MMA/ICMBio, Brasília, pp 105–119

Leão AR, Prates APL, Fumi M (2018b) Manguezal e as unidades de conservação. In: ICMBio – Instituto Chico Mendes de Conservação da Biodiversidade. Atlas dos Manguezais do Brasil. MMA/ICMBio, Brasília, pp 55–72

Löwy I (2017) Leaking containers: success and failure in controlling the mosquito *Aedes aegypti* in Brazil. Am J Public Health. Available in https://ajph.aphapublications.org/doi/epub/10.2105/AJPH.2017.303652. Accessed 23 Aug 2021

Machado W, Silva-Filho EV, Oliveira RR, Lacerda LD (2002) Trace metal retention in mangrove ecosystems in Guanabara Bay, SE Brazil. Mar Pollut Bull 44:1277–1280

Macieira RM, Oliveira LAS, Cardozo-Ferreira GC, Pimentel CR, Andrades R, Gasparini JL, Sarti F, Chelazzi D, Cincinelli A, Gomes LC, Giarrizzo T (2021) Microplastic and artificial cellulose microfibers ingestion by reef fishes in the Guarapari Islands, southwestern Atlantic. Mar Pollut Bull 167:112371

Maciel NC (1991) Alguns Aspectos da Ecologia do Manguezal. In: Braga RAP (ed) Alternativas de Uso e Proteção dos Manguezais do Nordeste. Série Publicações Técnicas. CPRH, Recife, pp 9–37

Maciel NC (2001) Legislação Ambiental e o Manguezal. In: Alves JR (org) Manguezais: educar para proteger. FEMAR/SEMADS, Rio de Janeiro, pp 35–45

Marangoni JC, Costa CSB (2009) Diagnóstico Ambiental das Marismas no Estuário da Lagoa dos Patos – RS. Atlantica 31(1):85–98

Metzger JP, Lewinsohn TM, Joly CA, Verdade LM, Martinelli LA, Rodrigues RR (2010) Brazilian law: full speed in reverse? Science 329(5989):276–277

Ministério da Transparência e Controladoria–Geral da União – MTCGU (2018) Portal da Transparência, Governo Federal. http://www.portaltransparencia.gov.br/receitas/. Accessed 30 May 2018

Ministério do Meio Ambiente – MMA (2010) Gerência de Biodiversidade Aquática e Recursos Pesqueiros. Panorama da conservação dos ecossistemas costeiros e marinhos no Brasil. MMA/SBF/GBA, Brasília

Ministério do Meio Ambiente – MMA (2021) Cadastro Nacional de Unidades de Conservação. Painel Unidades de Conservação Brasileiras. Brasília. https://antigo.mma.gov.br/areas-protegidas/cadastro-nacional-de-ucs.html

Onaga CA, Drumond MA (2007) Efetividade de gestão das unidades de conservação federais do Brasil. IBAMA, Brasília

Pacheco AA, Neves ACO, Fernandes GW (2018) Uneven conservation efforts compromise Brazil to meet the Target 11 of Convention on Biological Diversity. Persp Ecol Conserv 16:43–48

Pádua MTJ (2011) Do Sistema Nacional de Unidades de Conservação. In: Medeiros R, Araújo FFS (eds) Dez anos do Sistema Nacional de Unidades de Conservação da Natureza: lições do passado, realizações presentes e perspectivas para o futuro. Ministério do Meio Ambiente, Brasília, pp 23–36

Paludo D, Klonowski VS (1999) Barra de Mamanguape-PB: Estudo do impacto do uso de madeira de manguezal pela população extrativista e da possibilidade de reflorestamento e manejo dos recursos madeireiros. In: Série Cadernos da Reserva da Biosfera da Mata Atlântica, vol 16. Conselho Nacional da Reserva da Biosfera da Mata Atlântica, São Paulo

Pereira RS, Cordeiro RC, Sabadini-Santos E, Bidone ED (2021) Urbanization, environmental and socioeconomic trends in a coastal zone under an Anthropocene approach: 1940-2020 (Guanabara Bay hydrographic basin, Rio de Janeiro, Brazil). Rev Bras G&DR 17(2):194–209

Pinheiro MAA, Silva PPG, Duarte LFA, Almeida AA, Zanotto FP (2012) Accumulation of six metals in the mangrove crab *Ucides cordatus* (Crustacea: Ucididae) and its food source, the red mangrove *Rhizophora mangle* (Angiosperma: Rhizophoraceae). Ecotoxicol Environ Saf 81: 114–121

Pureza F, Pellini A, Pádua C (2015) Unidades de Conservação: fatos e personagens que fizeram a história das categorias de manejo. Matrix, São Paulo

RAMSAR (2021) Country profiles. The Convention on Wetlands. http://www.ramsar.org/wetland/brazil. Accessed 12 Apr 2021

Redford K (1991) The ecologically noble savage. Orion 9:24–29

Santana MFM, Moreira FT, Turra A (2017) Trophic transference of microplastics under a low exposure scenario: insights on the likelihood of particle cascading along marine food-webs. Mar Pollut Bull 121:154–159

Schaeffer-Novelli Y, Rovai AS, Coelho-Jr C, Menghini RP, Almeida R (2012) Alguns impactos do PL 30/2011 sobre os manguezais brasileiros. In: Souza G (org) Comitê Brasil em Defesa das Florestas e do Desenvolvimento Sustentável. Código Florestal e a Ciência: o que nossos legisladores ainda precisam saber. Comitê Brasil, Brasília, pp 18–17

Schaeffer-Novelli Y, Sierra EJS, Vale CC, Bernini E, Rovai AS, Pinheiro MAA, Schmidt AJ, Almeida R, Coelho C Jr, Menghini RP, Martinez DI, Abuchahla GMO, Cunha-Lignon M, Charlier-Sarubo S, Shirazawa-Freitas J, Cintrón-Molero G (2016) Climate changes in mangrove forests and salt marshes. Braz J Oceanogr 64(2):37–52

Schmidt AJ (2012a) Diagnóstico das principais ameaças à biodiversidade em ambientes de manguezal, com ênfase nas unidades de conservação. Technical report: Projeto para Conservação e Uso Sustentável Efetivos de Ecossistemas Manguezais no Brasil – PNUD BRA/07/G32. DIBIO, COMOB, Brasília

Schmidt AJ (2012b) Diagnóstico da capacidade de execução de monitoramentos da biodiversidade das UCs e seus parceiros. In: Schmidt AJ (ed) Minuta do Programa de Monitoramento da biodiversidade para as UCs federais com manguezais. Technical report: Projeto para Conservação e Uso Sustentável Efetivos de Ecossistemas Manguezais no Brasil – PNUD BRA/07/G32. DIBIO, COMOB, Brasília, pp 4–6

Schmidt AJ, Gomes TP (2016) Diagnóstico de Pesquisas Realizadas em Unidades de Conservação: Subsídio ao Programa de Monitoramento da Biodiversidade de Manguezais. Biodivers Bras 6(1):61–74

Schmidt AJ, Bemvenuti CE, Diele K (2013) Sobre a definição da zona de apicum e sua importância ecológica para populações de caranguejo-uçá *Ucides cordatus* (Linnaeus, 1763). Bol Tecn Cient CEPENE 19(1):9–25

Secretariat of the Convention on Biological Diversity – SCBD (2010) COP-10 Decision X/2. Secretariat of the Convention on Biological Diversity, Nagoya. https://www.cbd.int/doc/decisions/cop-10/cop-10-dec-02-en.pdf. Accessed 23 Aug 2021

Secretariat of the Convention on Biological Diversity – SCBD (2018) COP-14 Decision 14/34. Secretariat of the Convention on Biological Diversity, Sharm-El-Sheikh. https://www.cbd.int/doc/decisions/cop-14/cop-14-dec-34-en.pdf. Accessed 23 Aug 2021

Seeliger U, Costa CSB (1988) Impactos Naturais e Humanos. In: Seeliger U, Odebrecht C, Castello JP (eds) Os Ecossistemas Costeiro e Marinho do Extremo Sul do Brasil. Editora Ecoscientia, Rio Grande, pp 217–226

Silva PH, de Sousa FDB (2021) Microplastic pollution of Patos Lagoon, south of Brazil. Environ Challenge 4:100076

Spalding M, Kainuma M, Collins L (2010) World atlas of mangroves. Earthscan, London

Stevens S (2014) Indigenous peoples, national parks, and protected areas: a new paradigm: linking conservation, culture, and rights. University of Arizona Press, Tucson

Terborgh J, Peres CA (2002) O Problema das Pessoas nos Parques. In: Terborgh J, Van Schaik C, Davenport L, Rao M (eds) Tornando os Parques Eficientes: Estratégias para Conservação da Natureza nos Trópicos. Editora da Universidade Federal do Paraná and Fundação Boticário, Curitiba, pp 334–346

Tognella MMP, Falqueto AR, Espinoza HCF, Gontijo I, Gontijo ABPL, Fernandes AA, Schmildt ER, Soares MLG, Chaves FO, Schmidt AJ, Lopes DMS, Barcelos UD, D'Addazio V, Lima KOO, Pascoalini SS, Paris JO, Brites Júnior NV, Porto LA, Almeida Filho E, Oliveira CP, Leopoldo RVS, Leite S, Berribilli MP, Vieiras SFR, da Rosa MB, Sá F, Rodrigues Neto R, Ghisolfi RD, Castro MSM, Rigo D, Tosta VC, Albino J (2022) Mangroves as traps for environmental damage to metals: The case study of the Fundão Dam. Sci Total Environ 806(4):150452

van Bijsterveldt CEJ, van Wesenbeeck BK, Ramadhani S, Rave OV, van Gool FE, Pribadi R, Bouma TJ (2021) Does plastic waste kill mangroves? A field experiment to assess the impact of macro plastics on mangrove growth, stress response and survival. Sci Total Environ 756(2021): 143826

Vasconcelos ALS, Queiroz HM, Gabriel FA, Ferreira AD, Barcellos D, Andreote FD, Ferreira TO, Bernardino AF (2021) Processos biogeoquímicos no estuário do Rio Doce após o rompimento da barrage de Fundão em Mariana – MG. In: Vicente NMF, Sperber CF, Carneiro MAC (orgs) Dia D do Rio Doce. Um olhar científico sobre o maior desastre socioambiental do Brasil. Universidade Federal de Lavras, Lavras, pp 57–77

Viehman S, Pluym JLV, Schellinger J (2011) Characterization of marine debris in North Carolina salt marshes. Mar Pollut Bull 62(12):2771–2779

Zamprogno GC, Caniçali FB, Cozer CR, Otegui MBP, Graceli JB, da Costa MB (2021) Spatial distribution of microplastics in the superficial sediment of a mangrove in Southeast Brazil: a comparison between fringe and basin. Sci Total Environ 784:146963

Chapter 17
Contribution of Environmental Education to Mangrove Conservation

Renato de Almeida, Yara Schaeffer-Novelli, Jesus Manuel Delgado-Mendez, Guilherme Moraes de Oliveira Abuchahla, and Clemente Coelho-Jr

17.1 Education and Challenges for the Conservation of Mangroves

A worldwide study estimated that more than 120 million people would be living within less than 10 km of tropical coastal areas in 2015 (UNEP 2014). The same authors reported that the destruction rate of mangrove forests is three to five times higher than the estimated for other forest types. In Brazil, 13% of the mangrove cover was lost between 2000 and 2017 (MapBiomas 2018), mainly due to land conversion (i.e., port infrastructure, aquaculture, shrimp farming, and urban growth), pollution, and overexploitation of natural resources.

Recently, Rivera-Monroy et al. (2017) indicated that although countries with large mangrove areas allegedly acknowledged the social and economic value of this ecosystem, loss and degradation of mangroves and remaining adjacent coastal ecosystems continue unabated. In addition, the authors believe that the low percentage (7%) of mangrove settings classified as protected areas worldwide is inconsistent.

R. de Almeida (✉) · J. M. Delgado-Mendez
Centro de Ciências Agrárias, Ambientais e Biológicas, Universidade Federal do Recôncavo da Bahia, Cruz das Almas, BA, Brazil
e-mail: renato.almeida@ufrb.edu.br

Y. Schaeffer-Novelli
Instituto Oceanográfico, Universidade de São Paulo, São Paulo, SP, Brazil

IUCN Mangrove Specialist Group, Brasília, Brazil

G. Moraes de Oliveira Abuchahla
Leibniz Centre for Tropical Marine Research (ZMT), Bremen, Germany

C. Coelho-Jr
Instituto de Ciências Biológicas, Universidade de Pernambuco, Recife, PE, Brazil

Y. Schaeffer-Novelli et al. (eds.), *Brazilian Mangroves and Salt Marshes*, Brazilian Marine Biodiversity, https://doi.org/10.1007/978-3-031-13486-9_17

Because ecosystem functioning and services depend on conservation policies, the Brazilian government has opted for the creation of protected areas that allow for sustainable use by traditional populations (see Chap. 16). However, the criteria for sustainable use remain an issue. In Brazil, more than 36,000 ha of mangroves have been converted into shrimp farms (96.3% in landscape/seascape protected area – IUCN's Category V) (Freitas et al. 2018).

Since 1965, Brazilian mangroves are declared by law as Environmental Protection Areas (APAs). This means that the classification of a mangrove ecosystem as a protected area is merely a reiteration of the conservation status already provided by law. As much as 87% of all country's mangroves are encompassed within 6% of the 2071 protected areas, including the largest continuous mangrove area on the planet, which is located along the northern-northeastern coast, at the stretch of the states of Pará and Maranhão (Leão et al. 2018). Moreover, the creation of protected areas is only one of the current possible conservation strategies in the country (see Chap. 16). Official data in the Brazilian Mangroves Atlas (ICMBio 2018) indicate that there is 1,398,966.1 ha of mangroves along the Brazilian coast (Freitas et al. 2018).

Approximately 25% of the Brazilian population live in densely populated cities on the coast, which highlights the importance of mangrove functional diversity and its enormous socioecological importance. Mangrove's connectivity to adjacent ecosystems supports biodiversity and ecosystem services, in both land and sea. Currently, mangroves have become an ideal open classroom for learning about global ecological processes of adaptation, resilience, and biogeochemical cycles within the context of local features, presenting tangible examples between theory and practice. Carvalho (2004) suggests the adoption of environmental education programs in which human populations are recognized as part of the web of social, natural, and cultural connectivity, resulting from historical processes and whose role in problem identification, resolution, and mitigation of environmental problems must be active and based on knowledge.

An analysis of 37 reports by the United Nations (UN), World Bank (BIRD), and Intergovernmental Panel on Climate Change (IPCC) alongside Agenda 21 demonstrated that education has repeatedly been awarded high priority within the development agenda, and it is strongly correlated with all the 2015 UN's sustainable development goals (SDGs), except for SDG14 – Life below water (Vladimirova and Le Blanc 2016). This shows a relevant gap but also an opportunity to strengthen the role of ocean literacy (UNESCO 2017). All SDGs require reflection on sustainability, economic development, and resilience, including critical imaginative and innovative thinking about sustainable development in general and within each SDG.

It is well known that coastal ecosystems such as mangroves sustain complex interactions between marine and terrestrial environments, thereby supporting high biodiversity and the complex life cycle and connections of the marine food chain (coastal and oceanic). These interactions are still poorly understood, and the limited knowledge about biodiversity, ecosystem services, and the social importance of mangroves requires further research and environmental education programs that consider the importance of ocean literacy in a rapidly changing world (Ghilardi-Lopes et al. 2019a). Unfortunately, environmental education (EE) in

Brazil has been practically restricted to terrestrial environments (Berchez et al. 2016) but initiatives of coastal and marine environmental education are recently emerging (Ghilardi-Lopes and Berchez 2019).

Since the Rio de Janeiro Earth Summit (or Eco-92), the terms "sustainable development," "sustainable consumption," and "biodiversity" have implicated the need of promoting changes in the EE approach worldwide. Some countries have focused their efforts on social questions through social ecology (Ferreira 2002), focusing on biodiversity conservation to ensure sustainable development and allow sustainable consumption in areas with extreme poverty and famine issues. This reflects a partial understanding of what sustainability involves since it entails the use of resources in such a manner that regenerative capacity is not depleted.

The National Biodiversity Policy (Política Nacional de Biodiversidade) endorsed EE to restore respect and ethical awareness for biodiversity (Brasil 2002). Specific guidelines have been established to train educators on biodiversity, thereby reaffirming the importance of formal education on the topic.

A pioneering survey performed in 307 schools from different regions in Brazil (Loureiro et al. 2007) showed that biodiversity is one of the five most discussed themes in EE projects, including the umbrella themes of water (33.6%), waste and recycling (21.5%), pollution and basic sanitation (9.8%), health and nutrition (5.5%), and social and biological diversity (5.5%), among others (24.1%). This study also highlighted the importance of idealistic teachers acting as leaders, qualified teachers with higher education, and continuous training of teachers for effective EE in the school environment. This sample, although limited, is instructive. The complexity of Brazilian EE is reflected by the scope and diversity of institutions once Brazil is a country of continental proportions, divided into 27 administrative states (federal units) and 5570 municipalities, and the current school census detected more than 184,000 basic education schools (48.6 million students) and 2047 higher education institutions (8 million undergraduate students) (INEP/MEC 2018).

17.2 Different Views on Environmental Education in Brazil

The efforts of nongovernmental organizations (NGOs) and environmental activists led the discussion of environmental issues onto national and international agendas before the 1990s. In Brazil, EE was institutionalized and defined by the National Environmental Education Policy Act (NEEP; Federal Law No. 9795/1999) as "processes through which individuals and the community build social values, knowledge, and abilities, attitudes and competencies focused on environmental conservation, as an asset (human capital) of common use by the nation, essential to a healthy quality of life and sustainability." The institutionalization of the NEEP was an important legal landmark in Brazil, although it was only shaped at the turn of the century. Nevertheless, Layrargues (2002) considered the NEEP lacks important issues for its better functioning. In order to improve it, Layrargues and Lima (2014) incorporated the concept of *Campo Social* (field theory) into EE. They argued that

Table 17.1 The three pedagogical macrotrends of environmental education in Brazil, according to Layrargues and Lima (2014)

Conservationist
Linked to ecological principles and expressed by behavioral currents and value affection towards nature and sensory-perception activities outdoors.
Values a "green agenda" (biodiversity, protected areas, biomes, ecotourism, and agroecological experiences).
Adopted an ecological bias disregarding the social, political, economic, and cultural dimensions.
Limited potential for social transformation because it does not question the ruling social structure and only proposes sector reforms.
Pragmatic
Linked to free-market environmentalism resulting from a neoliberal hegemony. The market logic prevails over other social spheres, and this trend appropriates the "sustainable consumption" concept and is strongly influenced by education for sustainable development – ESD.
Values a "brown agenda" (solid waste, energy- and water-saving, carbon market, ecological footprint, certifications, and clean development mechanisms).
Considers a mechanism of adjustment to state reductions, thereby correcting "imperfections" of the production system based on consumerism, programmed obsolescence, and disposability.
Identifies results focused on a sustainable future but is limited by political realism, economic viability, and status quo maintenance, thereby decreasing the political confrontation of the environmental crisis.
Absence of contextual reflections, which would allow for the articulation of the causes and consequences of environmental problems.
Critical
It has a strong sociological and political bias with an emphasis on a critical revision of the grounds that allowed human domination and capital accumulation mechanisms; it attempts to identify political confrontations associated with inequality, social transformation, and socioenvironmental injustice.
Values the "citizen agenda" (democracy, participation, environmental justice, and conflict mediation).
Exposes contradictions in developmental and societal models.
Close to the complexity theory framework since reductionist solutions do not address environmental problems.

EE presented itself as a unified and homogenized concept to a wide unspecialized audience, thereby decreasing its diversity and the pedagogical, political, ethical, and epistemological characteristics that define its concepts and practices. On the other hand, environmental educators acknowledge the importance of the multiple natures of EE as well as of the numerous political-pedagogical trends that constitute it.

Layrargues and Lima (2014) identified three political-pedagogical macrotrends of EE in Brazil: conservationist, pragmatic, and critical (Table 17.1). In general, the conservationist and pragmatic macrotrends represent two evolutionary moments of the same lineage, whereas the critical macrotrend arises as an alternative. These three macrotrends should function as analytical models around which a plurality of pedagogical concepts in Brazilian EE is recognized. Despite being considered distinguished views on education, the three share many similarities.

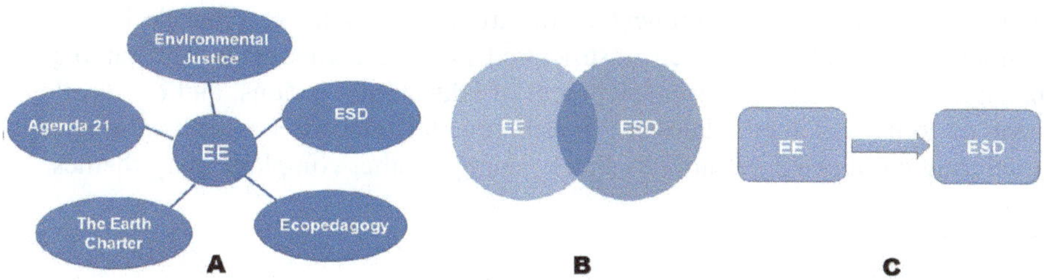

Fig. 17.1 Results of the survey conducted by Hesselink et al. (2000). (**a**) Education for Sustainable Development (ESD) is a subfield of Environmental Education (EE); (**b**) dialogue between EE and ESD; and (**c**) ESD represents an evolution of EE. (Adapted from Meira and Sato 2005)

The Rio+10 Summit, which took place in Johannesburg in 2002, highlighted the controversy associated with the goals and foundations of EE and Education for Sustainable Development (ESD). At that time, the study by Hesselink et al. (2000) for the IUCN served as a basis for understanding the worldwide perception of EE. A total of 50 participants from 25 countries answered an online questionnaire. The survey drew upon three different panoramas and showed a slight predominance towards a vision of ESD as an evolutionary stage or a new generation of EE (which had always been naturalist, apolitical, and lacking scientific rigor) (Fig. 17.1).

A work by UNESCO (2004) strongly differed from the analysis by Hesselink et al. (2000), highlighting that ESD should not be equated to EE. UNESCO also indicated that ESD covers EE and shows holistic and interdisciplinary characteristics, guided by values, critical thinking, and problem-solving, and is locally relevant and values participatory decision-making processes.

This specific position by UNESCO seems contradictory when compared to other documents produced by this same organization, previously defending EE as a transversal (non-disciplinary) approach, guided by the huge complexity and scale of the human relationship with the environment. This estrangement increased in Latin America and the Caribbean when González-Gaudiano (2004) interviewed 101 specialists from 17 countries and observed that 53% of them considered the transition from EE to ESD inconvenient for the following reasons: (1) Regionally, EE has social and economic elements that promote ESD but without the required institutional and political support, and (2) EE represents the loss of a symbolic capital constructed regionally with difficulty and with great transformative potential, that is, what we call prestige or honor that allows us to identify the social agents.

Despite the international debate, the Brazilian plurality of pedagogical concepts promoted an improved understanding of the origins and drivers that affected different EE trends. Formal EE was consolidated upon the establishment of the National Curriculum Guidelines for Environmental Education (NCGEE) by the National Education Council through Resolution No. 02/2012, which is a benchmark document of EE within the context of formal education.

Numerous Brazilian governmental campaigns (some emerging from social pressure) focused on different themes. Perhaps the most important element is the

National Program for Environmental Education (Programa Nacional de Educação Ambiental – PRONEA), which is structured in three components: (1) training of managers and educators, (2) development of educational actions, and (3) development of tools and methodologies. Environmental education mainstreaming and structuring culminated in public policies aiming at other complementary themes.

17.3 Mangroves and Environmental Education

Studies on mangroves as a theme within EE in Brazilian schools showed a great diversity of practical approaches (Ghilardi-Lopes et al. 2019b), frequently based on naturalistic observations, describing short-term experiences. Overall, schools focus on the identification of fauna and flora, as well as a superficial identification of environmental impacts on mangroves within the local context. Considering the pedagogical-epistemological dimension, a significant number of studies adopted a "before/after" methodological approach to evaluate the success or conceptual gains through the adopted interventions.

Most of the numerous perception studies aiming at understanding students' cognitive degree before exposure to contents were mediated by questionnaires, photographs, or drawings. Theoretical confusion appears to be generated by the different understandings of terms like perception, conception, and social representation. In general terms, these notions come from the field of social psychology and depend on the input associated with different contexts of human experience (Bronfenbrenner 2011) and pro-environmental beliefs and behaviors (Corral-Verdugo and Pinheiro 1999; Pol and Castrechin 2013), which were essentially absent or undervalued in the hereby considered studies. Most of these studies were published after 2005, when there was already a consolidated movement of national and regional environmental education meetings on mangroves (*Encontro Nacional de Educação Ambiental em Áreas de Manguezal* [ENEAAM] and *Encontro Regional de Educação Ambiental em Áreas de Manguezal* [EREAAM]), aiming to discuss mangroves in the context of EE (Table 17.2). These meetings introduced educational strategies typical of educommunication and were open to participation by community leaders and mangrove stakeholders. That led to the proposal of a set of educational practices aimed at improving expression abilities within the education community to allow the creation of better tools and environments for the process and the community.

The gap between events observed after 2008 was filled by meetings at the state level, especially in Maranhão and Alagoas. In 2004, at São Francisco do Sul, Santa Catarina State, the organizing committee drew closer to the Redmanglar International network at a moment of ideological unrest and need for rescaling goals due to the growing pressure from the shrimp farming industry to build new facilities close to or in mangrove areas. This highlighted the need for a more militant and critical EE focused on mangroves.

Table 17.2 Dates and locations of the national and regional environmental education meetings on mangroves areas (Encontros Nacionais e Regionais de Educação Ambiental em Áreas de Manguezal: ENEAAM/EREAAM, respectively) in Brazil

Events	City, State	Year
I ENEAAM	Maragojipe, BA	1993
II ENEAAM	Rio de Janeiro, RJ	1994
III ENEAAM	Cabedelo, PB	1995
IV ENEAAM	Serra, ES	1996
I EREAAM (Southeast)	São João da Barra, RJ	1997
I EREAAM (Northeast)	Ilhéus, BA	1997
V ENEAAM	Bragança, PA	1998
II EREAAM (Northeast)	Recife, PE	1999
VI ENEAAM	Parnaíba, PI	2000
II EREAAM (Northeast)	Maragojipe, BA	2001
III EREAAM (Northeast)	São Luís, MA	2003
VII ENEAAM	São Francisco do Sul, SC	2004
IV EREAAM (Northeast)	Recife, PE	2005
V EREAAM (Northeast)	Itaparica, BA	2007
VI EREAAM (Northeast)	Penedo, AL	2008

Brazilian states: *BA* Bahia, *RJ* Rio de Janeiro, *PB* Paraíba, *ES* Espírito Santo, *PA* Pará, *PE* Pernambuco, *PI* Piauí, *MA* Maranhão, *SC* Santa Catarina, *AL* Alagoas

Despite the first meeting occurring as early as 1993 (Table 17.2), it was not until 2007 that the Brazilian Association for Environmental Education in Mangrove Areas (EDUMANGUE) was created (Soffiati 2015). The association stressed the engagement in EE by academia, NGOs, diverse coastal stakeholders, and governmental institutions. The association's goals are as follows:

- To promote information exchange between parties, public institutions, organized society, and traditional communities.
- To support public and private institutions in implementing policies focused on the socioenvironmental and cultural sustainability of mangrove ecosystems in Brazil.
- To promote EE initiatives, scientific research, and training of cadres interested in the sustainability of mangroves.
- To propose and/or launch public and civil actions associated with related themes.

The meetings were promoted by researchers, NGOs, and environmentalists because of the need to overcome the perceived dichotomy between the so-called scientific and traditional knowledge gap concerning mangrove ecosystems. Hence, the inclusion and participation of traditional communities' representatives as lecturers became a core principle, which was previously considered to be incompatible within academic events. Therefore, ENEAAMs were expected to encourage the finding of new trends with the adoption of a reflective process combined with a strategy of teaching, research, and outreach in our universities. Unfortunately, EDUMANGUE lost its cohesion just after being created as a formal association (2007), probably due to a lack of effective articulation. Pedrini (2006) stated that these events acted as successful instruments of knowledge exchange but had little recorded significance in academic publications. As it is well known, academia does not yet reward publications aiming at EE practices or guidelines despite the efforts of some groups and authors.

17.4 The *Maravilhosos Manguezais do Brasil* Initiative

The Brazilian NGO Instituto BiomaBrazil and the Ministry of Environment have organized and implemented a Brazilian version of the NGO Mangrove Action Project's (MAP) education guide *Marvelous Mangroves* in the year 2008 (Almeida et al. 2008). The initiative has been named *Maravilhosos Manguezais do Brasil* (Marvelous Mangroves of Brazil), and it has been adopted as a tool for training teachers in mangrove-rich areas to improve knowledge on the ecosystem. This teaching guide has been distributed to educators during training courses that have theoretical and practical components.

A second stage of the course consisted of developing activities within the school context with monthly follow-ups and monitoring, eliminating the incidental character common in EE training projects. Each course consists of 100 hours of work associated with the teaching guide. Table 17.3 exposes quantitative data of this experience in Brazil.

The MAP has been developing this experience as an international experiment in more than 10 countries. In Brazil, actions included defining the target audience (basic education teachers) and monitoring in schools after the training course. Teachers were invited to participate in an organized teacher network,[1] which currently has already 520 members.

17.5 Final Remarks

Although Brazilian mangroves are protected by law, this has not guaranteed their conservation or the protection of their ecosystem services. Environmental education projects are as important as the creation of new protected areas. Currently, the lack of adequate attention to SDG14 makes educational programs focused on mangroves even more important for preventing and monitoring losses of marine biodiversity or detection and documentation of unsustainable patterns of consumption and production.

In Brazil, official EE documents adopt a critical and citizen-focused agenda, which helps to break away from a purely naturalist vision of biodiversity and incorporates political and cultural elements into the concept. For 15 years, EE in

[1] https://www.facebook.com/groups/151858038349284/

Table 17.3 Quantitative data from the Brazilian experience applying the teaching guide *Maravilhosos Manguezais do Brasil* in public schools

Location	Schools	Educators
Itapissuma (PE)	15	51
São José da Coroa Grande (PE)	10	40
Tamandaré (PE)	12	42
Japaratinga (AL)	11	40
Jequiá da Praia (AL)	10	21
Porto de Pedras (AL)	11	88
São Miguel dos Milagres (AL)	6	37
Maragojipe (BA)	14	31
Cariacica (ES)	26	91
Fundão (ES)	8	68
Cananéia (SP)	9	32
Total	132	541

Source: Instituto BiomaBrasil. States: *PE* Pernambuco, *AL* Alagoas, *BA* Bahia, *ES* Espírito Santo, *SP* São Paulo

mangrove areas has brought academic and traditional knowledge together based on rich experience with the so-called educommunication. Recent technical progress in EE has focused on perception studies, which are usually mediated by talks and process evaluations and centered on a cause-effect relationship (before and after), while guided visits to mangroves are still rare due to logistic difficulties by the school systems.

References

Almeida R, Coelho-Jr C, Corets E (2008) Guia Didático Os Maravilhosos Manguezais do Brasil. Papagaya Editora, Vitória

Berchez FAS, Ghilardi-Lopes NP, Correia MD, Sovierzoski HH, Pedrini AG, Ursi S, Kremer LP, Almeida R, Schaeffer-Novelli Y, Marques V, Brotto DS (2016) Marine and coastal environmental education in the context of global climate changes – synthesis and subsidies for ReBentos (coastal benthic habitats monitoring network). Braz J Oceanogr 64(SP2):137–156

BRASIL (2002) Decree no 4,339, August 22, 2002. Institui princípios e diretrizes para a implementação da Política Nacional da Biodiversidade

Bronfenbrenner U (2011) Bioecologia do desenvolvimento humano: tornando os seres humanos mais humanos (trans: Carvalho-Barreto A). Artmed, Porto Alegre

Carvalho ICM (2004) Educação Ambiental Crítica: nomes e endereçamentos da educação. In: Layrargues PP (ed) Identidades da educação ambiental brasileira. Ministério do Meio Ambiente, Brasília, pp 13–24

Corral-Verdugo V, Pinheiro JQ (1999) Condições para o estudo do comportamento pró-ambiental. Estud Psicol 4(1):7–22

Ferreira JG (2002) Biodiversity and environmental education: a contradiction? Koers 67(3): 259–269

Freitas DM, Ramos ALA, Sano EE, Borges KMR, Silvestre KS (2018) Mapas dos manguezais do Brasil. In: ICMBio – Instituto Chico Mendes de Conservação da Biodiversidade. Atlas dos Manguezais do Brasil. MMA/ICMBio, Brasília, pp 119–171

Ghilardi-Lopes NP, Barradas JI, Kremer LP (2019a) The importance of "ocean literacy" in the Anthropocene and how environmental education can help in its promotion. In: Ghilardi-Lopes NP, Berchez FAS (eds) Coastal and marine environmental education. Springer, Cham, pp 3–17

Ghilardi-Lopes NP, Berchez FAZ (2019) Coastal and marine environmental education. In: Turra A (ed) Brazilian marine biodiversity, Springer series, Springer Nature. Springer, Cham

Ghilardi-Lopes NP, Pimentel DS, Kremer LP, Almeida R, Meireles CP (2019b) Didactic materials as resources for the promotion of coastal and marine environmental education. In: Ghilardi-Lopes NP, Berchez FAS (eds) Coastal and marine environmental education. Springer, Cham, pp 119–133

González-Gaudiano E (2004) Encuesta latinoamericana y caribeña sobre la Educación para el Desarrollo Sustentable. In: Abstracts of Conferência Internacional de Educação para o Desenvolvimento Sustentável, Universidade do Minho, Braga, 19–22 May 2004

Hesselink F, Van Kempen PP, Wals A (2000) ESDebate. International debate on education for sustainable development. IUCN, Gland

Instituto Nacional de Estudos e Pesquisas Educacionais Anísio Teixeira/Ministério da Educação – INEP/MEC (2018) Censo Escolar 2017: notas estatísticas. http://portal.mec.gov.br/docman/janeiro-2018-pdf/81861-divulgacao-censo-2017-vi-pdf/file

Layrargues PP (2002) A conjuntura da institucionalização da Política Nacional de Educação Ambiental. OLAM: Ciência e Tecnologia 2(1):1–15

Layrargues PP, Lima GFC (2014) As macrotendências político-pedagógicas da educação ambiental brasileira. Ambiente & Sociedade 17(1):23–40

Leão AR, Prates APL, Fumi M (2018) Manguezal e as unidades de conservação. In: ICMBio – Instituto Chico Mendes de Conservação da Biodiversidade. Atlas dos Manguezais do Brasil. MMA/ICMBio, Brasília, pp 55–72

Loureiro CFB, Amorim EP, Azevedo L, Cossío MB (2007) Conteúdos, gestão e percepção da educação ambiental nas escolas. In: Mendonça PR, Trajber R (orgs) O que fazem as escolas que dizem que fazem Educação Ambiental? pp 35–81. Secretaria de Educação Continuada, Alfabetização e Diversidade, Brasília

MapBiomas (2018) Versão 3.0. http://mapbiomas.org. Accessed 10 Jan 2019

Meira P, Sato M (2005) Só os peixes mortos não conseguem nadar contra a correnteza. Rev Edu Pública 14(25):17–31

Pedrini AG (2006) A educação ambiental com a biodiversidade no Brasil: um ensaio. Ambiente & Educação 11:63–77

Pol E, Castrechin A (2013) Disrupción en la educación para la sostenibilidad? Rev Lat Am Psicol 45(3):335–349

Rivera-Monroy VH, Lee SY, Kristensen E, Twilley RR (2017) Introduction. In: Rivera-Monroy VH, Lee SY, Erik Kristensen E, Twilley RR (eds) Mangrove ecosystems: a global biogeographic perspective – structure, function, and services. Springer, Berlin and Heidelberg, pp 1–16

Soffiati A (2015) Pé no Mangue. Editora Autografia, Rio de Janeiro

United Nations Educational, Scientific and Cultural Organization – UNESCO (2004) United Nations decade of education for sustainable development 2005–2014. Draft: International Implementation Scheme. UNESCO, Paris

United Nations Educational, Scientific and Cultural Organization – UNESCO (2017) Ocean literacy for all – a toolkit. IOC manuals and guides 80. UNESCO, Paris

United Nations Environment Programme – UNEP (2014) The importance of mangroves to people: a call to action. In: van Bochove J, Sullivan E, Nakamura T (eds) United Nations environment programme. World Conservation Monitoring Centre, Cambridge

Vladimirova K, Le Blanc D (2016) Exploring links between education and sustainable development goals through the lens of UN flagship reports. Sustain Dev 24:254–271

Chapter 18
The Mangrove-Salt Marsh Complex: A Dynamic Landscape Approach

Gilberto Cintrón-Molero, Yara Schaeffer-Novelli, André S. Rovai, and Guilherme Moraes de Oliveira Abuchahla

18.1 Introduction

Although coastal wetlands share many attributes with inland types, they are exposed to a far more energetic environment where large amounts of energy are dissipated in constructive as well as destructive ways. Persistence in this environment requires adaptations at the species level to deal with coastal dynamism. At the population level, diaspore seeds and propagules with multiple dispersal characteristics provide rapid propagation and exploitation of substrates. This species richness translates into a gene pool at the landscape level and macroscales as a regional biogeographic pool. Along these lines of thought, it is appropriate to highlight that coastal wetlands are subject to disturbances and changes that span several spatiotemporal scales. Fast changes occur at the local level (hours, days). Slow change (10^3 years) is triggered by sea-level eustatic change, isostasy, and tectonic uplift or submergence. The third type of change involves catastrophic, rapid episodes such as high-magnitude floods and storms.

G. Cintrón-Molero (✉)
Instituto BiomaBrasil, São Paulo, SP, Brazil

Y. Schaeffer-Novelli
Instituto BiomaBrasil, Recife, Brazil

Instituto Oceanográfico/Insituto de Energia e Ambiente, Universidade de São Paulo, São Paulo, Brazil

A. S. Rovai
College of the Coast and Environment, Louisiana State University, Baton Rouge, LA, USA

G. Moraes de Oliveira Abuchahla
Leibniz Centre for Tropical Marine Research (ZMT), Bremen, Germany

Y. Schaeffer-Novelli et al. (eds.), *Brazilian Mangroves and Salt Marshes*, Brazilian Marine Biodiversity, https://doi.org/10.1007/978-3-031-13486-9_18

Depending on the scope of the disturbance, different levels of organization contribute to inducing regeneration of the landscape. At the local scale, community-level properties can cope with small frequent disturbances. Intermediate-level disturbances are addressed by landscape organizations such as self-organized changes in the configurations of mosaics. The most extreme disturbances, such as climate and sea-level changes, are addressed by dispersal from within or adjacent coastal environmental settings (CESs) that act as refuges and propagule source areas (Rovai et al. 2018; Twilley et al. 2018) (see also Chap. 3).

Wetlands have played a conspicuous role in the planet's history. Tropical peat and organic matter accumulations formed the large coal deposits of the carboniferous period. Wetlands were also the cradle of the earliest civilization. Marshes include swamps, bogs, and saltwater herbaceous intertidal formations in lagoons and estuaries where they often co-occur with mangroves. Although salt marshes occur along many of the world's shorelines particularly in middle to high latitudes, their current occurrence and extent are incomplete (Mcowen et al. 2017). They occupy the interface between terrestrial and marine habitats, forming ecotones between salt- and freshwater ecosystems.

In the 1960s, Teal's (1962) landmark paper on the energy flow in a marsh ecosystem helped to build broad generalizations about the structure and function of estuaries that recognized marshes as outwelling ecosystems. That paper expanded the perspective beyond the level of individual plant formation to the whole ecosystem and the broader landscape, having a major role in coastal and shelf productivity since outwelling from estuaries fuels phytoplankton productivity offshore (Twilley 1988). The transformative power of the salt marsh paradigm rapidly led to a generalized awakening about the importance of coastal wetlands in general. This recognition provided an early clue that wetland processes were best judged at the landscape geoecological level, the scale of the coastal environmental setting (CES) (Rovai et al. 2018; Twilley et al. 2018), and not at the individual plant community level. The recent recognition of wetlands as carbon sinks and their role in atmospheric CO_2 regulation has further spotlighted their importance.

The notable lack of information about salt marshes in Latin America was noticed by Costa and Davy (1992) and Junk et al. (2013). This is true as well of the salt marshes in Asia and Central and South America (Mcowen et al. 2017). Adam (2002) points out a clear difference in salt marsh diversity between different climatic zones, considering their distribution as arctic, boreal, temperate (Europe, Western North America, Japan, Australasia, and South Africa), dry coasts, West-Atlantic, and tropical. The latter two in which Brazil can be included: the West-Atlantic dominated by the smooth cordgrass *Spartina alterniflora* Loisel (syn. *Sporobolus alterniflorus* (Loisel.) P.M. Peterson & Saarela, while the tropical distribution dominated by the seashore dropseed *Sporobolus virginicus* (L.) Kunth and the seashore paspalum *Paspalum vaginatum* Sw. They co-occur with the saltwort *Batis maritima* L. and the shoreline purslane *Sesuvium portulacastrum* (L.) L., among others.

18.2 Assemblages, Complex Systems, or a *Continuum* of Coastal Marsh Ecosystems

The role of salt marshes has increasingly been recognized as ecosystemic, as an assemblage of interacting biological and physical processes. The emerging paradigm has taken into consideration particulate and dissolved organic matter (POM and DOM, respectively) dynamics and bacterial-microbial loops (see Chap. 7) that introduce organic carbon into the trophic web. These findings were later extended to mangrove ecosystems by Heald (1969), Odum (1970), and Odum and Heald (1975).

Teal (1962) had observed that most of the marsh *Spartina* spp. was not grazed upon in situ but decayed and became fragmented and then exported as detritus into the estuary. Later, Odum and de la Cruz (1967) and Mesquita (1983) observed that the detrital particles were promptly covered by bacteria forming bacterial/detrital aggregates, which then became incorporated into the estuarine food web. This became a dominant paradigm until Haines and Montague (1979) expanded the confirmatory evidence showing that the estuarine food web was far more broadly based and more complex than had been derived from previous gross oversimplifications. Using stable carbon isotopic ratio analysis, Haines and Montague (1979) and later researchers demonstrated that (1) algal-derived organic matter from microscopic blue-green algae and diatoms and edaphic and planktonic algae are a significant element of estuarine food webs (Sullivan and Moncreiff 1990); (2) vascular plant detritus may temporarily accumulate in situ where it is consumed by snails, its consumers, and associated trophic cascade; and (3) estuarine food webs, therefore, are far more complex, involving multiple sources of detritus and trophic pathways (Haines and Montague 1979; Domingos and Lana 2017).

There are serially concatenated as well as parallel pathways (networks) that can switch routings depending on prevailing conditions as during ecological succession or variations in environmental forcing functions. This is consistent with the view of marshes and mangroves as constituents of a larger highly interwoven, interactive estuarine system (Rovai et al. 2018). It highlights the need to recognize higher-level indirect as well as local, direct relationships. Advances in the understanding of complex systems' dynamics have been constrained by the dominance of a limited perspective based on parts and local scale rather than on whole entities.

A holological approach, on the other hand, recognizes higher-level functions that generate sustainability and long-term persistence (Ulanowicz 1997). Thus, the parts considered shaping an autocatalytic estuarine configuration that is not merely reactive to change but creates its own domain of influence and function. That approach is complementary to a traditional reductionist one, implying that ecological understanding requires a multiscalar perspective for framing causal processes.

18.2.1 Mangrove and Salt Marshes as a *Continuum*

West (1956) and others have recognized mangrove and salt marshes as a *continuum* vegetation unit that included mangroves and freshwater tidal swamps and mixed mangrove-*Acrostichum* spp. belt or patches that would correspond to a mangrove-salt marsh association. Schimper (1888) recognized a brackish vegetation formation adjacent to mangroves occupied by "semi-mangroves," in Asia, bordering the freshwater community inland. Generally, this zonation has been interpreted through the lens of a freshwater-saltwater succession. However, Clarke and Hannon (1969) recognized and described this formation as an integrated community rather than a transient seral community. Similarly, Walter (1977) considered this assemblage as a discrete community (ecosystem).

We expand in this line of thinking to suggest that the functional unit and its associated phenomena are "captured" by the geomorphological setting: a natural surface enclosing a mosaic (Rovai et al. 2018). The setting as a framework serves as a tangible discrete region that delimits aspects and process characteristics of the behavior of multiple interacting ecosystems. It also serves to integrate conceptual levels of observation and scale-independent entities (ecosystems and landscapes) into a scale-defined geographic unit. Furthermore, this perspective allows changing scales by adjusting the resolution (scope and grain), starting from a structurally defined entity (see Allen and Hoekstra 1990, 1992).

Here, we highlight that salt marshes are integral constituents of a coastal wetland geoecomorphological *continuum* where fresh- and saltwater marshes and mangrove forests self-assemble into landscape-level process-functional entities. Until now these areas have been considered aggregates of separate communities, whereas here we argue that a wetland *continuum* perspective is more realistic and helpful for understanding, management, and policymaking. See Chap. 3, Maps 1–17 show the location of mangroves, salt marshes, and salt flats in all Brazilian coastal states.

18.3 The Highly Dynamic Coastal Marshes

Coastal salt marshes have been defined in their most general sense as environments high in the intertidal zone where an often-muddy substrate supports halophyte plants (Allen and Pye 1992). They develop on the open coast, sheltered behind sandy barriers, and are also found in deltaic or non-deltaic estuaries where they occupy landforms between mean sea level and higher high water. The duration and frequency of tides influence the distribution although the configuration may be significantly modified by local site factors such as geomorphology and sediment inputs and transport. Site factors become particularly critical where salt marshes develop near ecological thresholds.

Salt marshes are highly dynamic formations vulnerable to erosion but capable of rapidly colonizing new sediments. Zonation is linked to flooding and salinity. At the higher intertidal elevations of depositional landform freshwater runoff, seepage and percolation reduce the salinity favoring glycophytes and freshwater marsh species, which become dominant (West 1977). The more halophytic species thrive at the lower marsh, where tidal mixing increases salinity and provides greater fluxes of nutrients and substrate aeration.

18.4 Salt Marshes in Brazil

In the context of Brazil, Junk et al. (2013) indicate that whereas there is considerable scientific information about the Brazilian mangroves (Brazil Blue Initiative 2018), the same is not true for other types of wetlands, particularly the ca. 121 km^2 coastal "marismas" (Portuguese for salt marshes) (Brazil Blue Initiative 2018) despite their presence along most of its coast. Only recently the importance of some of these marshes has been recognized by ornithologists as habitats for endangered bird species (Favretto 2018).

Much of the northern Brazilian coastal plains are dominated by the Amazon delta, which is occupied by assemblages of mangroves, marshes, and swamps that are prominent elements of the northern coastal landscape. In some environments, salt marshes form extensive monospecific meadows of *Spartina* spp. or can be associated with mangroves as in most of the tropical/subtropical Brazilian coastal landscape (Dansereau 1947; Lamberti 1969; Adaime 1978; Panitz 1992; Charlier-Sarubo et al. 2015; Schaeffer-Novelli et al. 2016). The most common genera throughout the country are *Spartina*, *Salicornia*, *Juncus*, *Paspalum*, *Cyperus*, *Batis*, *Sesuvium*, and *Bacopa* (Delaney 1962; Marangoni and Costa 2009).

In general, marshes are less prominent along the Brazilian coast, but they are locally important where freshwater seepage is significant and thrive behind mangrove fringes, or sandy barriers. The higher levels of these marshes occur where precipitation exceeds potential evapotranspiration (P > PET), which are subject to freshwater inputs from land drainage seeps and springs and salinities up to 20.

Tidal flushing creates topographic *catenas* within a tidal gradient. Technically these catenas are known as *haloseres* (or halophytic catenas). Some halophytes are facultative, but others persist only under the condition of low rainfall (P < 1300 mm year^{-1}) and high salinity where the high marsh becomes a distinctive morphoclimatic feature where salt accumulates, forming crusts and bare or succulent-covered areas (Guilcher 1979).

18.4.1 The *Apicum* Feature

Salt marshes are often integrated with mangroves where they develop best in the upper intertidal zone occupying spaces within a salinity range of 0–30. In arid or semiarid environments, mangroves merge into "apicuns," "salgados," or "areais" (Portuguese terms for salt pans, salt flats, or sabkhas).

The *apicum* (singular for *apicuns*) is dynamically coupled to the lower marsh and mangrove by haloseric successions, tidal flows, and animal migrations. The *apicum* is a feature of the mangrove ecosystem (just above the mean level of spring tides), landscapes typical of northeast Brazil and a dynamic feature of the upper intertidal zone between ~5° S and 15° S where the amplitude of spring tides is higher than 2.6 meters. Here, interstitial salinities reach values higher than 100 (Marius et al. 1987; Nascimento 1999; Meireles et al. 2007). These areas expand during periods of drought and contract depending on precipitation and long-term sea-level cycles (e.g., metonic, lunar cycle of 18.6 years). Mangroves are part of this dynamic halofacies dominating along creeks or as scrubby growth of the most tolerant species (*Avicennia* spp. and *Laguncularia racemosa*), which may be present as stunted individuals. This environment is dominated by pantropical salt-tolerant species such as *Sporobolus virginicus* and *Sesuvium portulacastrum* (Schaeffer-Novelli 2002).

The "bare" appearance of salt flats is deceiving. *Apicum* substrates are known to host bacteria, filamentous blue-green algae, microinvertebrate communities, scavengers, and predators such as fiddler crabs and birds (residents and long-distance migrants). During moist periods, algal mats may form, which break up and are dispersed by runoff during periods of heavy rain. Thus, these apparent bare flats export dissolved organic carbon (DOC) and dissolved organic matter (DOM) (Hogarth 2015).

The perception of *apicum* as a bare feature is also a result of short-term observations that ignore interannual or decadal oscillations in sea level and climate. The latter is not as static as previously thought, and several types of oscillations are intrinsic to the El Niño Southern Oscillation (ENSO) climate system and shifts in the Intertropical Convergence Zone (ITCZ) that can force adjustments in the configuration of the high marsh, which see-saws (fluctuates) in extent according to the strengths and periodicity of the local variability in climate (e.g., rain cycles associated with atmospheric oscillations; Kayano and Andreoli 2004). These sea level and atmospheric oscillations interact in either a constructive manner or destructively when they are out of phase causing expansion or contraction of "bare" areas along multidecadal scales (15–35 years).

18.5 The Gap for Salt Marshes: Discontinuity

A notable feature of the distribution of mangroves and salt marshes along the coast of Brazil is the absence of mangroves below 28° 30′ S and surprisingly, although not reported, the absence of marshes beyond this point (Table 18.1, see Chap. 3, Maps 16 and 17). This absence is a discontinuity because they reappear at 31° 48′ S (Isacch et al. 2006; Marangoni and Costa 2009) and extend to the southernmost estuarine portion of Patos Lagoon near 32° 07′ S and again reappear at the La Plata Estuary

Table 18.1 Austral limits for mangroves in Atlantic South America, where *PR* Paraná, *SC* Santa Catarina, *RS* Rio Grande do Sul (see Chap. 3, Maps 15, 16, and 17)

Latitudes	Sites	States	Sources
28° 56′ S	Araranguá River	SC	Chapman (1976)
27° S	Florianópolis	SC	Tomlinson (1986)
29° S	Araranguá	SC	Walter (1977)
28° 30′ S	Ponta Grossa River, Laguna	SC	Cintrón-Molero and Schaeffer-Novelli (1981)
28° 30′ S	Ponta Grossa River, Laguna	SC	Schaeffer-Novelli et al. (1990a)
33° S	Mangueira Lagoon	RS	Hogarth (1999)
27° 56′ S	Guarda do Embaú, Imbituba	SC	Spalding et al. (2010)
28° 28′ S	Ponta Grossa River, Laguna	SC	Soares et al. (2012)
25° 54′ S	Guaratuba Bay	PR	Scott (2014)
28° 28′ S (?)	Laguna	SC	Saintilan et al. (2014)
28° 30′ S	Laguna	SC	Ximenes et al. (2018)

(35° S, in Uruguay) from where they extend into austral latitudes (>40° S). Although some researchers consider the Patos Lagoon as an estuarine system (Lessa et al. 2018), tidal influence only reaches 60 km from the Rio Grande municipality (southern Patos Lagoon system) inlet and the estuarine portion of the lagoon is about 900 km^2 restricted to the southernmost part of the system (Odebrecht et al. 2010, 2017), limited to 10% of the system. Salt marshes that occur between 31° 48′ S and 32° 07′ S are indicators of saline estuarine environments (Marangoni and Costa 2009).

Some 70 km south of Laguna is the Araranguá River outlet (28° 56′ S), which is blocked by longshore transport of sediment (Siegle and Asp 2007) for periods longer than 6 months (Gabaglia 1916). In Araranguá, mangroves are absent (Chapman 1976; Soares et al. 2012; Ximenes et al. 2018); however, some brackish and freshwater species are present as *Hibiscus* spp. and *Dalbergia* spp. (Schaeffer-Novelli and Adaime 1979 pers. observation). At Tramandaí River (29° 56′ S), sediment core-sample records for palynological analyses reported the presence of both fresh- and saltwater marshes, e.g., *Salicornia gaudichaudiana*, *Acrostichum danaeifolium*, *Typha angustifolia*, and *Scirpus gigantescus* (Lorscheitter and Dillenburg 1998). The study registered episodes of salt marsh dominance associated with marine intrusions but no mangrove presence in the last 5000 years.

This physiographic limit has raised much interest in the scientific community because mangroves in other parts of the world reach more austral latitudes. Furthermore, the south of Laguna extends one of the world's largest lagoon systems (Patos Lagoon) and the climate type is similar to the climate in the latitudes elsewhere. For example, mangroves are still found in New Zealand in such latitudes and, where their expansion creates such great concern, national programs have been developed to remove them (ARC-Auckland Regional Council 2007).

As mentioned above, Brazil's mangrove forests or luxurious salt marshes are missing from most of Patos Lagoon, which extends from latitudes 30° S to 31° 48′ S, considered the world's largest choked lagoon (Kjerfve 1986). Most of its water is predominantly fresh to oligohaline, with fluvial and wind influences as main hydrologic drivers. The lagoon has a surface area of 10,360 km^2 (Odebrecht et al. 2010). P > PET in most of its watershed, resulting in freshwater predominance, which causes a head of surface-level considerably above sea level (Kjerfve 1986). This type of lagoon is typical of high-energy microtidal coasts, which becomes chocked by the extremely high littoral transport and high wave regime. Littoral sand transport chokes barrier openings and seals off the lagoon from the ocean except at its southern end. The connection to the Atlantic is maintained by jetties built early in the twentieth century to stabilize the inlet (Odebrecht et al. 2010).

Different hypotheses have been developed to explain this prominent discontinuity in mangroves, whereas the gap in salt marsh has been overlooked. Soares et al. (2012) suggest that mangrove dispersal is blocked by the longshore current described by Siegle and Asp (2007), suggested by the northward coastal orientation of the shoreline between Laguna (28° 30′ S) and the Araranguá River inlet (28° 56′ S).

Conventionally, latitudinal zonality (the use of latitude as proxy for temperature) has been used to explain the mangrove confinement to Laguna. Vegetation studies have emphasized the concept of climatic zonality. Many geomorphic, oceanographic, and meteorological factors have been suggested as causal agents for this notable absence; seawater and air temperature and the nearshore currents that flow sometimes in directions that would prevent dispersal are among those considered, as well as periodic frost events (Cintrón-Molero and Schaeffer-Novelli 1981; Schaeffer-Novelli et al. 1990b; Soares et al. 2012). What is notable in this ecological story is that not only mangroves are missing; salt marshes are missing as well except at the southern end in the Patos Estuary (Odebrecht et al. 2010; Fontoura et al. 2016; Odebrecht et al. 2017). Surprisingly, this discontinuity has remained poorly reported by researchers perhaps because the distinction between salt marshes and freshwater marshes needs a keenly trained eye and more than casual superficial observations. At ~32° 10′ S, *Spartina alterniflora* forms monospecific meadows where salinity is about 10 (Cunha et al. 2005).

18.6 The Freshwater Barrier to Mangrove and Salt Marsh Colonization

Immediately below the Serra Geral Mountain Range, orographically enhanced rain events (>1400 mm) contribute to the freshening of the narrow coastal strip (Siegle and Asp 2007) (see Chap. 3, Maps 16 and 17). This physiographic blockage is enhanced by freshwater inputs from high precipitation (>1200 mm), seepage, and over-land surface drainage. Here we suggest that the mangrove limit and the salt marsh gap that occurs in south Brazil is the result of a bottleneck created by a combination of physiographic factors and the regional-scale climatic feature of the South American continent. a feature that has just been recognized in the last decade (Carvalho and Cavalcanti 2016), the South American Monsoon System (SAMS), and its associated feature, the South American Convergence Zone (SACZ). These features, which are of continental and regional scales, produce abundant summer convective precipitation over land influencing soil moisture, river flows, large-scale flooding, and wetland hydroperiods (Cintrón-Molero and Schaeffer-Novelli 2019) (see Chap. 3).

SAMS is part of an integrated global climate system that includes other well-known and familiar monsoons: Asian, Australian, African/Indian, and North American. SAMS is South America's monsoon system, but it is somewhat different from others and perhaps for that reason has not been as widely recognized by nonspecialists. However, monsoonal characteristics have been recognized by Azevedo (1964) who suggested that the Amazon rainforest acted as an inland "sea" that feeds the Chaco Monsoon; Sioli (1984) suggested the presence of a large-scale branch of an aerial hydrological cycle as the agent bringing moisture to the La Plata Basin.

SAMS and SACZ are manifestations of an aerial "lake" and river that brings huge amounts of moisture from the Amazon Basin to Southeast Brazil and the La Plata Basin. This southernmost coastal segment receives drainage from that basin as well as abundant convective precipitation from the SACZ, which extends southwestward along the northeastern boundary of the basin during the summer months (Silva and Kousky 2012). Annual mean rainfall decreases from north to south and from east to west. A low-level jet runs along the Pampas syncline between 28° S and 34° S (Bridges 1990). This corridor is characterized by the deep convective activity of various sizes and intensities (Fig. 18.1).

In general, from 20° S to 25° S latitudes, summer rainfall is a distinctive signature of the mature SAMS and the SACZ keeping the chocked Patos Lagoon system almost fresh with minimal saltwater intrusion except near its southern end (Odebrecht et al. 2010).

It is widely recognized that mangroves are poor competitors under nonsaline conditions where they are replaced by glicophytes or brackish assemblages. According to Walter (1973), mangroves are azonal vegetation confined to saltwater tidal regions being absent even in tropical and subtropical regions where freshwater flows prevail over saline intrusions, such as the mouths of the Orinoco and the Amazon rivers (Huek 1978), and elsewhere such as coastal lakes. Salinity plays a key role in the growth and distribution of mangroves. Saltwater is an ecological requirement because although they are capable of growth in freshwater, saltwater prevents competition by non-halophytes. Despite their capacity to grow in freshwater for limited (time) periods, they are outcompeted by non-halophytes over longer (time) periods in nature. The presence of salt is a common feature of all mangroves although salinity may be only a seasonal or intermittent characteristic (Lugo et al. 1989).

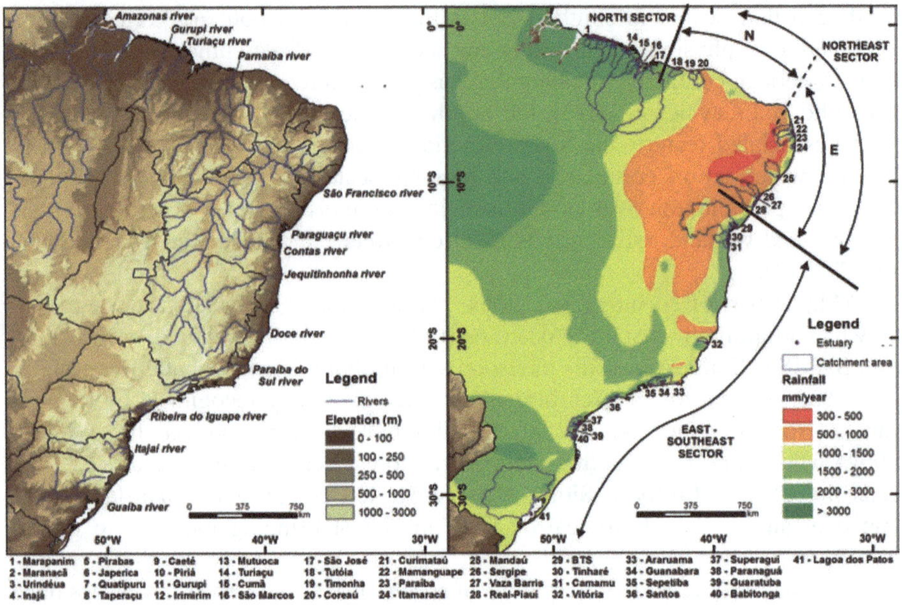

Fig. 18.1 Elevation map and the location of the Brazilian rivers with mean discharge higher than 100 m^3/s (left). Distribution of the mean annual rainfall in the continent, location of the main estuaries (>40 km^2), and their respective catchment areas (right). (Source: Lessa et al. 2018)

18.7 What to Expect from Climate Change?

Climate change has been an issue of increasing concern in the literature related to salt marshes and mangroves along the Brazilian coast (Schaeffer-Novelli et al. 2002; Soares et al. 2012; Schaeffer-Novelli et al. 2016; Ward et al. 2016; Servino et al. 2018; França et al. 2019).

Historically, the southern limit of Brazilian mangroves has been attributed to mangrove-salt marsh low-temperature limitations (Cintrón-Molero and Schaeffer-Novelli 1981; Schaeffer-Novelli et al. 1990b; Soares et al. 2012; Saintilan et al. 2014; Ximenes et al. 2018). The north-south orientation of Brazil's coast under the climatic change in temperature should favor the southward expansion of mangroves and encroachment on salt marshes. It is expected that these temperature changes will be manifested differently in separate coastal settings; for example, in northeast Brazil, episodic droughts would result in an expansion of *apicuns*.

Most climate projections predict enhanced seasonal amplitudes, early onset and late demise, and duration of SAMS (Soares and Marengo 2008), one of the drivers responsible for the bottleneck physiographic-climate-freshwater barrier mangrove's blockage below 28° 30′ S. This is consistent with the IPCC's Couple Module Intercomparison Project 5 – CMIP5 (Jones and Carvalho 2013). Nevertheless, in southeast Brazil, higher temperatures are expected to increase the frequency and intensity of storms and atmospheric moisture in this coastal setting (about 30° S to the south), where enhanced ITCZ and ENSO activity would increase rainfall and freshwater inputs would also alter salinity triggering changes in the configuration of mangrove-marsh ecotones and blocking mangrove and salt marsh expansion southward.

18.8 Marshes and Mangrove Forests Under a Wholistic Perspective: From Ecosystems to Landscape

Although we tend for convenience to see nature as an aggregate of separate entities (ecosystems), these do not act as autonomous parts; they interact and function as constituencies that are intrinsically interwoven into a complex trophic thicket that cannot be disentangled, a functional whole. This extreme level of connectivity is facilitated by overlapping environmental tolerances, active transport processes, species movements, and strong mixing energies due to ecosystem interactions (ecosystems are open systems). In fact, functional connectivity can be stronger than structural proximity once a complex tidal channel network (tidal creeks) provides for the widespread distribution of diaspores and nutrients throughout the area.

The relevance of higher-level organization to sustainability and endurance is that local systems may appear to be at a steady state, but as open systems, they are not in equilibrium. All ecosystems exist in a state of constant flux. Yet ecosystems persist; in fact, local variability builds long-term robustness. Robustness is generated by a hierarchically organized network of interactions that build resilience at different scales (Walker et al. 2005). Robustness is the combination of local resilience (engineering resilience), ecological resilience, mutualism, redundancy, and degeneracy (functional redundancy). These attributes together build landscape-level fitness and evolvability (Voigt et al. 2005).

The integrated system is a dialectical ensemble of interacting levels where the lower levels are closely linked to environmental change and act as testing and selecting grounds, while the higher levels provide a genetic pool that includes the biogeographic scale. The different levels act as an autocatalytic cycle that is self-enhancing. Thus, this autocatalytic configuration is not simply reactive to environmental change; it creates its own domain of influence (Ulanowicz 1997).

Ulanowicz introduced the notion of ascendence and "overhead" to describe the growth and development of ecological organizations. Using this perspective, we can infer that the local levels are characterized by ascendant attributes for growth and expansion, whereas the higher levels act as a reservoir of genetic information available for regenerative work. This corresponds to the overhead of the system. These attributes act in combination although at different spatiotemporal scales. This dialectical perspective helps to clarify how Clements (1916) and Gleason's (1939) considerations of community processes are complementary. Clements focused on the highest levels, whereas Gleason focused on the dynamics of local levels positing that at any instant the local community was a product of random opportunistic establishment of species available for invasion and self-assembly. There is no climax since the environment is always changing. The landscape at any one moment is "perfect" (*sensu* Phillips 2007).

18.9 Final Remarks

1. Salt marshes and associated systems (freshwater marshes and mangrove forests) form an integrated complex. A temporal and spatial *continuum* where interconnectivity and interrelationships restrain partitioning into isolated units.
2. The most appropriate perspective to understand marsh-mangrove dynamics is the landscape perspective comprised of the physiographic setting or CES (Thom 1984; Rovai et al. 2018; Twilley et al. 2018) (see Chap. 3).
3. At the extreme latitudinal range of distribution, local site factors can play highly influential roles in modifying environmental constraints buffering environmental stressors.
4. Freshwater-dominant coastal environments (such as Patos Lagoon) inevitably exclude mangrove plants acting as a biogeographic barrier to dispersal.
5. The north-south orientation of the southeast Brazil coast would favor southward expansion of marshes and mangroves in a warmer climate scenario, but warming is expected to increase atmospheric moisture in this part of Brazil, which will enhance ITCZ and ENSO activity that would increase freshwater inputs and enhance the blockage at the physiographic-freshwater barrier south of about 30° S.
6. These systems are nonequilibrium and constantly adjusting to internal and external drivers; variability is multiscalar and reflects oscillations in external forcing functions as well as internal self-organization.
7. Dynamic systems cannot be "segmented" or placed in "straitjackets"; they must remain open in order to adjust and reconfigure in response to the environmental orientors as an integrated whole consistent with Walter's (1973) hydrohaloseric *continuum* and more recently by Phillips' (2007) recognition of landscapes as "perfect landscapes" (also see Chap. 3).
8. Salt marsh and mangrove forests' spatial distribution and extents are important for calculating/estimating potential carbon storage and management of these areas in an integrated manner, not as separate units.

Coastal wetland ecology has underestimated the intertwining of connections within the system and to the higher-level landscape (Braswell 2015); what we suggest is an integration of observations and ideas of West (1956), Clarke and Hannon (1969), Walter (1968), and Thom (1984) who provided the notion of a geomorphic framework that entrains processes and function in a geographic discrete area.

References

Adaime RR (1978) Estudo da variação estacional do "standing-crop" e do repovoamento em um banco de *Spartina alterniflora* Loiseleur, 1807, complexo estuarino-lagunar de Cananeia. Bol Inst Oceanogr 27:1–43

Adam P (2002) Saltmarshes in a time of change. Environ Conserv 29(1):39–61

Allen TFH, Hoekstra TW (1990) The confusion between scale-defined levels, and conventional levels of organization in ecology. J Veg Sci 1:5–12

Allen TFH, Hoekstra TW (1992) Towards a unified ecology. Columbia University Press, New York

Allen JRL, Pye K (1992) Saltmarshes. Morphodynamics, conservation and engineering significance. Cambridge University Press, New York

Auckland Regional Council – ARC (2007) New Zealand mangroves. ARC, Auckland

Azevedo A (1964) Brasil: a terra e o homem, vol 1. Companhia Editora Nacional, São Paulo

Braswell A (2015) https://blogs.nicholas.duke.edu/river-center/redefining-salt-marsh-research-are-coastal-wetlands-end-members-or-part-of-a-continuum/. Accessed 26 Dec 2018

Brazil Blue Initiative (2018) http://www.mangrovealliance.org/wp-content/uploads/2018/05/Brazil-Blue-Initiative-Summary.pdf

Bridges EM (1990) World Geomorphology, 1st edn. Cambridge University Press, Cambridge

Carvalho LMV, Cavalcanti IFA (2016) The South American Monsoon System (SAMS). In: Carvalho J (ed) The monsoons and climate change: observations and modelling. Springer, New York City, pp 121–148

Chapman VJ (1976) Mangrove vegetation. Cramer, Vaduz

Charlier-Sarubo S, Cunha-Lignon M, Schaeffer-Novelli Y (2015) Protocolo de coleta de dados e monitoramento dos ecótonos entre manguezal e marisma e entre manguezal e vegetação de restinga. In: Turra A, Denadai MR (orgs) Protocolos de campo para o monitoramento de habitats bentônicos costeiros – Rede de Monitoramento de Habitats Bentônicos Costeiros – ReBentos. IOUSP, São Paulo, pp 104–111

Cintrón G, Schaeffer-Novelli Y (1981) Los manglares de la costa brasileña: revisión preliminar de la literatura. In: Informe Técnico preparado para la Oficina Regional de Ciencia y Tecnología para América Latina y el Caribe de Unesco y la Universidad Federal de Santa Catarina (ed/rev: Abuchahla GMO). Unesco, p 47. http://www.producao.usp.br/handle/BDPI/43826

Cintrón-Molero G, Schaeffer-Novelli (2019) The role of atmospheric-tropospheric rivers in partitioning coastal habitats and limiting the poleward expansion of mangroves along the southeast coast of Brazil. Int J Hydro 3(2):92–94

Clarke LD, Hannon NJ (1969) The mangrove swamp and salt marsh communities of the Sydney District II: the holocoenotic complex with reference to physiography. J Ecol 57(1):213–234

Clements FE (1916) Plant succession: an analysis of the development of vegetation. Carnegie I Wash 242:1–7

Costa CSB, Davy AJ (1992) Coastal Saltmarsh communities of Latin America. In: Seeliger U (ed) Coastal communities of Latin America. Academic Press, New York City, pp 179–199

Cunha S, Costa C, Asmus ML (2005) Production dynamics of *Spartina alterniflora* salt marshes in the estuary of Patos Lagoon (RS, Brazil): assimilation model approach. Braz J Aquat Sci Tech 9(2):75–85

Dansereau P (1947) Zonation et succession sur la Restinga de Rio de Janeiro. I. Halosere. Rev Can Biol 6:448–477

Delaney PJV (1962) Quaternary geologic history of the coastal plain of Rio Grande do Sul, Brazil. S Am Coast Stu Techn Rep 10(A):1–63

Domingos AM, Lana PC (2017) Detecting multiple states of trophic connectivity between Mangroves and Salt Marshes. Ecosystems 20:1179–1189

Favretto MA (2018) On the subtropical salt marshes of Brazil: discussions about a recently recognized ecosystem. In: Barnes D, Ellis C (eds) Salt marshes. Nova Science Publication Inc, Hauppauge, pp 49–68

Fontoura NF, Vieira JP, Becker FG, Rodrigues LR, Malabarba LR, Schulz VH, Möller OO, Garcia AM, Vilella FS (2016) Aspects of fish conservation in the upper Patos Lagoon basin. J Fish Biol 89(1):315–336

França MC, Pessenda LCR, Cohen MCL, Azevedo AQ, Fontes NA, Silva FB, Melo-Jr JCF, Piccolo MC, Bendassolli J, Macario K (2019) Late-Holocene subtropical mangrove dynamics in response to climate change during the last millennium. Holocene 29(3):445–456

Gabaglia FAR (1916) As fronteiras do Brasil. Typographia do Jornal do Commercio de Rodrigues & Co, Rio de Janeiro

Gleason HA (1939) The individualistic concept of the plant association. Am Midl Nat 92:92–110

Guilcher A (1979) Marshes and estuaries in different latitudes. Interdiscip Res 4:158–168

Haines EG, Montague CL (1979) Food sources of estuarine invertebrates analyzed using $^{13}C/^{12}C$ ratios. Ecology 60:48–56

Heald EG (1969) The production of organic Detritus in a South Florida Estuary. PhD dissertation, University of Miami

Hogarth PJ (1999) The biology of mangroves. Oxford University Press, New York City

Hogarth PJ (2015) The biology of mangroves and seagrasses. Oxford University Press, New York City

Huek K (1978) Los bosques de Sudmerica: ecología, composición y importancia económica. Sociedad Alemana de Cooperación Técnica, Bad Hersfeld, Eschborn

Isacch JP, Costa CSB, Rodrigues-Gallego L, Conde D, Scapa M, Gagliardini DA, Iribarne OO (2006) Distribution of saltmarsh plant communities associated with environmental factors along a latitudinal gradient of the South-West Atlantic coast. J Biogeogr 33(5):888–900

Jones C, Carvalho LMV (2013) Climate change in the South American Monsoon system: present climate and CMIP5 projections. J Climate 26:6660–6678

Junk WJ, Piedade MTF, Lourival R, Wittmann F, Kandus P, Lacerda LD, Bozelli RL, Esteves F, Nunes da Cunha AC, Maltchik L, Schöngart J, Schaeffer-Novelli Y, Agostinho AA (2013) Brazilian wetlands: their definition, delineation, and classification for research, sustainable management, and protection. Aquat Conserv 24(1):5–22

Kayano MT, Andreoli RT (2004) Decadal variability in northern, northeast Brazil rainfall and its relation to tropical sea surface temperature and global sea level pressure anomalies. J Geophys Res 109:C11011

Kjerfve B (1986) Comparative oceanography of coastal lagoons. In: Wolfe DA (ed) Estuarine variability. Academic, New York City, pp 63–81

Lamberti A (1969) Contribuição ao conhecimento da ecologia das plantas do manguezal de Itanhaém. Bol Fac Filosof Cienc Letras 23

Lessa GC, Santos FM, Souza Filho PW, Corrêa-Gomes LC (2018) Brazilian estuaries: a geomorphologic and oceanographic perspective. In: Lana PC, Bernardino AF (eds) Brazilian Estuaries. Springer, Cham, pp 1–37

Lorscheitter ML, Dillenburg SR (1998) Holocene paleoenvironments of the northern coastal plain of Rio Grande do Sul, Brazil reconstructed from palynology of Tramandaí lagoon sediment. In: Rebassa J, Salemme M (eds) Quaternary of South America and Antarctica Penninsula. Balkema AA, Rotterdam, pp 73–97

Lugo AE, Brinson MM, Brown S (1989) Synthesis and search of paradigms in wetland ecology. In: Lugo AE, Brinson MM, Brown S (eds) Forest wetlands: ecosystems of the world. Elsevier, Amsterdam, pp 447–460

Marangoni JC, Costa CSB (2009) Diagnóstico ambiental das marismas no estuário da Lagoa dos Patos – RS. Atlântica 31(1):85–98

Marius C, Archanjo D, Larque P (1987) Les sols de mangroves de la baie de Vitória (Brésil). Cahiers Orstom Série Pédologique 3:211–216

Mcowen C, Weatherdon L, van Bochove J-W, Sullivan E, Blyth S, Zöckler C, Santwell-Smith D, Kingston N, Martin CS, Spalding MD, Fletcher S (2017) Global map of saltmarshes. Biodivers Data J 5:e11764

Meireles AJA, Cassola RS, Tupinambá SV, Queiroz LS (2007) Impactos ambientais decorrentes das atividades de carcinicultura ao longo do litoral cearense, nordeste do Brasil. Mercator 12(6): 83–106

Mesquita HSL (1983) Suspended particulate organic carbon and phytoplankton in the Cananéia Estuary (25°S–48°W), Brazil. Océanographie Tropicale 18:55–68

Nascimento SA (1999) Estudo da importância do "apicum" para o ecossistema de manguezal. In: Sergipe, Relatório Técnico Preliminar. Governo do Sergipe, Aracaju

Odebrecht C, Abreu PC, Bemvenuti CE, Copertino M, Muelbert JH, Viera JP, Seeliger U et al (2010) The Patos Lagoon estuary, Southern Brazil; biotic responses to natural and anthropogenic impacts in the last decades (1979–2008). In: Kennish MJ, Paerl HW (eds) Coastal Lagoons: critical habitats of environmental change. CRC Press, Boca Raton, pp 437–459

Odebrecht C, Secchi ER, Abreu PC, Muelbert JH, Uilbein F (2017) Biota of the Patos Lagoon estuary and adjacent marine coast: long-term changes induced by natural and human-related factors. Mar Biol Res 13(1):3–8

Odum WE (1970) Pathways of energy flow in a South Florida Estuary. PhD Dissertation, University of Miami

Odum EP, de la Cruz AA (1967) Particulate organic detritus in a Georgia saltmarsh-estuarine ecosystem. Amer Ass Advan Sci Publ 83:383–388

Odum WE, Heald EJ (1975) Mangrove forests and aquatic productivity. In: Hasler AD (ed) Coupling of land and water systems. Springer, Berlin/Heidelberg/New York, pp 129–136

Panitz CMN (1992) Ecological aspects of a salt marsh ecosystem in Santa Catarina Island, Brazil. In: Seeliger U (ed) Coastal Plant Communities of Latin America. Elsevier, Amsterdam, pp 213–230

Phillips JD (2007) The perfect landscape. Geomorphology 84:159–169

Rovai AS, Twilley RR, Castañeda-Moya E, Riul P, Cifuentes-Jara M, Manrow-Villalobos M, Pagliosa PR (2018) Global controls on carbon storage in mangrove soils. Nat Clim Change 8(6):534–538

Saintilan N, Wilson N, Rogers K, Rajkaran A, Krauss KW (2014) Mangrove expansion and salt marsh decline at mangrove poleward limits. Glob Change Biol 20(1):147–157

Schaeffer-Novelli Y (2002) Manguezal, marisma e apicum. In: Fundação Biológica (ed) Avaliações e ações prioritárias para conservação da biodiversidade das Zonas Costeira e Marinha. MMA/SBF, Brasília

Schaeffer-Novelli Y, Mesquita HSL, Cintrón-Molero G (1990a) The Cananéia lagoon estuarine system, São Paulo, Brazil. Estuaries 13(2):193–203

Schaeffer-Novelli Y, Cintrón-Molero G, Adaime RR, Camargo TM (1990b) Variability of the mangrove ecosystem along the Brazilian coast. Estuaries 13(2):204–218

Schaeffer-Novelli Y, Cintrón-Molero G, Soares MLG (2002) Mangroves as indicators of sea level change in the muddy coasts of the world. In: Healy T, Wang Y, Healy J-A (eds) Muddy coasts of the world: processes, deposits and function. Elsevier Science BV, Amsterdam, pp 245–262

Schaeffer-Novelli Y, Soriano-Sierra EJ, Vale CC, Bernini E, Rovai AS, Pinheiro MAA, Schmidt AJ, Almeida R, Coelho-Jr C, Menghini RP, Martinez DI, Abuchahla GMO, Cunha-Lignon M, Charlier-Sarubo S, Shirazawa-Freitas J, Cintrón-Molero G (2016) Climate changes in mangrove forests and salt marshes. Braz J Oceanogr 64(2):83–98

Schimper AFW (1888) Die epiphytische Vegetation Amerikas. Bot Mitt Trop 2

Scott DB, Frail-Gauhtier J, Mudie PJ (2014) Coastal wetlands of the world, geology, ecology, distributions and applications. Cambridge University Press, Cambridge

Servino RN, Gomes LEO, Bernardino AF (2018) Extreme weather impacts on tropical mangrove forests in the Eastern Brazil Marine Ecoregion. Sci Total Environ 628:233–240

Siegle E, Asp NE (2007) Wave refraction and longshore transport patterns along the southern Santa Catarina coast. Braz J Oceanogr 55(2):109–120

Silva VBS, Kousky VE (2012) The South American Monsoon System: Climatology and Variability. In: Wang S-Y, Gillies RR (eds) Modern Climatology. Rijeka, Intech, pp 123–152

Sioli H (1984) The Amazon: limnology and landscape ecology of a mighty tropical river and its basin. Dr W Junk Publishers, Dordrecht

Soares WR, Marengo JA (2008) Assessments of moisture fluxes east of the Andes in South America in a global warming scenario. Int J Climatol 29(10):1395–1414

Soares MLG, Estrada GCD, Fernandez V, Tognella MMP (2012) Southern limit of the Western South Atlantic mangroves: assessment of the potential effects of global warming from a biogeographical perspective. J Coastal Res 101(10):44–53

Spalding MD, Kainuma M, Colins L (2010) World Atlas of Mangroves. Earthcan, London

Sullivan MS, Moncreiff CA (1990) Edaphic algae are an important component of marsh food webs from multiple stable isotope analysis. Mar Ecol Prog Ser 62:149–159

Teal HM (1962) Energy flow in the salt marsh ecosystem of Georgia. Ecology 43:614–625

Thom BG (1984) The Mangrove ecosystem: research methods. Department of Geography, University of New South Wales, Duntroon

Tomlinson PB (1986) Mangrove botany. Cambridge University Press, Cambridge

Twilley RR (1988) Coupling of mangroves to the productivity of estuarine and coastal waters. In: Jansson BO (ed) Coastal-offshore ecosystem interactions. Springer, Berlin, pp 155–180

Twilley RR, Rovai AS, Riul P (2018) Coastal morphology explains global blue carbon distributions. Front Ecol Environ 16(9):1–6

Ulanowicz RE (1997) Ecology, the Ascendent Perspective. Columbia University Press, New York City

Voigt CA, Mayo SL, Wang Z-G, Arnold F (2005) Directing the evolvable: utilizing robustness in evolution. In: Jen E (ed) Robust design: a Repertoire of biological, ecological and engineering case studies. Oxford University Press, New York City, pp 105–134

Walker BH, Peterson GD, Anderies JM, Kinzig A, Carpenter SR (2005) Robustness in ecosystems. In: Jen E (ed) Robust design: a Repertoire of biological, ecological and engineering case studies. Oxford University Press, Oxford, pp 173–189

Walter H (1968) Die Vegetation der Erde, vol II. Fischer, Jena/Stuttgart

Walter H (1973) Vegetation of the earth and ecological systems of the gosphere. Springer, Berlin

Walter H (1977) Vegetationszonen und Klima: die ökologische Glieregung der Biogeosphäre. Ulmer, Stuttgart

Ward RD, Friess DA, Day RH, Mackenzie RA (2016) Impacts of climate change on mangrove ecosystems: a region by region overview. Ecosyst Health Sustain 2(4):1–64

West RC (1956) Mangrove swamps of the Pacific Coast of Colombia. Ann Assoc Am Geogr 46(1): 98–121

West RC (1977) Tidal salt marsh and mangal formations of middle and South America. In: Chapman VJ (ed) Ecosystems of the world. Elsevier Scientific Publication, Amsterdam, pp 193–213

Ximenes AV, Ponsoni L, Lira CF, Dahdouh-Guebas F (2018) Does sea surface temperature contribute to determining range limits and expansion of mangroves in Eastern South America (Brazil)? Remote Sens 10(11):1787

Chapter 19
Coastal Zone Adaptive Management in the Context of Complex Adaptive Systems

Gilberto Cintrón-Molero, Yara Schaeffer-Novelli, Guilherme Moraes de Oliveira Abuchahla, and André S. Rovai

> Primum non nocere *"First do no harm" (Hippocrates of Kos, c. 460-377 BC)*

19.1 Introduction

Historically coastal zones have been important convergence zones of natural and social-ecological processes (see Chap. 14) and generators of history, tradition, and place value. Despite the emerging focus of resource management for the conservation and sustained delivery of ecological services, mangrove-salt marsh complex systems (or settings) are being lost at alarming rates due to urban, industrial, port, and aquaculture expansion (see Chap. 15). In part, those losses are triggered by economic growth perspectives that ignore or fail to consider the emerging notions of complex adaptive system (CAS) (*sensu* Levin 1995) dynamics in the coastal zone, and the pervasive interconnectivity that integrates coastal ecosystems into increasingly larger and more complex macroecosystems – from individual sites to coastal settings.

The destruction of coastal units that span whole landscapes has epistemic roots related to ignorance, as well as to the fundamental incoherence between worldviews based on the equilibrium theory (e.g., economy) and nonequilibrium dynamics

G. Cintrón-Molero (✉)
Instituto BiomaBrasil, São Paulo, SP, Brazil

Y. Schaeffer-Novelli
Instituto Oceanográfico/Instituto de Energia e Ambiente, Universidade de São Paulo, São Paulo, SP, Brazil

IUCN SSC Mangrove Specialist Group, São Paulo, SP, Brazil

G. Moraes de Oliveira Abuchahla
Leibniz Centre for Tropical Marine Research (ZMT), Bremen, Germany

A. S. Rovai
College of the Coast and Environment, Louisiana State University, Baton Rouge, LA, USA

Y. Schaeffer-Novelli et al. (eds.), *Brazilian Mangroves and Salt Marshes*, Brazilian Marine Biodiversity, https://doi.org/10.1007/978-3-031-13486-9_19

(ecology). Most decision-makers are not aware of this incompatibility and its consequences, resulting in a persistent fragilization of coastal systems at a time of rapid global environmental change when prudence and anticipatory principles were supposed to be the foundations of sustainability. Decision-makers are often deceived by the apparent profuse natural resilience and are not aware that this attribute is not permanent, but variable and eminently contextual. Effective resource management depends on how we perceive the way nature is organized and functions since there must be harmony and congruence between perception, understanding, and policies. Therefore, skillful management requires competence to interpret CAS behavior accurately. Without it, it is impossible to frame problems and design policies that are harmonious with natural processes.

The traditional managerial approach of authoritative top-down command-control is alien to the way CASs function and that explains to a large extent the degree and scope of coastal systems' deterioration. During times of accelerated change, policymaking must be highly flexible and command-control strategies become increasingly detrimental. Even when management policies have been well intended, the outcomes often have been harmful, or the unintended consequences intolerable. For example, the widespread use of breakwaters and defensive structures has resulted in accelerated coastal erosion and unplanned shoreline reconfiguration in many cases (Carter 1988).

Nature is too complex to be controlled. Control, as a foundation for management, is at best an illusory and not wise approach. Ecosystem collapse is more probable than sustainability, and this is a good reason for adopting a precautionary approach toward nature. Working against nature is a pyrrhic endeavor and a waste of human and social effort and resources. On the other hand, working with nature promotes win-win outcomes.

19.2 Coastal Ecosystem Features

Settings are regions of coincident physical and ecological limits; within the boundaries of settings, unique functional complexes become organized singularly since the complex is the result of a non-repeatable combination of agents, controls, and forcing functions to produce a highly improbable outcome considering interactions and contingency along its developmental trajectory. Key features of mangrove-salt marsh settings (including freshwater marshes) are their sensitivity and vulnerability to climate and sea-level changes (Schaeffer-Novelli et al. 2002), as well as multiple other non-climate contextual factors such as direct and indirect impacts of urban development in their periphery. Mangrove and salt marshes form large-scaled CASs (mangrove-salt marsh settings) and therefore a broad spectrum of responses can be expected in reaction to changes in environmental drivers (see Chap. 18). Because these systems are hierarchical, responses are also hierarchical, spanning a broad spatiotemporal domain with a diversity of site-level responses. The pervasive interconnectivity and interactivity indicate that these systems must be managed in an integrated fashion (Integrated Coastal Zone Management – ICZM), and not in a partial, fragmentary way (Ramsar Convention Secretariat 2010).

Complexity nevertheless makes specific predictions about managerial interventions or effects of environmental change impossible, particularly when long-term (strategic) policymaking is being attempted. Adaptive management (Holling 1978) is currently considered the most accepted way to deal with uncertainty. However, here we suggest that adaptive management, as currently implemented, has unrecognized limitations in that it contains implicit assumptions related to its application in steady-state systems in near equilibrium and the context of small-scale and spatial and temporal scales.

Holling's (1978) notion of adaptive systems has evolved since it was posited but the more recent perspective has yet to be widely understood or adopted by environmental managers, much less translated into regulatory processes (Folke 2006; Karkkainen 2006). The conventional view must be updated since it by itself may not provide adequate buffers and appropriate responses to deal with hierarchical organization and cross-scale dynamics, neither with the high level of uncertainty that results from the combination of complex multiscalar dynamics of CASs such as mangroves and salt marshes complexes as well as the diversity of habitats and fickle climate variability.

An updated management methodology or approach must be based on the following:

- Delaying and mitigation of effects (buffering)
- Restoration (recovery facilitation)
- Containment of impacts
- Increasing resilience (eliminating or reducing non-climatic stressors)
- Sustained foresight (attention to subtle change)

The abovementioned items entail anticipatory approaches based on awareness and early detection of undesirable effects. It involves a combined preemptive and regenerative approach that includes passive and active restoration (*sensu* Murray and Marmorek 2004). Policy responses to climate change must be based on a real-time methodology where predictions become less important than foresight, alertness, resilience, and versatility in terms of appropriate highly contextual responses to environmental change. Well-suited and timely responses are required because climate impacts cascade through a landscape and become manifested at different times and forms as modified by site factors (Schaeffer-Novelli et al. 2005a).

Unfortunately, despite increasing recognition that a system perspective is required for skillful resource management, many decision-makers continue to use obsolete thinking modalities that are excessively reductionistic and mechanistic. Many are not aware that coastal resources should be managed based on multiscalar considerations. A single spatiotemporal frame cannot provide a satisfactory framework for studying and understanding mangrove-salt marsh settings.

Holling's (1978) adaptive management notions have been steadily refined and the Panarchy theory (Gunderson and Holling 2002) provides an updated framework and can be used to apply a dynamic approach to problem framing that mimics the stratified nature of reality; reality must be understood considering the particulars and the whole simultaneously, taking in consideration the parts and the whole. This approach is grounded in holism and reductionism as complementary narratives for understanding whole system functions. In considering this "dynamic framing" (multiscalar management of coastal resources), we highlight that for convenience the coast can be divided into an infinite number of segments depending on the required grain and scope needed to frame a given problem (see Chap. 3).

The dynamic framing approach we suggest is multiscalar and grounded in Salthe's (1985) triadic approach, except we stress the importance of considering the focal level itself for an explanation of behavior. It is normally assumed that events in any hierarchical level are contingent upon lower-level mechanics and constrained by higher-level processes. The levels above and below are necessary to explore mechanisms and constraints that define behavior. The goal is to anticipate impacts while building propensities for increasing resilience as well as for harmonizing social and ecological systems. The goal is to promote self-organized management that requires the least manipulation by aligning management with natural processes. Coastal phenomena take place at various levels of detail, and the choice of a requisite level of detail (the level encompassing relevant processes) is a key criterion for mapping and assessing responses to environmental change and to frame management efforts. The selected minimum level of resolution must be able to "capture" the whole phenomenon.

The most strategic managerial option to deal with irreducible uncertainty and its potential consequences is prudence and caution as warning and preparedness provide important tactical advantages for containing and reducing impacts. Such a heuristic approach involves (1) minimizing the probability of errors in the interpretation of events, since misinterpreted evidence and poor judgments will become translated into dysfunctional decisions or policies, (2) delaying or hedging all decisions that involve an irrevocable commitment of resources, and (3) applying management measures to protect resilience and robustness at the setting scale (Janssen and Anderies 2007).

Here, system growth and diversity become manifested in "exploration" and constancy or persistence (Ulanowicz 1997). The outcome is evidenced in self-organization and ultimate persistence within dynamically changing geomorphic configurations. Biological and geomorphic forces work complementarily with the external energy regime and template constraints. Exploration leads to habitat exploitation of rapidly changing habitats, supported by the genomic aggregate of the system – its information resource or overhead. Overhead (as diversity and functional redundancy) is costly but required; it is complementary with the system ascendency (Ulanowicz 1997). That is, resources allocated to growth must be shared with what might appear to be redundant trophic structures that represent potential "strength-in-reserve" adaptability and evolvability.

Ecosystems are the prototypical example of Levin's (1990) CASs. This represents a new paradigm in the way we view ecosystems. Thus, mangrove-salt marsh ecosystems may be best viewed, studied, interpreted, and managed as CAS (Jørgensen 2012). Thus, these systems have been considered as "perfect" due to their uniqueness and fitness to local situations (Phillips 2007). Phillips' contention is that each landscape is indeed unique. Those systems are characterized by six properties: (1) a large number of components and dispersed nonlinear interactions, (2) hierarchical organization, (3) continuous adaptation, (4) perpetual novelty, (5) far-from-equilibrium dynamics, and (6) radical openness.

Openness and contingency are major sources of novelty, which makes a prediction and control very difficult if not impossible (Jørgensen 2012). It also makes these systems irreversible. Once altered they do not return to an earlier state but, if allowed and the context has not been significantly altered, they can regenerate. Since prediction and control are hallmarks of contemporary management, it is evident that CASs require new strategies for coping with irreducible uncertainty. Two principal attributes of complex adaptive systems are emergence and self-organization. Considering these coastal features as CAS is not merely giving them a new name; it has fundamental implications in how we view, study, interpret, and manage these systems.

19.3 Coastal Zone Landscape Management and Ecosystem Properties

In the presence of one or a few of the predicted climate changes and their consequences (sea-level rise, temperature rise, higher frequency, and recurrence of extreme events), it is possible that the characteristics of a system could change gradually or that the ecosystem could remain unchanged due to compensatory response, in terms of biomass and nutrient stores. Walker et al. (2004) indicated that ecological systems have three properties: resilience, adaptability, and transformability. The latter is defined as the ability to create a new system when various conditions (e.g., ecological and social-environmental drivers) deviate from the "normal" range of the system. It is known that complex systems hide latent damage and degrade "gracefully" only to undergo a sudden shift of state when a threshold is reached, or when a disturbance pushes the system beyond its residual resilience (Schaeffer-Novelli et al. 2005b; Alongi 2008).

Holling (1986) distinguished two types of resilience, engineering and ecological. Engineering resilience is defined as the speed of recovery of a system following shock. Changes in the speed of recovery can be used to measure this type of "homeostatic" resilience. In nonlinear systems subject to large disturbances, shifts can occur to an alternative state. Such complex systems are multi-stable. Ecological resilience recognizes that multiple states or regimes exist and therefore it is defined as the magnitude of disturbance that can be absorbed before a transition between states takes place (the size of the system's stability domain) (*sensu* Holling 1973). Because system shifts can result in different developmental trajectories, this type of resilience is homeorhetic rather than homeostatic. Holling (1973) has been the foundation from which the resilience perspective of social-ecological systems developed. Adaptability and transformability are the factors responsible for the persistence of systems such as mangrove and salt marshes, which have dominated the intertidal zone of the Brazilian coastal plains since the Post-Glacial Marine Transgression (Duke 1995).

What we see today represents a rare moment in the Quaternary in that we are in a high stand period. For the entire Quaternary the average position of the shoreline was located about 30 m below present, and more recently the entire coast of Brazil has experienced a slow 2–5-m sea-level fall in the last five to six thousand years (Dominguez 2009). Thus, these ecosystems are sensitive to changes in the external environment but have an extraordinary capacity for self-reconfiguration, either responding through adaptations to new conditions or exploring and exploiting new spaces that become available or promptly succumbing (Jimenez et al. 1985; Blasco et al. 1996; Schaeffer-Novelli et al. 2002; Menghini et al. 2011).

Mangroves' and salt marshes' potential for rapid accommodation to new conditions is the result of the typical set of plant species that share peculiar characteristics (Tomlinson 1986), which include:

- Broad tolerance to environmental factors
- Rapid growth
- Rapid maturation
- Continuous production of flowers and seeds/seedlings
- The release and dispersal of large numbers of diaspores
- Transport mechanisms that promote dispersal of reproductive structures over short and long distances by abiotic agents (e.g., currents and tides)

In the case of salt marshes, vegetative propagation should be added to this list of features (Dawes 1998). In general, these adaptations reflect the dynamism of the habitats these species occupy. Similarly, management must be aligned with the complexity and dynamics of the system being managed.

19.4 Managing Adaptive Systems

We consider that the key to sustainable and successful management is the consonance between natural processes and the boundaries of management units, taking ecosystem resistance and long-term resilience (i.e., allogenic resilience) from the whole context into consideration. The current destruction of tidal plains and wetlands represents an essentially irreversible process that eliminates natural regenerative assets and their future possibilities to accommodate sea-level rise, compromising the quality of adjacent marine systems and the delivery of ecosystem services to society.

A particular area of a mangrove forest or salt marsh cannot be evaluated independent of numerous factors that vary notably from site to site and the local social-ecological system perspective. This might or might not incorporate notions of adaptation and alignment with the system's self-organization as critical for the ability to persist as an integrated social-ecological system (Gunderson and Holing 2002). The broader setting-scale perspective provides the most appropriate framework for guiding and organizing the application of ecological knowledge within a social-ecological system.

Knowledge is always partial, provisional, and contextual and can be forgotten. Paradigms are provisional fallible frameworks that require constant refreshment. Abstract knowledge by itself does not lead to action. Knowledge must be transformed into awareness. The empowerment of local communities through the contextualization of scientific knowledge is an area of research under rapid evolution with important implications for sustainable development (see Chap. 17). The resilience approach provides a practical forum for the considerations of such socially relevant topics as vulnerability, ecological economics, and sustainability science and for generating interdisciplinary integrative science (Folke et al. 2002). This represents a major departure from the older adaptive management methodology, which was based on technocratic top-down action and lacked what is now considered essential local participation and creative stewardship.

The coastal zone should be considered as a priority space for anticipatory actions aimed at mitigation and preemptive adaptation in response to climate changes, regardless of whether these areas are natural or anthropogenic, according to Nicolodi and Peterman (2010). The emphasis on preemptive measures reflects the fact that the "back-loop" of the adaptive cycle is faster compared to the slow dynamics of system organization, growth, and assembly. Generally, systems can collapse faster than they can organize themselves. Order is built slowly and requires work; collapse can occur precipitously. The medium- and long-term monitoring of the dynamics of mangrove and salt marsh ecosystem forests should be considered as a priority for the timely identification of signs of imminent failure, in time to avoid sudden collapses triggered by cumulative brittleness and small disturbances on resilience-deficient systems (Kauffman and Donato 2012; Ellison 2012, 2015).

Another point that should be highlighted is the importance of monitoring animal groups with functional significance as "early warning" indicators of system performance deterioration. These proxies provide information associated with "health" and "vigor" of ecosystem functions under stress and increase the understanding about the type and scope of structural and functional changes and their influence on the composition of the associate fauna, including keystone species such as crabs, particularly the swamp ghost crab (*Ucides cordatus*) (Skilleter and Warren 2000; Alfaro 2010; Pinheiro and Almeida 2015; Schaeffer-Novelli et al. 2016). Such assessments should be performed along latitudinal gradients seeking generalizations that reduce "noise" due to site factors.

Although many attributes qualify mangrove and salt marsh as good indicators of changes in the relative sea level (Schaeffer-Novelli et al. 2002; Soares 2009), several site-specific characteristics must be considered in the prediction or interpretation of responses to expected changes, such as local tectonics, meteorological variability, local relative sea-level history, and impacts of human activities in hydrology, sediment supply, and changes in energy signatures (Thom 1982; Schaeffer-Novelli et al. 1990, 2002, 2005a; Jelgersma et al. 2002; Hadlich and Ucha 2009).

When mangroves and salt marshes respond to pulsing events (e.g., erosion, deposition, frost, droughts, floods, strong frontal systems), responses may reflect interaction with environmental legacies that had been unrecognized due to their latency accelerating the crossing of unknown thresholds (Schaeffer-Novelli et al. 2005b). Subtle responses may be missed if they are hidden within increased environmental noise. Climate change may be manifested as increased system instability.

Mangroves and salt marsh ecosystems may adapt and survive the climate and significant change and sea-level variations, but such adaptation does not depend on the individual properties of species but also on the elimination of non-climatic stressors that operate at local and regional scales (Schaeffer-Novelli et al. 1990, 2002, 2005a). No single scenario can adequately describe or anticipate the various possibilities and expectations regarding climate change and coastal dynamics. This entails that ample hedging and safety factors must be incorporated into policies, which must be structured to cope with increased variability and risk of failure. "Fail-safe" (able to return to a safe condition in the event of a failure) measures are required to protect regenerative potential embedded in exceptional landscapes/ecosystems. The propensity of failure increases when ignoring complexity and relying on simplistic mechanical-like approaches.

19.5 Primum Non Nocere

The CAS management approach addresses long-term protection of vital self-organizing processes while providing short-term rewards by safeguarding the sustained performance of services generated by climate-vulnerable coastal systems to society. This "fail-safe" approach nurtures win-win outcomes across the short and long terms and can be implemented at low cost if the real cost of business as usual is considered in terms of the loss of ecosystem services such as production, support, regulation, and cultural values (MEA 2005).

The management of CAS must be firmly grounded in sustainable stewardship that takes into consideration the capacity for self-organization and information storage capacity of the system. However, no matter how much these systems and their functions are studied, "unknown unknowns" (UUs; consequences that are impossible to predict or plan for) cannot be eliminated, and in any case, these systems do not behave as simple predictable mechanical entities. UUs can be features of fact-tail event distribution (unknown risks). For the contemporary resource manager and decision-maker, the first rule of management is *primum non nocere* ("first do no harm"); this is the oath taken by every medical doctor; it needs to be adopted by all decision-makers in the environmental arena. Just as important is that environmental decisions cannot be taken without consultation with knowledgeable stakeholder participation. Ecology and culture are intrinsically entangled at the social level; this is particularly true in coastal zones where social and geoecological processes interact to shape unique social-geoecological systems.

19.6 Final Remarks

Management systems are tied to the prevailing mindset or the way we "picture" reality. The outcome that may have been considered can turn out to be more adverse than the worst-case scenario. The recognition of complexity requires a revision of our mechanistic interpretation of nature. Management of natural systems is linked to the way nature is structured and the way we perceive nature is a function of our own subjective interpretation. This postmodern approach to management recognizes the complexity of reality, and because of this emergent awareness, it is necessary to revise the prevailing mindset and the ways we manage nature.

Mindsets, worldview institutions, and technologies are useful, but they become increasingly outdated and dysfunctional over time. They can become "competence traps" that can challenge long-term vitality becoming obstacles to ecosystem renewal or refreshment. Policy frameworks must be upgradable, not permanent features. Management is the regulation of change to ensure a desirable future. However, change currently is more destructive and irreversible than in the past. More than ever, management must be based on foresight, anticipation, and conservation of higher-level resilience. Latent self-organization is embedded in the landscape.

To the extent that a system can self-organize, it should allow ecosystems to operate far from equilibrium and maintain integrity within a "window of vitality." The purpose of virtuous management as CAS is to widen this "window" by preemptively building resilience. Change is a constant of nature, and management systems must be as dynamic as the context they intend to manage. The management of CAS must be grounded in sustainable stewardship that takes into consideration the following three broad principles:

1. Sustainable stewardship involves reserving for future generations the same options that are available now, keeping in mind that future challenges will be more complex than those faced today. Climate change and the impacts of increasing coastal squeeze will make systems more vulnerable putting a premium on the preservation of landscape-level, resilience, and robustness the drivers of regenerative capacity.
2. Sustainable stewardship is grounded in creative citizenship, the generation of a higher-level narrative through a dialectical process involving stakeholders where the ecological and social domains are fused into a narrative that is constructive of resilience and robustness. In such a dialectical process, a higher-level narrative emerges which is coherent with sustainable use and the creation of a social-ecological system where ecology and social processes act cooperatively rather than antagonistically. This is a product of citizen-based problem-solving.
3. Creation of an adaptive framework requires the capacity to educate and anticipate, have appropriate infrastructure for swift implementation (laws, regulations, etc.), and monitor and act in real time to contain and restore unexpected undesirable impacts to maintain the system in a state of high resilience and self-organizing capacity. Education and engagement of a motivated and concerned

public are recognized to be essential for building resilience solutions, and such a pool of people is the source of relevant creative and locally relevant and adapted solutions. Education creates and plays the critical role in screening and selection required for successful adaptation and maintenance of system fitness. Complexity entails novelty and change, which is incessant learning, relearning, adaptation, and evolutionary development. Change is open-ended and natural. Arresting change is unnatural.

We have explored mangrove forests and salt marsh behavior within a framework of CAS' principles that hopefully can lead to a better understanding despite difficulties in the prediction in their behavior and detailed understanding (Schaeffer-Novelli et al. 2016) of features, attributes, and functions that are still poorly recognized but that influence mangrove-salt marsh management and their future integrity. Coastal managers must think strategically (in a broad temporal and spatial perspective) but must be able to act tactically. Finally, dealing with CAS requires the capacity to unlearn the outdated and reframe problems in the context of complexity. Postmodern management is a matter of robust design, designing a sustainable future.

References

Alfaro AC (2010) Effects of mangrove removal on benthic communities and sediment characteristics at Mangawhai Harbour, northern New Zealand. ICES J Mar Sci 67:087–1104

Alongi DM (2008) Mangrove forests: resilience, protection from tsunamis, and responses to global climate change. Estuar Coast Shelf Sci 76:1–13

Blasco F, Saenger P, Janodet E (1996) Mangrove as indicators of coastal change. Catena 27:167–178

Carter RWG (1988) Coastal environments: an introduction to physical, ecological and cultural systems of coastlines. Academic, London

Dawes CJ (1998) Marine botany. Wiley, New York

Dominguez JML (2009) The coastal zone of Brazil. In: Dillenburg SR, Hesp PA (eds) Geology and geomorphology of Holocene Coastal Barriers of Brazil. Springer, Berlin, pp 17–51

Duke NC (1995) Genetic diversity, distributional barriers and rafting continents – more thoughts on the evolution of mangroves. Hydrobiologia 295:167–181

Ellison JC (2012) Climate change vulnerability assessment and adaptation planning for mangrove systems. World Wildl Fund, Washington, DC

Ellison JC (2015) Vulnerability assessment of mangroves to climate change and sea-level rise impacts. Wetl Ecol Manag 23:115–137

Folke C (2006) Resilience: the immerge of a perspective for social-ecological systems analyses. Glob Environ Change 16:253–267

Folke C, Carpenter SR, Elmqvist T, Gunderson LH, Holing CS, Walker BH, Bengtsso J, Berkes F, Colding J, Danell K, Falkenmark M, Gordon L, Kaspersson R, Kautsky N, Kinzig A, Levin SA, Mäler K-G, Moberg F, Ohlsson L, Olsson P, Ostrom E, Reid W, Rockström J, Savenije S, Svedin U (2002) Resilience and sustainable development: building adaptive capacity in a world of transformations. Science background paper commission by the Environmental Advisory Council of the Swedish Government in preparation for the World Summit on Sustainable Development. Ministry of the Environment, Stockholm

Gunderson LH, Holling CS (eds) (2002) Panarchy: understanding transformations in human and natural systems. Island Press, Washington, DC

Hadlich GM, Ucha JM (2009) Apicuns: aspectos gerais, evolução recente e mudanças climáticas globais. Rev Bras Geomorfolog 10(2):13–20

Holling CS (1973) Resilience and stability of ecological systems. Annu Rev Ecol Syst 4:1–23

Holling CS (1978) Adaptive environmental assessment and management. Wiley, New York

Holling CS (1986) The resilience of terrestrial ecosystems: local surprise and global change. In: Clark WC, Munn RE (eds) Sustainable development of the biosphere. Cambridge University Press, London, pp 292–317

Janssen MA, Anderies JM (2007) Robustness trade-offs in social-ecological systems. Int J Commons 1(1):43–65

Jelgersma S, Healy T, Marone E (2002) Relative sea level changes and some effects on muddy coasts. In: Healy T, Wang HJ-A (eds) Muddy coasts of the world: processes, deposits and function. Elsevier Science BV, Dordrecht, pp 83–97

Jimenez JA, Lugo AE, Cintrón G (1985) Tree mortality in mangrove forests. Biotropica 17:177–185

Jørgensen SE (2012) Introduction to systems ecology. CRC Press, Boca Raton

Karkkainen BC (2006) Panarchy and adaptive change: around the loop and back again. Minn J Law Sci Tech 7(1):59–77

Kauffman JB, Donato DC (2012) Protocols for the measurement, monitoring and reporting of structure, biomass and carbon stocks in mangrove forests. CIFOR, Bogor

Levin SA (1995) Ecosystems and the biosphere as complex adaptive systems. Ecosystems 1:431–436

MEA – Millennium Ecosystem Assessment (2005) Ecosystems and human well-being: synthesis. Island Press, Washington, DC

Menghini RP, Coelho-Jr C, Rovai AS, Cunha-Lignon M, Schaeffer-Novelli Y, Cintrón-Molero G (2011) Massive mortality of mangrove forests in southeast Brazil (Barnabé Island, Baixada Santista, State of São Paulo) as a result of harboring activities. J Coast Res 64(2):1793–1797

Murray C, Marmorek DR (2004) Adaptive management: a spoon full of vigour helps the uncertainty go down. In: Proceedings of the 16th annual meeting international conference, Society for Ecological Restoration, Victoria, 24–26 August 2004

Nicolodi JL, Petermann RM (2010) Mudanças climáticas e a vulnerabilidade da zona costeira do Brasil: Aspectos ambientais, sociais e tecnológicos. Revista de Gestão Costeira Integrada 10(2): 151–177

Phillips JD (2007) The perfect landscape. Geomorphology 84:159–169

Pinheiro MA, Almeida R (2015) Monitoramento da densidade e da estrutura populacional do caranguejo-uçá, *Ucides cordatus* (Linnaeus, 1763) (Brachyura: Ucididae). In: Turra A, Denadai MR (eds) Protocolos de campo para o monitoramento de habitats bentônicos costeiros – Rede de Monitoramento de Habitats Bentônicos Costeiros – ReBentos. IOUSP, São Paulo, pp 118–129

Ramsar Convention Secretariat (2010) Coastal management: wetland issues in integrated coastal zone management. Ramsar handbooks for the wise use of wetlands, vol 12, 4th edn. Ramsar Convention Secretariat, Gland

Salthe SN (1985) Evolving hierarchical systems: their structure and representation. Columbia University Press, New York

Schaeffer-Novelli Y, Cintrón-Molero G, Adaime RR, Camargo TM (1990) Variability of mangrove ecosystems along the Brazilian coast. Estuaries 13(2):204–218

Schaeffer-Novelli Y, Cintrón-Molero G, Soares MLG (2002) Mangroves as indicators of sea level change in the muddy coasts of the world. In: Healy T, Wang Y, Healy J-A (eds) Muddy coasts of the world: processes, deposits and function. muddy coasts of the world: processes, deposits and function. Elsevier Science, Amsterdam, pp 245–262

Schaeffer-Novelli Y, Cintrón-Molero G, Cunha-Lignon M, Coelho-Jr C (2005a) A conceptual hierarchical framework for marine coastal management and conservation: a "Janus-Like" approach. J Coast Res 42:191–197

Schaeffer-Novelli Y, Cintrón-Molero G, Primavera J, Quarto A (2005b) The mega tsunami of 26 December 2004: recognizing ecological lessons from a large-scale natural disaster. Indian Ocean Tsunami Fish 2(1):31–38
Schaeffer-Novelli Y, Soriano-Sierra EJ, Vale CC, Bernini E, Rovai AS, Pinheiro MAA, Schmidt AJ, Almeida R, Coelho-Jr C, Menghini RP, Martinez DI, GMO A, Cunha-Lignon M, Charlier-Sarubo S, Shirazawa-Freitas J, Cintrón-Molero G (2016) Climate changes in mangrove forests and salt marshes. Braz J Oceanogr 64(2):37–52
Skilleter GA, Warren S (2000) Effects of habitat modification in mangroves on the structure of mollusk and crab assemblages. J Exp Mar Biol Ecol 244:107–129
Soares MLG (2009) A conceptual model for the responses of mangrove forests to sea level rise. J Coast Res 56:267–271
Thom BG (1982) Mangrove ecology: a geomorphology perspective. In: Clough BF (ed) Mangrove ecosystems in Australia. Australian National University Press, Canberra, pp 3–16
Tomlinson PB (1986) The botany of mangroves. Cambridge University, Cambridge
Ulanowicz RE (1997) Ecology the ascendant perspective. Columbia University Press, New York
Walker B, Holling CS, Carpenter SR, Kinzig A (2004) Resilience, adaptability and transformability in social-ecological systems. Ecol Soc 9(2):5

Index

Y. Schaeffer-Novelli et al. (eds.), *Brazilian Mangroves and Salt Marshes*, Brazilian Marine Biodiversity, https://doi.org/10.1007/978-3-031-13486-9